T0215604

Hydrogeology

Groundwater science and engineering

This text combines the science and engineering of hydrogeology in an accessible, innovative style. As well as providing physical descriptions and characterizations of hydrogeological processes, it also sets out the corresponding mathematical equations for groundwater flow and solute/heat transport calculations. And, within this, the methodological and conceptual aspects for flow and contaminant transport modeling are discussed in detail. This comprehensive analysis forms the ideal textbook for graduate and undergraduate students interested in groundwater resources and engineering, and indeed its analyses can apply to researchers and professionals involved in the area.

Hydrogeology

Groundwater science and engineering

Alain Dassargues

CRC Press
Taylor & Francis Group
Boca Raton London New York

CRC Press is an imprint of the
Taylor & Francis Group, an **informa** business

CRC Press
Taylor & Francis Group
6000 Broken Sound Parkway NW, Suite 300
Boca Raton, FL 33487-2742

First issued in paperback 2020

© 2019 by Taylor & Francis Group, LLC
CRC Press is an imprint of Taylor & Francis Group, an Informa business

No claim to original U.S. Government works

ISBN-13: 978-1-4987-4400-3 (hbk)
ISBN-13: 978-0-367-65714-7 (pbk)

Library of Congress Cataloging-in-Publication Data

Names: Dassargues, Alain author.
Title: Hydrogeology : groundwater science and engineering / author,
Alain Dassargues.
Description: First Edition. | Boca Raton, Florida : Taylor & Francis, A
CRC title, part of the Taylor & Francis imprint, a member of the Taylor
& Francis Group, the academic division of T&F Informa plc, [2019] |
Includes bibliographical references and index. |
Identifiers: LCCN 2018009792 (print) | LCCN 2018018625 (ebook) |
ISBN 9780429894411 (Adobe PDF) | ISBN 9780429894404 (ePub) |
ISBN 9780429894398 (Mobipocket) | ISBN 9781498744003 (Hardback :
acid-free paper) | ISBN 9780429470660 (eBook)
Subjects: LCSH: Hydrogeology. | Groundwater flow.
Classification: LCC GB1003.2 (ebook) | LCC GB1003.2 .D37 2018 (print)
| DDC 551.49--dc23
LC record available at https://lccn.loc.gov/2018009792

Visit the Taylor & Francis Web site at
http://www.taylorandfrancis.com

and the CRC Press Web site at
http://www.crcpress.com

eResource material is available for this title at https://www.crcpress.com/9781498744003

Thanks!

Thanks to my wife, my three daughters and their companions, and my grand-children who have supported me in this writing experience. Sorry for the time that was captured by this book preparation and therefore not devoted to them!

Thanks to my colleagues within our research group "Hydrogeology and Environmental Geology" (University of Liège, Belgium), who have been very patient with me especially during the last two years while being more and more eager to discover the final result.

Thanks to many colleagues and friends with whom I discussed specific topics and contents of this book. Among others, I would especially like to thank Okke Batelaan, Luk Peeters, Craig Simmons, Philippe Renard, Daniel Hunkeler, Philip Brunner, John Molson, Jean-Michel Lemieux, and René Therrien for their fruitful exchanges of ideas during my half-year sabbatical in 2017. Particular thanks go to John, Jean-Michel, and René for having reviewed some of the chapters.

Thanks to Agathe Defourny for having greatly helped to draw the figures while she brilliantly pursued her Master in Geological Engineering at the University of Liège.

Thanks to the University of Liège and FNRS Belgium for the granted sabbatical half-year that allowed me to focus for a few months on the core topics of this book.

Thanks to the reader; I hope you will find in this book the theoretical and/or practical information you are looking for.

Contents

Foreword

Groundwater is an invisible but critical component of the hydrologic cycle. It represents the largest source of readily available drinking water on the planet and is therefore an essential resource for humans. Groundwater also provides baseflow to streams and plays a critical role in maintaining the health of freshwater ecosystems. Pressure on groundwater resources is increasing and there is growing evidence that extraction from major aquifers are gradually depleting groundwater reserves worldwide. Predicted climate variability, such as reduced precipitations or increased temperatures, may further exacerbate pressures on groundwater reserves. Emerging techniques to extract energy from geological materials are also creating challenges with respect to existing groundwater resources. Addressing these pressures and ensuring the sustainable development of groundwater resources represent a grand challenge for society and require a sound understanding of the physical and chemical processes that affect groundwater quantity and quality. Mathematical models that accurately represent these processes are also required to support sustainable development. Furthermore, a complete characterization of groundwater systems is not possible and, as a result, there will always be uncertainty associated with knowledge of these systems. This uncertainty should be accounted for when predicting groundwater quality or quality with mathematical models.

Hydrogeology: Groundwater Science and Engineering provides an up-to-date reference that combines the fundamental concepts and practical guidelines necessary to address challenges related to groundwater resources. The author, Professor Alain Dassargues, has been trained in both geosciences and engineering. He also has a long experience in teaching and conducting high-level groundwater research, as well as experience in solving practical problems in hydrogeology. His experience as a scientist and engineer has clearly influenced Professor Dassargues since the approach and content of the book are quantitative, which distinguishes it from other reference books on the topic. Physical and chemical concepts are presented rigorously, in mathematical terms, and an engineering approach is used to present methodologies and models required to solve practical problems. The book therefore combines the theoretical basis of groundwater science with the practical approach required by groundwater engineers.

The book presents groundwater in the context of the terrestrial water cycle and highlights its interactions with other components of the cycle, such as surface water or atmospheric water. It reviews the basic principles governing groundwater flow, defines the physical properties of geological materials that host groundwater, and presents the methods used to measure them. It also covers groundwater geochemistry and presents the fundamental concepts of advective-dispersive mass transport in groundwater, as well as energy transport. These fundamental concepts are called upon to cover a series of topics and challenges faced by groundwater specialists, such as managing water resources, investigating and remediating groundwater contamination, or applying physical principles to address land subsidence resulting from groundwater depletion. Advanced topics covered in detail also include unsaturated flow, salinization and density-dependent flow, and low-temperature geothermal systems.

Groundwater professionals now routinely use a variety of software such as numerical models to simulate groundwater flow and solute transport, geostatistical software for data analysis and interpolation, and Geographic Information Systems (GIS). Such software are now part of the hydrogeologist's toolbox and their underlying basic principles and concepts are presented in a series of chapters. The book also covers the critical topic of uncertainty in model parameters and model predictions. The author also takes the perspective of the groundwater practitioner and offers insightful advice on applying these emerging mathematical tools to solve practical problems.

The book will be a very valuable resource for an introductory course in hydrogeology, which could focus on the chapters that present the fundamental concepts. It is also well suited for an advanced undergraduate course or a graduate course that focuses on specific topics such as unsaturated flow, land subsidence, groundwater chemistry and contamination, numerical modeling, or statistical methods. The book will also be an extremely valuable reference for professionals who are looking for guidance in applying mathematical tools and want to learn the latest developments and concepts in the field.

René Therrien
Université Laval
Québec, Canada

Preface

If we fail to get the water, then it's ruin to the squatter,
For the drought is on the station and the weather's growing hotter,
But, we're bound to get the water deeper down

Banjo Paterson

Water does not belong to us; we only borrow it (modified from the quotation "We do not inherit the Earth from our ancestors; we borrow it from our children" of unknown origin). If we use or pollute water, the challenge is to restore and provide it back with at least the same quality and quantity as previously.

Hydrogeology refers to geology and hydrology engineering including the uncertainties of natural sciences and rigorous quantification techniques. Also, field work and theoretical background are complementary. All these opposite but complementary aspects can be considered the Yin and Yang of groundwater.[1]

"History outlines and defines the methods of science."[2] Hydrogeology is more a question of methods than of local results, even if the latter can be of paramount importance for the development and welfare of the local human society.

Even if pressure-wave propagation is quasi-instantaneous; groundwater flow through the pores and microfissures of the rock is slow, allowing a series of biochemical and physical processes to occur. The praise of slowness has its own meaning for groundwater.

As summarized by Jacob Bear (2011), "… modelling large-scale effects from small-scale influences is a delicate act. The starting point is the use of a magnifying glass to observe and understand what happens at points within the phases and on interphase boundaries and then, by employing homogenization of one kind or another obtain mathematical models that describe these phenomena in terms of measurable quantities in a domain regarded as a continuum."[3]

"Learn the rules like a pro so you can break them like an artist" (Pablo Picasso).

Sharing my modest knowledge and practice of groundwater science and engineering was a very interesting and important experience for me. Writing this book, I realized

---

[1] Lipson, D.S. 2017. The Yin and Yang of Groundwater. *Groundwater* 55: 287.
[2] Deming, D. 2016. The importance of History. *Groundwater* 54: 745.
[3] Bear, J. 2011. Response to the citation of the AGU 2010 Robert E. Horton Medal. *EOS* 92(6): 49.

that each chapter and even subchapter topic could be described at length in entire books or review papers. Consequently, one of the main challenge was undoubtedly to write an up-to-date comprehensive summary containing (hopefully) the most important information for the reader. Whatever the task, the beauty of the work lies in the satisfaction of the author in doing his best ...

Author

Alain Dassargues is Professor in Hydrogeology and Environmental Geology at the University of Liège in Belgium, and former Chair of the Belgian Chapter of the International Association of Hydrogeologists.

Chapter 1

General introduction

Water availability, along with land and food availability and other mineral resources, is an increasingly popular topic, a growing preoccupation as the world population continues to increase. This is not so much a global problem (de Marsily 2009) as it is a regional problem of availability to satisfy our needs for improving human health, food security, biodiverse natural ecosystems and effective energy production. Indeed, there are multiple feedback effects, interconnections, and couplings among these four main domains dependent on water resources. This relationship has been described as the "water—energy—food nexus" (Scanlon *et al.* 2017, Cai *et al.* 2018). The emergence of 'nexus thinking' comes from an increasingly perceptible understanding that natural resources may someday limit the development of our well-being and of our growing human communities. Consequently, win-win strategies must be developed for preserving environmental sustainability together with producing efficiency gains to balance the imposed growth from the demographic issue (Ringler *et al.* 2013). This problem is particularly true for freshwater (i.e., natural continental water with a limited ion content from brackish water and seawater) that is indeed essential for each of us.

1.1 Freshwater resources and groundwater resources

The global stock of water on Earth is currently estimated at approximately 1,387 million km^3. In fact, the amount is not what is important. More than 96.5% is seawater. Other saline waters are found at depth or in salty lakes for approximately 0.96%. These numbers mean that freshwater is found only in the remaining 2.54% of water on Earth. If we subtract water contained in ice caps and glaciers (1.75%), vapor in the atmosphere, soil moisture and permafrost (0.02%), the remaining "easy to use" freshwater is only 0.77% of water on Earth. In addition, the most amazing fact for the general public and the media is that out of this 0.77%, the share of rivers and lakes is less than 0.01%; rivers and lakes provide less than 1.3% of the directly valuable water. In other words, the ratio of fresh groundwater to fresh surface water is approximately 77 to 1. Data at the global scale are scarce or often result from inadequate upscaling techniques; therefore, these global estimates can be relatively inaccurate. Nevertheless, these figures clearly show the particular importance of groundwater for future generations in a world with a growing population. An additional critical question arises with the current climate change: Is the distribution of water changing between the different reservoirs (de Marsily 2009)?

Freshwater is quite unevenly distributed, inducing many local water availability issues. In arid zones, groundwater takes a critical importance while surface water may be very limited if they are not fed by seasonal glacier melt from mountainous regions. The renewability of groundwater reserves must also be considered. Again, in arid zones, water production from very old groundwater reserves, referred to as fossil groundwater (i.e., not renewed for thousands of years), automatically brings up the question of sustainable development. For example, in northern Mauritania, in the Sahara Desert, a water recharge of a few mm/year (i.e., on a multiannual basis and over the entire territory) would be enough to balance groundwater pumping needed for the current development of the region. However, determining whether these few mm/year of recharge actually occur under current climatic conditions is not a simple problem. This question already requires very extended and detailed hydrological study involving long periods of data measurement and accurate interpretations.

1.2 Anthropocentric vision

As mentioned above, research about water resources and especially about groundwater is most often motivated and supported by economic and human water needs. As noted by Hornberger *et al.* (2014), water sciences "have both pure and applied roots that stretch back to antiquity. The importance of water as a resource remains one of the central reasons for studying hydrological processes." Thus, hydrogeology refers not only to natural science but also to engineering. Hydrogeology is subject to the laws and uncertainties of the natural sciences while most often being used to ensure conservation of ecosystems and various needs of humankind. Engineering is involved because the social and economic importance of groundwater supplies requires not only quantified answers but also the uncertainty estimate of the answers. However, "groundwater is not visible in policy, governance and management matters as it needs to be" (Grabert and Kaback 2016), compared with its relative importance, for example, in the production of drinking water.

In a global DPSIR (Driver Pressure State Impact Response) approach, the drivers are clearly the growing population and climate-societal changes. Global and local water management systems must be efficient to provide a "water security" (Foster and MacDonald 2014). They are dealing with quantity and quality issues and renewable water resources. There is only a finite amount of water and there are no substitutes (Cosgrove and Loucks 2015) as desalinization of seawater is not yet based on sufficiently sustainable techniques to be used on a large scale. Thus, managing water, and especially water scarcity, is among the most pressing challenges humanity is facing today (Wheeling 2016).

When assessing water, especially groundwater needs, statements are often misleading and confusion exists between "used water," "consumed water" and "produced water," or "withdrawn water" (Simmons 2015). Water can be used many times, ensuring different successive functions or services with (recycled water) or without water treatment (reused water). Consumed water refer to water that is not (at least locally) recycled (i.e., evaporated, transpired, or transformed into food). Produced or withdrawn water is extracted from a source, and a part of it can actually be reinjected (recycled) or reused, while the other part is consumed (see above).

Large debates recur about privatization and marketing of water. This issue is not the scope here but indeed, theoretically, a cost assigned to water would ensure that water providers would be financially healthy and improve quality in their methods of managing water resources on behalf of all (Gautier 2008). At the same time, a minimal access to water should be guaranteed to everybody, and consequently, the role of local authorities is to determine the delicate equilibrium to be found.

Facing sustainability (i.e., defined as the ability to meet present needs without compromising those of future generations), the most difficult problem is that this concept seems quite opposite to the population growth, at least at current levels of resource consumption. Lower levels of water resources consumption (i.e., by fostering reused and recycled water) should be attained as soon as possible by technological progress.

1.3 Hydrogeology within hydrology

Hydrology is the scientific study of water (in general). Thus, it involves the study of movement, distribution, and quality of water on continents, including the water cycle, water resources, and environmental sustainability aspects. This study incorporates the properties of water, and the relationships between water and the biotic components of the environment. *Hydrogeology* is the study of groundwater hydrology, specifically taking geological conditions into account. Thus, hydrogeologists should master methods and scientific techniques from different sciences and engineering specialties: geology, physics, chemistry, biology, surface hydrology, hydraulics, fluid mechanics, and heat transfer as well as geostatistics, risk analysis, and geotechnical, numerical, and computational engineering. For fruitful interactions with other water and environment specialists and researchers, notions and understandings in civil and environmental engineering, and in water management, can also be essential.

1.4 Basics about groundwater: partially and fully saturated zones

Groundwater is water underground in pore spaces and fractures of the geological media. The part of the lithosphere in which each void (i.e., pores, fissures, caverns) is totally filled with water is referred to as the *saturated zone*. Above the saturated zone, a *partially saturated zone* is found where the void space contains water and air. Many terms are used, such as *unsaturated zone, variably saturated zone*, or *vadose zone*. Upwards, the uppermost layer is usually the soil moisture zone or root zone whose thickness varies according to the local combination of climatic, lithological, and topographical conditions. A capillary fringe can be distinguished overlying the saturated zone (Figure 1.1). In this fringe, water saturation is nearly reached but the capillary forces are greater than that of gravity. Groundwater filling the pores and fissures remains preferably attached to the solid matrix by capillary action. The thickness of this capillary fringe increases as the pore sizes and fissure apertures are small and uniform. Very high capillary fringes can give rise to significant sensitivity to any additional water infiltration: in this case, flood risk from groundwater is very high if groundwater levels rise rapidly up above the natural surface.

In a borehole or in a pumping well, the capillary fringe is most often not detected and thus actually neglected. The water level measured in the well gives the position of

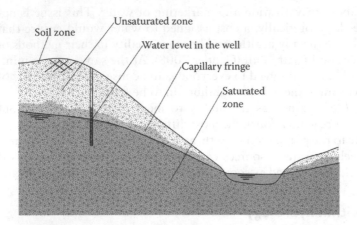

Figure 1.1 Above the water table, the capillary fringe is a zone of variable thickness where the medium is nearly saturated, but the capillary forces are greater than that of the gravity. This capillary fringe cannot be detected in a well. The measured water level in the well gives the position of the water table separating the saturated zone (i.e., where groundwater may flow by gravity) from the unsaturated zone.

the water table separating the saturated zone where groundwater may flow by gravity from the unsaturated zone (see Chapters 4 and 9).

1.5 Prospecting groundwater

If one considers the geological context in the upper zone of the Earth's crust, one can find groundwater (nearly) everywhere. Another story (i.e., challenge) is obviously to find groundwater in large enough quantity to be produced and meet local needs. For this production purpose, one is dependent on the two main properties of geological media influencing groundwater flow: their ability to store and conduct groundwater (see Chapter 4). Groundwater prospecting is clearly not within the scope of this book, and it can be very challenging.

Indeed, water dowsing is not relevant (more details in: Comunetti 1978, USGS 1988, Enright 1995, Betz 1995, Kölbl-Ebert 2012). Recently, hydrogeophysics has taken on much importance, providing the development and use of geophysical methods and inversion algorithms to image, monitor, and characterize subsurface properties and processes. Hydrogeophysics is currently used in environmental applications such as groundwater resources, contaminated sites, landfills, or shallow geothermal systems (among others: Telford *et al.* 1990, Rubin and Hubbard 2006, Vereecken *et al.* 2006, Nguyen *et al.* 2009, Robert *et al.* 2012, Binley *et al.* 2015, Moorkamp *et al.* 2016).

1.6 Content of this book

The content of this book is divided into 12 chapters.

Chapter 2 considers the water cycle and water balance equations with particular attention given to recharge assessments and drainable groundwater reserves. Precipitation and actual evapotranspiration terms are discussed with an emphasis on averaged assessments. The water table fluctuation and chloride mass balance methods are introduced for recharge assessments. The interpretation of recession hydrograph curves in terms of base flow is presented to estimate groundwater reserves that can be drained in a watershed.

The main terminology and different examples of groundwater occurrences are presented in Chapter 3, showing the importance of a thorough geological understanding to deduce the primary hydrogeological information before further investigations.

The main concepts and laws used to characterize groundwater flow in saturated geological porous media are described in Chapter 4. The effective drainable porosity (specific yield) is distinguished from the total porosity. The Bernoulli potential equation and Darcy's law are applied in different contexts of geological heterogeneity. The groundwater flow equations under steady-state and transient conditions are established in 2D and 3D and for confined and unconfined conditions. The transmissivity and storage coefficient are used when the Dupuit assumption of 2D horizontal groundwater flow can be accepted.

Hydraulic conductivity and (specific) storage coefficient measurements are described in Chapter 5. Various laboratory and in situ techniques, at different scales, are described to measure not only the transmissivity, saturated hydraulic conductivity (K), and/or intrinsic permeability (k) but also, under transient conditions, the storage coefficient and specific storage coefficient. Pumping test interpretation methods include Thiem's and Dupuit's method for steady-state conditions; the Theis, Cooper-Jacob, and Birsoy-Summers methods for transient conditions, recovery methods, and image well theory; and the Hantush and Neuman-Witherspoon methods.

Chapter 6 addresses the particular topic of land subsidence induced by pumping and drainage. Pore pressure changes may induce consolidation. It is often neglected or underestimated. Evolution of effective stress and water pressure at depth are explained for given drawdowns under confined and unconfined conditions. These processes clearly show how geomechanical aspects are fully coupled to groundwater flow with nonlinear evolution of specific storage coefficients and hydraulic conductivities. Examples of land subsidence processes in some of the most famous "sinking cities" are given.

Hydrochemistry and an introduction to groundwater quality are included in Chapter 7. Basic definitions and units of the most common physicochemical water characteristics are explained. Sampling and monitoring strategies are described and the conventional charts and diagrams used to represent and interpret the chemical compositions of groundwater are discussed. International groundwater quality standards are summarized.

Contaminant transport is the main topic of Chapter 8, where the mobility of dissolved (aqueous phase) and nonaqueous phase contaminants in groundwater is described. The different transport processes and related equations in heterogeneous geological media are derived. The main in situ remediation and natural attenuation techniques are introduced briefly. The assessment of transport times using isotopes and artificial and natural environmental tracers is explained. Groundwater ages and residence times are discussed in different contexts. Groundwater vulnerability and protection, conceptualized by process-based approaches, are presented.

The basic processes underlying partially saturated flow and transport are introduced briefly in Chapter 9. The specific assumptions and conventions are described, and emphasis is given to the nonlinear behavior of the parameters. The partially saturated flow and transport equations are derived.

Groundwater salinization is the topic of Chapter 10. If salts are dissolved, the groundwater density is increased, which influences flow and transport processes. Density-dependent groundwater flow is summarized, and a few typical occurrences of coastal seawater intrusion problems are presented. The concept of a sharp interface between seawater and freshwater is discussed and the Boussinesq approximation together with the usual assumptions for seawater intrusion calculations, is also presented.

Heat transfer in aquifers and shallow geothermy are the topics of Chapter 11. Heat is increasingly used not only as a tracer but also in relation to geothermal applications. New temperature measurement techniques, such as distributed-temperature-sensing (DTS), allow new perspectives in hydrogeology. The heat transfer processes are described, and the corresponding equations are introduced. Low-enthalpy shallow geothermal systems are described by highlighting the influence of hydrogeological conditions on their performances and environmental impacts.

The two last chapters of this book are dedicated to modeling. In Chapter 12, a complete methodology for groundwater flow and solute transport modeling is described step-by-step. Emphasis is given to the conceptual choices involving processes to be simulated, parsimony versus complexity, dimensionality, and specific choices linked to different hydrogeological contexts. Initial and boundary conditions for groundwater flow and solute transport simulations are discussed. Model design, data input, calibration, validation, sensitivity analysis, and inverse modeling are summarized in both deterministic and probabilistic frameworks. Geostatistical aspects and stochastic/probabilistic techniques together with uncertainty analysis are summarized for practitioners. These techniques help for treating the spatial and temporal variability found in geological formations and the different scales of heterogeneity.

Finally, the finite difference, finite element, and finite volume methods are presented in Chapter 13, keeping the mathematical description relatively simple. Explicit, implicit, Crank-Nicolson, and Galerkin time integration schemes are described and recommendations are given for the practitioner. For solute transport modeling, particular attention is given to solving advection-dominated problems with different methods, such as the Eulerian method with upstream weighting, TVD methods, Eulerian-Lagrangian methods, and random walk methods. Numerical Peclet and Courant numbers are defined to detect the actual numerical conditions or to adapt time steps. Multispecies reactive transport is addressed briefly as a coupled problem that can be simulated sequentially or in parallel.

References

Betz, H.D. 1995. Unconventional water detection: Field test of the dowsing technique in dry zones: Part 1. *Journal of Scientific Exploration* 9(I): 1–43.

Binley, A., Hubbard, S.S., Huisman, J.A., Revil, A., Robinson, D.A., Singha, K. and L.D. Slater. 2015. The emergence of hydrogeophysics for improved understanding of subsurface processes over multiple scales. *Water Resources Research* 51: 3837–3866.

Cai, X., Wallington, K., Shafiee-Jood, M. and L. Marston. 2018. Understanding and managing the food-energy-water nexus—opportunities for water resources research. *Advances in Water Resources* 111: 259–273.

Comunetti, A.M. 1978. Experimental investigation of the perceptibility of the artificial source for the dowsing agent. Progress report. *Experimentia* 35(3): 420–424.

Cosgrove, W.J. and D.P. Loucks. 2015. Water management: Current and future challenges and research directions. *Water Resources Research* 51: 4823–4839.

de Marsily, G. 2009. *L'eau, un trésor en partage (in French)*. Paris: Dunod.

Enright, J.T. 1995. Water dowsing: The Scheunen experiments. *Naturwissenschaften* 82: 360–369.

Foster, S. and A. MacDonald. 2014. The 'water security' dialogue: Why it needs to be better informed about groundwater. *Hydrogeology Journal* 22(7): 1489–1492.

Gautier, C. 2008. *Oil, water and climate. An introduction*. Cambridge: Cambridge University Press.

Grabert, V. K. and D.S. Kaback. 2016. Column theme: Groundwater management directions—Stewardship to sustain our water resources. *Groundwater* 54: 758.

Hornberger, G.M., Wiberg, P.L., Raffensperger, J.P. and P. D'Odorico. 2014. *Elements of physical hydrology* (2nd Edition). Baltimore: John Hopkins University Press.

Kölbl-Ebert, M. 2012. A history of diving rods. *Metascience* 21: 232–233.

Moorkamp, M., Lelièvre, P.G., Linde, N. and A. Khan (Eds.). 2016. *Integrated imaging of the earth: Theory and applications*. Hoboken NJ: John Wiley & Sons.

Nguyen, F., Kemna, A., Antonsson, A., Engesgaard, P., Kuras, O., Ogilvy, R., Gisbert, J., Jorreto, S. and A. Pulido-Bosch. 2009. Characterization of seawater intrusion using 2D electrical imaging. *Near-Surface Geophysics (Special Issue on Hydrogeophysics)* 7: 377–390.

Ringler, C., Bhaduri, A. and R. Lawford. 2013. The nexus across water, energy, land and food (WELF): Potential for improved resource use efficiency? *Current Opinion in Environmental Sustainability* 5(6): 617–624.

Robert, T., Caterina, D., Deceuster, J., Kaufmann, O. and F. Nguyen. 2012. A salt tracer test monitored with surface ERT to detect preferential flow and transport paths in fractured/karstified limestones. *Geophysics* 77(2): B55.

Rubin, Y. and S.S. Hubbard (Eds.). 2006. *Hydrogeophysics*. Volume 50. Dordrecht: Springer Science & Business Media.

Scanlon, B.R., Ruddell, B.L., Reed, P.M., Hook, R.I., Zheng, C., Tidwell, V.C. and S. Siebert. 2017. The food-energy-water nexus: Transforming science for society. *Water Resources Research* 53: 3550–3556.

Simmons, J. 2015. Water in the energy industry. *Geoscientist* 25(10): 10–15.

Telford, W., Geldart, L. and R. Sheriff. 1990. *Applied geophysics*. Cambridge: Cambridge University Press.

USGS. 1988. *Water dowsing. Open-File Reports Section*. Denver (CO): USGS publications.

Vereecken, H., Binley, A., Cassiani, G., Revil, A. and K. Titov (Eds.). 2006. *Applied hydrogeophysics*. Dordrecht: Springer.

Wheeling, K. 2016. The coming blue revolution. *EOS* 97(4): 29.

Chapter 2

Hydrologic balance and groundwater

2.1 Water cycle and balance assessments

The circulation of water through the atmosphere, land surface and subsurface, and rivers is referred to as the *water cycle* or *hydrological cycle* (Figure 2.1). The water cycle is an open system in which solar radiation is a constant source of energy. Many hydrology textbooks address this topic (e.g., Brutsaert 2005, Hornberger *et al.* 2014), and some of them focus explicitly on the importance of groundwater within hydrology (e.g., Todd and Mays 2005). Globally, evaporation is greater than precipitation at the sea/ocean surface; thus, the atmosphere carries water vapor from the oceans to the continents. The water that falls on land surfaces may return to the atmosphere through evaporation and transpiration (i.e., the water used by plants, animals, and humans that is provided back to the atmosphere). The rest penetrates the ground or flows into streams and rivers. The water that penetrates the soil flows deep underground (*groundwater recharge*) if it is not consumed by plants or humans.

Groundwater recharge can also be obtained, intentionally or not, from rivers, canals, excess irrigation, and artificial recharge ponds. Groundwater can flow in the subsurface for a very long time and travel long distances before either flowing back to the surface in springs, humid zones, rivers, and lakes, or flowing directly into the sea. Geological formations or layers can act in extremely varied ways relative to the transmission and storage of water (see Chapters 3 and 4). When it returns to the surface, groundwater emerges in the form of seepage zones, springs, or emergences in streams. In this way, groundwater is very important to sustain stream discharges during dry periods.

The processes of this water cycle can be summarized by the following words: precipitation, evapotranspiration, surface storage, surface flow (runoff), groundwater storage, and groundwater flow (Figure 2.1). Groundwater reserves and flows are thus quite dependent on other processes, particularly on their spatial distribution and temporal occurrences. In other words, all the "compartments" of the water cycle are deeply interdependent, spatially and temporally.

In this context of the water cycle, water balance assessments can be calculated for defined zones and during defined periods. The most basic balance equation can be written at the land surface (Figure 2.2) as:

$$P = ET + R + I \tag{2.1}$$

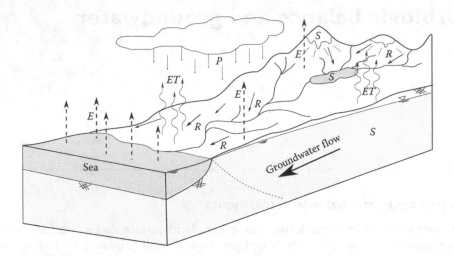

Figure 2.1 Hydrological or water cycle with seawater evaporation (*E*), precipitation (*P*), evapo-transpiration (*ET*), surface storage (*S*) and surface flow (runoff) (*R*), groundwater storage (*S*), and groundwater flow.

where *P* is the precipitation, *ET* is the evapotranspiration, *R* is the runoff (i.e., surface flow), and *I* is the infiltration. More details on these processes will be provided in the following sections. Values are usually expressed in "water depth" divided by a time period, such as mm/day, mm/month, or mm/year (or inches/day, inches/month, inches/year in the United States). Considering a volume (of water) per land surface unit divided by a duration, this provides a water height or column per time period [LT^{-1}].

Water balance, as used in hydrology, is most often expressed at the scale of a *drainage basin* or *watershed* (also referred to as *river basin* and *catchment*, respectively). The latter is related to a stream and has assumed impermeable boundaries corresponding to the water divides. However, the watershed definition is usually based on surface water flows (i.e., topographical water divide boundaries) more than on hydrogeological conditions. The surface water divide often does not correspond to the groundwater

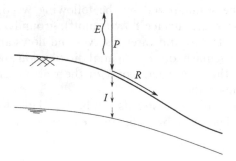

Figure 2.2 Processes playing a role in the water balance at the land surface.

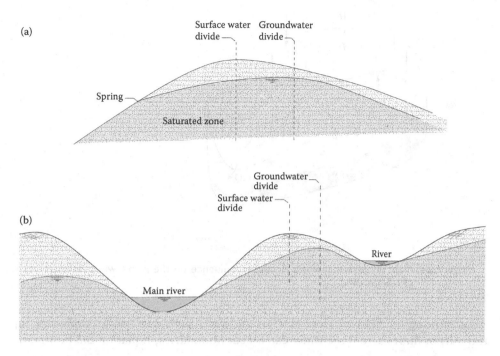

Figure 2.3 Topography combined to geological heterogeneity (a), or combined to different groundwater drainage base levels (rivers) (b), can induce nonzero groundwater fluxes at the catchment boundaries.

divide (Figure 2.3). For a given watershed, the water balance equation is a global approach for quantifying the water mass exchanges between the surface water, soil water, and groundwater reservoirs. The balance is simply expressed as a mass conservation principle:

Input = Output + Changes in reserves   (2.2)

Indeed, all the terms of this equation must be considered for the same time period. For a given watershed, Equation 2.2 is written as:

$$P = ET + Q + \Delta Res \qquad\qquad (2.3)$$

where Q [LT^{-1}] is the quantity of water leaving the watershed at the outlet calculated by the integration of the river discharge rate at the outlet during the considered period (Figure 2.4), and ΔRes [LT^{-1}] is the variation in the surface water and groundwater reserves. As stated previously, units of water height per time are used. This assumes implicitly that all the quantities of exchanged or stored water during the balance period, translated into water height per time, are considered to be uniformly distributed over the whole catchment.

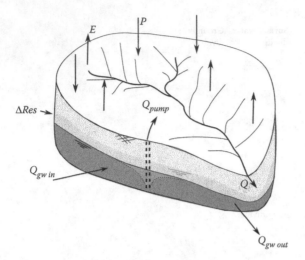

Figure 2.4 Main processes having a significant influence on the global water balance equation of a watershed.

Accounting explicitly for possible groundwater pumping and groundwater input and output fluxes across the water catchment boundaries, Equation 2.3 becomes:

$$P + Q_{gw\,in} = ET + Q + Q_{pumping} + Q_{gw\,out} + \Delta Res \qquad (2.4)$$

where $Q_{gw\,in}$ and $Q_{gw\,out}$ [LT^{-1}] state the groundwater input and output fluxes, respectively, across the watershed lateral underground boundaries (Figure 2.4), and $Q_{pumping}$ [LT^{-1}] is the pumped groundwater in the watershed. In practice, however, $Q_{gw\,in}$ and $Q_{gw\,out}$ are difficult to quantify, and Equation 2.4 is instead written as:

$$P = ET + Q + Q_{pumping} + \Delta Res + \varepsilon \qquad (2.5)$$

where ε is usually referred as to the water balance closure error. This error ε can be the result of inaccuracies or neglected groundwater exchanges fluxes. This makes a thorough interpretation of a balance study particularly difficult in water catchments where measurements are sparse and hydrogeological conditions are not well defined. While an unbalanced water budget could be interpreted or explained by possible groundwater fluxes that are not accounted for, the imbalance may also result from inaccuracies in the assessment of the other terms. On the other hand, a seemingly balanced water budget could also be the result of globally balanced errors committed during the assessment of each term including neglected groundwater fluxes $Q_{gw\,in}$ and $Q_{gw\,out}$ (e.g., an overestimation of *ET* balanced by an underestimation of *Q*, an over-estimation of *ET* balanced by neglecting $Q_{gw\,out}$, or an underestimation of *Q* balanced by neglecting $Q_{gw\,in}$).

Many human activities may lead to significant direct or indirect changes in the water balance of a watershed. Among others: deforestation, wetlands drainage, agriculture, irrigation, dams, industrial cooling, artificial inter-basin transfers, urbanization, mining, sluices, and leaking pipes. For example, the effects of increased groundwater pumping could be assessed at the watershed scale with accurate water budgets before a further detailed investigation on the hydrogeological conditions.

Understanding the impact of human activities on a water budget is often a very challenging topic. While the calculation of water budgets can be complex because of the different origins of errors, this is even more true for the interpretation of results. For example, a high recharge one year does not imply that the groundwater available for pumping will be increased by the same ratio: greater recharge will induce more drainage by the streams. These specific aspects are often discussed when trying to define "sustainable pumping" compared to "sustainable groundwater resources" in general (Loáiciga 2003, 2006, Devlin and Sophocleous 2005). Indeed, stream water allocations are very important, and dependent ecosystems must be considered during their estimation.

Water-budget analysis is a tool for operational water management. Most often, this analysis is still done by assuming a stationary local climate, while the current climate change should be considered for dynamic long-term analyses (Loáiciga 2017).

The different terms of the water balance equation will be described in the following paragraphs. However, detailed information about the surface processes associated with the water balance can usually be found in books about hydrology and will not be repeated here. On the other hand, there will be a particular emphasis on the assessment of groundwater recharge and drainable groundwater reserves (see Section 2.4), which are of great importance to groundwater studies.

2.2 Precipitation

Precipitation includes rain, drizzle, fog, mist, and dew, as well as snow, sleet hail, snow and ice pellets, freezing drizzle, and rain. The true spatial variability of precipitation is usually approximated by the extrapolation and interpolation of measurements made over time at only a few discrete points where precipitation gauges (or pluviometers) have been installed. Although the methods of obtaining reliable measurements for such different and variable processes will not be described here, we should note that measurements of local precipitation values are affected by (Dingman 1993):

- The high spatial and temporal variabilities of the processes relative to a limited number of gauges
- Local turbulence, and specific wind conditions
- Altitude, land use, orographic, and topographic effects
- Biases in rain gauges measurements resulting from dew, mist, fog condensation, and variable density snows

Discrete point measures are extrapolated and interpolated to provide averaged estimations at the regional scale. The combination of remote sensing data with discrete ground station measurements allows often improved estimates of P, particularly in "hard-to-reach" areas.

Precipitation heights corresponding to snow measurements that were taken with standard equipment for rain can be misleading. The snow water content can be determined by weighing snow cores from representative points within a catchment.

Snow cover controls can be performed by satellite images. In temperate and cold weather regions, snow precipitation is very important because the groundwater recharge is optimal during the snow melt periods when evapotranspiration is almost zero.

In practice, averaged precipitation values are used for balance water assessments. Time- and space-averaged values are calculated assuming that the averaging procedure does not bias the assessment. Unfortunately, the validation of the calculated values is difficult or nearly impossible in most cases. Long measurement periods are needed to establish significant time-averaged values. For example, the error in yearly averaged values is considered proportional to $1/\sqrt{t_t}$ where t_t is the total record duration in year (e.g., 20% for 25-year record, 14% for a 50-year record).

Time averaged values

Except for flood predictions, average precipitation values are used for daily, monthly, or even yearly periods. Averaging values over time impacts always the reliability of water balance terms. For example, for the same average value of P, the intensity of hydrological processes (e.g., runoff and infiltration) resulting from one month of low and continuous precipitation can be drastically different than that for a few intensive precipitation events. The time averaging of P creates smoothed values, while the reality can be considerably more variable over time. Considering the global water balance terms of Equation 2.1, one can easily understand that the time averaging of P tends to produce: (1) an overestimation of evapotranspiration (ET) resulting from the assumed water quantity available for evaporation processes during the whole period, (2) an underestimation of runoff (R) resulting from the relatively low value of the whole period replacing, by averaging, the real precipitation peaks, and (3) dependent on the (ET) conditions, a possible overestimation of infiltration (I) as $I = P - ET - R$ from Equation 2.1.

Spatial averaged values

As mentioned previously, the precipitation gauge network is always very limited compared to the actual variability of precipitation processes. Over a water catchment, the averaged values of the available discrete point measurements should normally account for the variable density of the point measure network, the type of land use in the catchment, the topography of the catchment, the slope orientation compared to the local prevailing wings that bring rain, and other possible influencing factors. Nevertheless, relatively simple spatial averaging procedures are still adopted in this water balance context. This results from the overall limited accuracy of these hydrological assessments compared to the true complexity of each process. In practice, the most common techniques use weighted averages, such as *Thiessen polygons*. In the Thiessen polygon method, the catchment is divided in subdomains around each of the n discrete measurement points. The Thiessen polygons are drawn around each of the points, and the weights (ω_i) are determined by the ratio:

$$\omega_i = \frac{a_i}{A} \rightarrow \quad \text{with} \quad \rightarrow i = 1, ..., n \rightarrow \quad \text{and} \quad \sum_{i=1}^{n} a_i = A \qquad (2.6)$$

where a_i represents the areas of the subdomains (polygons), A is the area of the whole catchment, and i is the index of each polygon (around each measurement point).

The spatial average $(\hat{P})$ is calculated as:

$$\hat{P} = \frac{1}{A} \sum_{i=1}^{n} a_i P_i \tag{2.7}$$

where P_i represents the measured values of precipitation at the n discrete measurement points.

Other techniques based on surface fitting or interpolation methods may be applied. The calculated surface therefore represents a smoothed model of the spatial variability of the measured P values. From an obtained map with contours of equal precipitation (referred to as *isohyets*), an areal average of P can be assessed. This can be done, for example, by considering the successive *isohyets* to serve as boundaries of subdomains within the catchment, where each subdomain is assigned a P value equal to the average of the values associated with its boundary isohyets (Dingman 1993).

There are multiple interpolation methods, and most of them are based on geostatistics to describe the spatial correlations between the measured data (see Chapter 12). For example, kriging and co-kriging, or even conditional and co-conditional stochastic simulations, allow the introduction of secondary data (e.g., land use or topographic information) to improve the description of the spatial variability of P. However, it is not rational to use techniques that are more elaborate than the quality of the measured data warrants. For example, it is often wiser to draw contours of equal precipitation (*isohyets*) manually, making use of physical and meteorological judgments to qualitatively account for different factors such as land use, elevation, topographic barriers, slope orientations, and prevailing rain origins.

2.3 Evapotranspiration

Evapotranspiration (*ET*) includes all the processes by which liquid and solid water become water vapor such as physical process of *evaporation* resulting from land-atmosphere energy exchanges under the influence of solar energy and dry winds and the plant *transpiration* processes in which soil water is absorbed by roots and transported through plants and evaporates from leaves, stems, and flowers. When water evaporates, heat energy (i.e., 590 calories per gram) is stored in the water vapor as latent heat. This energy mostly comes (directly or indirectly) from solar radiation. *ET* is most often the second largest component of the water balance (i.e., below precipitation). In arid and semi-arid zones, *ET* is nearly equivalent to P and is the most difficult water balance component to measure. This results from its wide spatial variability combined with its high temporal variability. The uncertainty affecting *ET* may often have a direct impact on the groundwater recharge assessments used to make critical groundwater management decisions. This uncertainty is particularly high in arid and semi-arid contexts where precipitation is limited (Van Camp *et al.* 2015).

It's not just hydrologists and hydrogeologists who use *ET* estimates. They are also used by plant physiologists, agronomists, and irrigation engineers who are interested in determining scientifically-based irrigation schedules. They use daily *ET* rates, while groundwater researchers are usually more interested in values averaged over weeks or months.

The measurement and quantification of the processes contributing to this upward flux of moisture into the atmosphere are quite complex. The estimation of averaged values (i.e., spatial and temporal averages) remains difficult and is based on data from multiple measurements.

It is far beyond the scope of this book to describe all these processes in detail and in a physically consistent way. After a brief description of the main factors influencing evaporation and transpiration, a few measurement techniques will be presented; then, empirical and physically-based estimation formulas will be proposed. If ET is not measured but is estimated by formulas, the key question becomes: how much ET actually occurs? As stated previously for precipitation (P), *evapotranspiration* (ET) is usually expressed in units of water height per time period [LT^{-1}]: mm/day, mm/month, or mm/year (or inches/day, inches/month, inches/year in the United States).

Evaporation

Solar energy is the main *evaporation* driver. Evaporation occurs when water is exposed to the air (i.e., at the land-surface, at the vegetation surface, in the soil, or in the unsaturated and shallow saturated zones), allowing water molecules to escape and form water vapor in the atmosphere. This process is active at all stages of the cycle but is very difficult to evaluate.

Evaporation depends on climatic variables (e.g., air temperature, air humidity, wind), and on landscape variables (e.g., surface temperature and roughness, humidity rate).

Transpiration

As stated previously, the water consumption by the plants or *transpiration* accounts for the uptake of water by roots and the subsequent loss as vapor through stomata in the leaves associated with the photosynthetic process. Transpiration depends on climatic variables (e.g., temperature, humidity, solar radiation), on landscape variables (e.g., soil type, slope, exposure angle, surrounding land use), and on plant growth characteristics (e.g., plant species, growth period, plant cycles).

ET measurements

Even if measurements of ET are difficult and complex, they are important and needed to validate values obtained by indirect and empirical methods. When evaluating their applicability to a site, users must be aware of the realistic uncertainty ranges of ET data resulting from the limitations, pitfalls, and high costs of the measurement techniques (Baker 2008). Direct point measurements and spatial measurements are very complementary and should be combined. Indirect methods include local and regional water balance investigations. A brief description of most of the techniques that are used, together with the required assumptions, is given by Shuttleworth (2008). One can practically delineate spatially different ET units or zones by mostly using land use information. Ideally, the ET rates should be measured during long periods in each of the dominant units.

Lysimeters

Measured values of evapotranspiration are often deduced from the soil water balance in lysimeters (Figure 2.5). This is the least expensive and thus most commonly used technique for *ET* measurements. An isolated soil sample for a given land use (i.e., different crops or vegetation) is used to measure the different terms of the local water balance. If precipitation (P), runoff (R), infiltration (I), and the difference in saturation ($S_f - S_i$) (i.e., the current saturation S_f compared to an initial value S_i) are measured, then *ET* can be obtained as follows:

$$ET = (S_f - S_i) + P - R - I \tag{2.8}$$

An undisturbed soil volume and high measurement accuracy are critical requirements to obtain a reliable estimate of the evapotranspiration during a given period. For example, a weighing lysimeter allows the continuous measurement of the gains and losses of saturation in the volume of soil. However, in the case of weighing lysimeters, the soil volume is usually smaller and more disturbed than the undisturbed natural soil column measured by a large lysimeter. In the latter type of lysimeter, drainage water is collected, and these data are often combined with miniaturized soil water pressure sensor measurements. Ideally, the chosen vegetation should match the existing vegetation in the immediate proximity of the lysimeter. Many questions can be asked about the representativeness of obtained values. In particular, significant scale and boundary effects (i.e., induced by the physical boundaries of the lysimeter) should be accounted for if the measured values are intended to be used at the catchment scale. For forested catchments, values from lysimeter measurements are poor representations of real forest conditions, while field conditions can be better represented for farm crops. In any case, lysimeter measurements are delicate, difficult, and take a long time, which makes them expensive.

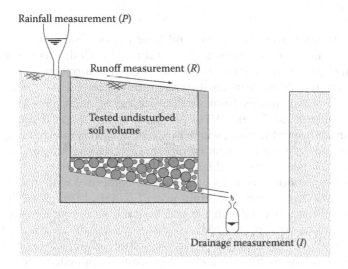

Figure 2.5 Schema of a lysimeter.

Micrometeorological fluxes methods

The *Bowen ratio-energy balance* (BREB) method is based on measurements of the net radiation above the surface and the soil heat flux below the surface. The difference between these measurements must correspond to the sensible heat (i.e., for heating the air) added to the latent heat (i.e., consumed by evaporation). The "Bowen ratio" between the sensible and latent heat is estimated from temperature and humidity measurements at two different elevations above the land surface. Thus, measurements require calibrated radiometers (for radiation), calibrated plates at depth supplemented by thermocouples in the surface layer (for soil heat flux), and a calibrated infrared gas analyzer or dew point hygrometer (for humidity gradients) (Baker 2008). All these instruments must be recalibrated periodically. Globally, BREB measurements are generally known as no better than 20% accurate (Meyers and Baldocchi 2005).

In the *eddy covariance* method, the vertical velocity and mixing ratio of eddies (parcels of air) are measured to deduce, from the mean covariance of these two variables, the vertical transport of water vapor through a given plane (Massman and Lee 2002). The needed humidity measurements are the most challenging to acquire, as is the case for the BREB method, and lead to the same level of accuracy. For both techniques, sensors should be located more than a few meters above the soil, particularly if the vegetation surface is rough and irregular. Unfortunately, they are not conventional or routine methods, as they require expensive installations and time-consuming operations on the long term.

In some very specific cases, daily gravity change data from superconducting gravimeters have allowed the direct measurement of *ET* in complex forested zones (e.g., Van Camp *et al.* 2016). Accurate and long-term gravity measurements can be used to infer the saturation variations of the partially saturated zone. However, many other factors influence the measured values, thus, they should be filtered out for reliable assessments.

Remote sensing methods

To provide reliable *ET* estimates over large areas and time periods, satellite-based maps can be combined with land-surface maps and be calibrated with discrete point measurements. This is an elegant way to scale-up site-based *ET* measurements.

Maps of vegetation cover are used to delineate different units where *ET* is expected to take different values. Thermal images from satellites integrate multiple factors affecting the *actual ET* (Allen *et al.* 2008). Most often, *ET* is estimated from an energy balance calculation at the land surface, where the sensible heat is inferred from the difference between air and evaporating surface temperatures, representing an estimate of the aerodynamic exchange resistance (Shuttleworth 2008).

Surface temperature and vegetation greenness maps may be combined with discrete point measurements in a statistical framework (Hultine *et al.* 2004). More generally, there is an extensive body of literature combining spatial *ET* data with point measurements (among many others: Cosgrove *et al.* 2003, Anderson *et al.* 2007, Allen *et al.* 2007a,b). However, satellite-based estimates are prone to multiple sources of random errors and systematic biases. Adequate filtering techniques must be applied; a description of these techniques is outside the scope of this book. Remote sensing algorithms can infer long-term water and energy budgets at the watershed and regional scale (Allen *et al.* 2008, Abiodun *et al.* 2018).

GRACE gravimetric satellite data may also be used. The change in soil water storage is derived (with a low resolution) from the measured changes in gravity detected by GRACE that essentially result from variations in the water table depth.

A Raman LIDAR may also be used for 3D measurements of water vapor concentrations, providing spatially resolved estimates of evapotranspiration. Eichinger *et al.* (2006) showed that this technique can be applied over an area of about a square kilometer, with a relatively fine (25 m) spatial resolution.

ET estimation equations

As direct measurements are generally expensive, time-consuming, and data-intensive, *ET* is often estimated using empirical or combined analytical equations (Woodhouse 2008). Considering the consumed energy for water passing into the atmosphere (i.e., *latent heat*), *ET* is dependent on the air temperature (relative to the surface), the wind velocity, the vegetation roughness, and the air dryness. These influences are integrated in "combined equations" that use measured climatic data (air temperature, air humidity, air velocity, and solar radiation) and landscape data (Allen 2008). Often, the proposed equations were originally developed for specific conditions (e.g., over crops or grass in flat and homogenous conditions).

The *ET* estimation is then performed in two steps. First, "*potential evapotranspiration*" (*PET*) is estimated under conditions of unlimited water availability. In the second step, the other water budget terms and the soil-water storage capacity are used to decrease the *PET* value to an *actual ET*. These calculations (see the following paragraphs) are far from free of errors and important approximations; in many situations, they are the best that can be done (Woodhouse 2008).

PET estimation

Although most of the following techniques were initially developed to be used with daily average values as needed in the fields of agronomy and irrigation engineering, we will focus on their applicability for hydrological balance studies, mostly considering monthly average values.

The Thornthwaite formula

An empirical formula providing an estimate of the potential evapotranspiration (*PET*) was developed by Thornthwaite (1948) based on correlations between the monthly mean temperature and evapotranspiration in zones where sufficient moisture was available for maintaining active transpiration. An average monthly *PET* (mm/month) is calculated as a complex function of the average monthly temperature and day length:

$$PET = 16\left(\frac{L_d}{12}\right)\left(\frac{N_m}{30}\right)\left(\frac{10T_m}{I_y}\right)^a \tag{2.9}$$

where T_m is the average monthly temperature (°C), N_m is the number of days in the month, L_d is the average day length (hours) of the month, I_y is the heat index of the year, which is calculated as follows: $I_y = \sum_{i=1}^{12} i_m$ where i_m the heat index of the month, and $i_m = (T_m/5)^{1.514}$, and $a = (6.75 \times 10^{-7})I_y^3 - (7.71 \times 10^{-5})I_y^2 + (1.792 \times 10^{-2})I_y + 0.49239$. The equation of a is often simplified as: $a = 0.016I_y + 0.5$ (Serra 1954).

Thornthwaite and Mather (1957) also published useful tables of values. This empirical formula cannot be validated rigorously but is widely used because of its simplicity, since it uses only temperature data. Let us reiterate that *PET* is defined as the evapotranspiration that occurs for ground covered by growing vegetation under unlimited water conditions. Generally, Thornthwaite's formula is known to underestimate *PET* in arid and semi-arid zones and overestimate *PET* in humid temperate areas (Pereira and Paes De Camargo 1989). The applicability of Thornthwaite's formula for monsoon climates has been checked by Kumar *et al.* (1987).

The Blaney-Criddle formula

The Blaney and Criddle (1950) formula provides the following estimate of a *reference ET* (i.e., for a "reference crop", which is considered actively growing green grass of 8–15 cm height):

$$ET_0 = p(0.46T_{mean} + 8.13) \tag{2.10}$$

where T_{mean} is the mean temperature (°C) during the considered period, and p is the mean percentage, for the period, of the annual daytime hours. In hydrology, the chosen period is usually one month, so the ET_0 unit is in mm/month.

As is also the case for the Penman-Monteith formula (see subsequent paragraphs), a crop coefficient can be associated with this reference value (ET_0) to determine more realistic values as the function of different vegetation types and growing seasons.

The Hamon formula

A monthly potential evapotranspiration value is estimated based on the mean air temperature and the possible hours of sunshine (Hamon 1961). This formula is written as (McCabe and Markstrom 2007):

$$PET = 13.97d\ D^2W_t \tag{2.11}$$

where *PET* is expressed in mm/month, d is the number of days in the month, D is the mean monthly hours of daylight in units of 12 hours, and W_t is a saturated water vapor density term calculated by $W_t = 0.0495\ e^{(0.062\ T_m)}$ where T_m the mean monthly air temperature (°C).

The Serra formula

A monthly potential evapotranspiration (mm/month) estimate is given by the Serra formula written as:

$$PET = 22.5\left(\frac{1-\varepsilon_m}{0.25}\right)\left(1-\frac{\tau^2}{1000}\right)e^{-0.0644T_m} \tag{2.12}$$

where T_m is the mean monthly temperature (°C), τ is the half-amplitude of the range between the extreme high and low temperatures of the month, and ε_m is the mean monthly air moisture.

The Penman-Monteith formula

A wide range of different methods are available to estimate PET or ET_0. Their results can be compared (Federer et al. 1996) but not rigorously validated. Some are temperature-based equations requiring only temperature and, possibly, daylight data (see previous paragraphs). These have shown strong limitations, particularly in arid or semi-arid regions (among others: Hashemi and Habibian 1979). A useful inter-comparison and evaluation of these methods can be found in Xu and Singh (2001). Other methods are mostly radiation-based equations (Priestley and Taylor 1972, Shuttleworth and Calder 1979). However, when radiation, temperature and humidity data are available, combination equations are recommended because they are known to be more reliable. This is the case for the Penman-Monteith equation, which uses net radiation, air temperature, wind speed, and relative humidity data. The initial equation developed by Penman (1948) was modified by Monteith (1965) to represent a reference evapotranspiration accounting for canopy conductance.

The relative accuracy of the Penman-Monteith approach in both arid and humid climates has been recognized and indicated by many studies from a wide range of locations and climates. This worldwide recognition has led to the formulation of a standardized equation that is recommended by the FAO (Food and Agricultural Organization). The FAO formulation (referred to as the FAO-56 procedure) of the Penman-Monteith equation is written as (Allen et al. 1998):

$$ET_0 = \frac{0.408\Delta(R_n - G) + \gamma(900/(T+273))u_2(e_s - e_a)}{\Delta + \gamma(1+0.34u_2)} \tag{2.13}$$

where ET_0 is the reference evapotranspiration (mm/day), Δ is the rate of change in the saturation vapor pressure with air temperature, the slope of the vapor pressure curve (kPa/°C), R_n is the net radiation at the crop surface (MJ/(m²day)), G is the soil heat flux density (MJ/(m²day)), γ is the psychrometric constant (kPa/°C), T is the mean daily air temperature at a 2 m height (°C), u_2 is the wind speed at a 2 m height (m/s), e_s is the saturation vapor pressure (kPa), e_a is the actual vapor pressure (kPa), and $(e_s - e_a)$ is the saturation vapor pressure deficit (kPa). Some of these required data are measured directly by weather stations. For the other parameters, the reader is advised to refer directly to the FAO Guidelines (Allen et al. 1998) to derive them from direct or empirical relationships. The Penman-Monteith *reference evapotranspiration* (ET_0) is defined as "the rate of evapotranspiration from a hypothetical reference crop with an assumed crop height of 0.12 m, a fixed surface resistance of 70 sec m⁻¹ and an albedo of 0.23, closely resembling the evapotranspiration from an extensive surface of green grass of uniform height, actively growing, well-watered, and completely shading the ground" (Allen *et al.* 1998, Irmak and Haman 2014).

Note that *PET* is not related to a specific crop. Instead, *PET* corresponds to the ET_0 of a grass lawn during its growing phase that has no water or nutritional restrictions (Verstraeten *et al.* 2008). A crop factor (K_c) should be used to correct *PET* for vegetation other than lawns:

$$PET = K_c ET_0 \qquad (2.14)$$

Since the evapotranspiration from well-watered agricultural crops may be 10%–30% greater than that from short green grass (Irmak and Haman 2014), it is important to give each vegetation cover unit a realistic but globally estimated crop factor. In hydrology, these corrections are not made at the parcel scale as is usually done in irrigation engineering. There is an extensive body of literature about crop factors in the agronomic and irrigation scientific communities. For example, a K_c larger than 1 is found for full-grown crops with close plant spacing and a large canopy height and roughness. On the other hand, for deciduous fruit trees cultivated without a ground cover crop, K_c values lower than 1 are found resulting from stomata on only the lower side of the leaf and the wide spacing of the trees (for mature trees, only 70% of the ground covered). A set of useful tables and curves can be found in the FAO-56 procedure document (Allen *et al.* 1998). A detailed example of the use of crop factors for wetlands can be found in Stannard *et al.* (2013).

In the same way, stress coefficients (K_s) may account for reductions in *ET* (compared to ET_0 or to $K_c ET_0$) resulting from a decrease in the plant transpiration induced by environmental stress (e.g., a lack of available water, air pollution, and soil salinization). Usually, native vegetation does not avoid recurring stressors of different origins. For crop factors, the ET_0 term of Equation 2.14 may be multiplied with an appropriate stress coefficient (Allen *et al.* 1998).

Actual ET estimation

THE THORNTHWAITE WATER BALANCE MODEL

Using *PET* values, a monthly simplified soil water balance model was developed by Thornthwaite and Mather (1955, 1957), and Mather (1969) for deducing the actual *ET* and monthly averaged values for groundwater recharge and runoff. This conceptual simplified model is known as the Thornthwaite water balance model. Assuming that the quantity of water in the soil (and unsaturated zone) can be conceptualized by a single water storage value (S_i) expressed in mm at the end of the month i with a given maximum water storage capacity (S_{max}), a monthly accounting procedure is adopted, as described in the flow sheet of Figure 2.6. Moisture is added to or subtracted from S_i depending on whether P_i (precipitation of the month i) is greater than or less than PET_i (average potential evapotranspiration of the month i). When S_i is increased, its value cannot be greater than S_{max}, and a possible water surplus is generated encompassing the groundwater recharge (R_{gw_i}) and runoff (R_i). On the other hand, when S_i is decreased, its value cannot be lower than zero and a possible water

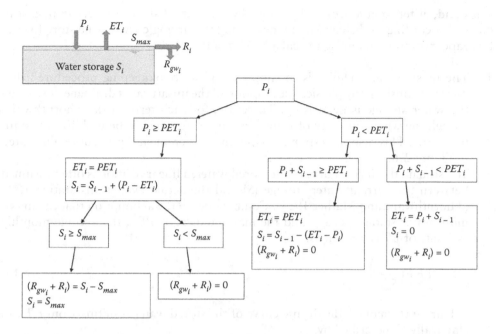

Figure 2.6 Conceptual schema and flow sheet of the Thornthwaite water balance model.

deficit is generated, causing ET_i (average actual evapotranspiration of the month i) to be lower than PET_i. This method was first applied for the water balance of the "root zone", and since then, it has been applied for the assessment of groundwater recharge at a whole watershed scale (Thornthwaite and Mather 1957).

Many slightly modified versions of this method have been applied in practice (Alley 1984). First, a choice must be made about the P_i (precipitation of the month i) that is used as input.

a. The total precipitation P_i is used as the input. In this case, when the soil water storage reaches its maximum capacity, any water surplus represents the sum of both the runoff generation (R_i) and the groundwater recharge (R_{gw_i}) for the month. This is the case illustrated in Figure 2.6.
b. An assumed value of the total runoff R_i (during the month i) is subtracted from P_i, and $P_i' = (P_i - R_i)$ is used as the input in the model. In this case, runoff is not considered in the model anymore; consequently, when the soil water storage reaches its maximum capacity, any water surplus only represents the groundwater recharge (R_{gw_i}).
c. A direct runoff value approximated as roughly as 5% of P_i is subtracted from P_i, and $P_i' = 0.95\,P_i$ is used as the input in the model (Wolock and McCabe 1999). In this case, when the soil water storage reaches its maximum capacity, any water surplus represents the sum of an additional generated runoff (R_i) and the groundwater recharge (R_{gw_i}). It is often proposed to divide this surplus value into equal parts (McCabe and Markstrom 2007). This choice can be considered to be very arbitrary, and one should find physical arguments to justify it.

Second, different practices of the Thornthwaite water balance model can be distinguished according to the chosen law describing the drying curve of the stored water by evapotranspiration during the calculated month.

1. The most classical choice is to simply adopt a "bookkeeping procedure" that does not consider any physical description of the unsaturated drainage flow when the water storage is decreasing. Therefore, for each term in the Thornthwaite model, the whole quantity of water (in mm) is assumed to be mobilized instantaneously. This case corresponds to the simple incremental procedure illustrated in Figure 2.6.

2. To estimate the drying curve of the stored water, a linear relation is often assumed between the current water storage (S) and the actual evapotranspiration (ET) (Thornthwaite and Mather 1955). Combined with the assumption that the maximum water storage is S_{max}, and the maximum ET is PET, the average monthly estimate of ET is written as:

$$ET = PET\left(\frac{S}{S_{max}}\right) \tag{2.15}$$

During the month, the drying curve of the stored water resulting from ET can classically be described by:

$$\frac{dS}{dt} = -ET \tag{2.16}$$

where S_{i-1} and S_i are the initial and final values during the month, respectively. Inserting Equation 2.15 into Equation 2.16, separating the variables and integrating both sides gives:

$$\ln\frac{S_i}{S_{i-1}} = -\frac{1}{S_{max}} PET_i \tag{2.17}$$

and

$$S_i = S_{i-1}e^{-(PET_i/S_{max})} \tag{2.18}$$

Thus, the actual evaporation (ET) for the month i is written as:

$$ET_i = P_i + S_{i-1}\left(1 - e^{-\frac{PET_i}{S_{max}}}\right) \tag{2.19}$$

In this case, when S_i decreases, it does not reach a zero value.

Applied to the same case study, these two ways of describing the drying of the stored water may lead to slight differences in the results in terms of calculated actual ET and water surplus (i.e., the sum of runoff generation (R_i) and groundwater recharge (R_{gw_i})) (Table 2.1, Figure 2.7).

Table 2.1 Thornthwaite balance model calculations for a typical hydrologic year in a temperate oceanic climate in northwestern Europe (as found typically near Brussels, Belgium). Given a maximum storage capacity of 100 mm, calculation results are shown for, respectively, the water storage (S_i), the actual evapotranspiration (ET_i), and the sum of runoff and groundwater recharge $(R_i + R_{gw_i})$ applying, respectively: (1) the "bookkeeping procedure," and (2) the negative exponential drying curve of the water storage as written in Equation 2.18. Slight differences can be detected in the results showing a slower decrease of the water storage resulting from the latter drying curve.

	P_i	PET_i	"Bookkeeping procedure" (1) mm/month			Negative exponential drying of the water storage (2) mm/month		
			S_i	ET_i	$R_i + R_{gw_i}$	S_i	ET_i	$R_i + R_{gw_i}$
January	70.9	8.5	100	8.5	62.4	100	8.5	62.4
February	77.4	21.4	100	21.4	56.0	100	21.4	56.0
March	54.9	38.8	100	38.8	16.1	100	38.8	16.1
April	29.4	67.7	61.7	67.7	0.0	68.2	61.2	0.0
May	76.3	81.2	56.8	81.2	0.0	64.9	79.6	0.0
June	42.7	107.4	0.0	99.5	0.0	34.0	73.6	0.0
July	71.7	103.1	0.0	71.7	0.0	24.8	80.9	0.0
August	49.3	81.4	0.0	49.3	0.0	18.0	56.1	0.0
September	89.1	68.3	20.8	68.3	0.0	38.8	68.3	0.0
October	117.8	57.2	81.4	57.2	0.0	99.4	57.2	0.0
November	90.7	14	100	14	58.1	100	14	76.1
December	114.6	15	100	15	99.6	100	15	99.6
Year Total	884.8			592.6	292.2		574.6	310.2

The Thornthwaite water balance model is conceptually very simple, but different factors must be discussed because they may significantly influence the results.

The maximum water storage capacity (S_{max}) is prescribed empirically based on qualitative data about the root zone where moisture evapotranspiration can occur. At the very least, a sensitivity analysis should be performed to assess its influence on results $(ET_i, R_i + R_{gw_i})$ of the chosen value. The balance calculations are preferably started during a period when an initial soil water storage can be assumed as the maximum water storage capacity (S_{max}).

Evaporation is assumed to occur at the potential rate as long as $(P_i + S_i) \geq PET_i$ even if, in reality, the actual evapotranspiration varies with plant root characteristics. This may lead to an overestimation of ET_i and consequently an underestimation of $R_i + R_{gw_i}$.

Despite these drawbacks, the method is known to produce reasonably reliable results at the monthly scale (among others: Alley 1984, Steenhuis and Van der Molen 1986, Varni and Usunoff 1999, Liebe *et al.* 2009, Sridhar and Hubbard 2010). Many authors have developed alternative or Thornthwaite-inspired methods that are more consistent with a daily step model and capture the nonlinearities of some processes (e.g., Pistocchi *et al.* 2008).

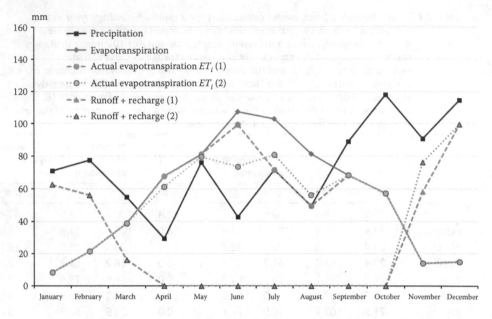

Figure 2.7 Graphical illustration of Table 2.1 results of the Thornthwaite water balance model. The calculated water storage values (*S_i*), actual evapotranspiration (*ET_i*), and water surplus (i.e., the sum of runoff (*R_i*) and groundwater recharge (*R_{gw_i}*)) are shown for the "bookkeeping" procedure (1), and for the negative exponential simulation of the water storage drying out (2).

THE TURC FORMULA

For large scale hydrology and catchments, a simple formula (Turc 1954) is used to provide yearly averaged values of actual evapotranspiration:

$$ET = \frac{P}{\sqrt{0.9 + (P/PET)^2}} \tag{2.20}$$

where *PET* is the average annual potential evapotranspiration approximated by a polynomial function of the mean annual temperature (T_y) written as (e.g., Meinardi 1994):

$$PET = 300 + 25T_y + 0.05T_y^3 \tag{2.21}$$

Initially, based on balance studies for more than 200 hydrological catchments in France, some practitioners and researchers argued that this simple formula was applicable to all hydrological circumstances if applied in large domains. Pistocchi *et al.* (2008) generalized this approach for describing actual *ET* at a monthly scale, while Lebecherel *et al.* (2013) proposed a regionalization approach for improved efficiency on ungauged catchments.

INTEGRATED MODELS

Currently, more and more "integrated hydrological models" include actual evapotranspiration and groundwater recharge calculations using process-based approaches involving the energy and water balances in the unsaturated zone (among others: VanderKwaak and Loague 2001, Panday and Huyakorn 2004, Ebel and Loague 2006, Kollet and Maxwell 2006, Maxwell *et al.* 2007, Maneta *et al.* 2008, Goderniaux *et al.* 2009, Delfs *et al.* 2013, Condon and Maxwell 2014, Niu *et al.* 2014, Hwang *et al.* 2015, Li *et al.* 2015).

2.4 Recharge

Introduction

Infiltration (i.e., water percolation into the subsurface) in the unsaturated zone is often considered as a mostly vertical process corresponding implicitly to a relatively uniform water flow in uniformly permeable layers. On the other hand, if layers or heterogeneities with contrasts in vertical permeability are found at depth, the water can move (for a certain distance) along a lateral component in the unsaturated zone. This hypodermic flow is usually referred to as *interflow* (Figure 2.8).

The remaining infiltration is considered as the *groundwater recharge*. Recharge is the main driver of any groundwater system, even if many aquifers contain "fossil" groundwater that infiltrated thousands of years ago. The assessment of recharge is essential for any water cycle and balance calculation and is often the first step in groundwater resource management. In this way, this topic has a central importance for humans and river-dependent ecosystems in the general water-food-energy nexus (Smerdon and Drewes 2017).

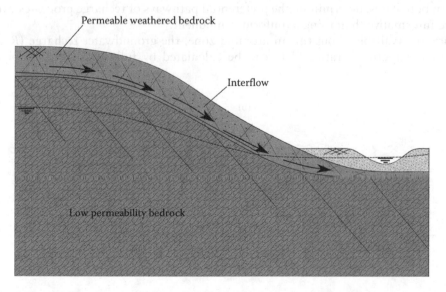

Figure 2.8 Interflow is defined as a hypodermic flow with a horizontal component occurring above a lower permeability geological formation.

There is a wide variety of recharge conditions. One can distinguish *diffuse recharge* derived from assumed uniform precipitation and/or irrigation conditions in large areas from *localized recharge* resulting from streams, lakes or any surface water body. Too often, only uniform rates of recharge are considered. If this important assumption is chosen, it may not affect further groundwater resource estimation too much; however, it may greatly affect contaminant transport predictions in areas with a highly variable input of water in the groundwater system (Scanlon *et al.* 1997). Some authors are also making a distinction between gross recharge and net recharge, where the latter is the gross recharge minus the water extracted by evapotranspiration from the saturated zone (Crosbie *et al.* 2010).

Methods for estimating recharge are the topic of many papers and books (e.g., Petheram *et al.* 2002, Scanlon *et al.* 2006, Healy 2010, Kim and Jackson 2012). They are generally classified into physical, tracer, and numerical modeling approaches (Scanlon *et al.* 2002). Any evaluation must be considered only as an approximate value resulting from the high temporal and spatial variability of the main influencing factors. Uncertainties should be considered accurately. To obtain reliable estimates, one should first check the applicability of the chosen method, obtain the data needed for the chosen method, and assess the expected spatial variability that is dependent on the groundwater system and overlying layers heterogeneities. Comparisons between the results from different methods can often be very informative, even if they theoretically do not guarantee increased reliability (Healy and Cook 2002, Somaratne *et al.* 2014). Bias can be corrected by combining ground and remote sensing techniques (e.g., Crosbie *et al.* 2015). Beyond the classical errors and uncertainties resulting from poor spatial coverage and sparse data, problems arise when the chosen conceptual model is not valid in the whole catchment. For example, a uniform diffuse recharge model can be far from the reality of the true recharge processes in karstic or fractured groundwater systems (Figure 2.9) (e.g., Guardiola-Albert *et al.* 2015, Alazard *et al.* 2016). In both cases, determining the preferential pathways of recharge processes can be more informative than using a uniform component.

If data are available about the unsaturated zone, the groundwater recharge (R_{gw}) and actual evapotranspiration (ET) can be calculated by fully integrated coupled

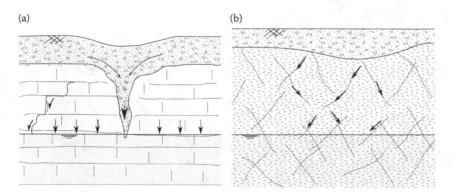

Figure 2.9 Actual recharge processes in karstic (a) and fractured (b) groundwater systems showing preferential pathways.

surface-unsaturated-saturated flow models (see previous section), or by 1D unsaturated flow models coupled to groundwater saturated flow models. ET and R_{gw} can be considered as a part of the output of a catchment scale groundwater model (e.g., Doble and Crosbie 2017). Except in these processes-based partially or fully coupled models, aquifer recharge is often considered essential data for the further modeling of groundwater resources (De Vries and Simmers 2002). Therefore, in this case, the recharge value should ideally be assessed by water balance and water table fluctuation methods and/or tracer techniques. Then, the recharge value can be prescribed as a boundary condition for further groundwater modeling (see Chapter 12).

As mentioned previously, the determination of groundwater recharge in arid and semi-arid zones is particularly difficult (Lerner *et al.* 1990). In these areas, the recharge flux has a magnitude close to those of the uncertainties and errors that commonly occur during the measurements and calculation of R_{gw} values.

The water balance method, as discussed in the previous section, allows estimates of the groundwater recharge (R_{gw}) to be obtained with the actual evapotranspiration (ET). Here, other methods for estimating recharge are briefly summarized: the water table fluctuation (WTF) method and the chloride mass balance (CMB) method. The latter can be qualified as a tracer technique. However, other tracer techniques used for recharge estimates will be addressed in Chapter 8, as most of them require a clear understanding of the solute transport processes affecting the different tracers and isotopes. They are often used for recharge history (e.g., Petersen *et al.* 2014) and are therefore related to the residence time of water in the groundwater system.

Water table fluctuation method

The *water table fluctuation* (WTF) method assumes that groundwater recharge is the only driver of a measured rise in the water table (in an unconfined aquifer, see Chapter 3) during a given time period Δt. This is expressed as:

$$R_{gw} = S_y \Delta h / \Delta t \tag{2.22}$$

where Δh is the change (rise) in the water table level during the period Δt, and S_y is the specific yield (defined in Section 4.2) equal to the effective (drainable) porosity of the medium. However, this method is based on the important assumption that the recharge is used exclusively to increase water storage and, accordingly, to increase the water table height. In other words, water level fluctuations are not supposed to affect other flow components, such as groundwater flow or evapotranspiration. This is particularly difficult to accept because the change in the water level is the main driver of groundwater flux changes (i.e., changes in lateral water discharge) in saturated media (see Chapter 4). Neglecting any increase in the lateral discharge resulting from a rise in the water level leads to the underestimation of the true recharge (i.e., the measured Δh is lower than the one in an assumed closed volume of the medium).

Moreover, it could be difficult to obtain a reliable representative value for the specific yield, especially for highly heterogeneous geological conditions. For all these reasons, the WTF method is only applied over short periods (i.e., heavy rain periods) in regions with a shallow water table where sharp variations in the water tables have been recorded (Scanlon *et al.* 2002).

Measured water table levels can also be partially biased by atmospheric pressure conditions, local and regional noncontinuous pumping or drainage, and evapotranspiration from the saturated zone (in arid and semi-arid zones) (Healy and Cook 2002, Sibanda *et al.* 2009). If local pumping is decreased in the influence zone during the period of the measured rise of the water table, the method may lead to an overestimation of the true recharge (i.e., the measured Δh will be higher than the one in an assumed closed volume of the medium during Δt). On the other hand, if a local pumping is increased, this may lead to an underestimation of the true recharge (i.e., the measured Δh is lower than the one in a assumed closed volume of the medium during Δt).

This method is used for interpreting relatively punctual (discrete) water table rises in long time series data to detect highly transient trends (Crosbie *et al.* 2005, Ordens *et al.* 2012).

An advantage of this method is that it integrates both preferential flowpaths and uniform recharge components. For example, a fractured bed-rock aquifer with minimal overburden cover could show a water table rise that exceeds the rise that would occur in a porous medium. This behavior results from the low value of the specific yield, possibly combined with inclined fractures exposed on or immediately adjacent to the outcrop (e.g., Miles and Novakowski 2016). In this case, an accurate measurement of the rapid water table response is possible for assessment of the regional recharge rates.

Chloride mass balance method

In the *chloride mass balance method*, one expresses the balance between the chloride content in rainfall (precipitation) and the chloride groundwater content in the unsaturated or saturated zone. Indeed, chloride from rainwater is concentrated in the infiltrating water by evapotranspiration processes. Thus, without any other source of chloride, the *Cl* concentration in groundwater relative to that in rainwater is assumed to be a measure of the proportion of the rainfall that has evaporated (Ordens *et al.* 2012). This can be expressed as:

$$R_{gw} = P\, C_{P+D} / C_{gw} \tag{2.23}$$

where P is the long-term average precipitation [LT^{-1}], C_{P+D} is the representative mean *Cl* concentration value (often a weighted average of measurements) in the precipitation including all dry depositions [ML^{-3}], and C_{gw} is the groundwater *Cl* concentration in the unsaturated or saturated zone. Initially reported by Schoeller (1941, 1962), this method was developed further by Eriksson and Khunakasem (1969).

This method is considered to provide reliable long-term estimates of R_{gw} (e.g., Alcalá and Custodio 2015), particularly in arid and semi-arid zones where other methods are especially difficult to implement. The method can not only be applied using the *Cl* profile in the unsaturated zone, but also using the measured *Cl* concentrations in the top of the saturated zone.

An extensive body of literature exists on the implicit and explicit assumptions that are required for the application of this method (e.g., Sibanda *et al.* 2009, Ordens *et al.* 2012, Somaratne 2015): (1) *Cl* is a conservative tracer, and its only source is atmospheric deposition (i.e., combined dry and rainfall deposition, with no sources/ sinks of Cl in the rock matrix), (2) there is no influence of the runoff in the recharge area (or its influence should be explicitly accounted for), (3) there are diffuse and steady-state conditions of the atmospheric solute input and recharge flux (i.e., a constant chloride profile in the unsaturated zone), and (4) there is 1D vertical recharge "piston" water flow in the unsaturated zone.

Practically, if no direct measurement of C_{P+D} is available, an empirical relation accounting for the dependency of the Cl content in the rainwater on proximity to the ocean is expressed as (Hutton 1976):

$$C_{P+D} = 35.45 \left(\frac{0.99}{d^{0.25}} - 0.23 \right) \tag{2.24}$$

where d is the distance from the coast line in km and C_{P+D} is expressed in mg/L. On the other hand, if C_{P+D} measurements are available, a weighted average chloride concentration (wet and dry) in the precipitation can be calculated using (Wood and Sanford 1995):

$$C_{P+D} = \frac{\sum_{i=1}^{n} P_i C_{pi}}{\sum_{i=1}^{n} P_i} \tag{2.25}$$

where n is the number of samples, P_i is the measured precipitation at the sample i location, and C_{pi} is the measured Cl concentration (including wet and dry processes) in the sample i.

Measurements of C_{gw} in Equation 2.23 have also been discussed extensively. Often, they are obtained by sampling conventional monitoring wells or water supply wells, or possibly by sampling a determined depth interval in the monitored well if a multiple level sampling scheme is applied (see Chapter 7) (Manna et al. 2016). As mentioned previously, C_{gw} can also be measured in the unsaturated zone, assuming a simplified chloride profile below the maximum depth of evapotranspiration (Figure 2.10). Indeed, if an unsaturated C_{gw} is used in Equation 2.23, the method does not account for the recharge supplied by preferential pathways such as macro-pores and fissures. Using detailed unsaturated flow and solute transport models (see Chapter 9), complex chloride profiles can be interpreted in more detail. This can be hugely important to assess recharge rates of the order of a few mm/year or lower in hyper arid deserts (e.g., Amundson et al. 2012, Bouhlassa et al. 2016).

Using the C_{gw} measured in the saturated zone guarantees to some extent an integration of the different recharge water fluxes (Somaratne 2015). However, if there is a large distance between the location where the water actually infiltrates and the sampling well, the chloride mixing processes in the saturated zone bias the results, as a "piston flow behavior" cannot be assumed anymore (see Chapter 8). The calculated recharge could be attributed to an infiltration area upgradient from the sampling well (Ordens et al. 2012). Corrections should be applied to describe the fate of chloride between the actual recharge zone and the sampling well using an accurate description of the groundwater flow (Chapter 4) and solute transport (Chapter 8).

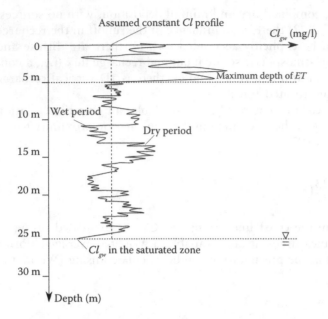

Figure 2.10 If an unsaturated Cl_{gw} is used for applying the chloride mass balance method (CMB), a simplified chloride profile is assumed below the maximum depth of evapotranspiration.

The CMB method has been widely used to estimate low recharge rates, and the maximum recharge rate that can be estimated by this method is estimated to be approximately 300 mm/year. This upper limit results from the uncertainties associated with the measurements of low Cl concentrations (i.e., associated with greater recharge rates) in comparison to other possible sources of chloride (Scanlon *et al.* 2002).

2.5 Base flow

Streamflow includes overland flow (or runoff) resulting from infiltration excesses (i.e., when the rate of rainfall exceeds the rate that water can infiltrate in the soil) or from saturation excesses (i.e., when the soil is fully saturated resulting from previous rainfall events), interflow (see Section 2.4), and base flow components from the geological formations drained by the stream. A description of streamflow and its associated measurement techniques is beyond the current scope of this section. Rather, the section focus is *base flow*, the groundwater discharge component (i.e., groundwater feeding the stream), with a very slow variation rate in comparison to the high frequency variations of the total streamflow resulting from precipitation events. Base flow is thus active during both dry and rainy periods. Because of the large groundwater reserves that can be drained by streams and the slowness of groundwater flow (see Chapter 4) in comparison to streamflow, the base flow changes only gradually over time.

In practice, base flow is often estimated using numerous conceptual and/or recursive filtering methods applied to a stream hydrograph (i.e., a time evolution diagram of the total discharge). However, these methods are difficult to validate (Figure 2.11).

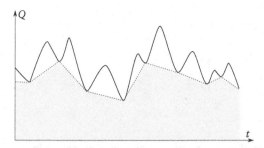

Figure 2.11 Schematic example of an empirical base flow separation assuming that an ensemble of minima in the observed hydrograph corresponds to the maximum envelope of the base flow.

This results from the empirical approaches that are used and the challenging inaccuracies of base flow measurements in the field. As mentioned previously (see Sections 2.3 and 2.4), "integrated hydrological models" using physically-based, surface water-groundwater flow simulations allow an accurate evaluation of the base flow to be obtained (e.g., Partington *et al*. 2012) if the available data enables an adequate calibration of the model (see Chapter 12).

During a period with no "excess precipitation" (i.e., when the assumption of no runoff and interflow components can be accepted) (Figure 2.12), the interpretation of the decreasing streamflow can be particularly interesting. For the interpretation of the corresponding *base flow recession*, a variety of methods have been developed to globally estimate (i.e., at the catchment scale) the groundwater reserves that can be drained by streams.

The earliest methods are based on a simple linear storage-discharge relationship between the groundwater domain, which is considered as a bucket reservoir, and the stream, which is considered as a limited outlet (Figure 2.13). According to Poiseuille's

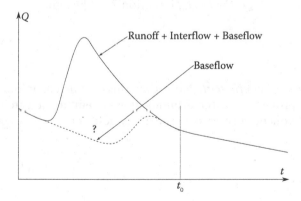

Figure 2.12 Typical hydrograph showing a peak resulting from runoff, interflow, and base flow as induced by heavy rainfall, followed by a long dry period and a base flow recession starting from t_0.

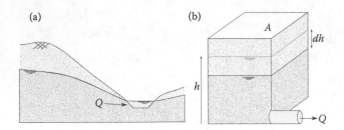

Figure 2.13 Schematic illustration of the reality (a), and of the Poiseuille's model (b) for representing base flow at the catchment scale. This conceptual model leads to the Maillet exponential reservoir model (Maillet 1905).

model for representing base flow at the catchment scale, the base flow discharge $[L^3 T^{-1}]$ is expressed as:

$$Q = Ch \tag{2.26}$$

where C $[L^2 T^{-1}]$ is a coefficient that is dependent only on geometry and properties of the reservoir and h is the water height in the reservoir (Figure 2.13b). During a period of time dt, a decrease in the water height dh is recorded, so that:

$$Qdt = Chdt = -Adh \tag{2.27}$$

where A is the effective area $[L^2]$ containing water in the reservoir (Figure 2.13b). Equation 2.27 is written as:

$$-\frac{C}{A}dt = \frac{dh}{h} \tag{2.28}$$

After integration, and given that $Q = Q_0$ for $t = t_0$, Equation 2.28 becomes:

$$\ln\frac{Q}{Q_0} = -\frac{C}{A}(t - t_0) = -\alpha(t - t_0) \tag{2.29}$$

where $\alpha = C/A$ $[T^{-1}]$ is the *recession coefficient*. Note that this coefficient is only dependent on the geometry and properties of the groundwater system at the catchment scale. This leads to the "exponential reservoir model" as described by Maillet (1905) and Horton (1933):

$$Q = Q_0 e^{-\alpha t} \tag{2.30}$$

with $t_0 = 0$.

Indeed, one problem of this method can be the choice of the arbitrary time t_0 (Figure 2.12). To help the practitioner, Fetter (2001) reported a rule of thumb from

Linsley *et al.* (1975) providing the number of days (D) after the hydrograph peak needed to obtain the end of the runoff and interflow components:

$$D = 0.827\ A^{0.2} \tag{2.31}$$

where A is the surface of the drainage basin expressed in km^2.

Practically, Equation 2.30 is used for assessing, at the catchment scale, the groundwater reserves that can be drained by the gauged stream. Based on different stream discharge measurements during the base flow recession period, a value can be deduced for the recession coefficient (α) (Figure 2.14). At the time t_0, the maximum groundwater reserves that can be potentially drained by the gauged stream (i.e., if the time is considered as infinite) can be found with (Figure 2.14):

$$\int_{t_0}^{\infty} Q\,dt = \int_{0}^{\infty} Q_0 e^{-\alpha t}\,dt = \frac{Q_0}{\alpha} \tag{2.32}$$

In the same way, the groundwater reserve variation (ΔRes) between two base flow measurements Q_1 and Q_2 at time t_1 and t_2, respectively, is given by (Figure 2.14):

$$\int_{t_1}^{t_2} Q\,dt = \frac{Q_1 - Q_2}{\alpha} = \Delta Res \tag{2.33}$$

One limitation of this recession curve interpretation is the fact that the true drainage of a groundwater system in a catchment is not similar to the emptying of a reservoir

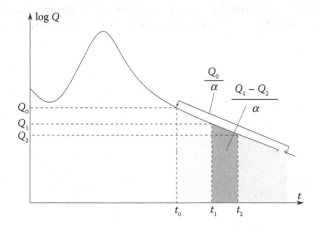

Figure 2.14 Practical use of the method for estimating the groundwater reserves that can be drained by the stream from time $t_0 = 0$, and for assessing the groundwater reserve variation between two measurements of the base flow (Maillet 1905).

with a horizontal water level. If we imagine a vertical cross section in a homogeneous medium, we can observe that the water level in the groundwater system (i.e., the water table, see Chapters 3 and 4) most likely describes a parabolic curve (Figure 2.13). Thus, using this linear storage-discharge relation leads to the subjective and unreliable fit of the recession line and the recession coefficient. The temptation is then great to use a multiple exponential reservoir model as follows (Figure 2.15) (Forkasiewicz and Paloc 1967):

$$Q(t) = Q_{01}e^{-\alpha_1 t} + Q_{02}e^{-\alpha_2 t} + \cdots + Q_{0n}e^{-\alpha_n t} \qquad (2.34)$$

where n is the total number of reservoirs. However, it is difficult, using these recession interpretation methods, to infer any value (or spatial heterogeneity) for the physical properties (see Chapter 4) of the groundwater system.

In their attempts to describe rainfall-runoff relationships, hydrologists have developed many methods to interpret hydrograph recessions. While their main aim is the prediction of the future behavior of a recession in a given catchment, there is controversy over the assumption of an existing time constant base flow recession coefficient (Thomas et al. 2015). Using a power law relationship between the global watershed storage and the base flow (Hall 1968), Brutsaert and Nieber (1977) proposed a combination with a water continuity equation (see Chapter 4) where the time variation of the storage can be expressed as a function of the base flow discharge, thus obtaining the relationship:

$$\frac{dQ}{dt} = -aQ^b \qquad (2.35)$$

where a and b are adjusted to experimental data. Many methods have been published for estimation of these parameters a and b varying (1) the recession segments in the hydrograph to be fitted, (2) the beginning time and end time of each recession, (3) the procedure used to fit Equation 2.35 to the true data, and (4) the approximation

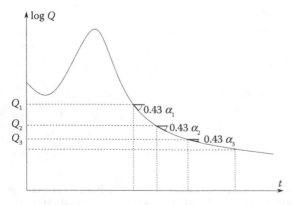

Figure 2.15 Interpretation of a base flow recession curve using a multiple exponential reservoirs model.

of the derivative dQ/dt at discharge Q (e.g., Roques *et al.* 2017). If $b = 1$ and $t = t_0$, Equation 2.35 is similar to Equation 2.30 with $a = \alpha$. Additionally, as proposed by Thomas *et al.* (2015), it could be necessary to explicitly account for seasonal effects or pumping impacts on the base flow response.

Beside these approaches, there is an extensive body of literature on analytical solutions describing hillslope subsurface flow. Most often, a link is made between the unconfined 2D steady state or transient groundwater flow equations (see Section 4.4) and simplified equations assuming the homogeneity of subsoil properties (hydraulic conductivity and drainable porosity, see Chapter 4) and different simplified geometries (e.g., Boussinesq 1877, Barnes 1939, Polubarinova-Kochina 1962, Henderson and Wooding 1964, Childs 1971, Beven 1981, Zecharias and Brutsaert 1988, Brutsaert 1994, Szilagyi *et al.* 1998, Verhoest and Troch 2000, Parlange *et al.* 2001, Troch *et al.* 2002, Mendoza *et al.* 2003, Troch *et al.* 2004, Rocha *et al.* 2007, Ajami *et al.* 2011, Thomas *et al.* 2015, Dralle *et al.* 2016, Ebrahim and Villholth 2016, Jepsen *et al.* 2016, Zhang *et al.* 2017). These approaches assume a simplified geometry and homogeneity to obtain information on the shallow groundwater availability, catchment recharge, or corresponding groundwater flow properties (see Chapter 4). Note that if the model is reduced to Equation 2.30, the recession coefficient α is known to be proportional to the hydraulic conductivity and inversely proportional to the drainage porosity.

References

Abiodun, O.O., Guan, H., Post, V.E.A. and O. Batelaan. 2018. Comparison of MODIS and SWAT evapotranspiration over a complex terrain at different spatial scales. *Hydrology and Earth System Sciences Discuss* 22: 2775–2794.

Ajami, H., Troch, P.A., Maddock III, T., Meixner, T. and C. Eastoe. 2011. Quantifying mountain block recharge by means of catchment-scale storage-discharge relationships. *Water Resources Research* 47: W04504.

Alazard, M., Boisson, A., Maréchal, J.C., Perrin, J., Dewandel, B., Schwarz, T., Pettenati, M., Picot-Colbeaux, G., Kloppman, W. and S. Ahmed. 2016. Investigation of recharge dynamics and flow paths in a fractured crystalline aquifer in semi-arid India using borehole logs: implications for managed aquifer recharge. *Hydrogeology Journal* 24(1): 35–57.

Alcalá, F.J. and E. Custodio. 2015. Natural uncertainty of spatial average aquifer recharge through atmospheric chloride mass balance in continental Spain. *Journal of Hydrology* 524: 642–661.

Allen, R.G. 2008. Why do we care about ET? *Southwest Hydrology* 7(1): 18–19.

Allen, R.G., Hendrickx, J.M.H., Toll, D., Anderson, M., Kleissl, J. and W. Kustas. 2008. From high overhead: ET measurement via remote sensing. *Southwest Hydrology* 7(1): 30–32.

Allen, R.G., Pereira, L.S., Raes, D. and M. Smith. 1998. Crop evapotranspiration—guidelines for computing crop water requirements. In *FAO Irrigation and drainage Paper 56*, Rome (Italy): Food and Agriculture Organization of the United Nations.

Allen, R.G., Tasumi, M. and R. Trezza. 2007a. Satellite-based energy balance for mapping evapotranspiration with internalized calibration (METRIC)—Model. *ASCE Journal of Irrigation & Drainage Engineering* 133: 380–394.

Allen, R.G., Tasumi, M., Morse, A. and R. Trezza. 2007b. Satellite-based energy balance for mapping evapotranspiration with internalized calibration (METRIC)—Applications. *ASCE Journal of Irrigation & Drainage Engineering* 133: 395–406.

Alley, W.M. 1984. On the treatment of evapotranspiration, soil moisture accounting, and aquifer recharge in monthly water balance models. *Water Resources Research* 20(8): 1137–1149.

Amundson, R., Barnes, J.D., Ewing, S., Heimsath, A. and G. Chong. 2012. The stable isotope composition of halite and sulphate of hyperarid soils and its relation to aqueous transport. *Geochim Cosmochim Acta* 99: 271–286.

Anderson, M.C., Kustas, W.P. and J.M. Norman. 2007. Upscaling flux observations from local to continental scales using thermal remote sensing. *Agronomy Journal* 99: 240–254.

Baker, J.M. 2008. Challenges and cautions in measuring evapotranspiration. *Southwest Hydrology* 7(1): 24–33.

Barnes, B.S. 1939. The structure of discharge-recession curves. *Transactions American Geophysical Union* 20(4): 721–725.

Beven, K. 1981. Kinematic subsurface storm flow. *Water Resources Research* 17(5): 1419–1424.

Blaney, H.F. and W.D. Criddle. 1950. Determining water requirements in irrigated areas from climatological irrigation data. In *Technical Paper No. 96*, Washington, DC: U.S. Department of Agriculture, Soil Conservation Service.

Bouhlassa, S., Ammary, B., Paré, S. and N. Safsaf. 2016. Integrating variations in the soil chloride profile and evaporativity for in-situ estimation of evaporation in arid zones: An application in south-eastern Morocco. *Hydrogeology Journal* 24(7): 1699–1706.

Boussinesq, J. 1877. Essai sur la théorie des eaux courantes du mouvement non permanent des eaux souterraines. [Essay on the theory of flowing waters and the transient movement of groundwater] in English. *Acad Sciences Institut France* 23: 252–260.

Brutsaert, W. 1994. The unit response of groundwater outflow from a hillslope. *Water Resources Research* 30(10): 2759–2763.

Brutsaert, W. 2005. *Hydrology: an introduction*. Cambridge: Cambridge University Press.

Brutsaert, W. and J.L. Nieber. 1977. Regionalized drought flow hydrographs from a mature glaciated plateau. *Water Resources Research* 13(3): 637–643.

Childs, E.C. 1971. Drainage of groundwater resting on a sloping bed. *Water Resources Research* 7(5): 1256–1263.

Condon, L.E. and R.M. Maxwell. 2014. Feedbacks between managed irrigation and water availability: Diagnosing temporal and spatial patterns using an integrated hydrologic model. *Water Resources Research* 50(3): 2600–2616.

Cosgrove, B.A., Lohmann, D., Mitchell, K.E., Houser, P.R., Wood, E.F., Schaake, J.C., Robock, A., Marshall, C., Sheffield, J., Duan, Q., Luo, L., Higgins, R.W., Pinker, R.T., Tarpley, J.D. and J. Meng. 2003. Real-time and retrospective forcing in the North American land data assimilation system (NLDAS) project. *Journal of Geophysical Research* 108(D22): 8842.

Crosbie, R.S., Binning, P. and J.D. Kalma. 2005. A time series approach to inferring groundwater recharge using the water table fluctuation method. *Water Resources Research* 41: W01008.

Crosbie, R.S., Davies, P., Harrington, N. and S. Lamontagne. 2015. Ground truthing groundwater-recharge estimates derived from remotely sensed evapotranspiration: a case in South Australia. *Hydrogeology Journal* 23(2): 335–350.

Crosbie, R.S., Jolly, I.D., Leaney, F.W. and C. Petheram. 2010. Can the dataset of field based recharge estimates in Australia be used to predict recharge in data-poor areas? *Hydrology and Earth System Sciences* 14: 2023–2038.

De Vries, J.J. and I. Simmers. 2002. Groundwater recharge: an overview of processes and challenges. *Hydrogeology Journal* 10: 5–17.

Delfs, J.-O., Wang, W., Kalbacher, T., Singh, A. and O. Kolditz. 2013. A coupled surface/subsurface flow model accounting for air entrapment and air pressure counterflow. *Environmental Earth Sciences* 69(2): 395–414.

Devlin, J.F. and M. Sophocleous. 2005. The persistence of the water budget myth and its relationship to sustainability. *Hydrogeology Journal* 13: 549–554.

Dingman, S.L. 1993. *Physical hydrology*. New York: McMillan Publishing Company.

Doble, R.C. and R.S. Crosbie. 2017. Review: Current and emerging methods for catchment-scale modelling of recharge and evapotranspiration from shallow groundwater. *Hydrogeology Journal* 25: 3–23.

Dralle, D.N., Karst, N.J. and S.E. Thompson. 2016. Dry season streamflow persistence in seasonal climates. *Water Resources Research* 52: 90–107.

Ebel, B.A. and K. Loague. 2006. Physics-based hydrologic-response simulation: Seeing through the fog of equifinality. *Hydrological Processes* 20(13): 2887–2900.

Ebrahim, G.Y. and K.G. Villholth. 2016. Estimating shallow groundwater availability in small catchments using streamflow recession and instream flow requirements of rivers in South Africa. *Journal of Hydrology* 541: 754–765.

Eichinger, W.E., Cooper, D.I., Hipps, L.E., Kustas, W.P., Neale, C.M.U. and J.H. Prueger. 2006. Spatial and temporal variation in evapotranspiration using Raman lidar. *Advances in Water Resources* 29(2): 369–381.

Eriksson, E. and V. Khunakasem. 1969. Chloride concentration in groundwater, recharge rate and rate of deposition of chloride in the Israel Coastal Plain. *Journal of Hydrology* 7: 178–197.

Federer, C.A., Vörösmarty, C. and B. Fekete. 1996. Intercomparison of methods for calculating potential evaporation in regional and global water balance models. *Water Resources Research* 32(7): 2315–2321.

Fetter, C.W. 2001. *Applied hydrogeology* (4th edition). Upper Saddle River (NJ): Pearson Education Limited.

Forkasiewicz, J. and H. Paloc. 1967. Le régime de tarissement de la foux-de-la-Vis. Etude préliminaire [First study on the depletion curve of the foux-de-la-Vis spring] in French. *Chronique d'Hydrogéologie* 3(10): 61–73.

Goderniaux, P., Brouyère, S., Fowler, H.J., Blenkinsop, S., Therrien, R., Orban, P. and A. Dassargues. 2009. Large scale surface–subsurface hydrological model to assess climate change impacts on groundwater reserves. *Journal of Hydrology* 373(1–2): 122–138.

Guardiola-Albert, C., Martos-Rosillo, S., Pardo-Igúzquiza, E., Durán Valsero, J.J., Pedrera, A., Jiménez-Gavilán, P. and C. Liñán Baena. 2015. Comparison of recharge estimation methods during a wet period in a karst aquifer. *Groundwater* 53(6): 885–895.

Hall, F.R. 1968. Base flow recessions: A review. *Water Resources Research* 4(5): 973–983.

Hamon, W.R. 1961. Estimating potential evapotranspiration. Proc. of the American Society of Civil Engineers. *Journal of the Hydraulic Division* 87(HY3): 107–120.

Hashemi, F. and M.T. Habibian. 1979. Limitations of temperature-based methods in estimating crop evapotranspiration in arid-zone agricultural development projects. *Agricultural Meteorology* 20(3): 237–247.

Healy, R.W. 2010. *Estimating groundwater recharge*. Cambridge, MA: Cambridge University Press.

Healy, R.W. and P.G. Cook. 2002. Using groundwater levels to estimate recharge. *Hydrogeology Journal* 10: 91–109

Henderson, F.M. and R.A. Wooding 1964. Overland flow and groundwater flow from a steady rainfall of finite duration. *Journal of Geophysical Research* 69: 1531–1540.

Hornberger, G.M., Wiberg, P.L., Raffensperger, J.P. and P. D'Odorico. 2014. *Elements of physical hydrology* (2nd Edition). Baltimore: John Hopkins University Press.

Horton, R.E. 1933. The role of infiltration in the hydrologic cycle. *Transactions American Geophysical Union* 14: 446–460.

Hultine, K.R., Scott, R.L., Cable, W.L., Goodrich, D.C. and D.G. Williams. 2004. Hydraulic redistribution by a dominant, warm-desert phreatophyte: Seasonal patterns and response to precipitation pulses. *Functional Ecology* 18: 530–538.

Hutton, J.T. 1976. Chloride in rainwater in relation to distance from the ocean. *Search* 7: 207–208.

Hwang, H.-T., Park, Y.-J., Frey, S.K., Berg, S.J. and E.A. Sudicky. 2015. A simple iterative method for estimating evapotranspiration with integrated surface/subsurface flow models. *Journal of Hydrology* 531(3): 949–959.

Irmak, S. and D.Z. Haman. 2014. Evapotranspiration: potential or reference? *University of Florida IFAS Extension: ABE* 343: 1–2.

Jepsen, S.M., Harmon, T.C. and Y. Shi. 2016. Watershed model calibration to the base flow recession curve with and without evapotranspiration effects. *Water Resources Research* 52: 2919–2933.

Kim, J.H. and R.B. Jackson. 2012. A global analysis of groundwater recharge for vegetation, climate, and soils. *Vadose Zone Journal* 11(1). doi:10.2136/vzj2011.0021RA.

Kollet, S.J. and R.M. Maxwell. 2006. Integrated surface–groundwater flow modeling: a free-surface overland flow boundary condition in a parallel groundwater flow model. *Advances in Water Resources* 29(7): 945–958.

Kumar, K.K., Kumar, K.R. and P.R. Rakhecha. 1987. Comparison of Penman and Thornthwaite methods of estimating potential evapotranspiration for Indian conditions. *Theoretical and Applied Climatology* 38: 140–146.

Lebecherel, L., Andréassian, V. and C. Perrin. 2013. On regionalizing the Turc-Mezentsev water balance formula. *Water Resources Research* 49: 7508–7517.

Lerner, D.N., Issar, A.S. and I. Simmers. 1990. Groundwater recharge: A guide to understanding and estimating natural recharge. In *IAH International Contributions to Hydrogeology 8*. Rotterdam: Taylor and Francis, Balkema.

Li, L., Lambert, M.F., Maier, H.R., Partington, D. and C.T. Simmons. 2015. Assessment of the internal dynamics of the Australian Water Balance Model under different calibration regimes. *Environmental Modelling and Software* 66: 57–68.

Liebe, J.R., van de Giesen, N., Andreini, M., Walter, M.T. and T.S. Steenhuis. 2009. Determining watershed response in data poor environments with remotely sensed small reservoirs as runoff gauges. *Water Resources Research* 45: W07410.

Linsley Jr., R.K., Kohler M.A. and L.H. Paulus. 1975. *Hydrology for engineers*. New York: McGraw-Hill.

Loáiciga, H.A. 2003. Sustainable groundwater exploitation. *International Geology Review* 44(12): 1115–1121.

Loáiciga, H.A. 2006. Comment on "The persistence of the water budget myth and its relationship to sustainability" by JF Devlin and M Sophocleous. Hydrogeology Journal 13: 549–554. 2005. *Hydrogeology Journal* 14: 1383–1385.

Loáiciga, H.A. 2017. The safe yield and climatic variability: Implications for groundwater management. *Groundwater* 55(3): 334–345.

Maillet, E. 1905. *Essais d'hydraulique souterraine et fluviale [Hydraulic tests in the subsurface and in rivers] in French*. Paris: Hermann.

Maneta, M.P., Schnabel, S., Wallender, W.W., Panday, S. and V. Jetten. 2008. Calibration of an evapotranspiration model to simulate soil water dynamics in a semiarid rangeland. *Hydrological Processes* 22(24): 4655–4669.

Manna, F., Cherry, J.A., McWhorter, D.B. and B.L. Parker. 2016. Groundwater recharge assessment in an upland sandstone aquifer of southern California. *Journal of Hydrology* 541(B): 787–799.

Massman, W.J. and X. Lee. 2002. Eddy covariance flux corrections and uncertainties in long-term studies of carbon and energy exchanges. *Agricultural and Forest Meteorology* 113: 121–144.

Mather, J.R. 1969. The average annual water balance of the world. In Proceedings of the Symposium on Water Balance in North America, Proceedings Series No. 7. Banff, Alberta, Canada: American Water Resources Association, 29–40.

Maxwell, R.M., Chow, F.K. and S.J. Kollet. 2007. The groundwater—land-surface—atmosphere connection: soil moisture effects on the atmospheric boundary layer in fully-coupled simulations. *Advances in Water Resources* 30(12): 2447–2466.

McCabe, G.J. and S.L. Markstrom. 2007. *A monthly water balance model driven by a graphical user interface.* U.S. Geological Survey Open-File report 2007, 1088.

Meinardi, C.R. 1994. Groundwater recharge and travel times in the sandy regions of the Netherlands, *PhD diss.*, Vrije Universiteit Amsterdam.

Mendoza, G.F., Steenhuis, T.S., Walter, M.T. and J-.Y. Parlange. 2003. Estimating basin-wide hydraulic parameters of a semi-arid mountainous watershed by recession-flow analysis. *Journal of Hydrology* 279: 57–69.

Meyers, T.P. and D.D. Baldocchi. 2005. Current micrometeorological flux methodologies with applications in agriculture. In *Micrometeorology in Agricultural Systems*, eds. J.L. Hatfield and J.M. Baker. Madison (WI): American Society of Agronomy, 381–396.

Miles, O.W. and K.S. Novakowski. 2016. Large water-table response to rainfall in a shallow bedrock aquifer having minimal overburden cover. *Journal of Hydrology* 541(B): 1316–1328.

Monteith, J.L. 1965. Evaporation and environment. In *The state and movement of water in living organism. 19th Symp.* Society of Experimental Bioloy, 205–234.

Niu, G.-Y., Paniconi, C., Troch, P.A., Scott, R.L., Durcik, M., Zeng, X., Huxman, T. and D.C. Goodrich. 2014. An integrated modelling framework of catchment-scale ecohydrological processes: 1. Model description and tests over an energy limited watershed. *Ecohydrology* 7(2): 427–439.

Ordens, C.M., Werner, A.D., Post, V.E.A., Hutson, J.L., Simmons, C.T. and B.M. Irvine. 2012. Groundwater recharge to a sedimentary aquifer in the topographically closed Uley South basin, South Australia. *Hydrogeology Journal* 20: 61–72.

Panday, S. and P.S. Huyakorn. 2004. A fully coupled physically-based spatially distributed model for evaluating surface/subsurface flow. *Advances in Water Resources* 27(4): 361–382.

Parlange, J.Y., Stagnitti, F., Heilig, A., Szilagyi, J., Parlange, M.B., Steenhuis, T.S., Hogarth, W.L., Barry, D.A. and L. Li. 2001. Sudden drawdown and drainage of a horizontal aquifer. *Water Resources Research* 37(8): 2097–2101.

Partington, D., Brunner, P., Simmons, C.T., Werner, A.D., Therrien, R., Maier, H.R. and G.C. Dandy. 2012. Evaluation of outputs from automated baseflow separation methods against simulated baseflow from a physically based, surface water-groundwater flow model. *Journal of Hydrology* 458–459: 28–39.

Penman, H.L. 1948. Natural evaporation from open water, bare soil and grass. *Proceedings Royal Society of London, Series A, Mathematical and Physical Sciences* 193(1032): 120–145.

Pereira, A.R. and Ã. Paes De Camargo. 1989. An analysis of the criticism of thornthwaite's equation for estimating potential evapotranspiration. *Agricultural and Forest Meteorology* 46(1): 149–157.

Petersen, J.O., Deschamps, P., Gonçalvès, J., Hamelin, B., Michelot, J.L., Guendouz, A. and K. Zouari. 2014. Quantifying paleorecharge in the Continental Intercalaire (CI) aquifer by a Monte-Carlo inversion approach of ^{36}Cl/Cl data, *Applied Geochemistry* 50: 209 221.

Petheram, C., Walker, G., Grayson, R., Thierfelder, T. and L. Zhang. 2002. Towards a framework for predicting impacts of land-use on recharge: 1. A review of recharge studies in Australia. *Australian Journal of Soil Research* 40: 397–417.

Pistocchi, A., Bouraoui, F. and M. Bittelli. 2008. A simplified parameterization of the monthly topsoil water budget. *Water Resources Research* 44: W12440.

Polubarinova-Kochina, P.Y.-A. 1962. *Theory of groundwater movement* (translated from Russian by R.J.M. De Wiest), Princeton (NJ): University Press.

Priestley, C.H.B. and R.J. Taylor. 1972. On the assessment of surface heat flux and evaporation using large-scale parameters. *Monthly Weather Review* 100(2): 81–82.

Rocha, D., Feyen, J. and A. Dassargues. 2007. Comparative analysis between analytical approximations and numerical solutions describing recession flow in unconfined hillslope aquifers. *Hydrogeology Journal* 15: 1077–1091.

Roques, C., Rupp, D.E. and J.S. Selker. 2017. Improved streamflow recession parameter estimation with attention to calculation of—dQ/dt. *Advances in Water Resources* 108: 29–43.

Scanlon, B.R., Healy, R.W. and P.G. Cook. 2002. Choosing appropriate techniques for quantifying groundwater recharge. *Hydrogeology Journal* 10: 18–39.

Scanlon, B.R., Keese, K.E., Flint, A.L., Flint, L.E., Gaye, C.B., Edmunds, W.M. and I. Simmers. 2006. Global synthesis of groundwater recharge in semiarid and arid regions. *Hydrol Process* 20: 3335–3370.

Scanlon, B.R., Tyler, S.W. and P.J. Wierenga. 1997. Hydrologic issues in arid, unsaturated systems and implications for contaminant transport. *Reviews of Geophysics* 35(4): 461–490.

Schoeller, H. 1941. L'influence du climat sur la composition chimique des eaux souterraines vadose (*in French*). *Bulletin Société Géologique de France* 11: 267–289.

Schoeller, H. 1962. *Les eaux souterraines (in French)*. Paris: Masson.

Serra, L. 1954. Le contrôle hydrologique d'un bassin versant. *IAHS, General Assembly Rome* 3(38): 349–357.

Shuttleworth, W.J. 2008. Evapotranspiration measurement methods. *Southwest Hydrology* 7(1): 22–23.

Shuttleworth, W.J. and I.R. Calder. 1979. Has the Priestley-Taylor equation any relevance to forest evaporation? *Journal of Applied Meteorology* 18: 639–646.

Sibanda, T., Nonner, J.C. and S. Uhlenbrook. 2009. Comparison of groundwater recharge estimation methods for the semi-arid nNyamandhlovu area, Zimbabwe. *Hydrogeology Journal* 17(6): 1427–1441.

Smerdon, B.D. and J.E. Drewes. 2017. Groundwater recharge: The intersection between humanity and hydrogeology. *Journal of Hydrology* 555: 909–911.

Somaratne, N. 2015. Pitfalls in application of the conventional chloride mass balance (CMB) in karst aquifers and use of the generalized CMB method. *Environmental Earth Sciences* 74: 337–349.

Somaratne, N., Smettem, K. and J. Frizenschaf. 2014. Three criteria reliability analyses for groundwater recharge estimations. *Environmental Earth Sciences* 72: 2141–2151.

Sridhar, V. and K.G. Hubbard. 2010. Estimation of the water balance using observed soil water in the Nebraska Sandhills. *ASCE J. Hydrologic Engineering* 15(1): 70–78.

Stannard, D.I., Gannett, M.W., Polette, D.J., Cameron, J.M., Waibel, M.S. and J.M. Spears. 2013. *Evapotranspiration from marsh and open-water sites at Upper Klamath Lake, Oregon, 2008–2010*. U.S. Geological Survey Scientific Investigations Report 2013–5014.

Steenhuis, T.S. and W.H. Van der Molen. 1986. The Thornthwaite-Mather procedure as a simple engineering method to predict recharge. *Journal of Hydrology* 84: 221–229.

Szilagyi, J., Parlange, M.B. and J.D. Albertson. 1998. Recession flow analysis for aquifer parameter determination. *Water Resources Research* 37: 1851–1857.

Thomas, B.F., Vogel, R.M. and J.S. Famiglietti. 2015. Objective hydrograph baseflow recession analysis. *Journal of Hydrology* 525: 102–112.

Thornthwaite, C.W. 1948. An approach toward a rational classification of climate. *Geographical Review* 38(1): 55–94.

Thornthwaite, C.W. and J.R. Mather. 1955. The water balance. *Publications in Climatology* VIII(1): 1–104.

Thornthwaite, C.W. and J.R. Mather. 1957. Instructions and tables for the computing potential evapotranspiration and the water balance. *Publications in Climatology* X(3): 311.

Todd, D.K. and L.W. Mays. 2005. *Groundwater hydrology* (3rd Edition). Phoenix: John Wiley & Sons.

Troch, P., van Loon, E. and A. Hilberts. 2002. Analytical solutions to a hillslope-storage kinematic wave equation for subsurface flow. *Advances in Water Resources* 25: 637–649.

Troch, P., van Loon, E. and A.H. Hilberts. 2004. Analytical solution of the linearized hillslope-storage Boussinesq equation for exponential hillslope width functions. *Water Resources Research* 40(8): W08601.

Turc, L. 1954. Le bilan d'eau des sols. Relation entre les précipitations, l'évaporation et l'écoulement (*in French*). *Annales Agronomiques* 5: 491–595, 6: 5–131.

Van Camp, M., de Viron, O., Pajot-Métivier, G., Casenave, F., Watlet, A., Dassargues, A. and M. Vanclooster. 2016. Direct measurement of evapotranspiration from a forest using a superconducting gravimeter. *Geophysical Research Letters* 43(10): 225–231.

Van Camp, M., Radfar, M. and K. Walraevens. 2015. A lumped parameter balance model for modeling intramountain groundwater basins: Application to the aquifer system of Shahrekord Plain, Iran. *Geologica Belgica* 18(2–4): 80–91.

VanderKwaak, J.E. and K. Loague. 2001. Hydrologic-response simulations for the R-5 catchment with a comprehensive physics-based model. *Water Resources Research* 37(4): 999–1013.

Varni, M.R. and E.J. Usunoff. 1999. Simulation of regional-scale groundwater flow in the Azul River basin, Buenos Aires Province, Argentina. *Hydrogeology Journal* 7(2): 180–187.

Verhoest, N.E.C. and P. Troch. 2000. Some analytical solutions of the linearized Boussinesq equation with recharge for a sloping aquifer. *Water Resources Research* 36(3): 793–800.

Verstraeten, W.W., Veroustraete, F. and J. Feyen. 2008. Assessment of evapotranspiration and soil moisture content across different scales of observation. *Sensors* 8(1): 70–117.

Wolock, D.M. and G.J. McCabe. 1999. Effects of potential climatic change on annual runoff in the conterminous United States. *Journal of the American Water Resources Association* 35: 1341–1350.

Wood, W.W. and W.E. Sanford. 1995. Chemical and isotopic methods for quantifying groundwater recharge in a regional, semiarid environment. *Ground Water* 33: 458–468.

Woodhouse, B. 2008. Approaches to ET measurement. *Southwest Hydrology* 7(1): 20–21.

Xu, C.-Y. and V.P. Singh. 2001. Evaluation and generalization of temperature-based methods for calculating evaporation. *Hydrological Processes* 15: 305–319.

Zecharias, Y.B. and W. Brutsaert. 1988. Recession characteristics of groundwater outflow and baseflow from mountainous watersheds. *Water Resources Research* 24(10): 1651–1658.

Zhang, J., Zhang Y., Song, J. and L. Cheng. 2017. Evaluating relative merits of four baseflow separation methods in Eastern Australia. *Journal of Hydrology* 549: 252–263.

Chapter 3

Groundwater terminology and examples of occurrences

3.1 Terminology

As mentioned previously, underground water circulation and storage can be seen as nothing more than hydraulics applied in a highly heterogeneous medium with spatial properties that are mostly influenced by the geological context. For groundwater flow, two main properties of the geological media will be considered: their ability to store and conduct groundwater (see Chapter 4). Therefore, contrasts between these two main properties resulting from geological and structural heterogeneities lead to an infinite number of different "hydrogeological contexts" in terms of water depth, natural groundwater springs or outlets, storage capacity, pumping and drainage efficiency, interconnection between different geological units, and interconnection with surface water bodies. The definition of this hydrogeological context related to surface/topographic and geological conditions is an important step in any hydrogeological study. The depth of the description needed is closely dependent on the scale, as well as the objectives of the study.

Hydrogeology does not escape partly anthropocentric terminology, which expresses physical realities related to human values and experience. The term *aquifer* (from the Latin words *aqua* [water] and *affero* [to bring], Kresic 2007) is defined as a geological entity made of unconsolidated or indurated rocks or a group of hydraulically connected geological formations that both stores and transmits water (i.e., whose pores and (micro)fissures are wide enough for water to be stored or circulated), which allows its saturated zone to yield useful quantities of water (e.g., Stone 1999).

On the other hand, an *aquitard* (from the Latin words *aqua* [water] and *tardo* [to delay], Kresic 2007) is a low permeability water bearing formation (i.e., with a high storage but a low transmission of water). *Aquiclude* (from the Latin words *aqua* [water] and *claudo* [to close], Kresic 2007) is also used and refers to an aquitard with a very low permeability (i.e., almost impermeable), and *aquifuge* refers to an impermeable formation that neither stores nor transmits water. These terms aquitard, aquiclude, and aquifuge have become increasingly difficult to distinguish among each other in many contexts. Therefore, they have been replaced by the more general term *confining unit*.

Most authors agree with these definitions with a few subtle variations (e.g., Freeze and Cherry 1979, Banton and Bangoy 1997, Stone 1999, Fetter 2001, Fitts 2002, Schwartz and Zhang 2003, Musy and Higy 2004, Todd and Mays 2005, Kresic 2007, 2008, Hornberger *et al.* 2014, Nonner 2015).

Thus, these definitions are relative and subjective. In reality, groundwater occurs everywhere but aquifers are considered only where groundwater can be extracted or drained for useful local needs (i.e., in relation to human and/or economic demands). Thus, a low permeable water bearing formation may qualify as an aquifer when, locally, all surrounding geological formations are less pervious, and the considered formation can be used for water supply. Clearly, the same formation would be considered as an aquitard if it was surrounded by more permeable layers (i.e., aquifers).

Groundwater flows in the subsurface under the action of gravity toward an exit point which has an altitude lower than the infiltration point. Groundwater can also be extracted by flowing or pumping wells. Resulting from the complex spatial variability of the main hydraulic properties, groundwater pathways can be complex and difficult to assess. Combining the heterogeneity from geological features with hydraulic laws may increase the number of unexpected hydrogeological contexts. This can be the case if the observer is excessively influenced by prevailing surface conditions.

Thus, understanding the local and regional geologic setting is a priority for any hydrogeological investigation. An aquifer can enclose multiple geological units or can only compose part of a geological unit. This is mostly dependent on the hydraulic properties (i.e., permeability and storage properties, see Chapter 4) of the subunits.

Typical aquifer lithologies are defined by unconsolidated sands and gravels, chalks, fractured (and possibly karstic) limestones and dolomites, siltstones and sandstones, or fractured metamorphic and plutonic rocks. However, a detailed lithology and fracturing state can be very important. For example, the nature of the cement and the filling rate of the pores between the solid grains of sandstone have an important influence on their hydraulic properties. The density of the microcracks in a porous chalk can also have a decisive influence on the hydraulic properties.

Low permeability geological units (such as aquitards and aquicludes) that bound an aquifer are defined as the *confining* units of the aquifer. Confining units should present contrasting permeabilities compared to the aquifer unit; otherwise, they may qualify as *semiconfining* units. Different hydrogeological contexts are illustrated in Figure 3.1.

A *confined aquifer* is overlain and underlain by confining units (Figure 3.2). The water level measured in a well is above the upper geological boundary of the aquifer unit. If the water level is above ground level, then the confined aquifer is also defined as an *artesian aquifer*, and any production well is defined as a *flowing artesian well* (Figure 3.3).

On the other hand, a *water table* or *unconfined* aquifer is defined as an aquifer which has a water table that can be considered as the upper boundary (Figure 3.4). The *water table* is defined roughly by all water levels that can be measured in shallow wells screened in the aquifer unit (Schwartz and Zhang 2003). Thus, the water table corresponds to a surface of zero water pressure. If the *capillary fringe* (see Chapter 1) is neglected, the water table can be considered as the boundary between the saturated part and the partially saturated part (or the unsaturated part) of the aquifer. In an unconfined aquifer, the water table is free to move according to the local recharge-discharge water balance. This is why the water table is referred to as a free surface. In French, the term "*aquifère libre*" meaning "free aquifer" is sometimes used to refer to an unconfined aquifer. The first unconfined aquifer encountered when drilling from the surface is commonly referred to as a *phreatic aquifer*. The term *phreatic zone* is sometimes (but rarely) used to refer to the saturated zone. The *perched aquifer,*

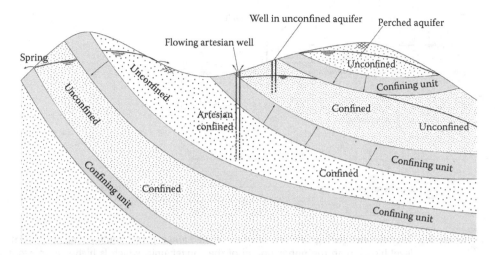

Figure 3.1 The heterogeneity of the geology (and of the dependent hydraulic properties) induces spatially variable hydrogeological conditions. The same aquifer may be unconfined, confined, or artesian dependent on the local context. Primary infiltration and discharge zones are also determined by the interplay between geologic, topographic, and hydraulic conditions.

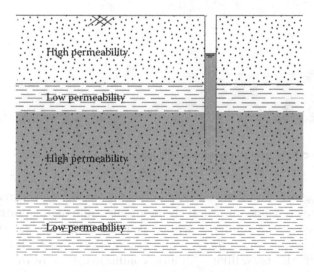

Figure 3.2 Typical hydrogeological context of a confined aquifer (i.e., with a water level higher than the upper bound of the aquifer unit).

which is typically unconfined, is defined when saturated conditions develop above a low permeability stratum, which is located above an unconfined aquifer with a larger extension (Figure 3.1). Accumulated groundwater flows laterally above the low permeability layer; if the topography intersects this limited perched saturated zone, seepage can occur, which forms a spring zone on the hillslope.

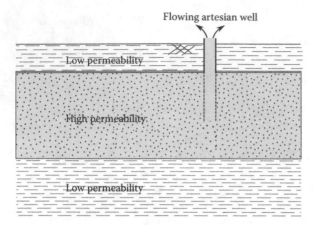

Figure 3.3 Typical hydrogeological context of a confined and artesian aquifer (i.e., with a water level higher than the upper bound of the aquifer unit, which is higher than ground level).

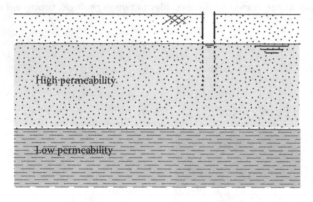

Figure 3.4 Typical hydrogeological context of an unconfined aquifer.

Often, there are hydrogeological conditions where aquifers are neither considered as completely confined or completely unconfined. If the bounding units of an aquifer are actually *leaky confining* layers, the aquifer is defined as a *semiconfined aquifer*. This type of aquifer exchanges *leakage water fluxes* with other aquifers through leaky confining units. Depending on the prevailing hydraulic conditions, downward or upward leakage fluxes occur (Figure 3.5) which can form important water recharge or discharge fluxes for the balance of the aquifer. A semiconfined aquifer should not be confused with a *partially confined aquifer*. The latter term refers to aquifers that are either confined during some time periods (not permanently; Figure 3.6) or in some places (not in the entire study area; Figure 3.7).

As mentioned previously, recharge preferably occurs in a recharge area where the aquifer unit crops out. However, slow recharge by downward, upward, or even lateral leakage through semiconfining or confining units can also be important (Fetter 2001).

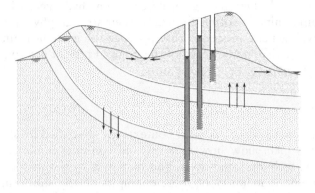

Figure 3.5 Downward and upward leakage fluxes that occur throughout the semiconfining units.

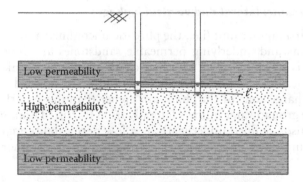

Figure 3.6 A partially confined aquifer resulting from time variations of the aquifer water level.

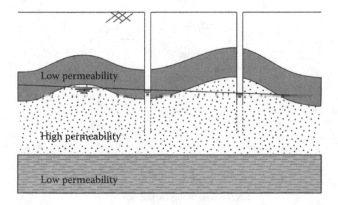

Figure 3.7 A partially confined aquifer resulting from spatial variations in the upper boundary
of the aquifer unit.

To obtain a clear relationship with their geological nature, some hydrogeological units (i.e., aquifers and confining units) are often defined by their stratigraphy name, relative depth, lithology, or even genetic processes that produce the geological unit (Stone 1999). This latter approach gives a good idea of the induced type of heterogeneity and spatial variability to be expected within the unit.

3.2 Examples of occurrences

In the following examples, different hydrogeological contexts are described by simplified vertical cross sections, which show how aquifers and confining units that are combined with topography, recharge conditions, and human activities may create specific groundwater systems. A majority of these examples clearly show that a detailed geological knowledge is needed (or is at least very useful) when starting hydrogeological investigations. Note that the reader will have a more detailed and thorough understanding of these illustrated case studies after having read Chapter 4.

Aquifers separated by confining units: lateral and vertical leakages

As described in Figure 3.8, the first aquifer unit (i.e., the phreatic unconfined aquifer) is composed of alluvial sediments and underlying permeable sandstones and limestones. This upper aquifer unit is separated from a deeper chalk aquifer by a leaky clay semiconfining layer. Most leakage groundwater fluxes are directed upward as the water levels in the chalk recharge zone of the chalk are higher than those in the upper aquifer. Chalk can receive groundwater inflows laterally from fractured granite bedrock. This chalk aquifer is also confined underneath by a low permeability marl through which very limited groundwater fluxes can occur from fractured metamorphic bedrock.

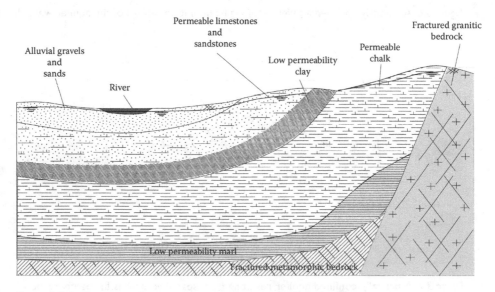

Figure 3.8 Different aquifer units separated by semiconfining layers.

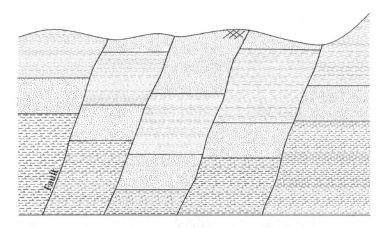

Figure 3.9 Sedimentary layers affected by horst/graben faults. Sandy aquifers may be partitioned due to the effects of the fault shifts combined with the alternating clay layers.

Partitioned aquifers resulting from recent horst/graben tectonics

As described in Figure 3.9, tectonic faults can divide aquifer layers into different compartments. A stack of initially horizontal sedimentary layers with alternating permeable sands and low permeability clays is partitioned. Note that when a fault shows walls with different lithologies, the weathered remains of the weakest (in terms of rheology) lithology provide most of the filling. Thus, clayey low permeability filling is found in faults separating clay from other geological material.

Aquifers in successive thrust faults and sheets

In the geological context of folded, fractured, and faulted sedimentary layers, which are characterized by successive thrust faults and sheets (as shown in Figure 3.10), faults can act as high permeability drains or low permeability screens, depending on the filling material. This leads to either connected aquifer units or disconnected aquifers of the same geological unit. Note that, usually, normal faults (i.e., resulting from traction stresses) are considered to be more permeable than inverse faults (i.e., resulting from compressive stresses) and transverse faults (resulting from shear stresses). However, when a clayey filling is found, the fault can be considered as a low permeability barrier (or screen) in any case.

Fractured bedrock and colluvium of variable lithology

Based on the geological context described in Figure 3.11, a spring appears on the hillside, which results from the permeability contrast between primarily sandy colluvium from the weathering of quartzite bedrock, and clayey colluvium from the weathering of shale.

Infiltration is lower in extremely low permeability shales. If these shales are not fractured, determination of the water level may be difficult and/or inaccurate for such a lithology (see Chapter 4).

1. Siltstones (aquifer)
2. Shales (aquitard)
3. Sandstones (aquifer)
4. Claystones (aquiclude)

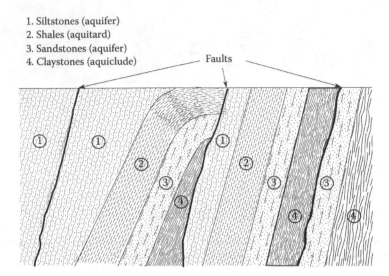

Figure 3.10 Successive thrust faults and sheets in a stratigraphic sequence connecting or disconnecting aquifer units.

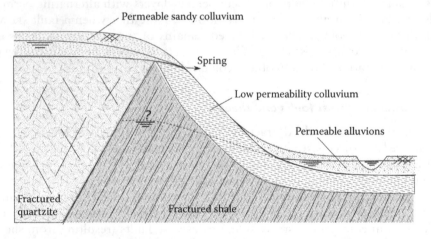

Figure 3.11 A lateral change in colluvium nature (i.e., from a permeable sand into a low permeability clay) is induced by a bedrock lithology change. As a result, a spring appears on the hillside.

Perched aquifer and heterogeneous bedrock aquifer

As described in Figure 3.12, a perched aquifer is defined by the saturation of sandy horizontal deposits above a geological unconformity at the top of low permeability eroded and weathered bedrock. When the bedrock is locally composed of high permeability karstic limestones, a perched aquifer is no longer observed. Therefore, the vertical infiltration occurs until it reaches the saturated zone and the water level is particularly low, which results from the global high permeability of karstic limestones combined with possible lateral drainage from streams.

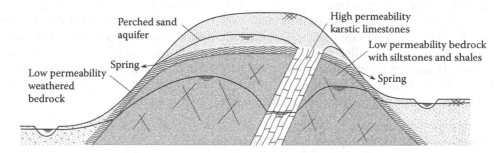

Figure 3.12 A perched sand aquifer above low permeability weathered bedrock. Laterally, the karstic limestone drains the deep regional aquifer, which induces lower water levels and deeper vertical infiltration.

Variable interactions between aquifers induced by human activities

Groundwater interactions between an alluvial aquifer and an underlying fractured bedrock are shown in Figure 3.13. If mining operations are active in bedrock formations, intensive pumping is needed to induce an important drawdown (Figure 3.13a). Thus, the alluvial aquifer loses groundwater into the fractured bedrock aquifer. When

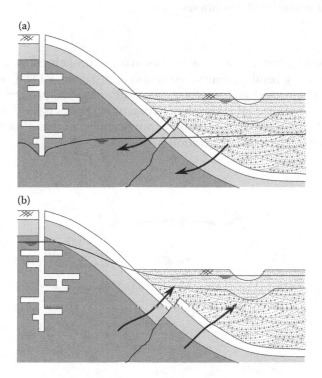

Figure 3.13 Groundwater interactions between an alluvial aquifer and the underlying fractured bedrock. The direction, values, and associated chemical concentrations of the main groundwater fluxes can be largely dependent on human-induced activities, such as (a) mining activities and intensive pumping in the bedrock aquifer and (b) the termination of mining activities.

mining and pumping activities are stopped, the increase in water levels in the bed-rock (Figure 3.13b) induces groundwater fluxes from the bedrock toward the alluvial aquifer. This change in exchanged groundwater fluxes is not only a matter of ground-water quantity as quality issues may also appear. For example, if coal mines are closed, rising groundwater levels are most often conjugated with sulfate and arsenic groundwater contamination from pyrite oxidation in the fractured exploited bedrock. As a result, this may increase the alluvial aquifer contamination.

A semiconfined to unconfined drained chalk aquifer

As described in Figure 3.14, groundwater is produced from drainage galleries in a chalk aquifer. Under natural conditions, this type of aquifer would be mostly semi-confined under a thick accumulation of loess. Under drainage conditions, this aquifer becomes mostly unconfined (under upgradient sediments), locally semiconfined under loess sediments, and confined under loamy alluvial sediments. The location of the upstream boundary of the chalk aquifer (i.e., the saturated zone in the chalk) is vari-able over time as function of the recharge conditions. Upstream ephemeral springs may become active in response to the seasonal increase in groundwater levels. The equilibrium between recharge and drainage determines current water levels and con-sequently the semiconfined or unconfined conditions.

Karstic groundwater system

In karstic limestone formations, groundwater flow occurs mainly through conduits and fractures that are enlarged as a result of limestone dissolution (Figure 3.15). Old abandoned conduit systems can also play an important role in drainage and preferen-tial flows in the unsaturated zone. The real geometry of conduit networks is hard to investigate and characterize. Sinking rivers and sinkholes provide most of the inflows into the aquifer system through preferential infiltration pathways.

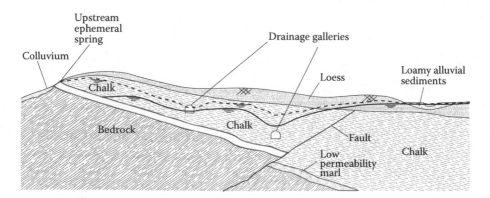

Figure 3.14 A semiconfined chalk aquifer becomes mostly unconfined as a result of local and regional drawdowns induced by drainage galleries. Downgradient, the aquifer becomes permanently semiconfined under a loess formation and locally confined under loamy alluvial sediments (the dashed line indicates high groundwater conditions).

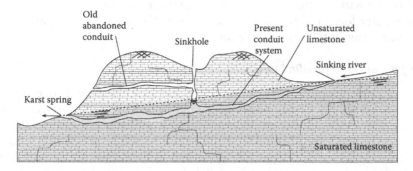

Figure 3.15 In a karstic limestone system, most of the groundwater flow occurs in conduits. Sinking rivers and sinkholes provide the main inflows in the aquifer system (Waltham 2009).

Groundwater flow direction in a limestone aquifer and the influence of the base water level

In a monoclinal limestone aquifer overlying low permeability shales (as described in Figure 3.16), the topography can be particularly misleading for assessing the groundwater flow direction. A spring appears on the hillside between the limestone layers and the low permeability shales (Figure 3.16a). If an alluvial aquifer and a drainage

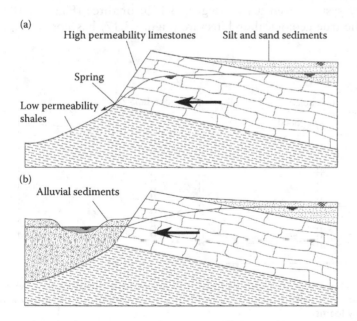

Figure 3.16 A limestone aquifer in monoclinal layers overlying low permeability shales. Hydrogeological contexts correspond to: (a) an aquifer outlet (spring) above the base level of local streams and (b) a main outflow influenced by the base water level of a stream and the corresponding alluvial aquifer.

stream are in contact with the limestone aquifer (Figure 3.16b), which imposes a relatively high base water level, the spring no longer exists. Groundwater flux may still exist from the limestones into the alluvial aquifer.

Importance of a clear and justified geological interpretation

When assessing aquifer geometry and the spatial variability hydraulic properties, a clear geological interpretation based on the understanding of identified geological processes is needed. For example, in a meandering alluvial sedimentary system, sandy deposits are either normally restricted, directly linked to main (present or old) stream channels. Loams and clays are preferentially deposited in flood basins and on levees (Walker 1976). This leads to permeable "elongated shoestring sand deposits" that are stratigraphically bound by low permeability loams and clays as shown schematically in the cross section of Figure 3.17. On the other hand, in a braided alluvial system, thicker and laterally more extensive sandy (permeable) bodies are found (Walker 1976).

Assuming that reality is illustrated by the cross section in Figure 3.17, we show in the next paragraph and in Figure 3.18 how misleading the interpretation can be when there is not a thorough understanding of the relevant geological processes.

First, we assume that only one borehole is available for the characterization of the alluvial aquifer. A rough and unprofessional (but unfortunately quite common) interpretation using general "pancake geology[1]" principles leads to misleading interpretation of the cross section as illustrated in Figure 3.18a. Compared to the reality as described in Figure 3.17, a set of uniformly horizontal layers alternating between sands and loams is interpreted from data of borehole 1. Without understanding the meandering sedimentary system, even if two (Figure 3.18b) or three (Figure 3.18c) boreholes are drilled, the true geological architecture (Figure 3.17) has not yet been correctly interpreted.

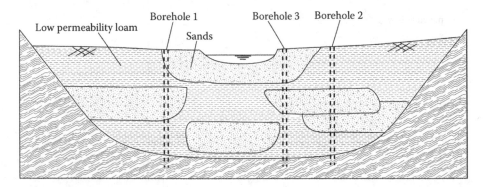

Figure 3.17 A typical architecture of alluvial deposits in a meandering fluvial system, which has "elongated shoestring sands deposits" that are stratigraphically bound by low permeability loams.

---

1 The term "pancake geology" is used here to refer to an unprofessional method of interpreting geological data based on the implicit assumption that layers are mostly horizontal, where correlations and extrapolations are mostly performed using straight lines.

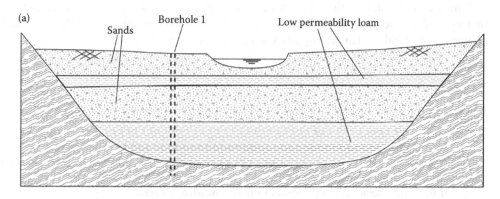

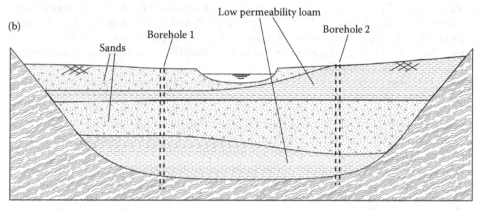

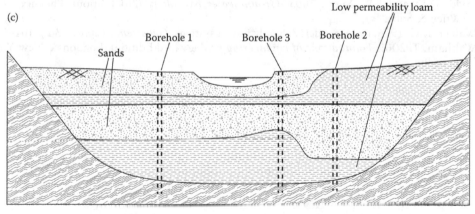

Figure 3.18 Unprofessional geological interpretation using one (a), two (b), and three (c) borehole(s). Using straight lines for joining and extrapolating borehole observed geological changes leads to erroneous interpretations of reality (as described in Figure 3.17).

This example shows the importance of a detailed understanding of geological processes for assessing the spatial variation in the lithology and, thus, the hydraulic properties needed for further hydrogeological studies.

References

Banton, O. and L.M. Bangoy. 1997. *Hydrogéologie. Multiscience environnementale des eaux souterraines (in French)*. Quebec, Canada: Presses de l'Université du Québec, AUPELF-UREF.

Fetter, C.W. 2001. *Applied hydrogeology* (4th Edition). Upper Saddle River (NJ): Pearson Education Limited.

Fitts, Ch.R. 2002. *Groundwater science*. London: Academic Press.

Freeze, R.A. and J.A. Cherry. 1979. *Groundwater*. Englewood Cliffs (NJ): Prentice Hall.

Hornberger, G.M., Wiberg, P.L., Raffensperger, J.P. and P. D'Odorico. 2014. *Elements of physical hydrology* (2nd Edition). Baltimore: John Hopkins University Press.

Kresic, N. 2007. *Hydrogeology and groundwater modeling* (2nd Edition). Boca Raton: CRC press, Taylor & Francis Group.

Kresic, N. 2008. *Groundwater resources. Sustainability, management, and restoration*. New York: McGraw Hill.

Musy, A. and C. Higy. 2004. *Hydrologie. Une science de la nature (In French)*. Lausanne, Switzerland: Presses polytechniques et universitaires romandes.

Nonner, J.C. 2015. *Introduction to Hydrogeology* (3rd Edition). Boca Raton: CRC press/ Balkema UNESCO-IHE.

Schwartz, F.W. and H. Zhang. 2003. *Fundamentals of Ground Water*. Wiley.

Stone, W.J. 1999. *Hydrogeology in practice. A guide to characterizing ground-water systems*. Upper Saddle River (NJ): Prentice Hall.

Todd, D.K. and L.W. Mays. 2005. *Groundwater hydrology* (3rd Edition). Phoenix: John Wiley & Sons.

Walker, R.G. 1976. Facies models 3. Sandy fluvial systems. *Geoscience Canada* 3(2): 101–109.

Waltham, T. 2009. *Foundations of engineering geology* (3rd Edition). London & New York: Spon Press.

Chapter 4

Saturated groundwater flow

4.1 Representative elementary volume (REV) concept

Loose sediments and hard rocks constituting the subsurface can be considered as forming what is called a *porous medium* characterized by the presence of a solid matrix and voids. These latter correspond to the pores and cracks of the rock, as well as fractures, and cavities that can be encountered in various geological formations. Fluid phases (e.g., air, water, oil) occupy these spaces scattered throughout the milieu. They may be interconnected or isolated. Generally, it will be assumed that a sufficient portion of these spaces is interconnected so as to make possible the movement of fluids and the associated transport of solutes.

In a porous medium, the different *phases* are unique domains separated by an interface or a transition zone. Each phase is assumed to form a *continuum*, even if it is actually not always the case, especially at the microscopic scale. Variables related to the phase (for example fluid pressure, solute concentration) and parameters related to its properties (i.e., viscosity, compressibility, density) are considered as taking a value in each point of the phase.

For example, macroscopically, at each point of a studied medium, a solute concentration in groundwater (e.g., for a given contaminant) can be found when microscopically this is not the case.

In fact, the variables and parameters are obtained by homogenizing (taking an average or "equivalent" value) the properties described at the microscopic level on a macroscopic representative elementary volume (REV). Notions of solid and fluid phases are conceptualized as continuous. This *continuity assumption* is fundamental for further quantification in hydrogeological science and engineering. The REV concept involves an integration in space that can be also understood as an "ensemble average" of a random function (de Marsily 1986, Dagan 1989).

Even if some authors consider that "the concept of a unique REV in natural heterogeneous sediments has not been conceptualized clearly or verified experimentally" (Molz 2015), the practical value of the reasoning is universally accepted. The idea is built on the fact that measurements are made on a sample or a "representative" volume of the medium to be characterized with respect to a defined process. This also means that the type of property or parameter in relation to a given study objective (and to studied processes) can have an influence on the size of the chosen REV and on the mean or "equivalent" value considered at the REV scale. Behind this concept, lies the fact that integration of the microscopic heterogeneities over the considered REV, yielding an equivalent value to be used at the REV scale should not bias the quantification of the studied process (Bachmat and Bear 1986, Bear and Verruijt 1987).

The REV concept is not only used in groundwater flow and solute transport problems but also in many other fields where quantification is needed for properties of a geological medium. For us, it is clearly an upscaling technique from microscopic fluid dynamics at the pore scale to a macroscopic (or megascopic) scale that is more realistic and practical in field-scale hydrogeology problems. Ideally (Figure 4.1), in order to avoid bias in the quantitative description of the reality, the REV must be large enough

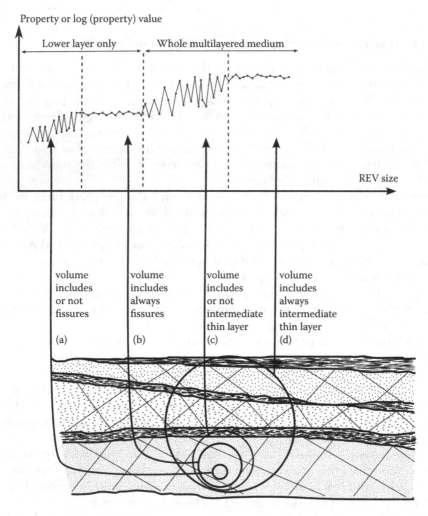

Figure 4.1 Scale of a considered porous medium and homogenizing actual heterogeneity in an REV. In a multilayer system including fracturation of some layers, (a) the REV may or may not include fissures producing highly variable values, (b) the REV always includes fissures producing less variable values, (c) the REV may or may not include intermediate thin layer producing highly variable values, (d) the REV always includes intermediate layer(s) producing more uniform values (Bear and Verruijt 1987).

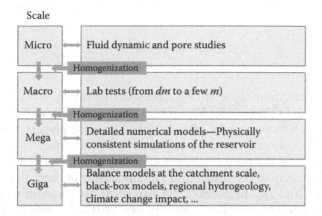

Scale

Figure 4.2 Scale involved for most of the methods used in hydrogeology, and upscaling property values by homogenization.

to be relevant with regards to the studied problem (avoiding the microscopic scale for a real case study) but including an acceptable level of homogeneity (i.e., a large number of microscopic heterogeneities) integrated in a reliable equivalent or average value. The REV must also be small enough to enable different property values from one zone to another within the studied domain, avoiding overly-smoothed values that would corrupt our description of the heterogeneities which have an influence on the governing processes.

Figure 4.2 shows how the REV and homogenization are systematically used by different hydrogeological study methods at different scales. The homogenization concept considers that whatever happens at the smaller scale is carried over to the larger scale in the form of lumped coefficients that will appear in the differential equations of the calculated process at that scale. To assess reliable values for properties at the chosen REV scale, the governing processes must be understood at the smaller scales. In hydrogeology, it also means that an actual fissured or fractured medium can be considered as an equivalent porous medium through the use of the continuity assumption and the REV concept.

4.2 Porosities

In any geological rock or soil, the solid does not occupy the entire volume or REV. In granular sediments, there are pores between the solid grains. In hardened and folded rocks, there are openings, cracks, and fissures. Some geological media have both pores and fissures.

Conceptually, in hydrogeology, even seemingly solid and compact rocks are often considered as porous to at least some degree.

Total porosity

The *total porosity* is defined as:

$$n = \frac{V_v}{V_t} \tag{4.1}$$

where V_v is the void volume and V_t the total volume. Total porosity can also be written as:

$$n = 1 - \frac{\rho_b}{\rho_s} \tag{4.2}$$

with ρ_b the dry bulk density of the medium ($\rho_b = M_s/V_t$), ρ_s the solid density ($\rho_s = M_s/V_s$), M_s is the solid mass, V_s is the solid volume, and $V_t = V_v + V_s$ in the considered REV. The pores (including all kind of fissures, cracks, channels, and bedding planes) may contain one or more fluids: for example, air, gas, water, or hydrocarbons. If more fluids are simultaneously present in the pore space, the capillary properties of each fluid relative to the solid will determine the distribution and behavior of these fluids at the pore scale (see Chapters 6 and 9). The pores of the geological environment may be connected or disconnected, regular or random, and may have particular geometric characteristics (planar, cellular, expanded cracks, …), depending on the texture[1] and the structure[2] of the rock. Essentially, the degree of interconnection of the pores will determine the permeability (see Section 4.4 and Chapter 5).

The porosity is the result of various physical and chemical processes related to the formation of the sediments or rock, including all interactions and changes throughout its history. In geology, the terms *primary porosity* and *secondary porosity* are distinguished chronologically to describe, respectively, the initial porosity of a sediment or rock, and additional porosity acquired posteriorly due to, for example, diagenesis, metamorphism, fracturing, or deformation. In hardened rocks, the secondary porosity, most often named fracture porosity or fissure porosity as opposed to interstitial or intergranular porosity, can have more influence on groundwater flow and the storage properties compared to the primary porosity (Figure 4.3). Hardened rocks can often be considered as dual porosity media (Figure 4.4), with fracture and interstitial porosities. However, in most cases, part of this interstitial porosity is reduced by cementing materials in the pore spaces.

In soil mechanics, a void ratio is used because the solid volume is taken as a constant reference (when solid compressibility is neglected with regards to the total volume compressibility, see Chapter 7) rather than the total volume:

$$e = \frac{V_v}{V_s} \tag{4.3}$$

where V_s is the solid volume in the REV.

As the total volume is the sum of the void volume and solid volume, we can write:

$$e = \frac{n}{1-n} \tag{4.4}$$

1 The texture is defined herein as a rock characteristic determined by the size, shape and arrangement of minerals in a rock on a microscopic scale. Example: a sandstone has a reticulated or granular texture with (rounded) particles, possibly welded together by a cement.
2 In contrast, the structure, considered at the macroscopic scale, describes the arrangement of rocks in their geological context. Example: a stratification stack of sedimentary layers is a structure.

Figure 4.3 Four examples of different porosities: intergranular porosity which is higher in a well sorted sediment (upper left) than in a poorly sorted sediment (upper right); in hardened rocks, fracture porosity (lower left), and fracture porosity enlarged by karstic dissolution (lower right).

and

$$n = \frac{e}{1+e} \tag{4.5}$$

It is understood that the concept of porosity is applied to a certain volume of the porous medium. In practice, the REV concept is always implicitly considered even though this is not necessarily acknowledged. Therefore, for the same geological formation, depending on the scale of consideration and thus the size of the considered REV, different porosity values can be measured (see Box 4.1).

As mentioned by de Marsily (1986), if a discrete porosity function was defined taking a value of 0 in the solid parts and 1 in the pores of the REV, the equivalent spatially integrated value for the REV would be expressed statistically by:

$$n = \int_V f(x)dV \tag{4.6}$$

with $f(x)$ a random function such that $f(x) = 1$ if $x \in V_v$ and $f(x) = 0$ if $x \in V_s$.

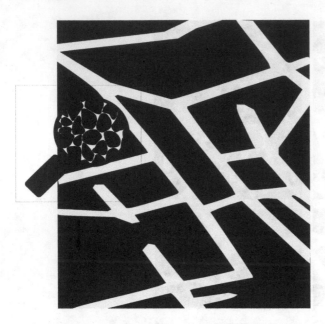

Figure 4.4 Dual porosity medium showing a fracture porosity together with an interstitial porosity.

Effective porosity

The porosity definition does not take into account the shape of the voids, nor the specific surface of the grains, whereas these two characteristics also influence the relationship between fluid(s) and solid matrix.

Now taking only water and air into consideration in the geological medium, and apart from the very small amount of water included within the rock minerals composition, groundwater in the pores and cracks can be immobile or mobile.

In fully saturated conditions, immobile water is composed of retention water (colloids and hygroscopic water, and water films), capillary water, and water contained in closed pores. Its quantity depends mainly on the specific surface of the solid matrix as it is retained by various surface physicochemical processes (e.g., adsorption and molecular attractions; Castany 1963). When groundwater flow is being studied, we are concerned with the mobile water porosity. However, the definition of this "mobile water porosity" is highly dependent on the existing pressure conditions in the geological medium, and on the relevant processes. As it will be discussed later (see Section 4.4 and Chapter 8), the "mobile water porosity" to be considered for groundwater flow is typically higher than the "mobile water porosity" acting in solute transport processes (Payne *et al.* 2008, Hadley and Newell, 2014).

In saturated conditions and for groundwater flow problems, an *effective porosity* is usually defined as approximating a kinematic or drainable porosity under the in situ pressure conditions:

$$n_e = \frac{V_m}{V_t} \tag{4.7}$$

where V_m is the volume of moving water in the REV. This effective porosity is usually quite less than the total porosity, except in well-sorted gravels and coarse sands. It is also named *specific yield* (S_y). On the contrary, the volume of water that is retained against drainage or immobile water is used to define the *retention capacity* or specific retention, expressed by:

$$S_r = \frac{V_{im}}{V_t} \tag{4.8}$$

where V_{im} is the volume of immobile water in the REV. The following relation can therefore be expressed:

$$n = S_y + S_r = n_e + S_r \tag{4.9}$$

while these definitions are conceived for saturated porous/fissured media, we clearly see the delicate and relative nature of the distinction between mobile and immobile water. Depending on the pressure conditions or hydraulic gradient in the medium (see Section 4.3), the proportion of mobile and immobile water can vary substantially. Kinematic porosity is in fact not measurable in absolute terms (Burger *et al.* 1985), so usually this effective porosity refers to the amount of fluid released by gravity drainage of a saturated rock after an undefined finite time. Actually, in an aquifer, the pressure or piezometric gradient (see Section 4.3) also influences the proportion

---

Box 4.1   Scale effect on REV porosity values:
Example of a Cretaceous chalk aquifer

At the *microscopic scale* (up to a few centimeters), the chalk consists of coccolith fossils of a few microns in length aggregated by diagenesis. The porosity may vary depending on localized $CaCO_3$ precipitation. The overall spatial distribution of voids seems relatively homogeneous. Samples of a few centimeters are apparently adequate REV to measure the pore porosity. The following values have been measured: $n = 0.40 - 0.42$ and $n_e = 0.35$.

At the *macroscopic scale* (up to a few decimeters) one can observe conjugate microcrack networks (or microfractures) and chalk layering. The effective drainable porosity is now mainly due to the porosity of cracks or fractures: $n = 0.42 - 0.45$ and $n_e = 0.01 - 0.03$.

At the *megascopic* scale (scale of field tests such as pumping tests, up to a few hundred meters), it is observed (through interpretation of tests and inverse modeling) that "homogenization" in a large REV ensures that all fractures, faults, and discontinuities should be considered as interconnected, which increases the effective porosity:

$$n = 0.42 - 0.45 \quad n_e = 0.05 - 0.10$$

Source: Data are relative to the Geer basin in Belgium, modified after Dassargues and Monjoie 1993.

of mobile water. Also, when pumping occurs, a forced flux is imposed on the aquifer. In a heterogeneous porous medium, possible "channeling" of the more mobile water increases, together with the contrast between the fraction of quasi-immobile water and the more mobile fraction.

The nature of the fluid can also play a role due to the influence of the size of molecules in relation to the size of the openings between the pores. Only the effective porosity to water will be considered here (unless otherwise stated).

Table 4.1 provides some representative values for effective porosity in different geological formations. These values must be considered with great caution because, as stated above, the proportion of water considered as mobile is dependent on pressure conditions and actual considered flow processes in relation to the scale of the REV (see Box 4.1).

Figure 4.5 shows the evolution of the porosity, effective porosity, and retention capacity in loose sediments as a function of the variation of grain size (Eckis 1934; Figure 4.5). In practice, it is more complex to assess these values because the shape of the grains, the packing, and the grain size distribution strongly influence the total and effective porosity values. There are many empirical relations linking grain size properties and especially grain size distribution of a sediment to its porosity. In geotechnique, a uniformity coefficient is defined as follows:

$$C_u = \frac{d_{60}}{d_{10}} \tag{4.10}$$

where d_{10} and d_{60} correspond respectively to the 10% and 60% lines of a cumulative particle size distribution curve (only 10%—by weight—or 60% of finer grains can be found in the sediment). Based on this coefficient, the sediment can be classified

Table 4.1 Typical porosity and effective porosity ranges (in %) for common lithologies

Lithology	n (%)	n_e (%)
Granite and gneiss	0.02–2	0.1–2[a]
Basalt	5–30	0.1–2[a]
Quartzite	0.5–2	0–2[a]
Shales	0.1–7.5	0.1–1[a]
Schists and slates	0.1–7.5	0.1–2[a]
Limestone and dolomite	0.5–15	0.5–14[a]
Chalk	0.5–45	0.5–15[a]
Sandstone, siltstone	3–38	3–25
Volcanic tuff	30–40	5–15
Gravels	15–25	5–25
Sands	15–35	5–25
Silts	30–45	5–15
Loams, loess, and clays	40–70	0.1–3

Source: Freeze and Cherry 1979, Fetter 2001.

Note
[a] Depends strongly on fractures, fissures.

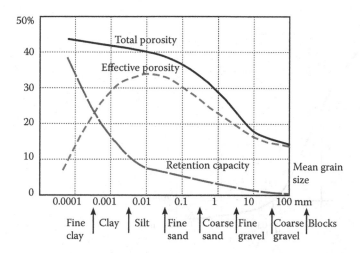

Figure 4.5 Total porosity, effective porosity, and retention capacity values in sediments according to the mean grain size (Eckis 1934, Castany 1963).

as a well sorted sediment with C_u lower than 4, or a poorly sorted sediment with C_u greater than 6 (Figure 4.6).

In practice, the total porosity is determined on undisturbed samples by using porosimeters (e.g., mercury porosimeter, air extraction porosimeter, expansion gas porosimeter), where the total volume and the volume of the voids are measured. Experiments on a laboratory scale to determine the effective porosity can be carried out by draining initially saturated undisturbed samples. Samples are often somewhat reworked especially in loose sediments. Some geotechnical tests, including oedometer tests, can help to determine indirectly the porosity. In compressible soils or porous media, the porosity is also highly dependent on the stress state (see Section 4.10 and Chapter 6).

In the case of a dual porosity rock, it is possible to assess the respective contribution of the pores and fissures to the total porosity using an index of continuity (I_C) determined by seismic measurements on a sample (Tourenq 1978, Denis et al. 1978):

$$I_C = \frac{v_{Lm}}{v_{Lc}} \tag{4.11}$$

with v_{Lm} and v_{Lc} respectively the measured and calculated (on the basis of mineral composition) longitudinal seismic velocities through the sample. Equation 4.11 has been used providing results from hundreds of experimental results on fissured sandstones, siltstones, limestones, and chalks. On the basis of these results, an experimental and empirical law was provided by Calembert et al. (1981):

$$v_{Lm} = \frac{v_{Lc}(100 - 1.6n_p - 22n_f)}{100} \tag{4.12}$$

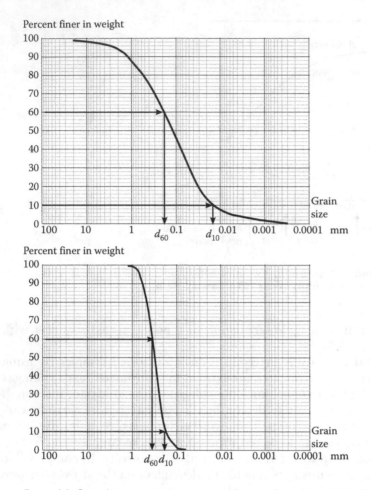

Figure 4.6 Cumulative grain size curves and geotechnical d_{10} and d_{60} dimensions of a sediment.

with n_p and n_f respectively the pore and fissure porosities. A useful corresponding abacus is shown on Figure 4.7.

It is also important to have accurate knowledge about the geological, structural, and tectonic conditions to assess, in 3D, the fracture network, orientations, directions, slopes, openings, density, roughness, and the degree of connectivity.

In the field, the total porosity can be derived from calibrated well-logging involving neutron measurements (geological formations are irradiated by a source of neutrons and the effects thereby produced are recorded by a receptor) providing a clear hydrogen content indicator. It is most often combined with gamma-ray and gamma-gamma logs (respectively clay content and density indicators) and possibly also with resistivity and sonic logs providing proxies to estimate porosity information. On a larger scale, the effective porosity can be determined as the specific yield of unconfined aquifers by interpretation and modeling of pumping tests (see Chapter 5).

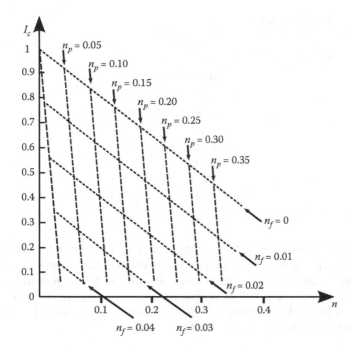

Figure 4.7 Experimental abacus to assess the respective parts of pore and fissure porosities in a dual porosity medium characterized by a total porosity and a continuity index (Calembert *et al.* 1981).

Water content

In a partially saturated media, the total porosity is partially filled with water so that the volumetric *water content* is defined as:

$$\theta = \frac{V_w}{V_t} \tag{4.13}$$

where V_w is the volume of water in the REV. If the grain size is small, the specific retention is high due to the high specific surface area, and the capillary fringe will be high between the saturated and partially saturated zones. Further developments about hydrodynamic conditions in the partially saturated zone will be provided in Chapter 9.

4.3 Piezometric heads

Introduction

Like in hydraulics, groundwater is assumed homogeneous and continuous. One can also assume that the properties of the smallest particles (i.e., water molecules) are the same throughout the entire groundwater system. In contrast with gases, liquids

Table 4.2 Representative values for density (in kg/m³)

Freshwater (at 4°C)	1×10^{-3}
Seawater (average value at the surface)	1.025×10^{-3}
Petrol	$0.660–0.760 \times 10^{-3}$
Fuels (various types)	$0.890–1.025 \times 10^{-3}$
Lamp oil	$0.790–0.820 \times 10^{-3}$
Benzene	0.88×10^{-3}
BTEX	$0.86–0.88 \times 10^{-3}$
Naphthalene (at 15.5°C)	1.145×10^{-3}
PCE	1.622×10^{-3}
Mercury	13.6×10^{-3}

have a very low compressibility, thus in hydrogeology, water can usually be considered incompressible in comparison with the bulk porous medium compressibility (see Section 4.10). For example, to decrease a volume of liquid water by 1%, a pressure of 25 MPa is needed corresponding to a 2,500 m-high column of freshwater (at 4°C).

Pressure (p) in liquids is defined as a force per unit of surface area, whereby the force (F) is considered perpendicular to the surface (A):

$$p = \frac{F}{A} \tag{4.14}$$

The IS unit of pressure is N/m² or Pascal (Pa) $[ML^{-1}T^{-2}]$.

The density of a liquid (ρ), also called the volumetric mass or the specific mass (M) per unit volume (V) is defined as:

$$\rho = \frac{M}{V} \tag{4.15}$$

with IS unit kg/m³ $[ML^{-3}]$. Representative liquid density values are given in Table 4.2.

The relative density of a liquid is a dimensionless number, defined as the ratio between the density of the liquid and the density of water at 4°C. For example, the relative density of seawater is equal to 1.025 on average at the surface.

At a depth z below the surface of a uniform density liquid, the fluid exerts a pressure that is equal to $\rho g z$, where g is the acceleration of gravity.

Simplified Bernoulli equation and piezometric head

An infinitesimal particle of fluid (e.g., water) below ground has a constant mechanical energy composed of potential and kinetic components (Bernoulli 1738). This simple form of Bernoulli's principle can be derived from the principle of conservation of energy and is valid for incompressible flow (Hubbert 1940).

Groundwater hydraulics is a special case of fluid mechanics for which flows are considered as almost always laminar (Burger et al. 1985). Exceptions are encountered in fissured rocks where turbulent flows can occur in open fissures with roughness and tortuosity and in granular sediments near strong pumping wells.

The total energy of a moving particle of fluid is therefore considered as an energy only due to altitude of the particle and pressure conditions because the kinetic energy is sufficiently low that it can be neglected. Bernoulli's equation is thus written accordingly, in water head expressed in m [L], which is used to define a *piezometric head* (Figure 4.8):

$$h = z + \frac{p}{\rho g} = h_g + h_p \tag{4.16}$$

where h_g is the elevation head and h_p is the pressure head.

The head h is also referred as groundwater hydraulic head, total head, piezometric level, groundwater level, and groundwater potential. As mentioned by Bear and Cheng (2010), in the mathematical sense, it is actually not a potential. The terms piezometric levels and groundwater levels can be confusing as there is no certainty about the actual reference plane (i.e., the geodetic datum or any local topographical reference surface).

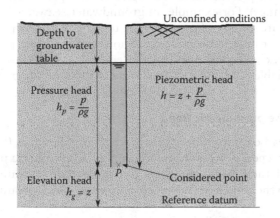

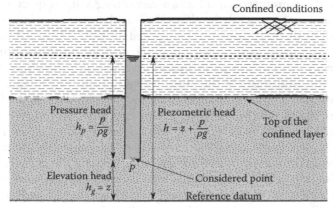

Figure 4.8 Definition of the piezometric head.

Equation 4.16 provides a direct relation between groundwater pressure and piezometric head:

$$p = \rho g (h - z) \tag{4.17}$$

where $(h - z)$ is the water-saturated depth or thickness above the considered location. For a varying water pressure at a given location of the porous medium, we can write:

$$\frac{\partial p}{\partial t} = \rho g \frac{\partial h}{\partial t} \tag{4.18}$$

In groundwater flow studies, water pressure or piezometric head can be used as the main variable of the problem. In hydrogeology practice, piezometric head is most often chosen as the main variable. It is the driving force for groundwater flow (Hubbert 1940), which is not especially the case with pressure (among others, Ingebritsen *et al.* 2006). However, it is important to observe that piezometric head values can be compared with each other only if groundwater is considered at the same temperature and with the same chemical composition (e.g., salt content). Otherwise the groundwater density can be affected (e.g., the pressure field is coupled to the spatial salinity distribution and its temporal change). For example, in groundwater seawater intrusion problems, it is crucial to measure salinity of groundwater together with any piezometric monitoring survey. Then, for interpretation, calculation, and modeling purposes, a choice must be made to work with pressure or with "equivalent freshwater piezometric head" as the main variable (Carabin and Dassargues 1999; see Chapter 10).

Practical measurements of the piezometric head

Note that the piezometric head definition needs three kinds of references (Chapuis 2007): a single elevation reference, a pressure reference (i.e., atmospheric pressure taken as 0 pressure), and (if the kinetic energy is not neglected) a groundwater velocity reference attached to the solid matrix/skeleton through which groundwater is flowing.

In practice, to measure the piezometric head at an elevation z with respect to a datum level (usually the regional mean sea level is taken as datum), a monitoring or observation well (usually called "piezometer") is needed. A pipe is inserted into the ground down to the measuring point where it is screened or perforated and surrounded by a gravel pack to enable a good hydraulic connection between the groundwater in the formation and water in the pipe. The piezometer well is indeed not pumped so that hydrostatic conditions can be assumed. With an *electric probe*, the "depth to water" from the top of the piezometer pipe can be measured (Figure 4.9a). The piezometric head is found by subtracting this "depth to water" from the elevation of the top of the piezometer pipe with regards to the reference datum. *Pressure sensors and transducers* can also provide pressure head measurements to which elevation of the point (z) is added to obtain the piezometric head (Figure 4.9b). These sensors are usually used in smaller diameter piezometers and in low hydraulic conductivity porous media.

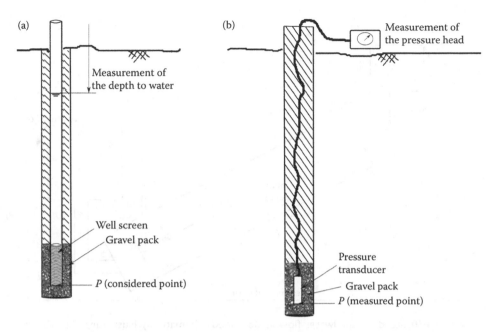

Figure 4.9 Piezometric head measurements (a) with a probe providing "depth to water" or (b) with a pressure transducer providing the pressure head (Chapuis 2007).

4.4 Darcy's law and hydraulic conductivity

Experimental Darcy's law

Any difference in piezometric head (energy per unit weight of water) at a given time is created by the resistance (due to friction) to groundwater flow in the complex network of tortuous paths of the considered porous medium. Testing different kinds of sand filters to provide water to the fountains of Dijon (France), Darcy (1856) proposed an experimental law to describe saturated flow in homogeneous sand columns. He observed that the flow rate was linearly proportional to the piezometric head difference, proportional to the cross-sectional area, and inversely proportional to the length of the considered sample. He formulated his empirical law in proposing a proportionality coefficient representing the resistance of the porous medium to the fluid motion. *Darcy's law* can be expressed (Figure 4.10) as:

$$Q = KA\frac{\Delta h}{L} \tag{4.19}$$

where Q (in m³/s) [L³T⁻¹] is the volumetric flow rate, $\Delta h/L$ [−] is the hydraulic gradient with Δh the head difference and L the length of the sample, A (m²) [L²] is the sample cross-sectional area, and K (m/s) [LT⁻¹] is the *hydraulic conductivity* of the saturated porous medium. K was originally referred to as the permeability coefficient.

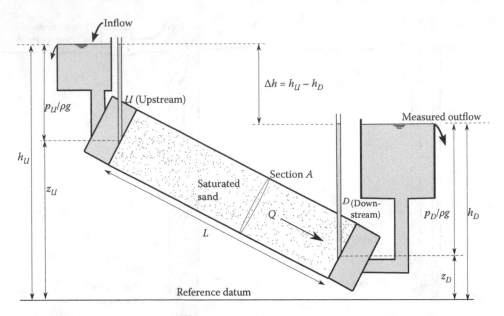

Figure 4.10 Steady groundwater flow as described schematically illustrating Darcy's law. The motion of water is directed from the higher upstream piezometric head toward the lower downstream piezometric head.

In this form, the Darcy's law is similar to Ohm's law of electricity, Fick's law of solute diffusion (see Section 8.2), or to Fourier's law of heat conduction. Indeed, before the advent of numerical models, many groundwater flow problems were solved using physical models involving electricity or heat conduction analogies.

Groundwater flow takes place from a higher piezometric head to a lower one ($\Delta h = h_U - h_D$) (Figure 4.10), with h_U and h_D respectively the upgradient and downgradient piezometric heads, as groundwater flow requires energy (Delleur 1999). It should be noticed that groundwater flow is not necessarily directed from a higher to a lower *pressure*. As shown in Figure 4.10, flow can be in the direction of increasing pressure ($p_D < p_U$) but is always in the direction of decreasing piezometric head. This phenomenological law, expressed in terms of piezometric heads, is valid for an incompressible fluid (i.e., with a constant density). If the study case involves varying density of water due to solute concentrations or (geo) thermal effects, the state variable should be the pressure (as mentioned previously in Section 4.3).

Specific discharge and velocities

One of the primary uses of Darcy's law is to calculate a *specific discharge,* dividing the flow rate by the cross-sectional area of the porous medium:

$$q = \frac{Q}{A} = K \frac{\Delta h}{L} \tag{4.20}$$

where q (in m/s) [LT^{-1}] is the specific discharge (or Darcy flux). It is also known inappropriately as the "Darcy velocity." Actually, groundwater flow takes place only in the pores of the medium, and indeed not through the solid/skeleton matrix.

As mentioned by Bear and Bachmat (1990), an average areal porosity can be considered equal to the volumetric porosity. The specific discharge must be divided by a porosity to obtain a velocity. In Section 4.2, different porosities were introduced: (a) the total porosity (n) including mobile, and immobile water; (b) the drainable effective porosity ($n_e = S_y$) (mobile water for flow); (c) the mobile water porosity (n_m) for solute transport. This last is often unfortunately also named "effective porosity." Although different groundwater velocities can be defined, all are averaged over the entire volume of the saturated porous medium considered as the chosen REV, and having units in m/s [LT^{-1}].

a. A global *mass-averaged velocity* of water (Bear and Cheng 2010) is defined by:

$$v_{avg} = \frac{q}{n} \tag{4.21}$$

b. An *effective velocity* relative to the mobile fraction of water in drainage and flow problems (i.e., balance equations for groundwater flow) is expressed by:

$$v_e = \frac{q}{n_e} \tag{4.22}$$

c. A mobile water velocity for solute transport named *transport velocity* or *advection velocity* (Payne *et al.* 2008) is expressed by:

$$v_a = \frac{q}{n_m} \tag{4.23}$$

Usually, $n_m < n_e < n < 1$ so that logically $v_a > v_e > v_{avg} > q$, where n_m is the mobile water porosity for transport.

Hydraulic conductivity and intrinsic permeability

One of the big challenges in hydrogeology comes from the fact that hydraulic conductivity values can vary over 13 orders of magnitude in common geological media. Table 4.3 provides indicative values for hydraulic conductivity in different geological formations. One of the reasons for the huge variations comes from the various joints, fractures, and karstic features that can be found under many different form, width, frequency, and intensity. These values must be considered with great caution because, as mentioned previously (see Section 4.2), the proportion of mobile water in a geological medium is dependent on head conditions. The groundwater motion process combined with a variety of fissures, fractures, and possibly karstic features that develop differently (orientation, density, aperture, interconnectivity) according to the lithology, produces a wide range of possible values, which are also related to the considered size of the REV (see Box 4.2).

Table 4.3 Typical hydraulic conductivity values (in m/s)

Lithology		K (m/s)
Granite and gneiss	with fissures	$1 \times 10^{-7} - 1 \times 10^{-4}$
	without fissures	$1 \times 10^{-14} - 1 \times 10^{-10}$
Basalt	with fissures	$1 \times 10^{-7} - 1 \times 10^{-3}$
	without fissures	$1 \times 10^{-12} - 1 \times 10^{-9}$
Quartzite	with fissures	$1 \times 10^{-7} - 1 \times 10^{-4}$
	without fissures	$1 \times 10^{-12} - 1 \times 10^{-9}$
Shales		$1 \times 10^{-13} - 1 \times 10^{-9}$
Schists and slates		$1 \times 10^{-9} - 1 \times 10^{-5}$
Limestone and dolomite	karstified	$1 \times 10^{-5} - 1 \times 10^{-1}$
	with fissures	$1 \times 10^{-9} - 1 \times 10^{-3}$
	without fissures	$1 \times 10^{-12} - 1 \times 10^{-9}$
Chalk		$1 \times 10^{-6} - 1 \times 10^{-3}$
Sandstone, siltstone	with fissures	$1 \times 10^{-5} - 1 \times 10^{-3}$
	without fissures	$1 \times 10^{-9} - 1 \times 10^{-5}$
Volcanic tuff		$1 \times 10^{-7} - 1 \times 10^{-3}$
Gravels		$1 \times 10^{-4} - 1 \times 10^{-1}$
Sands		$1 \times 10^{-6} - 1 \times 10^{-2}$
Silts		$1 \times 10^{-7} - 1 \times 10^{-4}$
Loams, loess and clays		$1 \times 10^{-13} - 1 \times 10^{-7}$

The hydraulic conductivity is in fact a specific parameter used to describe ground-water flow in a porous medium. Implicitly, groundwater is assumed to be freshwater taken at a 12°C–15°C temperature. Hydraulic conductivity depends on both the porous medium properties and the fluid properties. The relevant fluid properties are the density (ρ) and the dynamic viscosity (μ). The matrix properties are grain size fractions, grain shapes, pore distribution and shapes, intergranular porosity, and pore space tortuosity. These effects are all combined in a coefficient (k) called *intrinsic permeability* or *permeability* of the porous medium. The hydraulic conductivity is then expressed as:

$$K = \frac{k \rho g}{\mu} \tag{4.24}$$

where k is the intrinsic permeability (m^2) [L^2], μ is the dynamic viscosity (kg/(m.s)) [ML^{-1}T^{-1}], g is the acceleration of gravity (m/s^2) [LT^{-2}], and ρ is the density of water (kg/m^3) [ML^{-3}].

For decades, petroleum engineers use the unit *darcy* for the (intrinsic) permeability of a geological reservoir in place of m^2 or cm^2, with 1 darcy = 9.87×10^{-13} m^2. For fresh groundwater at a temperature of 20°C and under a pressure head of 1 m (or 0.98×10^4 Pa), 1 darcy is equivalent to 0.833 m/day (or 9.64×10^{-6} m/s $\cong 10^{-5}$ m/s). Via Equation 4.24, the hydraulic conductivity K depends on the fluid viscosity and

---

Box 4.2   Scale effect on hydraulic conductivity values: Example of a Cretaceous chalk aquifer

At the microscopic level (up to a few centimeters):
- $n = 0.40 - 0.42$ $n_e = 0.35$ (see Box 4.1)
- $K \cong 1.10^{-8}$ (m/s)

At the macroscopic scale (scale of laboratory tests, up to a few decimeters):
- $n = 0.42 - 0.45$ $n_e = 0.01 - 0.03$ (see Box 4.1)
- $1.10^{-5} \leq K \leq 1.10^{-4}$ (m/s)

At the megascopic scale (scale of pumping tests, up to a few hundred meters), it is observed that "homogenization" in a large REV ensures that all fractures, faults, and discontinuities should be considered as interconnected, resulting in an increase in the value of hydraulic conductivity:
- $n = 0.42 - 0.45$ $n_e = 0.05 - 0.10$ (see Box 4.1)
- $1.10^{-4} \leq K \leq 1.10^{-3}$ (m/s)

Source: Data are relative to the Geer basin in Belgium, modified after Dassargues and Monjoie 1993; see also Box 4.1, Section 4.2.)

---

density, which can vary with the temperature and/or total dissolved solids (TDS) content (see Chapter 7). As observed in Table 4.4, the first of these relationships is by far the one that has the most impact on the values of K. The K should therefore not be assumed constant when viscosity (and to a lesser extent density) can vary under the prevailing hydrogeological conditions. In practice, a sensitivity analysis must at least be performed to check if K variations would be significant for the studied problem.

3D Darcy's law

Darcy's law can be extended to a three-dimensional (3D) porous medium domain. Assuming a Cartesian coordinate system (with the z coordinates usually taken as vertical), the scalar gradient ($\Delta h/L$) in Equations 4.19 and 4.20 can be replaced by the piezometric or head gradient vector:

$$\boldsymbol{grad}\, h = \nabla h = \left(\frac{\partial h}{\partial x}, \frac{\partial h}{\partial y}, \frac{\partial h}{\partial z} \right) \tag{4.25}$$

In this text, the adopted typographic convention consists of representing vectors and tensors with boldface type. Also, the gradient operator will be noted ∇ ("nabla") in m^{-1} [L^{-1}]. In fact, the "nabla" will be used as a differential operator indicating taking a gradient (of a field variable) as well as taking a divergence (of a vector) (see Section 4.9). In 3D, $\nabla = ((\partial/\partial x),(\partial/\partial y),(\partial/\partial z))$.

Consequently, the specific discharge components can be expressed in each direction taking into account the possible anisotropy of hydraulic conductivity of the porous medium. In this last case, hydraulic conductivity depends upon the direction of

Table 4.4 Permeability (k) of 1 darcy converted to hydraulic conductivity K (m/s) depending on groundwater temperature and TDS (Total Dissolved Solids, see Chapter 7) which influence viscosity (kg/(m.s)) and density (kg/m³)

TDS (mg/L)	0	100	500	1000	10000	35000 (seawater)
T (°C)						
0	$\rho = 999.868$ $\mu = 1.79 \times 10^{-3}$ $K = 5.4 \times 10^{-6}$	$\rho = 999.950$ $\mu = 1.79 \times 10^{-3}$ $K = 5.4 \times 10^{-6}$	$\rho = 1000.278$ $\mu = 1.79 \times 10^{-3}$ $K = 5.4 \times 10^{-6}$	$\rho = 1000.687$ $\mu = 1.79 \times 10^{-3}$ $K = 5.4 \times 10^{-6}$	$\rho = 1007.980$ $\mu = 1.83 \times 10^{-3}$ $K = 5.3 \times 10^{-6}$	$\rho = 1028.131$ $\mu = 1.88 \times 10^{-3}$ $K = 5.3 \times 10^{-6}$
10	$\rho = 999.728$ $\mu = 1.31 \times 10^{-3}$ $K = 7.4 \times 10^{-6}$	$\rho = 999.807$ $\mu = 1.31 \times 10^{-3}$ $K = 7.4 \times 10^{-6}$	$\rho = 1000.122$ $\mu = 1.31 \times 10^{-3}$ $K = 7.4 \times 10^{-6}$	$\rho = 1000.514$ $\mu = 1.31 \times 10^{-3}$ $K = 7.4 \times 10^{-6}$	$\rho = 1007.527$ $\mu = 1.34 \times 10^{-3}$ $K = 7.3 \times 10^{-6}$	$\rho = 1026.979$ $\mu = 1.41 \times 10^{-3}$ $K = 7.05 \times 10^{-6}$
20	$\rho = 998.234$ $\mu = 1.00 \times 10^{-3}$ $K = 9.6 \times 10^{-6}$	$\rho = 998.310$ $\mu = 1.00 \times 10^{-3}$ $K = 9.6 \times 10^{-6}$	$\rho = 998.616$ $\mu = 1.00 \times 10^{-3}$ $K = 9.6 \times 10^{-6}$	$\rho = 998.997$ $\mu = 1.00 \times 10^{-3}$ $K = 9.6 \times 10^{-6}$	$\rho = 1005.820$ $\mu = 1.02 \times 10^{-3}$ $K = 9.5 \times 10^{-6}$	$\rho = 1024.790$ $\mu = 1.08 \times 10^{-3}$ $K = 9.2 \times 10^{-6}$
30	$\rho = 995.678$ $\mu = 0.80 \times 10^{-3}$ $K = 12.2 \times 10^{-6}$	$\rho = 995.753$ $\mu = 0.80 \times 10^{-3}$ $K = 12.1 \times 10^{-6}$	$\rho = 996.053$ $\mu = 0.80 \times 10^{-3}$ $K = 12.1 \times 10^{-6}$	$\rho = 996.427$ $\mu = 0.80 \times 10^{-3}$ $K = 12.1 \times 10^{-6}$	$\rho = 1003.122$ $\mu = 0.83 \times 10^{-3}$ $K = 11.7 \times 10^{-6}$	$\rho = 1021.755$ $\mu = 0.86 \times 10^{-3}$ $K = 11.5 \times 10^{-6}$
40	$\rho = 992.247$ $\mu = 0.65 \times 10^{-3}$ $K = 14.8 \times 10^{-6}$	$\rho = 992.322$ $\mu = 0.65 \times 10^{-3}$ $K = 14.7 \times 10^{-6}$	$\rho = 992.616$ $\mu = 0.65 \times 10^{-3}$ $K = 14.7 \times 10^{-6}$	$\rho = 992.988$ $\mu = 0.65 \times 10^{-3}$ $K = 14.7 \times 10^{-6}$	$\rho = 999.602$ $\mu = 0.72 \times 10^{-3}$ $K = 13.4 \times 10^{-6}$	$\rho = 1017.998$ $\mu = 0.74 \times 10^{-3}$ $K = 13.3 \times 10^{-6}$

measurement, and is no longer a scalar but a tensor K of nine components which can be written in matrix form as:

$$K = \begin{bmatrix} K_{xx} & K_{xy} & K_{xz} \\ K_{yx} & K_{yy} & K_{yz} \\ K_{zx} & K_{zy} & K_{zz} \end{bmatrix}$$

(4.26)

In the most complex 3D case, up to six distinct components are needed because the hydraulic conductivity tensor is positive definite. If the directions of the principal components of the hydraulic conductivity tensor are known and aligned with the selected coordinate system, only the three diagonal components of hydraulic conductivity are needed: K_{xx}, K_{yy}, K_{zz}. If the medium is assumed isotropic in the horizontal plane, there are only two remaining components: $K_{xx} = K_{yy} = K_h$ (horizontal hydraulic conductivity) and $K_{zz} = K_v$ (vertical hydraulic conductivity). Indeed, the anisotropy of the hydraulic conductivity is fully induced by the anisotropy of the permeability so that Equation 4.24 is generalized in 3D:

$$K = \frac{k\rho g}{\mu}$$

(4.27)

The generalized Darcy law can now be written in 3D:

$$q = -K \cdot \nabla h = -\frac{k\rho g}{\mu} \cdot \nabla h = -\frac{k}{\mu} \cdot (\nabla p + \rho g \nabla z)$$

(4.28)

A negative sign appears in Equation 4.28 since the specific discharge is directed to lower value of h, and the positive vertical z axis being directed toward the top.

As for any vector, the specific discharge has a direction (Figure 4.11) and its magnitude can be calculated by:

$$\|q\| = \sqrt{q_x^2 + q_y^2 + q_z^2}$$

(4.29)

In an anisotropic geological medium, the specific discharge vector is not necessarily collinear with the piezometric gradient vector.

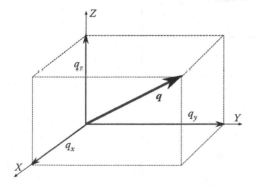

Figure 4.11 Specific discharge vector and its components in a 3D Cartesian coordinate system.

4.5 Heterogeneity: Upscaled, equivalent, and averaged hydraulic conductivity values

Heterogeneity of a geological medium is indeed a major source of complexity to quantify groundwater fluxes. Despite the different possibilities of measuring permeability (k) and hydraulic conductivity (K) at different scales in the laboratory or in the field (see Chapter 5), it is essential to understand how K values can be upscaled. In practice, choosing the adequate upscaling procedure to use effective, equivalent, or averaged K values dedicated for the studied process and in the considered medium is a crucial step. This is particularly true when using numerical techniques based on grid-cells or finite elements as applied in groundwater modeling. An upscaled K value usually refers to the K value of a larger volume of the medium (most often considered as the REV) given some finer scale observations or measurements.

If we consider groundwater flow through a porous medium, the upscaled K value on an assumed homogeneous REV will give, under the same boundary conditions, the same/equivalent flow as the actual heterogeneous medium the REV is representing (Ringrose and Bentley 2015). So, the upscaled values can be largely dependent on the considered process and the boundary conditions.

Equivalent averaged hydraulic conductivity values for flow parallel or perpendicular to stratified layers

For groundwater flow parallel to different layers of a stratified geological medium, applying Darcy's law, the equivalent K is an arithmetic average of the individual K values (Figure 4.12):

$$K_{eq} = \frac{\sum_{i=1}^{n} K_i d_i}{\sum_{i=1}^{n} d_i} \tag{4.30}$$

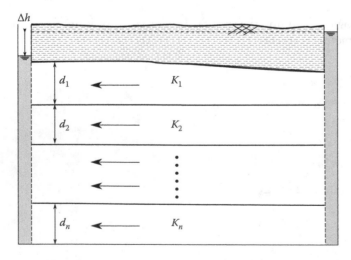

Figure 4.12 Groundwater flow parallel to different layers of stratification.

where the subscript i refers to the layer, and d_i the thickness of each layer. The value of K_{eq} is derived based on the equivalence of the total groundwater flow rate or the equivalence of the specific discharge (as the groundwater flow section is constant). Note that this averaged K_{eq} is only useful for groundwater quantity estimates. It should not be considered for solute transport (see Chapter 8). For example, calculations of contaminant arrivals would require considering the "worst case" scenario corresponding to the highest hydraulic conductivity.

For groundwater flow perpendicular to the layers, the specific discharge flowing through the stratified system must equal the flow through the equivalent homogeneous medium. The sum of the head drop across each layer must equal the piezometric head drop across the whole volume. Applying Darcy's law, the equivalent K is a harmonic average of the individual K values (Figure 4.13):

$$K_{eq} = \frac{\sum_{i=1}^{n} d_i}{\sum_{i=1}^{n} d_i/K_i} \tag{4.31}$$

in which it is assumed that the groundwater flow section remains constant (i.e., confined conditions). One can easily recognize similar laws in electricity for resistances in parallel and in series, as derived from Ohm's law.

On the basis of these two very simple examples, it is clear that the way of calculating equivalent values depends strongly on the final aim of the study. Another important point is the fact that the heterogeneity scale and problem scale must be considered when choosing the adequate REV size. In a similar approach as above, and still within a deterministic framework, upscaled equivalent K values and K tensor components can be calculated for various heterogeneous conditions including cross-bedded geological formations inducing crossflow between homogeneous sub-blocks (Durlofsky 1991, Pickup et al. 1994, Renard and de Marsily 1997, Ringrose and Bentley 2015).

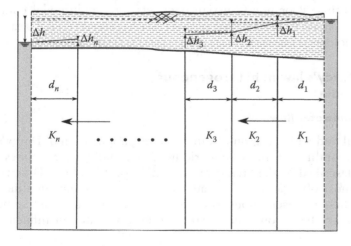

Figure 4.13 Groundwater flow perpendicular to different layers.

Geostatistically derived equivalent averaged hydraulic conductivity values

Much specialized literature is available on K_{eq} calculations under various heterogeneous conditions. Real geological media can be regarded has having a natural variability due to (1) structure/architecture properties at the macroscale and (2) texture/petrophysics properties at the microscale. As hydrogeologists, our main objectives aim to represent and quantify groundwater flow and associated transport processes at the macroscale (or megascale) with the help of the available data and measurements.

If not significantly affected by larger structural features, hydraulic conductivity at the macroscopic scale depends on microscopic properties that can be considered as stochastic quantities. Therefore, as stated by de Marsily (1986), K can be "regarded as a stochastic property and can be defined conceptually as a random function." As a consequence, probability distribution functions (pdfs), as well as average, variance, and covariance values can be calculated from spatially distributed data sets considered as statistically homogeneous. As developed and observed by many authors, hydraulic conductivity data are often found to approximately follow a log-normal distribution. This assumed Gaussian behavior of log (K) data, however, neglects the potential effect of geological structural architecture (i.e., faults, dominant fracture orientations, major bedding boundaries). More advanced geologically-based approaches (Ringrose and Bentley 2015) are recommended to capture these macroscale effects on the upscaled equivalent K components.

The mathematically correct average value (and equivalent value on an REV for an assumed uniform flow) for a log-normal distribution of hydraulic conductivity is the geometrical average of K (equal to the arithmetic average of the $\ln(K)$ values):

$$K_{eq} = \sqrt[n]{\prod_{i=1}^{n} K_i} \qquad (4.32)$$

where the subscript i refers to each local measurement of K.

This log-normal distribution of K is appealing and useful for further modeling (see Chapters 12 and 13). However, in practice, many data/measurements are needed to infer a Gaussian distribution. Poor non-Gaussian data sets (i.e., multimodal data) should be divided into several Gaussian distributions (with enough data in each distribution) based on geological knowledge.

4.6 Application of Darcy's law in heterogeneous and fractured media

Local and regional groundwater flow

When dealing with local and regional groundwater systems, application of Darcy's law allows deducing streamlines from piezometric head isocontours (or isolines). Typical situations are described in the literature using 2D horizontal or 2D vertical cross sections and most often assuming steady-state conditions and simplified boundary conditions. These representations can be useful and convenient for some interpretations but they are based on conceptual assumptions that can influence results.

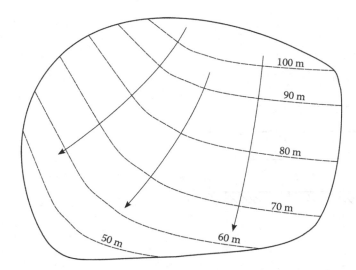

Figure 4.14 A piezometric map shows the equipotentials in a 2D horizontal view. Streamlines can often be drawn perpendicular to the equipotentials assuming an isotropic medium.

Piezometric maps are often used as a first assessment of the groundwater conditions in a specific aquifer. This 2D horizontal representation of the piezometric head spatial variation shows isolines of *h*, or *equipotentials*. Applying Darcy's law, local and general directions of groundwater flow can be deduced by tracing streamlines perpendicular to the *equipotential lines* assuming an isotropic medium (Figure 4.14). Most often, if there is no mention of the depth at which this horizontal section is drawn, an implicit assumption is taken that the vertical component of flow is negligible in comparison to horizontal flow components. This assumption, called the *Dupuit assumption* (Dupuit 1863) corresponds to:

$$\frac{\partial h}{\partial z} = 0 \tag{4.33}$$

Using Equation 4.16 and assuming no groundwater density variation with depth:

$$\frac{\partial p}{\partial z} = -\rho g \quad \text{and} \quad p = -\rho g z \tag{4.34}$$

This simplified situation is also known as a "hydrostatic" pressure distribution, which means that only vertical isosurfaces of head are represented by equipotentials in a horizontal 2D view.

In these 2D horizontal flow conditions, a change in the observed mapped piezometric head gradient can be caused by three potential reasons (Figure 4.15). Using Equation 4.19, if an increase of *K* occurs for unchanged groundwater flow cross-sectional area (*A*) and constant total flow rate in the aquifer (*Q*), it must be balanced

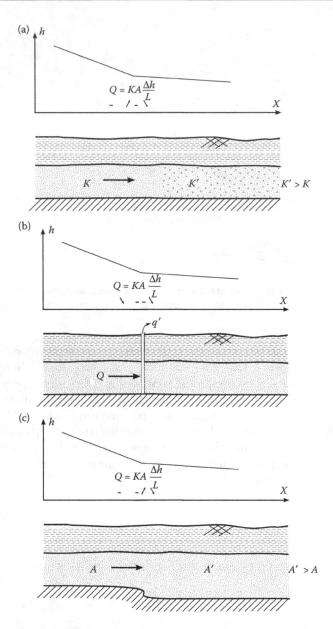

Figure 4.15 Piezometric head gradient can decrease locally for three potential reasons: (a) an increase of K; (b) a decrease of the total groundwater flow in the aquifer (Q) (e.g., due to pumping); and (c) an increase in the groundwater flow cross-sectional area.

by an increase in the gradient ($\Delta h/L$) (Figure 4.15a). On the other hand, if the total groundwater flow in the aquifer decreases due to a pumping well or drainage, it is balanced by a decrease of the head gradient for a constant K and A (Figure 4.15b). Finally, if the groundwater flow section is increased, it is balanced by a decrease of the head gradient if K and Q stay constant (Figure 4.15c).

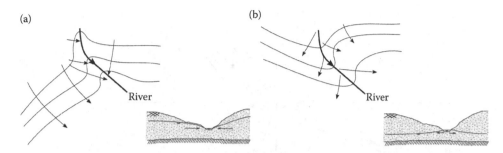

Figure 4.16 River—groundwater interactions as possibly detected by interpretation of a piezo-metric map: (a) converging streamlines toward a "gaining" or "draining" river; (b) diverging streamlines from a "loosing" or "feeding" river.

In practice, piezometric maps are drawn from interpolation of point-measured piezometric heads. Despite the existence of sophisticated mathematical or geostatistical methods to perform this type interpolation and contouring (e.g., among others, Hoeksema *et al.* 1989, Desbarats *et al.* 2002, Fasbender *et al.* 2008), many field hydrogeologists still interpolate and extrapolate piezometric heads by hand. This can be justified to the extent that it allows to implicitly take into account other factors in the contouring as, for example, the expected influence of the different geological formations (Figures 4.15 and 4.17), topographical effects, and local drainage or recharge from rivers (Figure 4.16). They integrate in the hand-made interpolations their own interpretation of those influences on local gradients. Indeed, the subjective character of the resulting map depends strongly on the density of available observation wells.

Two-dimensional piezometric maps are also used to assess *local and regional interactions between rivers and groundwater* (Figure 4.16). Assuming that the hydraulic conductivity of the river bed and embankments is of the same order as in the surrounding aquifer (i.e., that groundwater head will be largely influenced by the river's hydraulic level), groundwater streamlines will converge toward the "gaining" or "draining" river (Figure 4.16a). On the contrary, they will diverge from the river into the aquifer for a "loosing" or "feeding" river (Figure 4.16b).

Refraction of streamlines and of equipotentials at the interface between two media characterized by different hydraulic conductivities can be represented in the 2D horizontal plane or in a 2D vertical cross section (Figure 4.17). Although presented in 2D, it is actually valid in 3D. For $K_1 < K_2$ with both units assumed isotropic, the continuity of the 2D groundwater flow field is expressed by:

$$Q = -K_1 a \frac{\Delta h}{\Delta l_1} = -K_2 c \frac{\Delta h}{\Delta l_2} \tag{4.35}$$

Expressing $a = b \cos \alpha_1$, $\Delta l_1 = b \sin \alpha_1$, $c = b \cos \alpha_2$, $\Delta l_2 = b \sin \alpha_2$ (Figure 4.17) in Equation 4.35 gives:

$$\frac{K_1}{K_2} = \frac{\tan \alpha_1}{\tan \alpha_2} \tag{4.36}$$

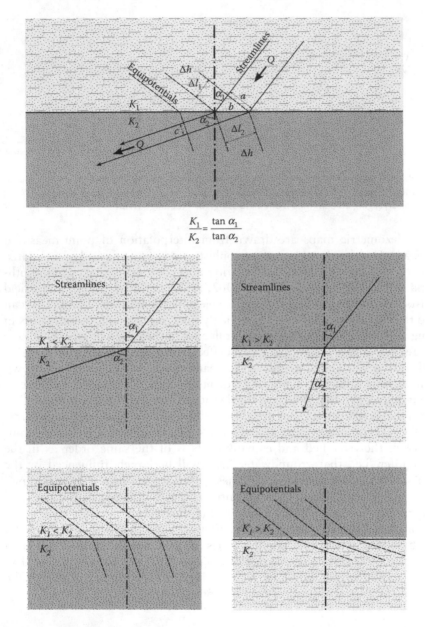

$$\frac{K_1}{K_2} = \frac{\tan \alpha_1}{\tan \alpha_2}$$

Figure 4.17 Refraction of streamlines and of equipotentials at the interface between two media characterized by different hydraulic conductivities ($K_1 < K_2$).

where α_1 and α_2 are the angles which the streamlines make with the normal to the boundary (Figure 4.17). Here, one can easily recognize a similar law as the optical refraction law.

Two-dimensional vertical cross sections are often used to show diagrams (called flownets) describing flow systems using groundwater streamlines and equipotentials.

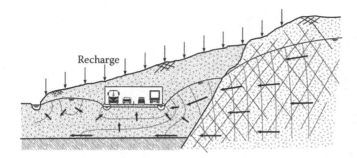

Figure 4.18 2D vertical cross section schematically showing groundwater flow directions and equipotentials influenced by recharge, topography, civil works (tunnel), and geological conditions.

This approach assumes that flow components not in this vertical plane can be neglected (i.e., transverse flow is negligible).

In the past, many authors (among others, Tóth 1962, 1963, Freeze and Witherspoon 1967) have used analytical approaches to calculate flownets to show how the water table controls regional groundwater flow and locations of discharge and recharge areas. However, the results are of limited interest and can even be misleading because of the important assumptions underlying these calculations. For example, the water table is often chosen (prescribed empirically on the basis of assumed similarities with the topographic surface) instead of prescribing a recharge to the system. The medium is assumed isotropic and homogeneous with impermeable boundaries. Buoyancy-driven flow (induced by variations in water density due to changes in groundwater temperature or composition) is neglected.

Modern numerical modeling tools now allow for better understanding and description of heterogeneous groundwater flow systems. Using 2D numerical models, Yang *et al.* (2010) have shown that under some conditions involving temperature differences, groundwater flow can be influenced by buoyancy effects at low water table gradients (<0.0005). Two-dimensional vertical cross sections are also very useful to schematically describe a groundwater system with geological heterogeneity and how it interacts with topography, lakes, wetlands, rivers, and infrastructure (Figure 4.18). These systems must indeed be simulated on a case-by-case basis to check their accuracy and realism with regards to the actual hydrogeological conditions and conceptual model (i.e., hydrogeological parameters and stress factors).

Hydraulic conductivity and groundwater flow in fractured rocks

Considered individually, a natural fracture will most often have an irregular shape and variable aperture and roughness. Despite this observation, the most common way to describe groundwater flow in a single fracture is based on the parallel plate model derived by application of Poiseuille's law which provides a *hydraulic conductivity of the fracture* (Figure 4.19):

$$K_f = \frac{\rho g}{12\mu} a_f^2 \tag{4.37}$$

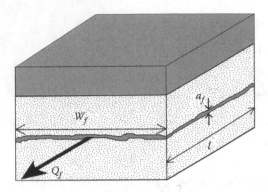

Figure 4.19 Groundwater flow in a fracture along the direction *l* (Ringrose and Bentley 2015).

where a_f (in m) [L] is the fracture aperture (assumed constant). The volumetric water flow Q_f (in m³/s) [L³T⁻¹] along the axis *l* of a fracture presenting a flow section A_f is written (Rausch *et al.* 2002):

$$Q_f = \left(\frac{\rho g}{12\mu} a_f^2 \right) A_f \frac{\partial h}{\partial l} \tag{4.38}$$

$$A_f = a_f w_f \tag{4.39}$$

where w_f is the fracture width (m) [L]. As a result, the so-called *"cubic-law"* is highly dependent on fracture aperture:

$$Q_f = \left(\frac{\rho g}{12\mu} a_f^3 \right) w_f \frac{\partial h}{\partial l} \tag{4.40}$$

In practice, a transmissivity T_f (m²/s) [L²T⁻¹] of the fracture is often used, determined by multiplying the hydraulic conductivity by the aperture:

$$T_f = \left(\frac{\rho g}{12\mu} a_f^3 \right) \tag{4.41}$$

The rock matrix can be assumed porous or not and the fracture filling can be assumed with a relative high porosity (almost 1) where open fractures are considered.

In practice, fracture aperture is hard to measure and generally is only indirectly and globally measured by pressure data (Prandtl probes and Halliburton tests results). General approaches for groundwater flow calculations through fractured media can be divided into three methods (Ringrose and Bentley 2015):

- Matrix and fractures are taken together within the REV and effective/equivalent *K's* are assumed to be adequately upscaled at the REV scale, referred to as the *equivalent porous medium approach* (EPM).

- A *discrete fracture network approach* (DFN) with a matrix considered impervious or not, requiring the explicit fracture network geometry (from the most detailed field knowledge).
- *Dual domain or dual-continuum approaches* (i.e., fracture network and rock matrix) where the REV concept is applied for both domains. Separate equations are written and exchange terms must be defined. A *dual porosity approach* corresponds to the simplified case where hydraulic conductivity is assumed to have a role only in the fracture network.

Groundwater flow in a fracture network does not necessarily follow Darcy's law as the flow can be partially turbulent (see Section 4.7). If the EPM approach is used, all the conceptual assumptions used in an REV are adopted (see Section 4.1) and equivalent hydraulic conductivity calculations can be done (Equations 4.30 and 4.31) considering the combined effect of flow through the matrix and fractures together (i.e., arithmetic or harmonic averages for groundwater flow respectively parallel or perpendicular to the fracture direction). For example, for groundwater flow parallel to continuous fractures of aperture a_f and a mean distance (thickness) of the rock matrix between fractures of d_m, the calculated equivalent hydraulic conductivity is:

$$K_{eq} = \frac{a_f}{a_f + d_m} K_f + \frac{d_m}{a_f + d_m} K_m \qquad (4.42)$$

where K_m is the matrix hydraulic conductivity. If $a_f \ll d_m$ then Equation 4.42 is reduced to:

$$K_{eq} \cong \frac{a_f}{d_m} K_f + K_m \qquad (4.42a)$$

This method is applicable only at a larger scale of observation as it does not describe in detail the specific discharges, velocities, and heads in any of the individual fractures. The REV can be very large or even ill-defined if fractures have irregular apertures and are widely spaced (Fitts 2002). Most often, a tensor K_{eq} describes at the large REV scale, the anisotropy induced by two or more fracture systems. Many techniques to calculate upscaled/equivalent hydraulic conductivity tensor are now available (among others, Chen *et al.* 2015). Note that, as mentioned by de Marsily (1986), the principal components of the K tensor most often do not correspond to the fracture system directions. The principal components can be calculated as a function of the respective directions of the considered fracture systems and the corresponding assessed K_f values for each component (Maini and Hocking 1977, Singhal and Gupta 2010).

4.7 Limitations of the validity of Darcy's law

Darcy's law is an experimental law that has been generalized, and fortunately it has been validated by practice and observations in many situations. However, it has some limitations to describe groundwater flow in very low or very high permeability media. As water flows through the complex interconnected spaces made of pores or micro-fissures of the geological medium, the microscale high variations of directions and velocities are overlooked in favor of equivalent (volume-averaged) values in the REV (Fitts 2002, Bear and Cheng 2010). Darcy's law is confirmed when groundwater flow can be considered as mainly laminar. The *Reynolds number (Re)* is used classically in fluid mechanics as a criterion to distinguish between laminar and turbulent flows. It can be interpreted as a ratio between inertial and viscous forces on the moving water. By analogy with pipe flow, the *Re* number in a porous medium can be defined as:

$$Re = \frac{\rho q d}{\mu} \tag{4.43}$$

where d is a characteristic length assumed to represent the flow cross section in the pore space of the medium. Various definitions of d (m) [L] are proposed in the literature aiming to obtain a length that can be easily measurable in practice. The most commonly used definitions are d = the mean grain-size, or $d = d_{10}$ where d_{10} is based on the particle size distribution (see Section 4.2) and $d = \sqrt{k/n}$ where the ratio of intrinsic permeability to the porosity is assumed to represent a type of hydraulic radius of the pore space. As mentioned by many authors (among others, de Marsily 1986, Bear and Cheng 2010), in spite of the different definitions, it is commonly admitted that Darcy's law is valid if the *Re* number is between 1 and 10 ($1 < Re < 10$). In practice, for example in highly karstified or fissured media and if the REV is not large enough to produce acceptable average values with an EPM approach, groundwater flow is assumed to be nonlinearly laminar for $10 < Re < 100$ and turbulent for $Re > 100$ (Bear 2007). As shown in Figure 4.20, calculations based on Darcy's law would systematically produce an overestimation of the specific discharge (q) for very low as for very high hydraulic conductiviy values.

Actually, for high K, energy is lost in turbulence (more frictional losses) in comparison with a full laminar flow. Consequently, a higher gradient ($\Delta h/L$) should be applied to obtain the same discharge through the considered medium (i.e., karstified/fractured).

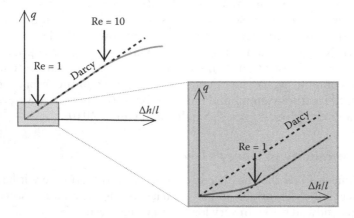

Figure 4.20 Deviations from Darcy's law for groundwater discharge calculations in very high and very low permeability media.

In engineering geology, for civil and mining engineering underground works like tunnels and engineered cavities, it is well known that the measured "dynamic pressure" (p_{dyn}) in fractures can be lower than the total water pressure (p) due to *head losses* linked to frictional effects between groundwater and the fracture walls (influenced by water viscosity, velocity, fracture roughness and fracture aperture, width, and length). These head losses usually contribute to increased stability of underground tunnels and cavities. However, flow is needed to have them mobilized (Maréchal and Perrochet 2003): thus, stability is improved on the condition that there is active drainage.

Figure 4.21 shows how drainage conditions in a tunnel may help in decreasing the head water pressure by mobilizing head losses in addition to a possible drawdown of the water table. If drainage is stopped, head losses cease instantaneously and the dynamic water pressure increases until recovering to the static water pressure.

At the other extreme, in very low permeability media and especially in compact clays, a higher gradient than foreseen by Darcy's law is necessary to obtain active groundwater flow (Figure 4.20). This was observed by Jacquin (1965) and other authors interested in very low permeability compact clays for clay liners and possible future storage

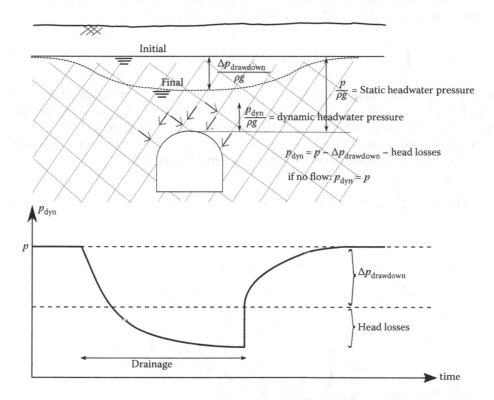

Figure 4.21 Head losses induced by active drainage in a tunnel help to decrease the (dynamic) head water pressure. When the drainage is stopped, head losses cease instantaneously and the dynamic water pressure increases until recovering to the static water pressure.

of radioactive wastes in natural clay layers. This is due to different interactions between water and the structured clay sheets (i.e., electrical forces, non-Newtonian behavior of water in capillary spaces, inverse electro-osmotic fluxes). However, recognizing the difficulty in determining an alternative law (Wagner and Egloffstein 1990), a typical conservative reasoning was implicitly adopted. Since applying Darcy's law will systematically overestimate the specific discharge through the low permeability medium, engineers are most likely to be on the safe (conservative) side. If permeability tests are conducted, interpreted values of permeability and unsaturated relative permeability must be made with extreme caution as a result of non-Darcian flow behavior (Liu 2014).

4.8 Transmissivity concept

Often, the vertical dimension can be considered as much smaller than the horizontal dimensions of an aquifer. The assumption that an average value of the hydraulic conductivity can be applied over the vertical dimension (*"depth-averaged" conditions*), is often adopted in confined and phreatic aquifers of large 2D extent. Implicitly, under these conditions, the Dupuit assumption (see Section 4.6) is chosen, as groundwater flow is mostly horizontal.

In a confined aquifer, the *transmissivity* of the aquifer is defined by the integral over the layer thickness of the hydraulic conductivity (Figure 4.22):

$$T(x,y) = \int_0^{b(x,y)} K(x,y,z)dz = K_{avg}(x,y)b(x,y) \tag{4.44}$$

where $T(x, y)$ is the transmissivity (m²/s) $[L^2T^{-1}]$ at a point (x, y) in the 2D horizontal plane, $b(x, y)$ is the thickness of the confined aquifer and $K_{avg}(x, y)$ the averaged hydraulic conductivity over the vertical dimension of the considered point.

In an unconfined aquifer, the transmissivity is obtained similarly from the integral over the vertical saturated thickness of the aquifer (Figure 4.23):

$$T(x,y) = \int_0^{h(x,y)} K(x,y,z)dz = K_{avg}(x,y)h(x,y) \tag{4.45}$$

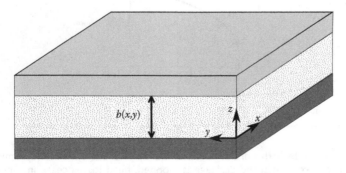

Figure 4.22 Transmissivity definition in a confined aquifer.

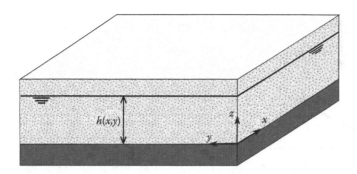

Figure 4.23 Transmissivity definition in an unconfined aquifer.

where $h(x, y)$ is the saturated thickness in the unconfined aquifer. Assuming that the datum level is the bottom of the aquifer, $h(x, y)$ is equal to the piezometric head at this point. Note however that $h(x, y)$ is a variable, a function of time under transient conditions. In other words, under transient local or regional perturbations (pumping, drainage, recharge), $h(x, y)$ and consequently $T(x, y)$ will also vary. As indicated by many authors, the Dupuit assumption is acceptable when unconfined flow is shallow and the slope of the free surface is small (<0.001). In summary, this implies that groundwater flow is dominantly horizontal across any vertical cross section, and that the specific discharge is considered constant over depth and can be calculated using the small (<0.001) free surface slope as the piezometric gradient (Delleur 1999). Equations 4.33 and 4.34 describing the hydrostatic pressure distribution are used corresponding (as mentioned previously in Section 4.6) to vertical head isosurfaces represented by equipotentials in a horizontal 2D view.

As a 2D parameter, anisotropy of the transmissivity is described only in the horizontal dimension by a 2D tensor T. The general Darcy's law in 2D can be therefore written as:

$$q = -T \cdot \nabla h \tag{4.46}$$

where the components of q are expressed in m²/s [L^2T^{-1}].

4.9 Equations of the steady-state groundwater flow (saturated conditions)

Darcy's law leads to a groundwater flow equation. In the continuum approach, the balance equation must be written for the mass of water under saturated conditions. First it will be expressed under steady-state flow conditions. Consequently, there is no change of water storage as function of time in the saturated geological medium. In a continuum, the classical mathematical tool to express the continuity (e.g., water mass conservation) with an Eulerian approach (i.e., with a fixed reference axis system) is to express the "divergence" of the motion vector. For our case, the negative divergence of the Darcy flux represents the excess of inflow in a "control box" over outflow, per

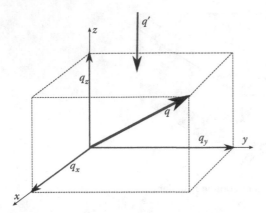

Figure 4.24 Groundwater mass conservation for a volume of saturated geological medium under steady-state conditions.

unit volume of porous medium and per unit time. For saturated groundwater flow, it is written:

$$-div(\rho q) = -\nabla \cdot (\rho q) = -\rho q' \qquad (4.47)$$

where $div(q) = \nabla \cdot q = ((\partial q_x/\partial x),(\partial q_y/\partial y),(\partial q_z/\partial z))$ in the 3D saturated medium (Figure 4.24), q' (s^{-1}) [T^{-1}] is the water flow rate per unit volume of the geological medium that is injected ($q' > 0$) or withdrawn ($q' < 0$). The term q' is often called the "source" or "sink" term, also referred to as the "stress" ("perturbation") term of the groundwater flow equation.

Entering the Darcy law Equation 4.28 in the Equation 4.47 leads to the following 3D equation:

$$\nabla \cdot (\rho K \cdot \nabla h) + \rho q' = 0 \qquad (4.48)$$

and

$$\nabla \cdot \left(\rho \frac{k}{\mu} \cdot (\nabla p + \rho g \nabla z) \right) + \rho q' = 0 \qquad (4.49)$$

The different terms of these equations are in units of kg/(m^{3}s) [ML^{-3}T^{-1}] expressing the mass of flowing water per volume of porous medium per second. Equation 4.47 written in indicial notation is:

$$\frac{\partial}{\partial x_i} \left(\rho K_{ij} \frac{\partial h}{\partial x_j} \right) + \rho q_i' = 0 \qquad (4.50)$$

Note that this equation can be written under different simplified forms. The most classical is the following:

$$\frac{\partial}{\partial x} \left(K_{xx} \frac{\partial h}{\partial x} \right) + \frac{\partial}{\partial y} \left(K_{yy} \frac{\partial h}{\partial y} \right) + \frac{\partial}{\partial z} \left(K_{zz} \frac{\partial h}{\partial z} \right) + q' = 0 \qquad (4.51)$$

where the density is assumed constant and the principal directions of the K tensor are known and aligned with the selected coordinate system. Here, the different terms are in s^{-1} expressing the volume of flowing water per volume of porous medium per unit time.

If the medium can be considered as isotropic and homogeneous and if there is no source/sink term, Equation 4.51 is reduced to Laplace's equation:

$$\nabla^2 h = \frac{\partial^2 h}{\partial x^2} + \frac{\partial^2 h}{\partial y^2} + \frac{\partial^2 h}{\partial z^2} = 0 \qquad (4.52)$$

In a 2D vertical cross section, assuming flow components only exist in this vertical plane, Equation 4.51 is reduced to:

$$\frac{\partial}{\partial x}\left(K_{xx}\frac{\partial h}{\partial x}\right) + \frac{\partial}{\partial z}\left(K_{zz}\frac{\partial h}{\partial z}\right) + q' = 0 \qquad (4.53)$$

where the terms of this equation are in units of $m^3/(m^2\ m\ s)$ or s^{-1} $[T^{-1}]$ expressing the volume of flowing water divided by the transverse surface area of the porous medium, per second and per unit width.

In a 2D horizontal plane, considering the Dupuit assumption, Equation 4.51 is reduced to:

$$\frac{\partial}{\partial x}\left(T_{xx}\frac{\partial h}{\partial x}\right) + \frac{\partial}{\partial y}\left(T_{yy}\frac{\partial h}{\partial y}\right) + q'' = 0 \qquad (4.54)$$

where q'' is the water flow rate per unit area of the geological medium that is injected ($q'' > 0$) or withdrawn ($q'' < 0$). Note that all the terms of this equation are in $m^3/(m^2\ s)$ or m/s $[LT^{-1}]$ expressing the volume of flowing water per unit cross-sectional area of the porous medium, per second.

4.10 Storage variation under saturated conditions

If we want to express the balance equation under *transient conditions*, the decrease or increase of groundwater mass per unit volume of porous medium per unit time $\partial(\rho n)/\partial t$ must be introduced in Equation 4.47:

$$\nabla \cdot (\rho q) - \rho q' = -\partial(\rho n)/\partial t \qquad (4.55)$$

The negative sign in front of the right-hand side term comes from the fact that decreased water storage leads to more groundwater available for flow. This storage term is then globally positive or negative corresponding to respectively more or less groundwater available for flow, and also to, respectively, a decrease or an increase of stored water mass per unit volume of saturated medium and per unit time ($kg/(m^3 s)$) $[ML^{-3}T^{-1}]$.

Specific storage coefficient or specific storativity definition

For our convenience, we should like to express this right-hand member of Equation 4.55 in a form showing explicitly the role of the piezometric head variation in time on this change of groundwater storage. The desired relationship leads automatically to the definition of the *specific storage coefficient* (S_s) which is the volume of groundwater gained or released per unit volume of saturated porous medium and per unit of h change (m^{-1}) [L^{-1}]:

$$\frac{\partial(\rho n)}{\partial t} = \rho S_s \frac{\partial h}{\partial t} \tag{4.56}$$

Geological media and especially porous media are deformable, that is, the volume compressibility allows deformations under applied geomechanical stresses. Under saturated conditions, the possible change of stored volume (i.e., the specific storage coefficient) is obviously linked to this volume compressibility. In fact, piezometric variations will induce geomechanical stress changes, inducing deformation and storage changes. The stress state in a porous medium must therefore be described.

Effective stress and Terzaghi concept

In a porous granular medium, the concept of *effective stress* was first introduced in soil mechanics by Terzaghi (1943), defined as the stress that is transmitted directly from grain to grain at the contact surfaces in a granular porous medium. This effective stress noted σ', added to the pressure p of the groundwater gives the total stress σ (Figure 4.25):

$$\sigma = \sigma' + p \tag{4.57}$$

To write Equation 4.57, the solid grain compressibility is neglected. Many authors have discussed this concept and contributed to the theoretical background of deformation in saturated porous media. When a saturated porous medium is loaded, compaction (i.e., deformation produced by rearrangement of the grains or rearrangement within the solid matrix texture) can take a long time and is called *consolidation*. This delayed compaction is due to the slow outflow of the interstitial water (transient flow) and can be described most often by elastic or elastoplastic rheological laws.

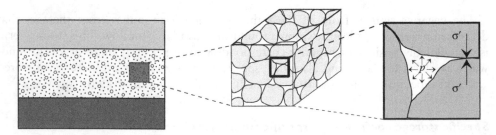

Figure 4.25 REV in a saturated porous medium, and microscopic vertical cross section describing the stress distribution in the medium.

Geotechnical engineers attribute all consolidation to the effective stress change. They express the volume compressibility of a porous medium with the relation:

$$-\frac{1}{V}\frac{\partial V}{\partial t} = \alpha\frac{\partial \sigma'}{\partial t} \tag{4.58}$$

where α is the volume compressibility of the porous medium (Pa^{-1}) [M^{-1}LT^2]. The negative sign comes from the fact that for an increasing effective stress, a decrease of total volume is expected. Note that, at a given time, the compressibility of the medium is dependent on the effective stress level: $\alpha(\sigma')$. This compressibility and its evolution can be measured with oedometer tests that are drained consolidation tests. In theory, the volume deformations are not only due to the porosity decrease of the medium but also to solid compressibility (β_s) and water compressibility (β_w). Some authors, such as Biot (1941), have accordingly proposed that:

$$\sigma = \sigma' + \left(1 - \frac{\beta_s}{\alpha} - \frac{\beta_w}{\alpha}\right)p \tag{4.59}$$

Then, neglecting the water compressibility, Equation 4.59 becomes:

$$\sigma = \sigma' + \left(1 - \frac{\beta_s}{\alpha}\right)p \tag{4.60}$$

where $\left(1 - \frac{\beta_s}{\alpha}\right)$ is often called the Biot coefficient. In practice, this coefficient is not easily characterized, requiring consolidation tests in undrained conditions in a tri-axial cell. However, this coefficient often remains very close to 1 for most natural soils (Verruijt 1982).

Specific storage coefficient and development of the mass balance equation

As shown in Equation 4.56, the rate of change of the mass of flowing groundwater per unit volume of porous medium is:

$$\frac{\partial(\rho n)}{\partial t} = n\frac{\partial \rho}{\partial t} + \rho\frac{\partial n}{\partial t} \tag{4.61}$$

Normally the water density depends on pressure, temperature, and solute concentration:

$$\rho = \rho(p, T, C) \tag{4.62}$$

At this point, isothermal conditions and groundwater of constant composition are assumed to restrict the discussion to the influence of the pressure. If the water compressibility is defined by:

$$\beta_w = \frac{1}{\rho}\frac{\partial \rho}{\partial p} \tag{4.63}$$

then the first term on the right-hand side of Equation 4.61 is written:

$$n\frac{\partial \rho}{\partial t} = n\frac{\partial \rho}{\partial p}\frac{\partial p}{\partial t} = n\rho\beta_w \frac{\partial p}{\partial t} \qquad (4.64)$$

If we assume that the total stress remains unchanged (i.e., $\partial\sigma = 0$; see Chapter 6) then taking the derivative of Equation 4.57 gives:

$$\frac{\partial \sigma'}{\partial t} = -\frac{\partial p}{\partial t} \qquad (4.65)$$

One must recall that

$$n = \frac{V_p}{V} \quad \text{and} \quad (1-n) = \frac{V_s}{V} \qquad (4.66)$$

where V_p and V_s are respectively the pore and solid volumes in the REV.

The second term on the right-hand side of Equation 4.61 can be expanded as:

$$\rho\frac{\partial n}{\partial t} = \frac{(V_p + V_s)(\partial V_p/\partial t) - V_p((V_p + V_s)/\partial t)}{V^2} \qquad (4.67)$$

and simplified using Equations 4.66 and if the solid compressibility is neglected $(\partial V_s/\partial t \cong 0)$, then:

$$\rho\frac{\partial n}{\partial t} = \frac{V_s(\partial V/\partial t)}{V^2} = \rho\frac{(1-n)}{V}\frac{\partial V}{\partial t} = \rho\frac{(1-n)}{V}\frac{\partial V}{\partial \sigma'}\frac{\partial \sigma'}{\partial t} \qquad (4.68)$$

Using Equation 4.65, and the definition of the volume compressibility (Equation 4.58), gives:

$$\rho\frac{\partial n}{\partial t} = \rho(1-n)\alpha\frac{\partial p}{\partial t} \qquad (4.69)$$

With Equations 4.64 and 4.69, Equation 4.61 becomes:

$$\frac{\partial(\rho n)}{\partial t} = \rho\{n\beta_w + (1-n)\alpha\}\frac{\partial p}{\partial t} \qquad (4.70)$$

The first term vanishes if the water compressibility is also assumed negligible $(\beta_w \cong 0)$ with regards to the volume compressibility.

Now, entering this water storage variation in the mass balance equation, and expressing that the Darcy specific discharge is here in fact a relative discharge with respect to a moving solid, Equation 4.55 can be rewritten as:

$$\nabla \cdot \rho(q_r + nv_s) - \rho q' = -\rho\{n\beta_w + (1-n)\alpha\}\frac{\partial p}{\partial t} \qquad (4.71)$$

where v_s is the solid's velocity vector (taken at the REV scale), q_r is the relative specific discharge vector, and $(q - nv_s) = q_r$. This solid's velocity is an additional variable in the problem that is eliminated from the equation expressing the solid's mass balance equation written as follows:

$$-\nabla \cdot \rho_s(1-n)v_s = \frac{\partial\{(1-n)\rho_s\}}{\partial t} \tag{4.72}$$

where ρ_s is the solid mass density (kg/m³) [ML⁻³]. If the solid mass density can be considered constant (i.e., the solid compressibility β_s negligible) and using Equation 4.69, this solid's balance equation is reduced to:

$$\nabla \cdot v_s = \alpha \frac{\partial p}{\partial t} \tag{4.73}$$

By replacing $\nabla \cdot v_s$ in Equation 4.71 by its value here above, we obtain:

$$\nabla \cdot \rho(q_r) - \rho q' = -\rho\{n\beta_w + n\alpha + (1-n)\alpha\}\frac{\partial p}{\partial t} = \rho(\alpha + n\beta_w)\frac{\partial p}{\partial t} \tag{4.74}$$

Transforming $\partial p/\partial t = \rho g \, \partial h/\partial t$ we obtain:

$$\nabla \cdot \rho(q_r) - \rho q' = -\rho^2 g(\alpha + n\beta_w)\frac{\partial h}{\partial t} \tag{4.75}$$

Finally, using the definition of the *specific storage coefficient* as expressed by Equation 4.56, we obtain:

$$S_s = \rho g(\alpha + n\beta_w) \tag{4.76}$$

Note that many assumptions were needed to obtain this relation: (1) isothermal conditions, (2) a uniform groundwater composition, (3) Terzaghi's concept and volume compressibility definition, (4) a constant total stress, and (5) a negligible solids' compressibility compared to volume compressibility. Most often, the water compressibility β_w is also assumed negligible compared to volume compressibility so that the *specific storage coefficient* (or specific storativity) in m⁻¹ [L⁻¹] can be directly written in relation to volume compressibility (Pa⁻¹) [M⁻¹LT²]:

$$S_s = \rho g\alpha \tag{4.77}$$

This relation shows that S_s depends directly on the volume compressibility of the whole porous medium (α). Table 4.5 gives general ranges of compressibility values. The compressibility varies with the effective stress level (σ') that will influence the groundwater pressure (p), so that a full coupling exists between groundwater flow and storage and the geomechanical stress-strain state of the saturated porous medium REV. This coupling could be further strengthened by also considering the variation of permeability due to consolidation. Until now, this variation has been neglected. This

Table 4.5 General ranges of volume compressibility (Pa^{-1}) for different porous media

Lithology	Volume compressibility α (Pa^{-1})
Highly organic alluvial clays and peats, underconsolidated clays	1.5×10^{-6}–1×10^{-6}
Normally consolidated alluvial clays	1×10^{-6}–3×10^{-7}
Clays of lake deposits/outwash, normally consolidated clays at depth, weathered marls	3×10^{-7}–1×10^{-7}
Tills and marls	1×10^{-7}–5×10^{-8}
Over-consolidated clays	5×10^{-8}–1×10^{-8}
Sand	5×10^{-7}–1×10^{-9}
Gravel	5×10^{-8}–1×10^{-10}
Fractured rock	5×10^{-8}–1×10^{-10}
Hard rock	5×10^{-9}–1×10^{-11}

(Freeze and Cherry 1979, Carter and Bentley 1991). Note that volume compressibility of a given porous medium is highly dependent on the effective stress.

effect will be further explained and developed in Chapter 6 as it underlies most of the physical processes involved in land subsidence.

4.11 Equations of the transient groundwater flow

When S_s (Equation 4.76) is introduced in Equation 4.75, we obtain the final *equation of saturated transient groundwater flow*:

$$-\nabla \cdot \rho(q_r) + \rho q' = \rho S_s \frac{\partial h}{\partial t} \tag{4.78}$$

3D groundwater flow equations

Again, as previously done, entering Darcy's law (Equation 4.28) in this equation leads to:

$$\nabla \cdot \rho(K \cdot \nabla h) + \rho q' = \rho S_s \frac{\partial h}{\partial t} \tag{4.79}$$

and

$$\nabla \cdot \rho\left(\frac{k}{\mu} \cdot (\nabla p + \rho g \nabla z)\right) + \rho q' = \rho S_s \frac{\partial h}{\partial t} \tag{4.80}$$

The different terms of these equations are in kg/(m³s) [ML^{-3}T^{-1}] expressing the mass of flowing water per volume of porous medium per second. Equation 4.80 written in indicial notation is:

$$\frac{\partial}{\partial x_i}\left(\rho K_{ij} \frac{\partial h}{\partial x_j}\right) + \rho q_i' = \rho S_s \frac{\partial h}{\partial t} \tag{4.81}$$

Similarly to Section 4.9, this equation can be written under different simplified forms. The most classical is the following:

$$\frac{\partial}{\partial x}\left(K_{xx}\frac{\partial h}{\partial x}\right)+\frac{\partial}{\partial y}\left(K_{yy}\frac{\partial h}{\partial y}\right)+\frac{\partial}{\partial z}\left(K_{zz}\frac{\partial h}{\partial z}\right)+q'=S_s\frac{\partial h}{\partial t} \tag{4.82}$$

where the density is assumed constant and the principal directions of hydraulic conductivity are known and aligned with the selected coordinate system. This equation is the logical development of Equation 4.51 under transient conditions. The different terms are in s^{-1} [T^{-1}] expressing the volume of flowing water per volume of porous medium per unit time.

If the medium can be considered as isotropic and homogeneous and if there is no source/sink term, it is reduced to the equation:

$$\nabla^2 h=\frac{\partial^2 h}{\partial x^2}+\frac{\partial^2 h}{\partial y^2}+\frac{\partial^2 h}{\partial z^2}=\frac{S_s}{K}\frac{\partial h}{\partial t} \tag{4.83}$$

2D vertical groundwater flow equations

In a 2D vertical cross section, assuming only flow components in this vertical plane, Equation 4.82 is reduced to:

$$\frac{\partial}{\partial x}\left(K_{xx}\frac{\partial h}{\partial x}\right)+\frac{\partial}{\partial z}\left(K_{zz}\frac{\partial h}{\partial z}\right)+q'=S_s\frac{\partial h}{\partial t} \tag{4.84}$$

where the terms of this equation are still in s^{-1} [T^{-1}] expressing the volume of flowing water divided by the transverse surface area of the porous medium, per second and per unit width.

One should note that all these equations can be applied only under strictly saturated conditions. In the unsaturated zone and consequently for unconfined aquifer conditions, the actual storage change is far greater than in the saturated zone (see below).

Storage coefficient

If one considers the groundwater flow in a 2D horizontal plane, considering the Dupuit assumption, an integral of the specific storage coefficient (or storativity) is to be calculated on the layer thickness:

$$S(x,y)=\int_0^{b(x,y)} S_s(x,y,z)dz=S_{s_{avg}}(x,y)b(x,y) \tag{4.85}$$

where $S(x, y)$ is the *storage coefficient* (-) at a point (x, y) in the 2D horizontal plane, $b(x, y)$ is the thickness of the confined aquifer and $S_{s_{avg}}(x, y)$ is the averaged specific storage coefficient over the vertical dimension of the considered point. In practice, under "depth-averaged" conditions, most hydrogeologists use:

$$S=S_s b \tag{4.86}$$

This approach and the required assumptions are similar in all respects to what was highlighted in Section 4.8 for the transmissivity definition. The storage coefficient is the volume of groundwater gained or released per unit area of saturated porous medium and per unit change in h (dimensionless) [-].

In an unconfined aquifer, the corresponding equation is:

$$S = S_s h \tag{4.87}$$

where h is the saturated thickness in the unconfined aquifer. As in Section 4.8, assuming that the datum level is the bottom of the aquifer, the thickness h is equal to the piezometric head at this point.

In this unconfined case, it is obvious that only a small part of the storage change in the REV is considered. The main contribution to the storage coefficient value comes from the drainable porosity (or specific yield or effective porosity, see Section 4.2) as desaturation or resaturation of the porous medium occurs as a function of h variations. The storage coefficient in unconfined conditions can therefore be written as:

$$S = n_e + \int_0^h S_s dz = n_e + S_s h \tag{4.88}$$

Values of the volume compressibility are so low (see Table 4.5) that even integrated over a saturated thickness of hundreds (or thousands) of meters, the result remains negligible compared to effective porosity (specific yield) values. In most cases, Equation 4.88 therefore reduces to:

$$S \cong n_e = S_y \tag{4.89}$$

As illustrated in Figure 4.26, the storage coefficient under unconfined and confined conditions have therefore very different physical meanings. To be very clear about the link between the storage coefficient and the specific yield in unconfined aquifers, note that the specific yield, assumed equal to the drainable effective porosity (Equation 4.7 in Section 4.2), was defined as the ratio between the volume of drainable water and the total volume of the porous medium. The storage coefficient in unconfined conditions is defined as the ratio between the volume of drainable water and the unit surface area of the aquifer multiplied by the unit change of h.

2D horizontal groundwater flow equations in confined conditions

In a 2D horizontal plane, considering the Dupuit assumption, Equation 4.79 is reduced to:

$$\nabla \cdot (\boldsymbol{T} \cdot \nabla h) + q'' = S \frac{\partial h}{\partial t} \tag{4.90}$$

where T is the transmissivity tensor of order 2 in the horizontal plane.

As for Equation 4.54 (steady-state Section 4.9) q'' is the water flow rate per unit surface area of the geological medium that is withdrawn ($q'' < 0$) or injected ($q'' > 0$).

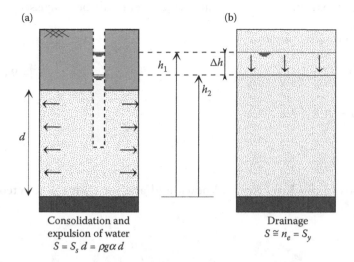

(a)

h_1

Δh

h_2

d

Consolidation and
expulsion of water
$S = S_s\, d = \rho g \alpha\, d$

(b)

Drainage
$S \cong n_e = S_y$

Figure 4.26 The storage coefficient reflects very different physical processes: (a) consolidation and groundwater expulsion in confined conditions; (b) drainage of the medium in unconfined conditions.

Note that all terms of this equation are in m³/(m² s) or m/s [LT⁻¹] expressing the volume of flowing water per surface area of the considered porous medium, per second.

In indicial form, Equation 4.90 can be written as:

$$\frac{\partial}{\partial x_i}\left(T_{ij}\frac{\partial h}{\partial x_j}\right) + q_i'' = S\frac{\partial h}{\partial t} \qquad (4.91)$$

where $i = 1, 2$.

As previously, if the main principal directions of hydraulic conductivity are known and aligned with the selected coordinate system, Equation 4.91 can be written under the following form:

$$\frac{\partial}{\partial x}\left(T_{xx}\frac{\partial h}{\partial x}\right) + \frac{\partial}{\partial y}\left(T_{yy}\frac{\partial h}{\partial y}\right) + q'' = S\frac{\partial h}{\partial t} \qquad (4.92)$$

2D horizontal groundwater flow equations in unconfined conditions

In unconfined conditions, we cannot write a 3D groundwater flow equation without taking into account partially saturated groundwater flow (see Chapter 9). In saturated conditions, the 2D horizontal groundwater flow equation can be written as follows:

$$\nabla\cdot(T(h)\cdot\nabla h) + q'' = n_e\frac{\partial h}{\partial t} = S_y\frac{\partial h}{\partial t} \qquad (4.93)$$

where $T(h)$ is the transmissivity tensor in the horizontal plane whose components are depending on the piezometric head. This is clearly what numerical engineers call a "nonlinearity" of this T parameter. Solving such a problem will be discussed in Chapters 12 and 13.

Other forms of the same equations are:

$$\frac{\partial}{\partial x_i}\left(T_{ij}\frac{\partial h}{\partial x_j}\right) + q_i'' = n_e\frac{\partial h}{\partial t} = S_y\frac{\partial h}{\partial t} \tag{4.94}$$

in indicial form, where $i = 1, 2$, or:

$$\frac{\partial}{\partial x}\left(T_{xx}\frac{\partial h}{\partial x}\right) + \frac{\partial}{\partial y}\left(T_{yy}\frac{\partial h}{\partial y}\right) + q'' = S\frac{\partial h}{\partial t} \tag{4.95}$$

if the principal directions of the K tensor are known and aligned with the selected coordinate system.

References

Bachmat, Y. and J. Bear. 1986. Macroscopic modelling of transport phenomena in porous media, part 1: The continuum approach. *Transport in Porous media* 1(3): 213–240.

Bear, J. 2007. *Hydraulics of groundwater*. New-York: Dover.

Bear, J. and Y. Bachmat. 1990. *Introduction to modeling phenomena of transport in porous media*. Dordrecht: Kluwer.

Bear, J. and A.H.D. Cheng. 2010. *Modeling groundwater flow and contaminant transport*. Dordrecht: Springer.

Bear, J. and A. Verruijt. 1987. *Modeling groundwater flow and pollution*. Dordrecht: Reidel Publishing Company.

Bernoulli, D. 1738. *Hydrodynamica*. Basel: Sumptibus Johannis Reinholdi Dulseckeri.

Biot, M.A. 1941. General theory of three-dimensional consolidation. *Journal of Applied Physics* 12: 155–164.

Burger, A., Recordon, E., Bovet, D., Cotton, L. and B. Saugy. 1985. *Thermique des nappes souterraines (in French)*. Lausanne: Presses polytechniques romandes.

Calembert, L., Monjoie, A., Polo-Chiapolini, C. and C. Schroeder. 1981. Géologie de l'ingénieur et mécanique des roches (1st part) (in French). *Annales des Travaux Publics de Belgique* 6: 543–571.

Carabin, G. and A. Dassargues. 1999. Modeling groundwater with ocean and river interaction. *Water Resources Research* 35(8): 2347–2358.

Carter, M. and S.P. Bentley. 1991. *Correlations of Soil Properties*. London: Pentech Press.

Castany, G. 1963. *Traité pratique des eaux souterraines (in French)*. Paris, Bruxelles, Montréal: Dunod.

Chapuis, R.P. 2007. *Guide des essais de pompage et leurs interprétations (in French)*. Québec: Bibliothèques et archives nationales du Québec.

Chen, T., Clauser, Ch., Marquart, G., Willbrand, K. and D. Mottaghy. 2015. A new upscaling method for fractured porous media. *Advances in Water Resources* 80: 60–68.

Dagan, G. 1989. *Flow and Transport in Porous Formations*. Heidelberg, Berlin, New York: Springer-Verlag.

Darcy, H. 1856. *Les fontaines publiques de la ville de Dijon* (in French). Paris: Dalmont.

Dassargues, A. and A. Monjoie. 1993. Chalk as an aquifer in Belgium. In *Hydrogeology of the chalk of north-west Europe*, ed. R.A. Downing, M. Price and G.P. Jones. Oxford: Oxford University Press, 153–169.

de Marsily, G. 1986. *Quantitative hydrogeology: Groundwater hydrology for engineers.* San Diego: Academic Press.

Delleur, J.W. 1999. *The handbook of groundwater engineering.* Boca Raton: CRC Press.

Denis, A., Panet, M. and C. Tourenq. 1978. L'identification des roches par l'indice de continuité (in French). In *Proc. of the 4th Int. Cong. of the Int. Soc. of Rock Mechanics.* Montreux, 2, 95–98.

Desbarats, A., Logan, C., Hinton, M. and D. Sharpe. 2002. On the kriging of water table elevations using collateral information from a digital elevation model, *Journal of Hydrology* 255: 25–38.

Dupuit, J. 1863. *Estudes théoriques et pratiques sur le mouvement des eaux dans les canaux découverts et à travers les terrains perméables (in French)* (2nd Edition), ed. Dunod Paris.

Durlofsky, L.J. 1991. Numerical calculations of equivalent grid block permeability tensors for heterogeneous porous media. *Water Resources Research* 27(5): 699–708.

Eckis, R. 1934. Geology and ground-water storage capacity of valley fill, South Coastal Basin Investigation: California Dept. Public Works. *Division Water Resources Bull* 45, 273.

Fasbender, D., Peeters, L., Bogaert, P. and A. Dassargues. 2008. Bayesian data fusion applied to water table spatial mapping. *Water Resources Research* 44: W12422.

Fetter, C.W. 2001. *Applied hydrogeology* (4th Edition). Upper Saddle River (NJ): Pearson Education Limited.

Fitts, C.R. 2002. *Groundwater science.* London: Academic Press.

Freeze, R.A. and J.A. Cherry. 1979. *Groundwater.* Englewood Cliffs (NJ): Prentice Hall.

Freeze, R.A. and P.A. Witherspoon. 1967. Theoretical analysis of regional groundwater flow: 2. Effect of water-table configuration and subsurface permeability variation. *Water Resources Research* 3(2): 623–634.

Hadley, P.W. and C. Newell. 2014. The new potential for understanding groundwater contaminant transport. *Groundwater* 52(2): 174–186.

Hoeksema, R., Clapp, R., Thomas, A., Hunley, A., Farrow, N. and K. Dearstone. 1989. Cokriging model for estimation of water table elevation, *Water Resources Research* 25: 429–438.

Hubbert, M.K. 1940. The theory of ground-water motion. *The Journal of Geology* 48(8) Part 1: 785–944.

Ingebritsen, S., Sanford, W. and C. Neuzil. 2006. *Groundwater in geologic processes* (2nd Edition). New York: Cambridge University Press.

Jacquin, C. 1965. Interactions entre l'argile et les fluides. Ecoulement à travers les argiles compactes. Etude bibliographique (in French). *Revue de l'Institut Francais du Petrole* 20(10): 1475–1501.

Liu, H.H. 2014. Non-Darcian flow in low-permeability media: Key issues related to geological disposal of high-level nuclear waste in shale formations. *Hydrogeology Journal* 22: 1525–1534.

Maini, T. and G. Hocking. 1977. An examination of the feasibility of hydrologic isolation of a high level waste repository in crystalline rocks. In *Invited Paper Geologic Disposal of High Radioactive Waste Session, Ann. Meet Geol. Soc. Am.* Seattle, Washington.

Maréchal, J.C. and P. Perrochet. 2003. New analytical solution for the study of hydraulic interaction between Alpine tunnels and groundwater. *Bulletin de la Société Géologique de France* 174(5): 441–448.

Molz, F. 2015. Advection, dispersion and confusion. *Groundwater* published online, DOI: 10.1111/gwat.12338

Payne, F.C., Quinnan, A. and S.T. Potter. 2008. *Remediation hydraulics.* Boca Raton: CRC Press/ Taylor & Francis.

Pickup, G.E., Ringrose, P.S., Jensen, J.L. and K.S. Sorbie. 1994. Permeability tensors for sedimentary structures. *Mathematical Geosciences* 26(2): 227–250.

Rausch, R., Schäfer, W., Therrien, R. and Ch. Wagner. 2002. *Solute transport modelling: An introduction to models and solution strategies.* Berlin-Stuttgart: Gebr. Borntraeger Science Publishers.

Renard, Ph. and G. de Marsily. 1997. Calculating equivalent permeability: A review. *Advance Water Resources* 20: 253–278.

Ringrose, Ph. and M. Bentley. 2015. *Reservoir model design: A practitioner's guide.* Dordrecht: Springer.

Singhal, B.B.S. and R.P. Gupta. 2010. *Applied hydrogeology of fractured rocks* (2nd Edition). Dordrecht: Springer Science & Business Media.

Terzaghi, K. 1943. *Theoretical soil mechanics.* London: Chapman and Hall.

Tóth, J. 1962. A theory of groundwater motion in small drainage basins in Central Alberta. *Journal of Geophysical Research* 67: 4375–4387.

Tóth, J. 1963. A theoretical analysis of groundwater flow in small drainage basins. *Journal of Geophysical Research* 68: 4795–4812.

Tourenq, C. 1978. *Les essais de granulats en France.* Budapest: R.I.L.E.M. Int. Colloquium on granular materials, 367–377.

Verruijt, A. 1982. *Theory of groundwater flow* (2nd Edition). London: MacMillan.

Wagner, J.F. and Th. Egloffstein. 1990. Advective and /or diffusive transport of heavy metals in clay liners. In *Proc. of the 6th Int. IAEG Congress,* ed. D.G. Price, 1483–1490.

Yang, J.W., Feng, Z.H., Luo, X.R. and Y.R. Chen. 2010. Numerically quantifying the relative importance of topography and buoyancy in driving groundwater flow. *Science China Earth Sciences* 53: 64–71.

Chapter 5

Hydraulic conductivity measurements

5.1 Introduction

Many different investigation methods have been developed to measure saturated hydraulic conductivity (K) and/or intrinsic permeability (k) in the laboratory and in the field, and at various scales. Generally, a limited accessibility to the geological medium hampers the collection of many samples or performing sufficient field measurements to quantify, at the desired scale, the heterogeneity of the geological medium. Moreover, the ability to characterize the subsurface K heterogeneity is not only dependent on the density and scale of laboratory and in situ measurements but also on an adequate understanding of the geological processes conditioning the spatial structure of the geological medium.

Usually, laboratory and field methods are addressing different scales of investigation. They can be considered as complementary because small-scale variability in K has implications at a larger scale (de Marsily *et al.* 2005, Huysmans and Dassargues 2009, Ronayne *et al.* 2010, Rogiers *et al.* 2013). Field scale is considered as the adequate scale for most applications involving groundwater flow (e.g., water supply, well capture zone delineation, dewatering) and contaminant transport. What is measured is always an "effective" hydraulic conductivity as lab and field experiments have, in any case, specific boundary conditions influencing measurements. Indeed, experimental conditions should be chosen in a way to decrease as much as possible this influence. It is always relevant to compare "equivalent values" (see Section 4.5) from in situ large-scale measurements with numerically upscaled values resulting from small-scale tests results. This upscaling procedure consists, for example, in using conditioned stochastic simulations (see Sections 12.6–12.7) of a (large) number of porous media having the same probability of occurrence.

On the basis of extensive studies involving many measurements in different conditions and particularly in fractured rocks, some authors propose interpolation procedures for hydraulic conductivity values developed on the distance between field measurements and the tested aquifer volumes (e.g., Schultze-Makuch *et al.* 1999, Niemi *et al.* 2000, Lemieux 2002, Neuman and Di Federico 2003, Nastev *et al.* 2004).

5.2 Laboratory tests

Core drilling samples and outcrop samples can be tested in a laboratory. Various methods can be applied from grain size indirect K assessments to measurements in

Darcy's law based permeameters. However, in all cases, a fundamental requirement to obtain realistic results is to test, as far as possible, undisturbed samples.

Empirical relations based on grain size distribution

Various empirical relations exist to obtain indirectly an estimation of the saturated hydraulic conductivity (K) from grain-size data. This is probably the easiest way to obtain a first approximation for K values in unconsolidated sediments. Traditionally, most methods used only a single grain-size related parameter in empirical relations, each of them being more or less validated on some specific sediments. Typically the pioneering Hazen equation (Hazen 1892) expresses a relation between K and the 10th percentile grain-size diameter (d_{10}) (grain size for which 10% of the sample in weight is passing):

$$K = Cd_{10}^2 \tag{5.1}$$

where C is an empirical adapted coefficient. Since then, numerous other attempts were made to improve the assessment of hydraulic conductivity of soils and sediments by this type of relations. Other particle size parameters are used as: the geometric mean grain-size diameter (d_m), the uniformity coefficient $C_u = d_{60}/d_{10}$ (see Equation 4.10) (a sediment showing a $C_u < 4$ is considered as well sorted), the x-intercept of a $d_{50} \sim d_{10}$ straight line (Cronican and Gribb 2004). A main limitation is that the whole information encompassed by the entire grain-size distribution is mostly omitted or neglected. On the other hand, a model that includes more information from the entire grain-size distribution is the Kozeny-Carman equation (Kozeny 1927, Carman 1938). This equation, known as the KC equation, was linking, initially, the intrinsic permeability k in m² to the specific surface of solids (S_{SP}) in m²/kg [L²M⁻¹]:

$$k = C_{KC} \frac{e^3}{\rho_s^2 S_{SP}^2 (1+e)} \tag{5.2}$$

where C_{KC} is an empirical coefficient, ρ_s is the solid mass density (kg/m³) [ML⁻³], and e the void ratio (see Equation 4.3). The main difficulty lies in the determination of the soil specific surface (Chapuis and Aubertin 2003), so that many authors have replaced the specific surface by data from grain size analysis ending up, for example, to the following relation (Carrier 2003) to estimate the hydraulic conductivity:

$$K = 1.10^5 C_{KC} \frac{e^3}{(1+e)} \left(\sum \frac{f_i}{D_{av\,i}} \right)^{-2} \frac{1}{SF^2} \tag{5.3}$$

where SF is a grain shape factor, f_i are the grain-size fractions, $D_{av\,i}$ is the average particle size in each of the ith fraction, calculated practically with:

$$D_{av\,i} = D_{li}^{0.404} D_{si}^{0.595} \tag{5.4}$$

where D_{li} and D_{si} are respectively the largest and smallest grain-size diameters of the ith fraction assuming the particle size distribution as log-linear in each fraction. Different values for SF are proposed: spherical 6.0; rounded 6.1; worn 6.4; sharp 7.4; and angular 7.7 (Fair and Hatch 1933) or rounded 6.6; medium angularity 7.5; angular 8.4 (Loudon 1952). In general, the relations based on the Kozeny-Carman equation were reworked many times and appeared to be highly discussed or contradicted especially in clays.

More recently, new methods were developed tending to use the whole information encompassed by the entire grain-size distribution using artificial neural networks (ANNs) as nonlinear regression technique taking bulk density, organic matter, and carbonate content into account (Schaap *et al.* 1998). These methods can be also combined in a generalized likelihood uncertainty estimation (GLUE) approach (Rogiers *et al.* 2012).

Constant head permeameter

Applying Darcy's law, laboratory hydraulic conductivity measurements are performed most often on cylindrical samples from borings. Ideally, the samples are undisturbed, however, depending on the drilling conditions, the samples are disturbed to some degree. The saturated sample is placed between two highly porous and permeable disks and a constant piezometric head difference is maintained during the whole test (Figure 5.1a), so that steady-state can be assumed. Perfect impermeable lateral boundaries are assumed. Rearranging Darcy's law (Equation 4.19), one can deduce a value for an equivalent hydraulic conductivity at the scale of the whole sample and including the effect of the boundary conditions. K_{eq} is found as function of the measured flow-rate ($Q_{measured}$), the known sample section (A) and length (L), and the constant applied piezometric head difference (h):

$$K_{eq} = \frac{Q_{measured} L}{A h} \tag{5.5}$$

For low hydraulic conductivity media, this permeameter test is unreliable because low flow-rate are to be measured, so that evaporation losses at the outflow or, accidental leakage along the vertical walls of the permeameter can become nonnegligible.

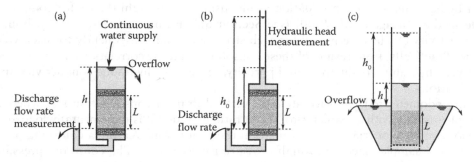

Figure 5.1 Constant head (a) and falling-head (b) permeameters used in laboratory. A falling-head permeameter can also be used as a fairly basic but robust field device (c) (Lewis 2016).

Falling head permeameter

When a low or very low hydraulic conductivity is expected, measurement of the flow-rate or the water volume flowing through the sample during a given period of time is done upstream of the sample in a vertical graduated tube. Then, hydraulic head is measured as a function of time and the discharge flow-rate is decreasing. During a time dt a volume of water Qdt was flowing as well through the sample, as through the tube section (Figure 5.1b), so that:

$$Qdt = -adh = K\frac{A}{L}hdt \qquad (5.6)$$

where a is the tube section and $-dh$ is the decrease in hydraulic head during dt. Integrating this equation and considering that at the initial time t_0 the hydraulic head difference is h_0, leads to:

$$K = -\frac{a}{A}\frac{L}{(t-t_0)}\ln\frac{h}{h_0} \qquad (5.7)$$

If the hydraulic conductivity is so low that only extremely low discharge flow rates can be obtained, higher gradients must be used in pressurized cells as performed in soil mechanics (oedometer and triaxial tests).

Based on this falling head permeameter principle, a simple field method was recently proposed by Lewis (2016) to assess near-surface saturated hydraulic conductivity of undisturbed as well as of repacked samples from cuttings or disturbed samples. The sample is (re)packed in a thin-walled rigid metal tube placed vertically in a container full of water (Figure 5.1c). After full saturation of the sample, the rest of the tube is fulfilled by water and a kind of falling head permeameter is actually in place.

Oedometer and isotropic tests

The aim here is not to describe extensively these geomechanical consolidation tests but by back analyzing consolidation results using the Terzaghi theory for compressible media (Delage *et al.* 2000) (see Section 6.2; Figures 6.4 and 6.5), hydraulic conductivity can be determined for each stress state and consequently for each void ratio. Practically, from results of these consolidation tests, correlations are produced between hydraulic conductivity and porosity, or between hydraulic conductivity and void ratio.

An oedometer test is a saturated 1D vertical drained consolidation test where a sample (usually with a diameter-to-height ratio of about 3) is incrementally loaded.

Isotropic 3D (or triaxial) consolidation tests consist in creating a confining pressure for a sample placed in a high-pressure isotropic cell. The confining pressure is incrementally increased, and in drained conditions, the determination of the hydraulic conductivity at various void ratios can be deduced (Parent *et al.* 2004; Deng *et al.* 2011).

5.3 Slug tests

In a slug test, the water level in the well is perturbed nearly instantaneously by injection or extraction of a volume of water. Then, the water level in the well returns to its initial (assumed in equilibrium) value. Water levels are recorded accurately until stabilization.

This test can even be performed without adding or extracting water by the use of a closed cylinder or by using pressurized air (Greene and Shapiro 1995). Starting from a static water level in the well (Figure 5.2a), the cylinder is lowered and creates an instantaneous rise of the water level equivalent to a true slug of added water. The lowering of the water level is then followed until stabilization (Figure 5.2b). When the level is stabilized, the cylinder is rapidly withdrawn creating an instantaneous decrease of the water head in the well so that the rising back of the head in the well can be followed until a new equilibrium is reached (Figure 5.2c).

In fact, the actual test must be directed in another way. As recommended by Butler (1997), the first direction of the induced flow should be ideally from the formation into the well ("slug-out test"). The flow from the well into the formation ("slug-in test") could induce a partial clogging of the surrounding aquifer by transported fine particles. Water level measurements should be recorded by pressure transducers to be able to capture the rapid change in head in the well. Practically, it is advised to know approximately the expected initial head displacement in order to place the transducer sufficiently below the top of the water column. In doing so, interference can more easily be avoided between the slug cylinder, and the cable and transducer.

Advantages: The method is very simple and quick and needs few equipment. The technique can be performed in small diameter wells and does not require any pumping. Slug tests were used previously only in low permeability media, but now they are also performed in all other media to avoid, in particular, pumping contaminated groundwater. Repeating slug tests in many wells allow mapping of the heterogeneity of the *K*-field.

Disadvantages: The tested volume of medium is small as the duration is short. For the rising head test, the screen must be entirely in the saturated zone. The measured

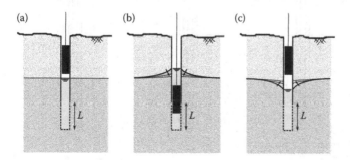

Figure 5.2 The double procedure of a slug test. Using a closed cylinder, an instantaneous rise of the water level is induced (a) and measurements follow the lowering until stabilization (b). After withdrawing of the cylinder, corresponding to an instantaneous decrease of the water level, measurements follow the rising back up until stabilization (c). As far as possible, it is recommended to start with this last step.

hydraulic conductivity is representative of the direct surrounding of the well. Detailed information about the well construction and equipment is required to avoid bias analysis of results. Effective screen length, and an "effective" radius of the well including the filter pack thickness (at least if the gravel-sand pack has—as expected—a greater permeability than the geological formation) must be known or estimated. Storage coefficient values can be deduced but are hardly reliable and long-term effects are not detected. The method is known as providing mostly a horizontal hydraulic conductivity value.

Interpretation

There are many methods used to interpret slug-test data. They are described in many papers and books (among others, Hvorslev 1951, Cooper *et al.* 1967, Bouwer and Rice 1976, Bouwer 1989, Butler 1997), without mentioning numerous works about corrections to account for different effects as partial penetration of the well, double porosity, elastic storage in the aquifer, and partially submerged screens and skin effects (i.e., effects of the altered disturbed zone in the immediate vicinity of the well screen and gravel pack). However, in practice, the aim is more to assess an order of magnitude and "the choice of the method results in little variation compared to the typical spatial variability of hydraulic conductivity" (Fitts 2002). The most used interpretation method, developed by Hvorslev (1951), is very similar to the falling head permeameter solution (Equation 5.7) and provides a value for the horizontal hydraulic conductivity:

$$K_h = -F \frac{d^2}{(t - t_0)} \ln \frac{h}{h_0} \tag{5.8}$$

where F is a shape factor depending on the geometry of the well equipment (Figure 5.3) expressed in m^{-1} [L^{-1}], d is the diameter of the well casing [L], t is the time and h is the

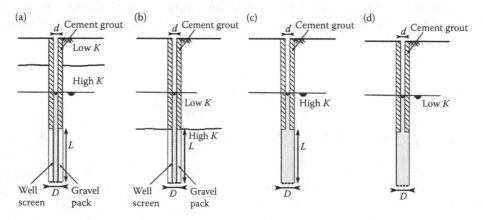

Figure 5.3 Different cases of slug-test configurations to characterize the shape factor F (Equation 5.10) of the Hvorslev method: (a) the high K zone is thicker than the screen length, (b) the screen length is equal to the thickness of the high K tested zone, (c) the pipe is not screened in a high K medium, and (d) the pipe is not screened in a low K medium (Fitts 2002).

measured head in the well with regards to the static level considered as $h = 0$. A first practical problem comes from the fact that actual ($\ln h$, t) plots are often not so linear as expected. The early nonlinear behavior can be resulting from skin effects, screens that are not totally in the saturated zone and irregularities in the gravel pack around the screens (perforated or slotted pipes) and late time nonlinear effects can be due to relative inaccuracy of measurements. So, it is advised to calculate the slope coefficient of the ($\ln h$, t) line for intermediate times (i.e., between t_1 and t_2). This slope coefficient is equal to $-Fd^2/K_h$ using:

$$(t_2 - t_1) = -\frac{Fd^2}{K_h}\ln\frac{h_2}{h_1} \tag{5.9}$$

Then, the remaining problem is to determine a value for the shape factor F. To keep it simple, and as developed initially by Hvorslev and presented by Fitts (2002), four situations are described (Figure 5.3) and correspond to four different values:

$$F = \frac{1}{8L}\ln\left[\frac{mL}{D} + \sqrt{1 + \left(\frac{mL}{D}\right)^2}\,\right] \quad \text{(Figure 5.3a)}$$

$$F = \frac{1}{8L}\ln\left[\frac{2mL}{D} + \sqrt{1 + \left(\frac{2mL}{D}\right)^2}\,\right] \quad \text{(Figure 5.3b)} \tag{5.10}$$

$$F = \frac{\pi m}{11D} \quad \text{(Figure 5.3c)}$$

$$F = \frac{\pi m}{8D} \quad \text{(Figure 5.3d)}$$

where D is the diameter of the well including the gravel pack or backfill around the well, L is the screen length and

$$m = \sqrt{\frac{K_h}{K_v}} \tag{5.11}$$

with K_h and K_v respectively the horizontal and vertical hydraulic conductivity values. Most often, an assumption must be taken to estimate the horizontal-to-vertical hydraulic conductivity anisotropy factor for the tested volume of medium, allowing to assess the m value. Finally, using Equation 5.9, K_h can be calculated. If the vertical hydraulic conductivity is decreased, the horizontal component must be adapted somewhat to compensate the effect, but the sensitivity of K_h final result to this m value is not too high.

Studies that have compared results of slug-test data to those interpreted by different methods (e.g., Fritz et al. 2016) have confirmed that it is quite unrealistic to seek a greater accuracy by using exact but complex analytical interpretation methods. Slug tests measurements are critically affected by actual well-aquifer conditions that are not always perfectly known. Moreover, as mentioned previously, the natural variability of hydraulic conductivity values remains quite huge compared to differences in interpreted results. In highly permeable aquifers or in wells with a high water column above the screen, oscillatory head response is often obtained after perturbation. Van der Kamp (1976) developed an interpretation method assuming a value for the storage coefficient. Finally, it must be noted that, in general, but especially in high-conductivity aquifer layers, estimation of storage coefficient values using slug-test data could be highly unreliable (Cardiff *et al.* 2011).

5.4 Pumping tests

Pumping tests are used to answer the following direct and basic questions:

*   What is the production discharge rate that can be pumped in a given well?
*   What are the values of the main hydraulic properties of the aquifer?
*   What is the impact (i.e., induced drawdown in the aquifer) of a future pumping at a given flow-rate?
*   What is the efficiency of the well (theoretical vs. actual productivity)?
*   Are there aquifer limits, and if so, what are their nature (recharge, impermeable) and position?

This information can be useful directly or indirectly for any issues addressing environmental risks, long-term trends, groundwater overexploitation, induced land subsidence, seawater intrusion, humid zones, and dependent ecosystems conservation.

There are numerous books and papers in the hydrogeological literature about pumping test interpretation. In this chapter, only the main and more useful methods are described. Consequently, many detailed and accurate existing analytical solutions will not be recalled for two main reasons:

*   Nowadays, pumping test results can also be easily interpreted by numerical modeling techniques (see Chapters 12 and 13), as building a local model of a given situation does not take so much time with the existing pre- and postprocessors of many groundwater flow programs.
*   The strong assumptions for interpretation of a pumping test by analytical solutions and the true natural variability of the hydrogeological parameters make it cumbersome and useless to interpret results with too much detail and accuracy.

A pumping test consists in extracting groundwater from a well and measuring the aquifer response as function of time through measurements in monitoring wells and in the pumping well itself. There are many possible configurations involving steady-state or transient measurement conditions, step-by-step increase of pumping rate, recovery timing, and number and location of available observation wells. Each configuration leads to using another analytical solution. Moreover, the type of aquifer (i.e., confined or unconfined), the type of overlying layer (considered as leaky or not), the length of the screen zone in relation to the aquifer thickness, the well-equipment influence (screen, filter pack), and the presence of heterogeneity in the influenced zone can theoretically change or modify the adopted analytical solution to be used for an accurate interpretation of the measured results. So, this could become tedious to address every actual pumping situation. Only the most used and robust analytical solutions are proposed covering most of the practical cases. Remarks will be made with references to specialized works for more specific cases.

Design, procedures, and measurements

Before starting a pumping test, geological and hydrogeological information must be collected concerning possible lithological and structural features that can influence groundwater flow, preexisting piezometric gradient, infiltration conditions, and other pumping wells in the vicinity. A pretest "forward problem" can be solved to predict roughly piezometric head spatial distribution with the projected pumping rate. The analytical solutions presented below can be used to predict drawdown for the projected pumping rate and an assessed value of hydraulic conductivity. This calculation can be very helpful to design the pumping test itself. If the hydraulic conductivity is completely unknown, a slug test could be conducted beforehand.

Most of the following recommendations to design a pumping test aim (a) to ensure reliable and significant measurements, and (b) to match as possible to the assumptions required for analytical interpretation of results.

Location of the well must be far enough from other hydrogeological influences (e.g., other pumping wells or drainage). Diameter of the well should allow the use of the adequate pumps in relation with the foreseen discharge rates. The depth of the well should be chosen to have a fully penetrating well in the tested confined aquifers. If used, screens and associated filter packs should be chosen maximizing the saturated screen length in relation with the geological conditions and the expected drawdown of the piezometric heads. Indeed, increasing the length of the saturated screen will tend to reduce the vertical component of groundwater (i.e., neglected in most of the analytical solutions).

A steep and narrow or a wide and flat drawdown/depression cone is respectively expected in low and high permeability media. It must be taken into account to decide about distances of the observation wells (piezometers) from the pumping well. Small diameter piezometers are accurate and less expensive but diameters can also be chosen as a function of the possibility to use each well as a possible future pumping well for further measurements of local heterogeneity in hydraulic parameters of the aquifer.

First, an initial *static water level* is measured in all wells to define initial conditions. Then, measurements of the induced drawdowns are made maintaining a constant pumping rate (Figure 5.4). If the measured drawdown is stabilized as function of time, the corresponding values will be used for interpretation under steady-state conditions. All other measurements (i.e., before reaching stabilized values) will be used for interpretation under transient conditions. Indeed, *recovery* measurements (i.e., rising back of the heads after the pumping was stopped) are transient. Very often, step-drawdown tests (Figure 5.5) are conducted to assess optimum future production pumping rate of the well. Logically, and as observed in Figure 5.5, the measured drawdowns take more and more time to reach a stabilized value with the increasing considered discharge rate.

Interpretation of pumping test results can be qualified as an "inverse problem" using all measurements (i.e., piezometric heads and pumping rate) under steady-state or transient conditions to calculate back the hydraulic conductivity or transmissivity,

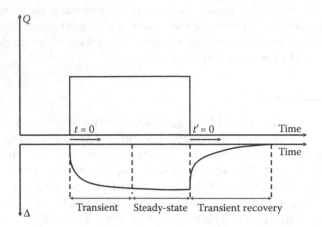

Figure 5.4 Measured drawdown (Δ) function of time for a constant pumping rate. Transient data are measured before a stabilization of piezometric heads corresponding to steady-state data (if the pumping rate is not too high). During the recovery phase, the rising back up of the heads toward the initial static piezometric level also corresponds to transient data.

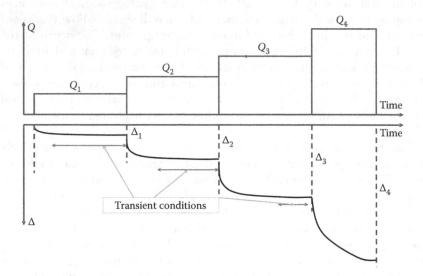

Figure 5.5 Measured drawdown function of time for a step-drawdown pumping test. For each pumping step, the measured drawdown takes more and more time to reach a stabilized value.

the specific storage coefficient or storage coefficient, and a few important additional data as the influence radius and the critical discharge rate of the pumping well. Note that interpretation of pumping tests yields parameters averaged over a domain of uncertain extent (Pechstein *et al.* 2016) with the important assumption of homogeneity of this influenced domain. Consequently, equivalent hydraulic conductivity values from a pumping test can be time dependent as function of the heterogeneity scale.

Interpretation of steady-state data

Thiem's method for confined aquifers under steady-state conditions

For confined and steady-state conditions, Thiem (1906) proposed a 2D analytical solution for radial groundwater flow to the well with the help of the following additional assumptions:

- The aquifer is homogeneous, isotropic, and of uniform thickness over the influenced zone.
- There is no initial piezometric gradient.
- The induced groundwater flow is only horizontal.
- The aquifer has an infinite areal extension.
- The well is screened over the entire thickness of the aquifer (full penetrating well).
- There is no leakage or infiltration from or to the aquifer.
- Delayed storage variation effects are neglected (i.e., groundwater removed from storage is assumed to be discharged instantaneously with decline of piezometric head).
- The storage in the well can be neglected (the diameter of the well is small).

The 2D horizontal groundwater flow Equation 4.54 can be written under isotropic conditions, without any leakage or infiltration and in radial coordinates. It becomes:

$$T\left(\frac{\partial^2 h}{\partial r^2} + \frac{1}{r}\frac{\partial h}{\partial r}\right) = 0 \tag{5.12}$$

Note that radial or polar coordinates are usually expressed:

$$\begin{cases} x = r\cos\theta \\ y = r\sin\theta \end{cases} \tag{5.13}$$

with r the radial distance from the pumping well, θ the angle of the considered direction. It vanishes from the equation resulting from the considered isotropic conditions.

Under steady-state, the constant discharge of the well (Q) must be balanced by the Darcy's specific discharge toward the well, integrated on the total flow surface at any distance of the well (a cylinder of radius r and height b equal to the aquifer thickness) (Figure 5.6):

$$Q = 2\pi r b K \frac{dh}{dr} = 2\pi r T \frac{dh}{dr} \tag{5.14}$$

Note that the negative sign of Darcy's law vanishes as the discharge occurs in the negative r direction. Rearranging and integrating this equation leads to:

$$h = \frac{Q}{2\pi T}\ln r + Cst \tag{5.15}$$

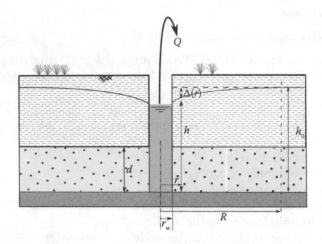

Figure 5.6 Pumping test in a confined aquifer under steady-state conditions.

At large distance, mathematically, the drawdown is never zero but tends to zero. In practice, it reaches a zero value resulting from many features supplying water to the aquifer (i.e., creating actual boundary conditions at some distance). If the initial static level is h_0, the boundary condition can be written $h = h_0$ equivalent to no drawdown at a distance corresponding to the influence radius (R) from the pumping well.

$$h_0 = \frac{Q}{2\pi T} \ln R + Cst \qquad (5.16)$$

At a distance r from the pumping well, the measured piezometric head h shows that a drawdown $\Delta(r) = (h_0 - h)$ is observed with regards to initial conditions. It gives:

$$\Delta(r) = (h_0 - h) = \frac{Q}{2\pi T} \ln \frac{R}{r} \qquad (5.17)$$

If the piezometric head is measured in the pumping well (at a distance r_w) and at some point, close to the well (at a distance r):

$$\Delta(r_w) - \Delta(r) = (h - h_w) = \frac{Q}{2\pi T} \ln \frac{r}{r_w} \qquad (5.18)$$

Finally, between two points located at distances r_1 and r_2 from the pumping well, it gives:

$$\Delta(r_1) - \Delta(r_2) = (h_2 - h_1) = \frac{Q}{2\pi T} \ln \frac{r_2}{r_1} \qquad (5.19)$$

The Thiem's equation is often written as function of the constant discharge rate as follows:

$$Q = \frac{2\pi T(\Delta(r_1) - \Delta(r_2))}{\ln(r_2/r_1)}$$

(5.20)

Or using Equation 5.17:

$$Q = \frac{2\pi T(\Delta)}{\ln(R/r)}$$

(5.21)

A direct application is often used in practice, transforming Equation 5.17 in:

$$\Delta(r) = \frac{0.366\,Q}{T}\log R - \frac{0.366\,Q}{T}\log r$$

(5.22)

Therefore in a (log r, Δ) diagram, measured drawdown at different distances can be reported and the slope of the best fitted line (Figure 5.7) will provide (0.366 Q/T) and thus T can be deduced.

The fitted line can be prolonged until crossing the log r-axis at zero drawdown, providing a rough[1] assessment of the influence radius (R) of the pumping well (Figure 5.7). The Thiem equation was tested with regards to the effects of its simplifications (Tügel et al. 2016) by comparison with results from a detailed 3D numerical model. It has been shown that the most critical simplification is related to the assumed uniformly distributed horizontal groundwater flow toward the well. On the other hand,

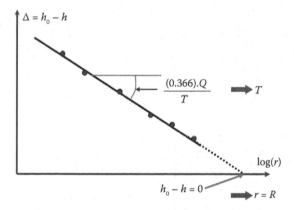

Figure 5.7 Application of the Thiem's solution for transmissivity determination: in a (log r, Δ) diagram, the angular coefficient of the best fitted line provides (0.366 Q/T) from which T can be deduced.

---

1 This is only a rough assessment resulting from the large uncertainty linked to the fitting of a straight-line on observed drawdown values. Usually, only a few points are available (if not only two) including the pumping well itself.

the influences of an initial hydraulic gradient and of the gravel pack on the drawdown were found negligible.

Dupuit's method for unconfined aquifers under steady-state conditions

The same type of interpretation was published previously by Dupuit (1863) for steady-state conditions in unconfined aquifers. The same list of assumptions (as in confined conditions) is adopted. However, the horizontal groundwater flow assumption is particularly questionable, as the saturated thickness changes. Practitioners recognize this assumption as acceptable only if the drawdown is limited to a quarter of the initial saturated thickness in the unconfined aquifer ($\Delta = h_0 - h < 0.25h_0$).

The Darcy's specific discharge toward the well, integrated on the total flow surface at any distance of the well, is now expressed as function of h as aquifer saturated thickness (replacing b in the cylinder surface at the distance r from the pumping well) (Figure 5.8):

$$Q = 2\pi r h K \frac{dh}{dr} \tag{5.23}$$

As previously, the negative sign of Darcy's law vanishes as the discharge occurs in the negative r direction. Rearranging and integrating this equation leads to:

$$h^2 = \frac{Q}{\pi K} \ln r + Cst \tag{5.24}$$

As stated previously, a zero drawdown is considered at the distance corresponding to the influence radius:

$$h_0^2 = \frac{Q}{\pi K} \ln R + Cst \tag{5.25}$$

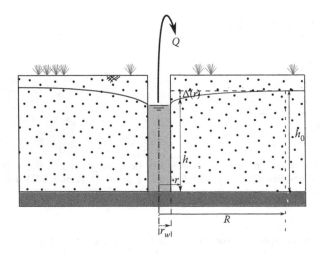

Figure 5.8 Pumping test in an unconfined aquifer under steady-state conditions.

At any point at a distance r from the pumping well:

$$(h_0^2 - h^2) = \frac{Q}{\pi K} \ln \frac{R}{r} \tag{5.26}$$

Using the measured drawdown in the unconfined aquifer, $\Delta = (h_0 - h)$ and consequently $(h_0^2 - h^2) = 2h_0\Delta - \Delta^2$, Equation 5.26 written as function of the constant discharge rate becomes:

$$Q = \frac{\pi K(2h_0\Delta - \Delta^2)}{\ln(R/r)} \tag{5.27}$$

One can observe that if $\Delta' = \Delta - \Delta^2/2h_0$, we can choose to express this Equation 5.29 as follows:

$$Q = \frac{2\pi Kh_0(\Delta')}{\ln(R/r)} \tag{5.28}$$

That is similar to Equation 5.21, with a transmissivity calculated over the initial saturated thickness $T = Kh_0$. So, in unconfined aquifers and provided that the drawdown is limited (see above), one can use equations established for confined conditions replacing Δ by Δ' (Kruseman and de Ridder 1994).

If the piezometric head is measured in the pumping well (at a distance r_w) and at some point close to the well (at a distance r), Equation 5.26 is written as:

$$(h^2 - h_w^2) = \frac{Q}{\pi K} \ln \frac{r}{r_w} \tag{5.29}$$

And between two points located at distances r_1 and r_2 from the pumping well:

$$(h_2^2 - h_1^2) = \frac{Q}{\pi K} \ln \frac{r_2}{r_1} \tag{5.30}$$

As a function of the constant discharge rate, the Dupuit's equation is most often written as:

$$Q = \frac{\pi K(h_2^2 - h_1^2)}{\ln(r_2/r_1)} \tag{5.31}$$

In practice, a direct application is made of Equation 5.26 in:

$$(h_0^2 - h^2) = \frac{0.73\,Q}{K} \log R - \frac{0.73\,Q}{K} \log r \tag{5.32}$$

Therefore in a $(\log r, (h_0^2 - h^2))$ diagram, the slope of the best fitted line (Figure 5.9) will provide $(0.73\,Q/K)$ and thus K can be deduced. The fitted line can be prolonged

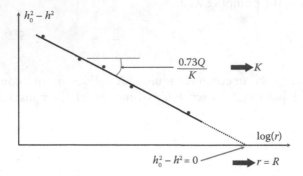

Figure 5.9 *Figure 5.9* Application of the Dupuit's solution for hydraulic conductivity determination: in a $(\log r, (h_0^2 - h^2))$ diagram, the angular coefficient of the best fitted line provides $(0.73\ Q/K)$ from which K can be deduced.

until crossing the log r-axis at zero drawdown, providing a rough[2] assessment of the influence radius (R) of the pumping well (Figure 5.9).

Remarks

1. It is always more reliable to use observed drawdown values from observation wells than in the pumping well itself (influenced by head losses in the well). Thus, it is advised to have a minimum of two observation wells in order to fit the line on measured values (Figure 5.10a).
2. If measurements are available in many observation wells, different lines can often be fitted to experimental points in relation to a considered direction or in relation to a particular zone (Figure 5.10b), which highlights possible local anisotropy and/or heterogeneity in the influenced medium.

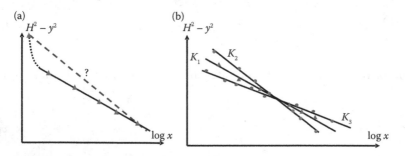

Figure 5.10 (a) Possible bias of a fitted line including measurement value in the pumping well; (b) Possible anisotropy or heterogeneity of the medium as detected by measurements in many observation wells.

---

2 This is only a rough assessment resulting from the large uncertainty linked to the fitting of a straight-line on observed drawdown values. Usually, only a few points are available (if not only two) including the pumping well itself.

Characteristic curve of a well, well yield, and specific capacity

When the purpose of the study is rather focused on the production of the well, a *characteristic curve* (Q, Δ) (Figures 5.11a and 5.12a) is constructed on the basis of stabilized drawdown obtained for a step-drawdown test with increasing constant pumping rates (Figure 5.5). The *well yield* is equivalent to the *recovery yield* (i.e., the pumping rate balanced by the feeding of the well from the aquifer) when a stabilized or steady-state drawdown is recorded in the well. It means that a drawdown taking a lot of time to be stabilized is showing that the pumping rate is not far from the critical well yield. A *specific-capacity* value can also be defined for the well considering the pumping rate (yield) divided by the stabilized (i.e., steady-state) drawdown (Figures 5.11b and 5.12b):

$$q_\Delta = Q/\Delta \tag{5.33}$$

PUMPING WELL IN A CONFINED AQUIFER

For confined aquifers and for limited drawdowns (i.e., laminar flow conditions), the drawdown can be considered as directly proportional to the pumping rate. Using Equation 5.17 gives then:

$$\Delta(r) = (h_0 - h) = \frac{Q}{2\pi T} \ln\frac{R}{r_w} = C_1 Q \tag{5.34}$$

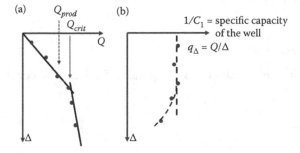

Figure 5.11 Characteristic curve (a) and specific-capacity evolution (b) for a pumping well in confined conditions.

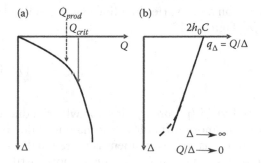

Figure 5.12 Characteristic curve (a) and specific-capacity curve (b) for a pumping well in unconfined conditions.

where $C_1 = (1/2\pi T)\ln(R/r_w)$ is a laminar-flow coefficient considered globally as constant. Then, for greater drawdowns, a turbulent flow component influences also the measured drawdown (Castany 1963):

$$\Delta(r) = (h_0 - h) = \frac{Q}{2\pi T}\ln\frac{R}{r} = C_1 Q + C_2 Q^n \tag{5.35}$$

where C_2 is the turbulent flow coefficient. Most often, $n = 2$ is considered (Jacob 1947). The characteristic curve of a well in a confined aquifer describes a straight line as long as drawdowns are moderate and becomes parabolic for larger drawdowns (Figure 5.11a).

Using Equations 5.34 and 5.35, the specific-capacity is constant equal to $1/C_1$ as far as drawdowns stay moderate (Figure 5.11b). Then, this specific-capacity is decreasing linearly for larger drawdowns:

$$q_\Delta = \frac{1}{C_1 + C_2 Q^{n-1}} \tag{5.36}$$

Therefore, a critical well yield (pumping rate) can be assessed when larger drawdowns are observed inducing increasing turbulent effects in the water flow. A reasonable and safe production yield or pumping rate is usually accepted adopting $Q_{production} = 0.75\, Q_{critical}$ (Figure 5.11a).

PUMPING WELL IN AN UNCONFINED AQUIFER

For unconfined aquifers, using $(h_0^2 - h^2) = 2h_0\Delta - \Delta^2$, Equation 5.26 can be considered as a parabolic (Q, Δ) equation:

$$Q = C(h_0^2 - h^2) = 2h_0 C\Delta - C\Delta^2 \tag{5.37}$$

where $C = (1/\pi K)\ln(R/r_w)$ is considered as a constant. The characteristic curve of a well in unconfined conditions describes a parabola (Figure 5.12a). Indeed, if the drawdowns become larger, turbulent effects are increased.

The specific capacity, as defined in Equation 5.33, varies in a linear way tending to zero (i.e., as Δ is increasing more than Q) (Figure 5.12b):

$$q_\Delta = \frac{Q}{\Delta} = 2h_0 C - C\Delta \tag{5.38}$$

It means also that there is a critical well yield (pumping rate) for which induced drawdowns become prohibitive. It is however more difficult to estimate its value on a characteristic curve than in confined conditions. An assessment is more reliable on the basis of a specific-capacity curve. If an assessment is made, this is more justified to choose a production yield in a safer (i.e., more conservative) way than in confined conditions $Q_{production} < 0.75\, Q_{critical}$ (Figure 5.12a).

Interpretation of transient data

Using data recorded under transient conditions, another set of methods can be applied to determine hydraulic conductivity values of the influenced zone of saturated medium. The discharge rate being maintained constant, measurements of the varying piezometric heads as function of time provide the possibility to determine also specific storage coefficient values (or storage coefficient if integrated over the saturated thickness).

Theis method

In confined and transient conditions, Theis (1935) proposed a 2D analytical solution for radial groundwater flow to the well with the help of the same assumptions as previously:

- The aquifer is homogeneous, isotropic, and of uniform thickness over the influenced zone.
- There is no initial piezometric gradient.
- The induced groundwater flow is only horizontal.
- The aquifer has an infinite areal extension.
- The well is screened over the entire thickness of the aquifer (full penetrating well).
- There is no leakage or infiltration from or to the aquifer.
- Delayed storage variation effects are neglected (i.e., groundwater removed from storage is assumed to be discharged instantaneously with decline of piezometric head).
- The storage in the well can be neglected (the diameter of the well is small).

The 2D horizontal groundwater flow Equation 4.90 can be written under isotropic conditions, without any leakage or infiltration and in radial coordinates. It becomes:

$$T\left(\frac{\partial^2 h}{\partial r^2} + \frac{1}{r}\frac{\partial h}{\partial r}\right) = S\frac{\partial h}{\partial t} \tag{5.39}$$

where T is the transmissivity (m²/s) and S the storage coefficient (-) (see Section 4.11).

Applying a change of variable with the dimensionless variable $u = r^2 S/4Tt$, the Theis solution is expressed as:

$$\Delta(r,t) = (h_0 - h) = \frac{Q}{4\pi T} W(u) \tag{5.40}$$

where $W(u)$ is referred as to the "well function" by hydrogeologists. As explained by Renard (2005), this "well function" is known by mathematicians as the *Exponential integral function E1* obtained by using Laplace transforms, and expressed mathematically as an infinite series:

$$W(u) = \int_{r^2 S/4Tt}^{\infty} \frac{e^{-u}}{u}\,du = -0.577216 - \ln u + u - \frac{u^2}{2(2!)} + \frac{u^3}{3(3!)} - \frac{u^4}{4(4!)} + \dots \tag{5.41}$$

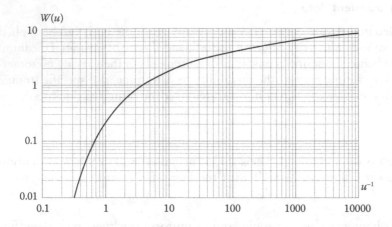

Figure 5.13 Theis curve or well function expressed in a $(u^{-1}, W(u))$ diagram. Practically, in confined conditions, the observed drawdown $\Delta(r, t)$ function of (r^2/t) evolves in a similar way.

where t is the time since the constant pumping rate Q is applied starting from a previously stabilized situation. This exponential integral function E1 is tabulated in mathematical handbooks and drawn in the hydrogeological literature as $(u^{-1}, W(u))$ diagrams (Figure 5.13) presented as the "Theis transient curve."

Practically, bringing measured data in a $(\log(r^2/t), \log \Delta)$ diagram, the resulting curve can be superimposed and fitted on the Theis curve. Thus, for any point in the diagram, one value of $\Delta(r, t)$ corresponds to one value of $W(u)$, and similarly one value of $1/u$ corresponds to one value of t/r^2. These two relations allow to determine the two unknowns: T and S.

Theoretically, the Theis method was developed for confined aquifers. The assumption of an immediate storage variation in the saturated porous medium is only reasonable in confined aquifers.

Neuman's method

In *unconfined conditions* (Figure 5.14), at early times, water is coming (as for confined conditions) from the elastic behavior of the medium, but at later times, water comes mainly in a delayed way from the drainage porosity (see Section 4.2) (Neuman 1972) and automatically involving a more important vertical flow component. The main contribution of water storage from the drainage is decreasing with time and even becoming negligible (i.e., the vertical component vanishes) so that very late drawdown can again follow a Theis curve. Interpretation of early drawdowns will provide a storage coefficient that is similar to confined conditions (i.e., integration of the specific storage over the saturated thickness). The determined value is only valid for elastic storage and not as long-term storage of the unconfined aquifer. As mentioned by Kruseman and de Ridder (1994), and without entering too much in detail, it has been shown that the late-time drawdown curve may be interpreted by the Theis method and provide reasonably realistic values for the specific yield of the unconfined aquifer (in fact, specific yield or effective drainage porosity added to the elastic storage this last being quite lower than the specific yield).

log(Δ)

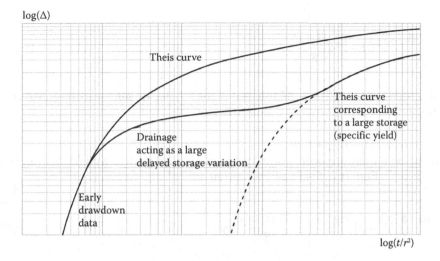

Figure 5.14 Modified curve as obtained in unconfined conditions. The important delayed storage variation due to drainage is reducing the drawdown (Neuman 1972).

In practice, if the observation well is far enough from the pumping well, confined conditions can be applied if the drawdown is limited compared to the initial saturated thickness ($\Delta = h_0 - h < 0.25 h_0$) to avoid the vertical component of the groundwater flow becoming too large. A better approach is also to use the correction $\Delta' = \Delta - \Delta^2/2h_0$ mentioned in Equation 5.28 converting observed drawdown in "equivalent confined conditions" drawdown. The Theis method for confined aquifers can then be applied on the late-time measured drawdown.

The method known as the Boulton (1963) method for interpretation of pumping tests in unconfined conditions consists in fitting the final measured drawdown to obtain an estimation of T and the global storage coefficient (i.e., specific yield + elastic storage). Then the early-time measured drawdowns are used to fit the Theis curve, keeping T unchanged and adjusting for the assessment of S (i.e., elastic storage only). Afterward, the specific yield (S_y) can be deduced.

Cooper-Jacob methods

Simplified methods were developed by Cooper and Jacob (1946) from the Theis method. These methods can be used for confined aquifers but they are in practice often used also for unconfined aquifers under certain conditions (see below).

For $(Tt/r^2 S) > 10$ or low values of u (i.e., for short distances r, long times t, and/or large T/S ratios), the higher order terms become negligible and Equations 5.40 and 5.41 can be approximated by:

$$\Delta(r,t) = (h_0 - h) = \frac{Q}{4\pi T}(-0.577216 - \ln u) = \frac{Q}{4\pi T}\ln\left(\frac{2.25\,Tt}{r^2 S}\right) \tag{5.42}$$

or

$$\Delta(r,t) = \frac{(0.183)Q}{T} \log\left(\frac{2.25\,Tt}{r^2 S}\right) \tag{5.43}$$

Practically, depending of the available data, four types of semi-log diagrams are used to interpret transient drawdown measurements.

a.   *(log t, Δ) diagram—single well*
      If drawdown measurements are only available in one single well (including the pumping well), Equation 5.43 can be written as:

$$\Delta(t) = \frac{(0.183)Q}{T} \log\left(\frac{2.25\,T}{r^2 S}\right) + \frac{(0.183)Q}{T} \log t \tag{5.44}$$

The slope of the best fitted line (Figure 5.15a) on the experimental points provides $0.183Q/T$ and T can be deduced. The fitted line can be extended until crossing the $\log t$-axis at zero drawdown, providing a t_0 value for which $2.25T/r^2 S = 1/t_0$ and the value of S can be deduced[3].

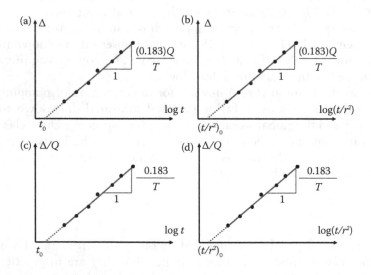

Figure 5.15 Application of the Cooper-Jacob method for different pumping rate and drawdown measurements conditions: (a) for one pumping rate and drawdown measurements in one well, (b) for one pumping rate and drawdown measurements in multiple wells, (c) for a step-drawdown test with drawdown measurements in one well, and (d) for a step-drawdown test with drawdown measurements in multiple wells.

---

3 A large uncertainty can be the result of the straight-line fitting.

b. *(log t/r², Δ) diagram—multiple wells*

If drawdown measurements are available in different wells (including possibly the pumping well), Equation 5.43 can be written as:

$$\Delta(r,t) = \frac{(0.183)Q}{T}\log\left(\frac{2.25\,T}{S}\right) + \frac{(0.183)Q}{T}\log\left(\frac{t}{r^2}\right) \tag{5.45}$$

As stated previously, T can be deduced from the slope of the best fitted line (Figure 5.15b) on experimental points. The fitted line can also be extended until crossing the $\log(t/r^2)$-axis at zero drawdown, providing a $(t/r^2)_0$ value for which $2.25\,T/S = (r^2/t)_0$ and the value of S can be deduced[4]. If different observation wells are available, one can choose to use this method integrating all the available transient drawdown measurements in one single diagram. The obtained equivalent values can be considered as more "robust" (i.e., based on more data) taking the assumption that the influenced medium can be considered as sufficiently homogeneous to obtain significant equivalent values of the parameters T and S. An alternative could be to use a separate diagram (method a) for each data set from each observation well. This can be used to reveal possible local heterogeneity between the pumping and observation wells.

c. *(log t, Δ/Q) diagram for a step-drawdown test—single well*

In the case of a step-drawdown test, with drawdown measurements only available in one single well (possibly the pumping well), Equation 5.43 can be written as:

$$\frac{\Delta}{Q}(t) = \frac{0.183}{T}\log\left(\frac{2.25\,T}{r^2 S}\right) + \frac{0.183}{T}\log t \tag{5.46}$$

The method is similar to method (a) and the slope of the best fitted line (Figure 5.15c) provides $0.183/T$. The fitted line can be extended until crossing the $\log t$-axis, providing $2.25\,T/r^2 S = 1/t_0$. This method integrates in a same diagram the transient drawdown values observed for different pumping rates. An alternative could be to use a separate diagram (method [a]) for each pumping rate and check if the interpreted values evolve with the rising pumping rate. If this is the case, this may indicate heterogeneity of the medium revealed by an increased volume influenced by greater pumping rates.

d. *(log t/r², Δ/Q) diagram for a step-drawdown tests—multiple wells*

Finally, for a step-drawdown test with drawdown measurements in different wells (including possibly the pumping well), Equation 5.43 can be written as:

$$\frac{\Delta}{Q}(r,t) = \frac{0.183}{T}\log\left(\frac{2.25\,T}{S}\right) + \frac{0.183}{T}\log\left(\frac{t}{r^2}\right) \tag{5.47}$$

The slope of the best fitted line (Figure 5.15d) provides $0.183/T$, and this line can be extended until crossing the $\log(t/r^2)$-axis, providing $2.25\,T/S = (r^2/t)_0$ and the value of S. As mentioned previously, this method integrates all transient drawdown measurements in one single diagram. The deduced values are more

---

4 A large uncertainty can be the result of the straight-line fitting.

"robust" integrating all the effects of local possible heterogeneities in an equivalent value. The alternative choice could be to draw separate diagrams for each pumping rate and for each observation well (methods a, b, and c). This can be used to reveal possible local heterogeneity between the pumping and observation wells, and possibly also to reveal heterogeneity of an increased volume of medium influenced by greater pumping rates.

In practice, the Cooper-Jacob methods are also used in *unconfined conditions* if the drawdowns are limited ($\Delta < 0.25h_0$). Rigorously, this should not be the case, but the induced inaccuracies can most often be considered as negligible compared to the natural variability of the parameter values, and relative to the chosen assumptions. In practice, comparisons of interpreted values applying or not the correction $\Delta' = \Delta - \Delta^2/2h_0$ reveal usually very similar obtained values.

Superposition principle

The groundwater flow equation in confined conditions is mathematically linear as the parameters (i.e., T and S) are considered constant (most often) and not dependent on the main variable h. Note that this is not the case in unconfined conditions as $T = Kh$ (see Equation 4.45). Therefore, a problem involving multiple stresses can be solved by adding the solutions calculated for each individual stress. This *superposition principle* can be applied in the time to interpret data in varying pumping rate conditions, as well as in the spatial domain accounting for the effect of actual boundary conditions or heterogeneity of the influenced medium. In practice, this principle can be used to solve many different problems such as the impact of pumping on an existing groundwater flow field, pumping from different wells, boundary effects by using the image-well theory, and time-varying step-incremental pumping. As mentioned previously, nowadays, it is more and more easy to build a local numerical model and obtain results for various scenarios varying parameters and stress factors.

Even if the superposition principle is theoretically not applicable in *unconfined conditions*, analytical methods using this principle are actually accepted for pumping test in unconfined aquifers if the drawdown is limited ($\Delta = h_0 - h < 0.25h_0$). A classical way of linearizing the unconfined equations is also to use (as previously) $\Delta' = \Delta - \Delta^2/2h_0$.

Interpretation of step-incremental pumping test data

If the pumping rate is risen step-wise with Q_1 between t_0 and t_1, Q_2 between t_1 and t_2, ... Q_n between t_{n-1} and t_n, the superposition principle is applied to the Cooper-Jacob approximation of the Theis equation by adding the effect of each increasing of pumping rate $\Delta Q_i = Q_i - Q_{i-1}$ (Birsoy and Summers 1980):

$$\Delta_n(r,t) = \frac{(0.183)Q_n}{T} \log\left[\frac{2.25\,T}{r^2 S}(\beta_{t(n)}(t - t_n))\right] \qquad (5.48)$$

with

$$\beta_{t(n)}(t - t_n) = (t - t_1)^{\Delta Q_1/Q_n} \cdot (t - t_2)^{\Delta Q_2/Q_n} \cdots (t - t_n)^{\Delta Q_n/Q_n} \qquad (5.49)$$

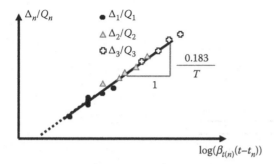

Figure 5.16 Application of the Birsoy-Summers method for step-incremental pumping: fitting a line on the experimental points measured for the pumping rates allows an estimation of T.

where t_i is the starting time of the ith pumping period with Q_i, accordingly, $(t - t_i)$ is the time since the ith pumping period started. Practically, a semi-log $(\log\beta_{t(n)}(t - t_n)$, $\Delta_i/Q_i)$ diagram (or a $(\log\beta_{t(n)}(t - t_n)/r^2$, $\Delta_i/Q_i)$ diagram for multiple wells measurements) is used to fit a line (Figure 5.16) on the experimental points. The slope of the fitted line provides $0.183/T$ and T can be deduced.

Solutions can also be found for *interrupted step-drawdown tests* (i.e., with periods of no pumping) as described in Kruseman and de Ridder (1994) from Birsoy and Summers (1980).

Interpretation of recovery data

When the pumping is stopped, piezometric heads in the tested aquifer are rising back up to return to their initial level: this is the recovery period. Using these transient data is very easy and can be very useful to obtain a reliable value of transmissivity (see below).

The superposition principle is applied to provide an analytical solution of the recovery part of a pumping test. In the Theis and Cooper-Jacob equations, an additional term is considered describing the actual stop of the pumping by the addition of an equivalent water injection rate applied from the moment of the pumping stop (Figure 5.17).

Using the Cooper-Jacob equations, if $T(t - t')/r^2 S > 10$, the approximated resulting or residual drawdown is expressed:

$$\Delta_{res}(r, t, t') = \frac{(0.183)Q}{T} \log\left(\frac{2.25\,Tt}{r^2 S}\right) + \frac{(0.183)(-Q)}{T} \log\left(\frac{2.25\,Tt'}{r^2 S}\right) \tag{5.50}$$

where t is the time from the pumping start, t' is the time from the pumping stop. Equation 5.50 can be simplified in:

$$\Delta_{res}(t, t') = \frac{(0.183)Q}{T} \log\left(\frac{t}{t'}\right) \tag{5.51}$$

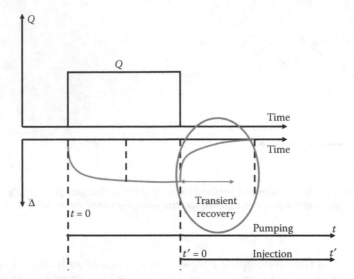

Figure 5.17 During recovery, piezometric heads rise back up toward the initial static piezometric level. Residual drawdown (Δ_{res}) are measured as function of two time-scales corresponding to t and t'.

In a (Δ_{res}, log t/t') diagram, the slope of the best fitted line (Figure 5.18) on the experimental points provides $0.183Q/T$ and T can be deduced. This method is often applied when drawdown measurements are available only in the pumping well. This allows avoiding most of the head losses (i.e., turbulent flow), skin effects (i.e., by clogged screen and high K gravel pack), and most of the well-bore storage effects (except for very large diameter wells). Note that with increasing time, the ratio t/t' is decreasing. One can observe that Equation 5.51 is not dependent on the value of the storage coefficient.

Banton and Bangoy (1996) proposed a method allowing to deduce the storage coefficient S if the recovery is measured in a minimum of one additional observation

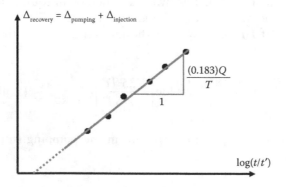

Figure 5.18 Application of the Cooper-Jacob method on recovery data. Using the superposition principle, the slope of the best fitted line allows an estimation of T.

well. The exponential integral function E1 or well function series (Equation 5.41) is approximated taking one additional term into account, so that:

$$\Delta(r,t) = \frac{Q}{4\pi T}(-0.577216 - \ln u + u) \tag{5.52}$$

As the truncation error is lower than in the Cooper-Jacob method, application can be done with less drastic conditions than the one mentioned for Equation 5.42. During recovery, the residual drawdown can be expressed by:

$$\Delta_{res}(r,t,t') = \frac{(0.183)Q}{T} \log\left(\frac{t}{t'}\right) + \frac{r^2 SQ}{16\pi T^2}\left(\frac{1}{t} - \frac{1}{t'}\right) = A(t,t') + B(t,t')r^2 \tag{5.53}$$

where $A(t, t')$ and $B(t, t')$ are respectively the intercept on the Δ_{res}-axis and the slope of the experimental straight lines in a (r^2, Δ_{res}) diagram. There are as many lines as measurement times (i.e., t from the start of the pumping, t' since the pumping stop). A is in fact similar to $\Delta_{res}(t, t')$ of Equation 5.51 and T can be deduced, as previously, from the slope of a fitted line in a (log t/t', Δ_{res}) diagram. If T is known, $-SQ/16\pi T^2$ is then the slope of the fitted experimental line in a $((1/t')-(1/t)$, $B)$ diagram as $B = (SQ/16\pi T^2)((1/t) - (1/t'))$ and the storage coefficient can be deduced.

If the recovery is following a step-drawdown test (i.e., with step rising pumping rates), using the Birsoy and Summers (1980) method:

$$\frac{\Delta_{res}(t,t')}{Q_n} = \frac{0.183}{T} \log\left(\beta_{t(n)}\frac{(t-t_n)}{(t-t')}\right) \tag{5.54}$$

with

$$\frac{\beta_{t(n)}(t-t_n)}{(t-t')} = \frac{1}{(t-t')}\left[(t-t_1)^{\Delta Q_1/Q_n} \cdot (t-t_2)^{\Delta Q_2/Q_n} \cdots (t-t_n)^{\Delta Q_n/Q_n}\right] \tag{5.55}$$

where t_i is the starting time of the ith pumping period with Q_i, accordingly, $(t - t_i)$ is the time since the ith pumping period started, and t' is the time from the pumping stop. Practically, a semi-log (log$\beta_{t(n)}(t - t_n)/(t - t')$, Δ_i/Q_i)diagram is used to fit a line (Figure 5.19) on the experimental points. The slope of the fitted line provides $0.183/T$ and T can be deduced.

Deviations due to aquifer boundaries or heterogeneities (Image well theory)

The superposition principle is applied in the spatial domain to take into account the effect of actual boundary conditions or heterogeneity in the influenced medium, through the use of the *image well* theory. If the groundwater flow is clearly not any-more radially convergent to the pumping well, the measured drawdown values deviate from the previous theories. The tested aquifer is never infinite so this is useful to investigate how boundaries will influence the drawdown response. Two cases are considered: one corresponds to a boundary with a more permeable zone or ultimately

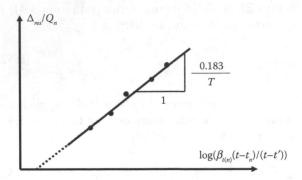

Figure 5.19 Application of the Birsoy-Summers method for the recovery phase after a step-incremental pumping.

a surface water body providing as much water as needed, the second corresponds to a less permeable zone or ultimately a totally impermeable medium.

a.   *Influence of a high hydraulic conductivity zone*
      In this case, the high hydraulic conductivity zone supplies more groundwater in the influenced zone and to the pumping well. The measured drawdown values are lower than expected. To check this effect, the situation may be represented by the extreme situation corresponding to a constant head boundary condition at a given distance from the well. Indeed, this boundary condition assumes that as much groundwater flows into the domain as needed to maintain a piezometric head unchanged at a given distance from the pumping well. This is similar to the effect of an infinitely large reservoir of water (i.e., a lake or large river) in direct contact with the influenced zone. Then, the superposition principle may be applied with an image well located opposite from the real well relative to a symmetry plane corresponding to the boundary (Figure 5.20). A constant head boundary condition is then expressed on this boundary by adding the influence of the image well where $-Q$ (i.e., an injection rate equivalent to the discharge rate in the pumping well) is applied. The Theis method (Equation 5.40) expressing the resulting drawdown gives:

$$\Delta_{res}(r,r',t) = \frac{Q}{4\pi T}\left[W\left(\frac{r^2 S}{4Tt}\right) - W\left(\frac{r'^2 S}{4Tt}\right) \right] \tag{5.56}$$

Using the Cooper-Jacob approximation:

$$\Delta_{res}(r,r') = \frac{0.183Q}{T}\log\left(\frac{r'^2}{r^2}\right) \tag{5.57}$$

where r is the radial distance from the measured observation well to the pumping well (as previously), and r' is the radial distance from the measured point to the

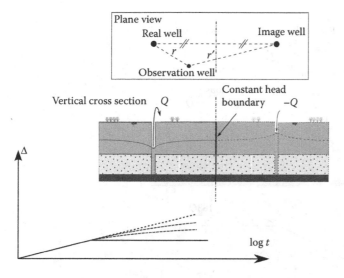

Figure 5.20 Use of the image well theory and the superposition principle to detect the influence of a high hydraulic conductivity zone (ultimately a constant head boundary condition) on the observed drawdown during a pumping test.

image well. Note that this extreme boundary condition induces that late draw-down data are not depending anymore on time. A stabilized situation is observed. Theoretically, if the geometry of a true boundary is known, r' could be calculated and T deduced directly from Equation 5.57. Usually, this reasoning is rather used to understand how a zone of high K influences the curve of observed drawdowns (or the experimental line in a Cooper-Jacob diagram) (Figure 5.20).

b. *Influence of a low hydraulic conductivity zone*

In this case, the low hydraulic conductivity zone supplies less groundwater to the influenced zone and the pumping well. The measured drawdown values are greater than expected. To check this effect, the situation may be represented by the extreme situation corresponding to an impermeable boundary condition at a given distance from the well, similar to the effect of a very low hydraulic conductivity lithology in direct contact with the influenced zone. As previously, an image well, located opposite from the real well relative to a symmetry plane corresponding to the boundary (Figure 5.21), is added with a pumping rate Q (i.e., equivalent to the discharge rate in the pumping well). The Theis method (Equation 5.40) expressing the resulting drawdown gives:

$$\Delta_{res}(r,r',t) = \frac{Q}{4\pi T}\left[W\left(\frac{r^2 S}{4Tt}\right) + W\left(\frac{r'^2 S}{4Tt}\right)\right] \tag{5.58}$$

and using the Cooper-Jacob approximation:

$$\Delta_{res}(r,r') = \frac{0.183Q}{T}\left[2\log\left(\frac{2.25\,Tt}{r^2 S}\right) + \log\left(\frac{r'^2}{r^2}\right)\right] \tag{5.59}$$

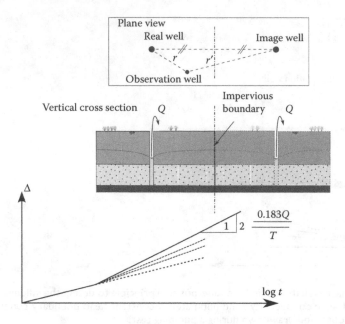

Figure 5.21 Use of the image well theory and the superposition principle to detect the influence of a low hydraulic conductivity zone (ultimately an impermeable boundary condition) on the observed drawdown during a pumping test.

where r is the radial distance from the measured observation well to the pumping well (as previously), and r' is the radial distance from the measured point to the image well. Practically, in a (log t, Δ) diagram, the slope of the straight line fitted on the late-drawdown measurements is multiplied by 2. This is the case in the theoretical and extreme situation of a perfectly impermeable boundary. Usually, the reasoning is rather applied to understand how a zone of low K influences the curve of observed drawdowns (or the experimental line in a Cooper-Jacob diagram) (Figure 5.21).

Hantush and Neuman-Witherspoon methods for leaky confining layer

Pumping can induce or increase leakage through the layers bounding (above and below) the concerned aquifer. If this leakage becomes significant and if the "source aquifer" piezometric head is considered as not affected by the leakage, solutions proposed by Hantush and Jacob (1955) can be applied to solve the convergent radial flow equations including this leakage term and without change of storage in the confining layer (neglected compressibility). Later, Hantush (1960) also proposed a solution taking the variation of the storage coefficient of the leaky confining layer into account.

To keep it simple, most of the assumptions used for the previous methods are required except that we consider a strictly vertical leakage through a low-permeability layer located above the pumping aquifer (Figure 5.22a). In this case, the piezometric head in the "source" aquifer or in located above the aquitard is assumed not to be affected ($h_s = constant$). The 2D radial convergent groundwater flow Equation 5.39 can be written, with the additional leakage expresses as a source term:

(a)

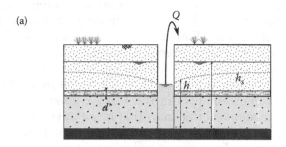

(b)

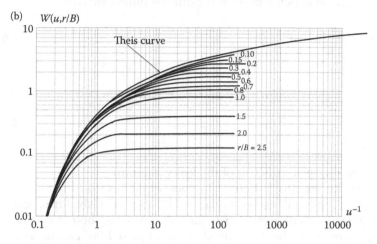

Figure 5.22 Influence of a leaky confining layer on pumping test results (a); Hantush curves for different values of the r/B ratio (b).

$$T\left(\frac{\partial^2 h}{\partial r^2} + \frac{1}{r}\frac{\partial h}{\partial r}\right) = S\frac{\partial h}{\partial t} + \left(\frac{K'}{b'}\right)(h_s - h) \tag{5.60}$$

where K' are the hydraulic conductivity and b' the thickness of the low permeability layer (aquitard). The proposed solution is similar to the Theis analytical solution, but now with a well function dependent on a second dimensionless variable r/B (Hantush and Jacob 1955):

$$\Delta(r,t) = (h_0 - h) = \frac{Q}{4\pi T} W'\left(u, \frac{r}{B}\right) \tag{5.61}$$

with $1/B = \sqrt{K'/Tb'}$ in m^{-1} [L^{-1}]. Values for $W'(u,(r/B))$ are tabulated but Walton (1962) also produced the different curves (for different values of r/B) in a log-log diagram (Figure 5.22b) allowing fitting graphically experimental data on theoretical curves. Practically, bringing measured data in a $(\log(r^2/t), \log\Delta)$ diagram, the resulting curve can be superimposed and fitted on one of the Walton curves. Depending on the chosen theoretical curve, a value of r/B is assessed providing a ratio between K' and T for a known thickness of the aquitard. Then, classically, as for the Theis method, for any point in the diagram, one value of $\Delta(r, t)$ corresponds to one value of $W'(u, r/B)$, a value of $1/u$ corresponds to one value of t/r^2. It allows to determine T and S. Consequently, this interpretation method allows to obtain an estimation of the K' value, or at least of K'/b'. Logically, if r/B becomes very low (<0.01), a convergence toward the Theis solution is observed.

In a similar way, Hantush (1960) proposed a solution which accounts for the variation of the storage (thus the compressibility) in the leaky confining layer. Equation 5.60 is now written with an aquitard thickness that is not anymore constant (i.e., $(\partial/\partial z)(h_s - h)$ is replacing $(h_s - h)/b'$ in the source term). Solutions are found using a well function dependent on three dimensionless variables:

$$\Delta(r,t) = (h_0 - h) = \frac{Q}{4\pi T} W''\left(u, \frac{r}{B}, \beta\right) \tag{5.62}$$

with

$$\beta = \frac{r}{4}\sqrt{\frac{K'S_s'}{TS}} \tag{5.63}$$

where S_s' is the specific storage coefficient of the leaky aquitard.

Practically, different families (for different β values) of curves (for different values of r/B) are available in a log-log diagram (Figure 5.23). As previously, fitting graphically experimental data on theoretical curves (i.e., choosing the best β family and the best r/B curve), T and S can be determined, K' or at least K'/b' can be assessed from r/B, and finally S_s' can also be assessed using the value of β. This solution can be applied for the following range of u values:

$$\frac{(r/B)^4}{80\beta^2} \leq u \leq \frac{(r/B)^4}{(1.6)\beta^2} \tag{5.64}$$

Note that the fitting of experimental curves on the theoretical curves becomes very subjective and relatively inaccurate leading to a rough assessment of the hydraulic parameters.

In fact, if the storage variation is taken into account, it means that a drawdown is measured in the confining layer when the drawdown in the overlying source aquifer is still negligible. Neuman and Witherspoon (1969a,b) developed a solution based on the ratio of the drawdowns as measured at the same elapsed time t and at the same distance r from the pumping well:

$$\frac{\Delta'(r,t)}{\Delta(r,t)} = f(u, u') \tag{5.65}$$

where $u' = r^2 S'/4T't$ with $T' = K'd'$ and $S' = S_s'd'$. From the measured ratio Δ'/Δ and a previously determined u value for the pumped aquifer, a u' value can be deduced from the published curves (Neuman-Witherspoon nomogram) (Figure 5.24).

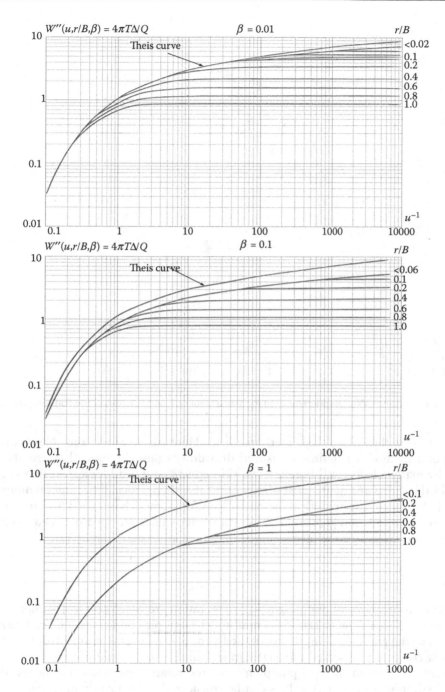

Figure 5.23 Hantush method for a leaky confining layer with storage variation: different families of curves according to β values and the r/B ratio.

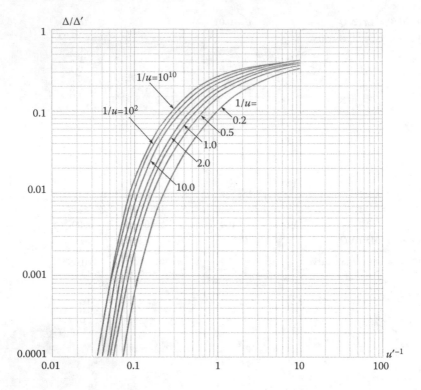

Figure 5.24 Neuman-Witherspoon method for assessment of T' and S' values of the confining layer.

However, for these two last methods (i.e., Hantush method and Neuman-Witherspoon method) early-time measured drawdowns must be used that are also affected by many effects (well-bore effects, skin effects, and head losses) so that the obtained values for S and S' are most often not reliable. In practice, it becomes tedious and unreliable still to use analytical methods. Nowadays, models are used to simulate pumping tests in such complex conditions. Parameter values are then deduced from the calibration process (or the inverse modeling procedure) (see Chapter 12).

Anisotropy

All the previous methods of pumping test interpretation are based on the assumption that the groundwater flow is 2D horizontal under isotropic conditions (i.e., radially convergent to the pumping well). Unfortunately, the possible vertical anisotropy that is the most frequently observed especially in stratified layers, could not be detected or taken into account in classical pumping test interpretation methods. The deduced effective hydraulic conductivity values can be considered consequently as horizontal values.

Anisotropy of hydraulic conductivity in the horizontal plane is not exceptional and can be detected if the pumping test is performed with measurements in a minimum of two observation wells.

In practice, two cases can be distinguished depending on the knowledge that we may have about the principal components of the K tensor.

Under homogeneous conditions, if *the two main horizontal anisotropy directions are known*, the reference axis system (x, y) can be chosen adequately following the two principal components of the T tensor. Then the ellipse of equal drawdown around the pumping well can be reduced to a circle using the following variable change (Figure 5.25):

$$x' = \sqrt{\frac{T}{T_x}}\, x \quad \text{and} \quad y' = \sqrt{\frac{T}{T_y}}\, y \tag{5.66}$$

where $T = \sqrt{T_x T_y}$ and T_x is the transmissivity in the main direction of anisotropy parallel to x, T_y is the transmissivity in the main direction of anisotropy parallel to y. Using Equations 5.66, the 2D groundwater flow equation (simplified from Equations 4.92 or 4.95) becomes:

$$T_x \frac{\partial^2 h}{\partial x^2} + T_y \frac{\partial^2 h}{\partial y^2} = S \frac{\partial h}{\partial t} \tag{5.67}$$

Equation 5.67 is then transformed in an equivalent isotropic equation:

$$T \frac{\partial^2 h}{\partial x'^2} + T \frac{\partial^2 h}{\partial y'^2} = S \frac{\partial h}{\partial t} \tag{5.68}$$

Practically, the Theis solution can be written as:

$$\Delta(r', t) = \frac{Q}{4\pi T}\, W\!\left(\frac{r'^2 S}{4 T t}\right) \tag{5.69}$$

with $r'^2 = x'^2 + y'^2$

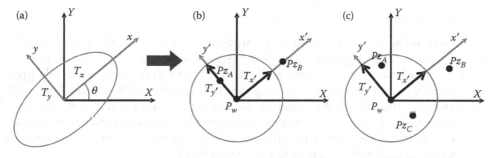

Figure 5.25 For interpretation of a pumping test in the well (P_w) in anisotropic conditions, the axis system (x, y) is chosen following the principal components of the T tensors and a change of variable is performed (a). Then the Theis solution is applied using measured data from 2 piezometers (Pz_A and Pz_B) if they are located in the directions of the principal components (b), or from 3 piezometers (Pz_A, Pz_B, and Pz_C) if they are located anywhere (c).

If two observation wells (Pz_A and Pz_B) are adequately placed (Figure 5.25) along the main axes of anisotropy x and y, we have $r'^2 = x'^2 = (T/T_x)x^2$ along x, and $r'^2 = y'^2 = (T/T_y)y^2$ along y, and the Theis solution gives:

$$\begin{cases} \Delta_A(x,t) = \dfrac{Q}{4\pi T}\, W\left(\dfrac{x^2 S}{4T_x t}\right) \\[2ex] \Delta_B(y,t) = \dfrac{Q}{4\pi T}\, W\left(\dfrac{y^2 S}{4T_y t}\right) \end{cases} \qquad (5.70)$$

and T_x, T_y, and S can be deduced as previously (see Theis method).

If *the two main horizontal anisotropy directions are unknown*, an additional unknown angle θ must be determined between the used reference system and the reference axis system (x, y) that follows the two principal components of the T tensor (Figure 5.25) so that:

$$\begin{cases} x = X\cos\theta + Y\sin\theta \\ y = -X\sin\theta + Y\cos\theta \end{cases} \qquad (5.71)$$

As stated previously, the change of variable described by Equations 5.66 is applied and the Theis solution as in Equation 5.69 is used with:

$$r'^2 = \frac{T}{T_x}(X\cos\theta + Y\sin\theta)^2 + \frac{T}{T_y}(-X\sin\theta + Y\cos\theta)^2 \qquad (5.72)$$

A minimum of three observation wells (Pz_A, Pz_B and Pz_C) are needed to apply Equation 5.69 on the measured drawdowns as function of time in each of them ($\Delta_A(r'_A,\theta,t), \Delta_B(r'_B,\theta,t), \Delta_C(r'_C,\theta,t)$), so that T_x, T_y, S, and θ can be deduced.

Partially penetrating well

If the screen length is not equal to the entire aquifer thickness, the pumping induces vertical components of groundwater flow close to the pumping well. Due to the deficit of the groundwater feeding of the well, the measured drawdown is actually greater than expected with the classical assumptions of the analytical solutions. The possible anisotropy of the saturated medium is the key influencing point and normally only three dimensional analytical solutions or numerical models can consider these effects. A frequent case is the *partial penetrating well* where the open borehole or the well screen corresponds to only a part of the aquifer thickness.

In practice, Hantush (1964) demonstrated that measured drawdowns in observation wells at a distance $r > 1.5d\sqrt{K_v/K_h}$ can be interpreted according to the methods seen previously. As K_v and K_h are most often not known at this stage, the effect of a partially penetrating well is most often neglected if the observation well is located at a distance $> 1.5\,d$. If this is not the case, seemingly surprising observations can be measured. For example, possible greater drawdown in a more distant well than in a closer one. The actual combination of screen lengths and relative positions of pumping and observation well screens can be very determining in predicting these "surprising effects." Despite numerous efforts to find appropriate corrections in analytical solutions (Huysman 1972), detailed interpretation in such conditions should be done by the use of 3D numerical models.

Well-bore effect

One of the basic assumptions of the previous methods is that water storage in the pumping well would be negligible. Indeed, if the pumped well has a large diameter, this is not any more the case, and a *well-bore effect* should be taken into account. A part of the pumped water is flowing from the well itself. This part becomes more important as the hydraulic conductivity of the porous medium is low.

For a fully penetrating well in confined conditions, a curve fitting method is proposed by Papadopoulos and Cooper (1967):

$$\Delta(r,t) = (h_0 - h) = \frac{Q}{4\pi T} W''' \left(u, \alpha, \frac{r}{r_c} \right) \tag{5.73}$$

where r_w is the effective radius of the well (i.e., open hole or internal well screen radius), $\alpha = Sr_w^2/r_c^2$ with r_c the well radius in the unscreened part if different than r_w (i.e., in the casing where the water level changed) and W''' is a well-function given in published tables and different family curves. However, the curve-fitting is particularly delicate. Indeed, for long pumping time, the late-time measured drawdown should no longer be influenced by the well storage effect. Papadopoulos and Cooper (1967) determined that this is surely the case if $t > 250 r_c^2/T$.

For a partially penetrating well in unconfined conditions, a curve-fitting method is proposed by Boulton and Streltsova (1976) with a well-function dependent on seven dimensionless parameters. Let alone the fitting procedure, it becomes quite too intricate to obtain values for this well-function (i.e., some are proposed by Kruseman and de Ridder [1994]). Chapuis (1992) proposed to use the classical methods after correction of the timescale: subtracting the time needed to empty the well-storage (neglecting during this initial period the feeding from the aquifer). The corrected pumping time is written as:

$$t_{corrected} = \frac{V - V_w}{Q} \tag{5.74}$$

where V_w is the emptied well-volume, V is the total volume of pumped water, Q is the constant pumping flow rate (as previously).

Head losses in the pumping well

When a drawdown is measured in a pumping well, the head decrease not resulting from the aquifer reaction is attributed to *head losses*. A part of them are due to the well-bore *skin effects* (i.e., no proper development and equipment of the bore-hole after drilling operations). *Skin effects* are resulting from a filter cake, drilling additives, or fines on the borehole walls (i.e., *well-bore effects*), a clogged screen or a screen of low conductivity with a high conductivity gravel pack (Houben 2015). Many of these effects can actually be combined inducing additional measured draw-down in the pumping well that is not representative of the drawdown in the aquifer. Skin and well-bore effects can also induce an increase of the water volume to be emptied before really influencing the aquifer. If the skin region can be considered of infinitesimal thickness and without storage capacity, Agarwal *et al.* (1970) proposed a modification to the Papadopoulos and Cooper method using an additional dimen-sionless variable *SF*, named the skin factor:

$$SF = \frac{(T - T_s)}{T_s} \ln\left(\frac{r_s}{r_w}\right) \tag{5.75}$$

where T_s is the transmissivity of the "skin cylinder" around the well, r_s is the radius of the well including the "skin cylinder" (i.e., the gravel pack). Logically, a positive skin factor $(T_s < T)$ indicates a clogged well and a negative skin factor $(T_s > T)$ a gravel pack with a greater hydraulic conductivity than the porous medium. In practice, a log-log curve fitting method is used with the dimensionless time versus dimensionless drawdown. Different families of curves are available as function of the skin factor *SF*.

Other head losses can be corresponding to the *screen and well interior head losses* (i.e., resulting from vertical water flow in the well and turbulence). According to the review on that topic by Houben (2015), effects of these additional head losses can be considered most often as quite lower than the skin effects.

5.5 Other measurements methods

Lugeon test or packer test

Lugeon (1933) developed a packer test aiming mainly to assess the grouting need in soils and rock foundations at dam sites. Consequently, this test was conceived to esti-mate hydraulic conductivity of rock masses. The Lugeon test can be considered as a constant head or constant pressure step test involving an isolated portion of the bore-hole between two inflatable packers. Different increasing pressures steps are tested until reaching a predefined maximum pressure (i.e., that must be lower than the nor-mal confinement stress expected at the considered depth, in order to be sure to avoid any hydraulic fracturing). Then, decreasing pressure steps are also performed. At each pressure step, the averaged flow rate that was necessary to maintain the pressure constant is measured. A local hydraulic conductivity is expressed in Lugeon units, 1 Lugeon being empirically defined when a flow rate of 1 liter/minute per meter of test interval is found. If the hydraulic conductivity is assumed constant in the medium, the flow rate (Q) is increased to balance the increase of pressure so that:

$$\text{Lugeon Value} = \frac{Q/L}{p/p_0} \tag{5.76}$$

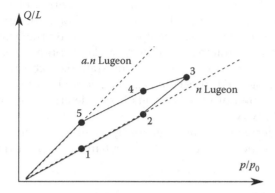

Figure 5.26 Interpretation of a Lugeon test: in practice, due to actual test conditions far from a perfect laminar flow, an experimental "pressure loop" is obtained (instead of a line) in a $(p/p_0, Q/L)$ diagram. As illustrated for 5 pressure steps, a same pressure ratio does not correspond to one single value of the Q/L ratio leading possibly to different values of measured hydraulic conductivity in Lugeon units.

where p is the averaged pressure during the considered pressure-step, p_0 is the initial pressure, L is the tested length between the packers. In practice however a kind of "pressure loop" (Quinones-Rozo 2010) is often found if the Lugeon test data are taken in a $(p/p_0, Q/L)$ diagram (Figure 5.26). Without entering into details, the flow can be considered as essentially laminar if this pressure loop describes nearly a line. If this is not the case (most often the case) different effects can be invoked as turbulence effects, dilation of fissures in the investigated medium, washing-out of the filling particles, and void filling with fines (Houlsby 1976). The investigated volume is often considered as not very large with a maximum influenced distance of about $3L$. If this volume is considered as homogeneous and isotropic (that is actually not the case), 1 Lugeon is empirically corresponding to a hydraulic conductivity (K) value of about $1.3 \ 10^{-7}$ m/s.

A development of this Lugeon test has led to the *Ménard pressio-permeameter* test, where the tested zone between packers is surrounded by two other injection zones to force a plane radial water flow around the injection well. Then, a horizontal radial water flow can be assumed and the hydraulic conductivity can be estimated using:

$$K = \frac{Q}{2\pi L((p/\rho g) - (p_0/\rho g))}\left(\frac{1}{2} + \ln\frac{L}{r}\right) \tag{5.77}$$

where r is the diameter of the injection well. As previously, Q is the measured flow rate for a test pressure p, p_0 is the initial pressure, L is the tested length between the packers.

Lefranc method

When a pumping test is not possible, an interesting alternative can be to inject or pump water in a drilled volume to assess the hydraulic conductivity around this cavity. Most often, this *Lefranc test* can be performed during the execution of the drilling itself, the volume (cavity) being created at the basis of the borehole, bounded by the bottom and a portion of the borehole walls. The flow generated by the head difference is assumed to take place only through the cavity as a clay/bentonite seal is installed around the upper casing. As for other tests, a steady-state (i.e., constant head) or a transient (i.e., falling or rising head) procedure can be adopted (Rat *et al.* 1968). The main issue in the interpretation of measurements lies in determining the actual volume and form of the tested cavity.

Under steady-state flow, and considering only one value for hydraulic conductivity (i.e., homogeneous and isotropic equivalent value) in the immediate vicinity of the cavity, Darcy's law is written as:

$$\Delta = (h_0 - h) = \frac{Q}{KC} \tag{5.78}$$

where C is a "cavity coefficient" in m [L] dependent on the cavity geometry. For each tested flow rate, a stabilized measured rise or drawdown of the piezometric head is measured. A (Q, Δ) diagram can be produced providing an experimental line from which $1/KC$ can be deduced. For a cavity considered rather of spherical shape with a diameter D, $C = 2\pi D$. For a cylinder of height L and diameter D, $C = 2\pi L/\ln(2L/D)$ if the ratio L/D is greater than 2.

Under transient flow conditions, $(h_0 - h)$ is measured as function of time for a given flow rate Q, so that:

$$Q = S\frac{dh}{dt} = KCh \tag{5.79}$$

with S the assumed homogeneous value of the storage coefficient in the immediate vicinity of the cavity. Integration gives:

$$\ln h - \ln h_0 = \frac{KC}{S}(t - t_0) \tag{5.80}$$

In a $(\log(h/h_0), t - t_0)$ diagram, the slope of the experimental line corresponds to $(KC/2.3\,S)$.

However, in practical conditions, many deviations can be found from an ideal experimental straight line. This may be resulting from many interference effects as for example a part of the flow bypassing the sealing, clogging of the cavity walls by fines, and a partially turbulent flow.

Jacob and Lohman solution for an artesian flowing well test

If the well is naturally flowing under confined artesian conditions in the aquifer, a flow outlet is arranged by capping at a given level above the ground (the method can be used also under water constant pressure between packers). The outlet elevation is considered as constant and the decreasing discharge is measured as function of time.

This is thus a constant drawdown—variable discharge test. Derived from a thermal analogy, Jacob and Lohman (1952) provided a mathematical solution:

$$Q(t) = 2\pi T(h_0 - h)G(\alpha) \tag{5.81}$$

where $(h_0 - h) = \Delta$ is the drawdown from the initial piezometric head h_0, $G(\alpha)$ is a complex integral (endless series) dependent on the value of the dimensionless parameter $\alpha = Tt/Sr_w^2$, r_w being the radius of the well. It has been observed that this solution can be approximated by the Cooper-Jacob method (method [d], Equation 5.47) with the same accepted assumptions and for $t > 10^5$. In a $(\log t/r_w^2, \Delta/Q)$ diagram, the slope of the best fitted line (Figure 5.15d) provides $0.183/T$, and this line can be prolonged until crossing the $\log(t/r_w^2)$-axis, providing $2.25\ T/S = (r^2/t)_0$ and the value of S (a large uncertainty can result from this fitting). Wendland (2008) has shown that the interpreted values of hydraulic conductivity and above all of storage coefficient could be largely erroneous due to friction head losses (about 20% of error in T values and 5 orders of magnitude for S values). He also demonstrated the importance to consider only interpreted values from long-term drawdown tests.

Inverse auger hole, infiltrometer, or Porchet method

This method was developed mainly to measure the saturated hydraulic conductivity with a test performed above the water table (i.e., in an initially partially saturated medium) like in an infiltrometer. This method, described as the Porchet method in the French specialized literature, consists simply to fill with water a given hole (most often excavated from the ground surface), and measuring the water head decrease with time. Since the true tested zone is not known, the interpretation is based on the (important) assumption that a one-meter gradient can be supposed to apply Darcy's law under saturated conditions. For a cylindrical auger hole of radius r, and at any time t:

$$Q = KA = K(2\pi rh + \pi r^2) = -\pi r^2 \frac{dh}{dt} \tag{5.82}$$

This can be written as:

$$K\left(h + \frac{r}{2}\right)dt = -\frac{r}{2}dh \tag{5.83}$$

After integration, taking into account that $h = h_0$ at time $t = t_0$ and using log instead of ln, it gives (van Hoorn 1979):

$$t = \frac{1.15r}{K}\log\left(h_0 + \frac{r}{2}\right) - \frac{1.15r}{K}\log\left(h + \frac{r}{2}\right) \tag{5.84}$$

Using a semi-log diagram $(t, \log(h + (r/2)))$ allows to fit a line on the measured experimental points and deduce the slope coefficient $1.15\ r/K$ providing an order of magnitude for the saturated hydraulic conductivity of the tested zone of soil. However, the test conditions do not allow us to have a clear view on this true involved zone.

Field-based air permeameter measurements

Air permeameters measurements on outcrops or on laboratory samples can provide additional useful information at relatively small scale compared to pumping and slug tests. This technique can be very useful to analyze aquifer heterogeneity combining measurements and K-proxy data at different scales (Possemiers *et al.* 2012, Huysmans *et al.* 2013, Rogiers *et al.* 2013, Rogiers *et al.* 2014a, 2014b) helping significantly in reservoir analog and fault zones studies. Since the 1960s, air permeability measurements with a hand-held device have been performed (Bradley *et al.* 1972). They are cheap, rapid, nondestructive, repeatable, and pose often less practical problems than water flux methods. They have been used to characterize loose sediments as well as indurated rocks, on natural outcrops, in quarries, on soils, or even on borehole cores (Figure 5.27).

Figure 5.27 In situ air permeameter measurement: the device is pressed against a sand outcrop face. The plunger is depressed to create a vacuum, causing air to flow from the unsaturated sand into the device. Gas flow rate and pressure are monitored by a pressure transducer and analyzed by the microprocessor unit.

It consists in an annulus through which a gas is released from (or injected into) the porous medium. Leakage is avoided using a compressible impervious ring at the probe tip. The gas flow and pressure data are recorded. The gas permeability is deduced from empirical relations or analytical solutions of modified forms of Darcy's law including a geometrical factor to account for the size and the form of the permeameter. There are so many relations that it is useless here to provide a list of them. Each of them must be adapted to the actual test conditions and devices. They should also be corrected to account for gas slippage and high velocity, and depth of investigation (often estimated between four and two times the probe inner radius). Among others, an example is the relation proposed by Goggin et al. (1988):

$$k_a = \frac{2\mu p_1 Q_1}{(p_1^2 - p_2^2)aG_0} \tag{5.85}$$

where k_a is the air permeability $(m^2)[L^2]$, μ is the air dynamic viscosity $(kg/(m.s))$ $[ML^{-1}T^{-1}]$ at atmospheric pressure, p_1 and p_2 are the injection and outflow air pressures (Pa) $[ML^{-1}T^{-2}]$ respectively, Q_1 is the volumetric airflow rate (m^3/s) $[L^3T^{-1}]$, a is the radius of the seal aperture (m) $[L]$, and G_0 is a dimensionless "Goggin" geometrical factor. If we consider that the intrinsic permeability of the medium is the same for air and for water, the air permeability can be translated under saturated hydraulic conductivity using Equation 4.24:

$$K = \frac{k_a \rho g}{\mu} \tag{5.86}$$

where (see Section 4.4) μ is the water dynamic viscosity $(kg/(m.s))$ $[ML^{-1}T^{-1}]$, ρ is the density of water (kg/m^3) $[ML^{-3}]$. The assumed equality of the air and water intrinsic permeability actually neglect the air slippage at the interface with solid matrix, and the influence of the partial saturation of the tested medium (i.e., atmospheric pressure does not act as a true fluid continuum in the medium). Totally dry sediment conditions are rarely found in situ, so that empirical equations have been developed to deduce the saturated hydraulic conductivity values from the measured k_a (among others, Davis et al. 1994, Loll et al. 1999, Iversen et al. 2003). To avoid seal quality problems and possible effects of external weathering of the medium, the development and use of a small drill hole minipermeameter probe has been developed and applied (Dinwiddie et al. 2003).

References

Agarwal, R.W., Al-Hussainy, R. and H.J.J. Ramey. 1970. An investigation of wellbore storage and skin effect in unsteady liquid flow. SPE Journal 10: 279–290.

Banton, O. and L. Bangoy. 1996. A new method to determine storage coefficient from pumping test recovery data. Ground Water 34(5): 772–777.

Birsoy, Y.K. and W.K. Summers. 1980. Determination of aquifer parameters from step tests and intermittent pumping data. Ground Water 18: 137–146.

Boulton, N.S. 1963. Analysis of data from non-equilibrium pumping tests allowing for delayed yield from storage. *Proceedings of the Institution of Civil Engineers*. 26: 469–482.

Boulton, N.S. and T.D. Streltsova. 1976. The drawdown near an abstraction well of large diameter under non-steady conditions in an unconfined aquifer. *Journal of Hydrology* 30(1–2): 29–46.

Bouwer, H. 1989. The Bouwer and Rice slug test—an update. *Ground Water* 27: 304–309.

Bouwer, H. and R.C. Rice. 1976. A slug test for determining hydraulic conductivity of unconfined aquifers with completely or partially penetrating wells. *Water Resources Research* 12: 423–428.

Bradley, V.W., Duschatko, R.W. and H.H. Hinch. 1972. Pocket permeameter: Handheld device for rapid measurement of permeability. *Bulletin American Association Petroleum Geologists* 56: 568–571.

Butler, J.J. 1997. *The design, performance, and analysis of slug tests*. Boca Raton: CRC Press.

Cardiff, M., Barrash, W., Thoma, M. and B. Malama. 2011. Information content of slug tests for estimating hydraulic properties in realistic, high-conductivity aquifer scenarios. *Journal of Hydrology* 403: 66–82.

Carman, P.C. 1938. The determination of the specific surface of powders. *Journal of the Society of Chemical Industry, Transactions* 57: 225.

Carrier, W.D. 2003. Goodbye, Hazen; Hello, Kozeny-Carman. *Journal of Geotechnical and Geoenvironmental Engineering* 129(11): 1054–1056.

Castany, G. 1963. *Traité pratique des eaux souterraines (in French)*. Paris: Dunod.

Chapuis, R.P. 1992. Using Cooper-Jacob approximation to take account of pumping well pipe storage effects in early drawdown data of a confined aquifer. *Ground Water* 30(3): 331–337.

Chapuis, R.P. and M. Aubertin. 2003. On the use of the Kozeny-Carman equation to predict the hydraulic conductivity of soils. *Canadian Geotechnical Journal* 40: 616–628.

Cooper, H.H., Bredehoeft, J.D. and I.S. Papadopulos. 1967. Response of a finite-diameter well to an instantaneous charge of water. *Water Resources Research* 3: 263–269.

Cooper, H.H. and C.E. Jacob. 1946. A generalized graphical method for evaluating formation constants and summarizing well filed history. *Transactions of the AGU* 27: 526–534.

Cronican, A. and M. Gribb. 2004. Literature review: Equations for predicting hydraulic conductivity based on grain-size data. Supplement to technical note entitled: Hydraulic conductivity prediction for sandy soils. *Ground Water* 42(3): 459–464.

Davis, J.M., Wilson, J.L. and F.M. Phillips. 1994. A portable air-minipermeameter for rapid in situ field measurements. *Ground Water* 32(2): 258–266.

de Marsily, G., Delay, F., Goncalves, J., Renard, P., Teles, V. and S. Violette. 2005. Dealing with spatial heterogeneity. *Hydrogeology Journal* 13(1): 161–183.

Delage, P., Sultan, N. and Y.J. Cui. 2000. On the thermal consolidation of Boom clay. *Canadian Geotechnical Journal* 37(2): 343–354.

Deng, Y.F., Tang, A.M., Cui, Y.J. and X.L. Li. 2011. Study on the hydraulic conductivity of Boom clay. *Canadian Geotechnical Journal* 48: 1461–1470.

Dinwiddie, C.L., Molz, F.J. and J.W. Castle. 2003. A new small drill hole minipermeameter probe for in situ permeability measurement: Fluid mechanics and geometrical factors. *Water Resources Research* 39(7): 1–13.

Dupuit, J. 1863. *Etudes théoriques et pratiques sur le mouvement des eaux dans les canaux découverts et à travers les terrains perméables (in French)* (2nd Edition). Paris: Dunot.

Fair, G. M. and L.P. Hatch 1933. Fundamental factors governing the streamline flow of water through sand. *Journal of American Water Works Association* 25: 1551–1565.

Fitts, Ch. R. 2002. *Groundwater science*. London: Academic Press.

Fritz, B.G., Mackley, R.D. and E.V. Arntzen. 2016. Conducting slug tests in mini-piezometers. *Groundwater* 54: 291–295.

Goggin, D.J., Chandler, M.A., Kocurek, G.A. and L.W. Lake. 1988. Patterns of permeability in eolian deposits: Page Sandstone (Jurassic), Northeastern Arizona. *SPE Formation Evaluation* 3: 297–306.

Greene, E.A. and A.M. Shapiro. 1995. *Methods of conducting air-pressurized slug tests and computation of type curves for estimating transmissivity and storativity.* U.S. Geological Survey Open-File Report 95–424.

Hantush, M.S. 1960. Modification of the theory for leaky aquifers. *Journal of Geophysical Research* 65: 3713–3725.

Hantush, M.S. 1964. Hydraulics of wells. In *Advances in Hydroscience*, ed. V.T. Chow. New York: Academic Press, 281–432.

Hantush, M.S. and C.E. Jacob. 1955. Non steady radial flow in an infinite leaky aquifer. *Transactions of the American Geophysical Union* 36: 95–100.

Hazen, A. 1892. Some physical properties of sands and gravels, with special reference to their use in filtration. In: *24th Annual Rep., Massachusetts State Board of Health*, Pub. Doc. no. 34, 539–556.

Houben, G.J. 2015. Review: Hydraulics of water wells—head losses of individual components. *Hydrogeology Journal* 23: 1659–1675.

Houlsby, A.C. 1976. Routine interpretation of the Lugeon water-test. *Quarterly Journal of Engineering & Geology* 9: 303–313.

Huysman, L. 1972. *Groundwater recovery.* New York: Winchester Press.

Huysmans, M. and A. Dassargues. 2009. Application of multiple-point geostatistics on modelling groundwater flow and transport in a cross-bedded aquifer (Belgium). *Hydrogeology Journal* 17(8): 1901–1911.

Huysmans, M., Orban, P., Cochet, E., Possemiers, M., Ronchi, B., Lauriks, K., Batelaan, O. and A. Dassargues. 2013. Using multiple point geostatistics for tracer test modeling in a clay-drape environment with spatially variable conductivity and sorption coefficient. *Mathematical Geosciences* 46(5): 519–537.

Hvorslev, M.J. 1951. *Time lag and soil permeability in ground water observations.* U.S. Army Corps of Engineers Waterways Experimentation Station. Bulletin 36.

Iversen, B.V., Moldrup, P., Schjonning, P. and O.H. Jacobsen. 2003. Field application of a portable air permeameter to characterize spatial variability in air and water permeability. *Vadose Zone Journal* 2(4): 618–626.

Jacob, C.E. 1947. Drawdown test to determine effective radius of artesian well. *Transactions of the ASCE* 112(2312): 1047–1070.

Jacob, C.E. and S.W. Lohman, 1952. Nonsteady flow to a well of constant drawdown in an extensive aquifer. *Transactions of the American Geophysical Union* 33: 559–569.

Kozeny, J. 1927. Ueber kapillare Leitung des Wassers im Boden (in German). Wien, *Akad. Wiss.* 136(2a): 271.

Kruseman, G.P. and N.A. de Ridder. 1994. *Analysis and evaluation of pumping test data* (2nd Edition). Publication 47. Wageningen: International Institute for Land Reclamation and Improvement.

Lemieux, J.M. 2002. *Caractérisation multi-approche à petite échelle de l'écoulement de l'eau souterraine dans un aquifère carbonaté fracturé et implications pour les changements d'échelle.* (in French), MSc thesis, Canada: Laval University.

Lewis, J. 2016. A simple field method for assessing near-surface saturated hydraulic conductivity. *Groundwater* 54(5): 740–744.

Loll, P., Moldrup, P., Schjønning, P. and H. Riley. 1999. Predicting saturated hydraulic conductivity from air permeability: Application in stochastic water infiltration modeling. *Water Resources Research* 35(8): 2387–2400.

Loudon, A. G. 1952. The computation of permeability from simple soil tests. *Geotechnique* 3: 165–183.

Lugeon, M. 1933. *Barrage et Géologie (in French)*. Paris: Dunod.

Nastev, M., Savard, M.M., Lapcevic, P., Lefebvre, R. and R. Martel. 2004. Hydraulic properties and scale effects investigation in regional rock aquifers, south-western Quebec, Canada. *Hydrogeology Journal* 12(3): 257–269.

Neuman, S.P. 1972. Theory of flow in unconfined aquifers considering delayed response of the water table. *Water Resources Research* 8(4): 1031–1045.

Neuman, S.P. and V. Di Federico. 2003. Multifaceted nature of hydrogeologic scaling and its interpretation. *Reviews of Geophysics* 41(3): 1014.

Neuman, S.P. and P.A. Witherspoon. 1969a. Theory of flow in a confined two aquifer system. *Water Resources Research* 5: 803–816.

Neuman, S.P. and P.A. Witherspoon. 1969b. Applicability of current theories of flow in in leaky aquifers. *Water Resources Research* 5: 817–829.

Niemi, A., Kontio, K., Kuusela-Lanthinen, A. and A. Potteri. 2000. Hydraulic characterization and upscaling of fracture networks based on multiple-scale well test data. *Water Resources Research* 36(12): 3481–3497.

Papadopoulos, I.S. and H.H.J. Cooper. 1967. Drawdown in a well of large diameter. *Water Resources Research* 3(1): 241–244.

Parent, S.E., Cabral, A., Avanzi, E. and J.G. Zornberg. 2004. Determination of the hydraulic conductivity function of a highly compressible material based on tests with saturated samples. *Geotechnical Testing Journal* 27(6): 1–5.

Pechstein, A., Attinger, S., Krieg, R. and N. K. Copty. 2016. Estimating transmissivity from single-well pumping tests in heterogeneous aquifers. *Water Resources Research* 52: 495–510.

Possemiers, M., Huysmans, M., Peeters, L., Batelaan, O. and A. Dassargues. 2012. Relationship between sedimentary features and permeability at different scales in the Brussels Sands. *Geologica Belgica* 15(3): 156–164.

Quinones-Rozo, C. 2010. Lugeon test interpretation, revisited. In *Collaborative Management of Integrated Watersheds. 30th Annual USSD Conference Sacramento*, California, April 12–16, 2010, 405–414.

Rat, M., Laviron, F. and J.C. Jorez. 1968. Essai Lefranc (in French). In *Proc. of the JELPC— Journées d'études d'Hydraulique des sols*, Paris, 27–29 November 1968, 56–66.

Renard, Ph. 2005. *Quantitative analysis of groundwater field experiments*. Switzerland: University of Neuchatel.

Rogiers, B., Beerten, K., Smeekens, T., Mallants, D., Gedeon, M., Huysmans, M., Batelaan, O. and A. Dassargues. 2013. The usefulness of outcrop analogue air permeameter measurements for analysing aquifer heterogeneity: Quantifying outcrop hydraulic conductivity and its spatial variability. *Hydrological Processes* 28: 5176–5188.

Rogiers, B., Mallants, D., Batelaan, O., Gedeon, M., Huysmans, M. and A. Dassargues. 2012. Estimation of hydraulic conductivity and its uncertainty from grain-size data using GLUE and artificial neural networks. *Mathematical Geoscience* 44(6): 739–763.

Rogiers, B., Vienken, T., Gedeon, M., Batelaan, O., Mallants, D., Huysmans, M. and A. Dassargues. 2014a. Multi-scale aquifer characterization and groundwater flow model parameterization using direct push technologies. *Environmental Earth Sciences Journal* 59: 1303–1324.

Rogiers, B., Winters, P., Huysmans, M., Beerten, K., Mallants, D., Gedeon, M., Batelaan, O. and A. Dassargues. 2014b. High resolution hydraulic conductivity logging of borehole cores using air permeability measurements. *Hydrogeology Journal* 22(6):1345–1358.

Ronayne, M.J., Gorelick, S.M. and C. Zheng. 2010. Geological modeling of submeter scale heterogeneity and its influence on tracer transport in a fluvial aquifer. *Water Resources Research* 46(10): 1–9.

Schaap, M.G., Leij, F.J. and M.T. van Genuchten. 1998. Neural network analysis for hierarchical prediction of soil hydraulic properties. *Soil Science Society of America Journal* 62: 847–855.

Schultze-Makuch, D., Carlson, D.A., Cherkauer, D.S. and P. Malik. 1999. Scale dependency of hydraulic conductivity in heterogeneous media. *Ground Water* 37: 904–919.

Theis, C.V. 1935. The relation between the lowering of the piezometric surface and the rate and duration of discharge of a well using groundwater storage. *Transactions of the AGU* (2): 519–524.

Thiem, G. 1906. *Hydrologische Methoden (in German)*. Leipzig: Gebhardt.

Tügel, F., Houben, G.J. and T. Graf. 2016. How appropriate is the Thiem equation for describing groundwater flow to actual wells? *Hydrogeology Journal* 24(8): 2093–2101.

van der Kamp, G. 1976. Determining aquifer transmissivity by means of well response tests: The underdamped case. *Water Resources Research* 12(1): 71–77.

van Hoorn, J.W. 1979. Determining hydraulic conductivity with the inverse auger hole and infiltrometer methods. In *Proc. of the Int. Drainage Workshop*, ed. J. Wesseling. Wageningen: ILRI Publication 25, 150–154.

Walton, W.C. 1962. Selected analytical methods for well and aquifer evaluation. Urbana: Illinois State Water Survey Bulletin 49.

Wendland, E. 2008. Friction loss correction in flowing well discharge tests. *Water Resources Research* 44: W01428.

Chapter 6

Land subsidence induced by pumping and drainage

6.1 Introduction

Currently, much attention is paid to sea level rise, but the problem of land subsidence induced by anthropogenically changed (fluid) groundwater conditions can be far more significant locally (Baeteman 2010, Showstack 2014). The "sinking" regions most often correspond to densely populated areas located where compressible loose sediments are found. They form underconsolidated (see Section 6.2) and compressible layers made of recent coastal sediments, particularly estuarine, deltaic, and lacustrine sediments. Venice, Mexico, Bangkok, Shanghai, Changzhou, Jakarta, Manila, New Orleans, Houston, Tokyo, Ho Chi Minh City, and Hanoi are only a few examples of the numerous "sinking cities." An excellent review of the most challenging cases, main processes, and milestone papers of sinking cities was provided by Gambolati and Teatini (2015). For a long time, groundwater pumping or drainage has been known to induce land subsidence (Poland and Davis 1969). The Terzaghi and Biot theories were developed early (Biot 1941, Terzaghi 1943) (see Section 4.10). However, only in the last few decades have causal links between pumping or drainage and subsidence been clearly shown (Gorelick and Zheng 2015). This land subsidence hazard is still mostly underestimated or ignored. For example, civil engineering works involving excavation and groundwater pumping (for building foundations or underground parking) in alluvial deposits include the consolidation of the loamy/peaty parts of the deposits. This is common in circumstances where very local land subsidence is observed with very damaging differential settlements, highly dependent on the actual heterogeneity of the alluvial deposits.

The transient groundwater flow equation (see Chapter 4, Equations 4.77 and 4.79) in saturated media implies a direct relation between groundwater flow conditions and geomechanical consolidation processes. This coupling is expressed with the value of the specific storage coefficient (S_s) function of the volume compressibility of the porous medium. Another coupling can be invoked as consolidation clearly reduces the porosity, permeability, and compressibility of the concerned REV of porous medium.

6.2 Effective stress and water pressure variations in depth

Recent unconsolidated or semi-consolidated deposits often form a succession of layers that can be considered, from a hydrogeological point of view, as

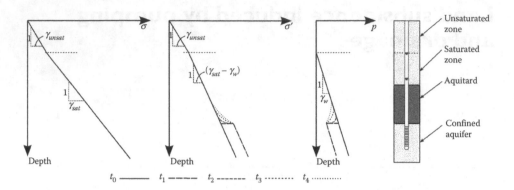

Figure 6.1 Evolution of the total stress (σ), effective stress (σ'), and pore pressure (p) as a function of depth (z) in a *confined aquifer* and in the confining unit above (solid lines represent the initial state, assumed to be in equilibrium, and dotted lines represent the dynamic states after the lowering of the piezometric head in the aquifer) by applying the Terzaghi's concept (Dassargues 1995).

semi-confined or confined aquifer systems (Poland 1984). They consist of silty sands or sands of high permeability and low compressibility, interlayered with clayey aquitards characterized by low permeability and high compressibility. Applying Terzaghi's concept (see Section 4.10, Equation 4.57), the lowering of the piezometric head (resulting from pumping) in a *confined aquifer* (Figure 6.1) induces additional effective stress first directly in the concerned aquifer and then in the compressible confining layers, with a delay dependent on their characteristics. During this stage, the total stress can be assumed constant as the propagation of the pressure decrease through the aquitard is very slow, and the saturation of the upper layers can possibly be maintained by the recharge from the surface (Dassargues 1995).

In detail and as schematized in Figure 6.1, the *initial conditions* (considering the medium to be homogeneous) are as follows:

- For the total stress, in the unsaturated zone, the increase in σ with z is linearly proportional to $\gamma_{unsat} = \rho_{unsat}\, g$ where γ_{unsat} and ρ_{unsat} are the specific weight and the mass of the unsaturated medium, respectively. In the saturated zone, the increase of σ with z is linearly proportional to $\gamma_{sat} = \rho_{sat}\, g$ where γ_{sat} and ρ_{sat} are the specific weight and the mass of the saturated medium, respectively (with $\gamma_{sat} > \gamma_{unsat}$).
- For the pore pressure, p is considered 0 in the unsaturated zone and it increases linearly in proportion to $\gamma_w = \rho\, g$.
- For the effective stress, in the unsaturated zone, the increase in σ' is similar to the increase in σ (i.e., $p = 0$). In the saturated zone, the increase in σ' with z is linearly proportional to $(\gamma_{sat} - \gamma_w) = (\rho_{sat} - \rho_w)\, g$ where γ_w is the specific weight of water (with $\gamma_{sat} > \gamma_{unsat} > (\gamma_{sat} - \gamma_w)$).

After the *lowering of the piezometric head* in the confined aquifer (Figure 6.1),

- The total stress distribution can be considered unchanged, as it does not affect the saturation of the medium above.
- The pore pressure is decreased immediately in the confined aquifer in accordance with $\Delta p = \rho g \Delta h$, where part of Δp progressively enters the aquitard; this is clearly a transient effect resulting from the low permeability of this compressible layer.
- The effective stress is unchanged where pore pressures did not change; however, an increase in the effective stress is observed where an equivalent decrease in the pore pressure occurred in accordance with $\Delta \sigma' = -\Delta p$ when $\Delta \sigma = 0$.

The change in pore pressure and the equivalent increase in effective stress in the aquifer can induce consolidation processes locally where the aquifer contains clay, loam, or peat lenses. The slow propagation of the pore pressure variation in the confining low permeability layer automatically induces an equivalent increase in the effective stress of this compressible layer, starting a drained consolidation process. This corresponds to a highly transient behavior that can take years. After a very long time passes, one could theoretically expect that the induced Δp (or part of it) will reach the top of the confining layer, and then, in the case of unconfined conditions (see below) a possible decrease in the saturated zone thickness could be observed.

In an *unconfined aquifer*, the lowering of the piezometric head due to pumping or drainage (Figure 6.2) induces additional effective stresses directly in the concerned aquifer. Consequently, consolidation can occur if compressible (loam, clay, peat) lenses are in the aquifer.

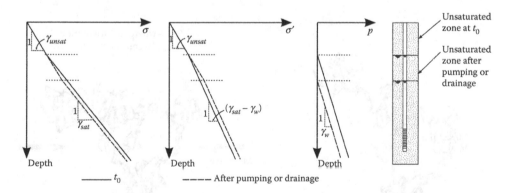

Figure 6.2 Evolution of the total stress (σ), effective stress (σ'), and pore pressure (p) as a function of depth (z) in an *unconfined aquifer* (solid lines represent the initial state, which is assumed to be in equilibrium, and dotted lines represent the dynamic states after the lowering of the piezometric head in the aquifer) by applying Terzaghi's concept (Dassargues 1997).

In detail and as schematized in Figure 6.2, the *initial conditions* of the unconfined aquifer are similar to those of the confined aquifer. After the *lowering of the piezometric head* (Figure 6.2),

- For the total stress distribution, the depth of the unsaturated conditions increases; thus, at depth, the total stress is decreased compared to the total stress of the initial conditions.
- The zone where $p = 0$ is deeper and the pore pressure decreases immediately in the aquifer in accordance with $\Delta p = \rho g \Delta h$.
- The effective stress is increased because of Terzaghi's concept, where the decrease in pore pressure (Δp) is balanced by a decrease in the total stress ($\Delta \sigma$) and an increase in the effective stress ($\Delta \sigma'$).

Consolidation processes can thus locally occur in unconfined aquifers if some parts of the aquifers contain clay, loam, or peat lenses. Consolidation can occur also in the underlying compressible layers, occurring as far as the pore pressure variation can propagate.

6.3 Coupling groundwater flow and geomechanical aspects in porous media

Many authors have observed that clay minerals tend to orient their sheets orthogonally to the direction of the main applied stress, thus developing a kind of microstructural anisotropy (Rieke and Chilingarian 1974, Delage and Lefebvre 1984; Figure 6.3). Clayey soils and loose sediments have a geomechanical behavior that is more often qualified as nonlinear elasticity with progressive plasticity and viscosity (Dassargues 1998). This complex behavior leads, in practice, to the choice

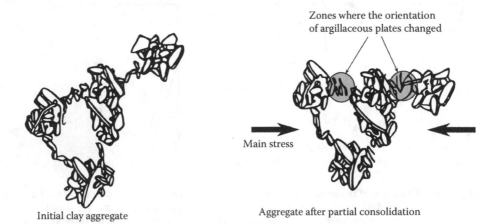

Zones where the orientation of argillaceous plates changed

Main stress

Initial clay aggregate

Aggregate after partial consolidation

Figure 6.3 Preferential orientation of argillaceous plates orthogonal to the main stress (Dassargues 1997).

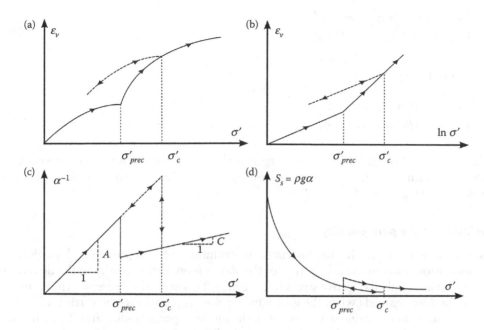

Figure 6.4 Schematic results of oedometer tests in a (σ', ε_v) diagram (a), in a linearized ($\ln \sigma', \varepsilon_v$) diagram (b), in a (σ', α^{-1}) diagram (c), and in a (σ', S_s) diagram (d), showing the evolution of the specific storage coefficient value during a consolidation process (Dassargues 1998).

of rheological models based on experimental laws rather based on combinations of theoretical models. Elasto-visco-plastic laws can be established from experimental results. For practical reasons, 1D (instead of 3D) relations based on vertically measured strain are preferred. With this assumption, oedometer tests provide consolidation results plotted in (σ', ε_v) diagrams (Figure 6.4a), where ε_v is the relative vertical deformation volume of the porous medium. This allows the determination of a volume compressibility value for each effective stress level (see Section 4.10, Equation 4.58). Note that the volume compressibility one could calculate based on this (σ', ε_v) curve is not constant but dependent on the value of σ' and on the preconsolidation effective stress σ'_{prec} (i.e., the highest effective stress that was previously applied to the porous medium). In the following considerations, only "primary consolidation" will be considered. The "secondary consolidation," mostly involving the long-term viscous behavior of the consolidating porous medium will be neglected.

Variation of the specific storage coefficient

Consequently, to characterize the porous medium with constant coefficients, one uses ($\ln \sigma', \varepsilon_v$) diagrams (Figure 6.4b). A "swelling constant" A and a "compression constant" C can be defined for the considered elastic stress-strain behavior ($\sigma' < \sigma'_{prec}$) and for the plastic behavior ($\sigma' \geq \sigma'_{prec}$), respectively, leading to:

$$\begin{cases} d\varepsilon_v = d\sigma'/(A\,\sigma') & \sigma' < \sigma'_{prec} \\ d\varepsilon_v = d\sigma'/(C\,\sigma') & \sigma' \geq \sigma'_{prec} \end{cases} \tag{6.1}$$

Thus (Figure 6.4c):

$$\begin{cases} \alpha(\sigma') = 1/(A\,\sigma') & \sigma' < \sigma'_{prec} \\ \alpha(\sigma') = 1/(C\,\sigma') & \sigma' \geq \sigma'_{prec} \end{cases} \tag{6.2}$$

The specific storage variation ($S_s = \rho g \alpha$) is thus expressed as a noncontinuous $1/\sigma'$ function, with the noncontinuity corresponding to the preconsolidation effective stress value (Figure 6.4d).

Variation of the permeability

As mentioned above, the microstructural evolution of compressible clays during consolidation leads to the orientation of the sheets more orthogonally to the direction of the vertical applied effective stress (Figure 6.3). This evolution increases the tortuosity of the flow channels when the groundwater flow is parallel to the vertical effective stress. In addition, it induces a decrease in the intrinsic permeability and the hydraulic conductivity. Many reports have been published linking k or K to the porosity or the void ratio (see Chapter 5). To describe the decrease in the permeability during a consolidation process, it may be logical to adopt a similar approach considering the elastic and plastic rate of k variations for consolidation at $\sigma' < \sigma'_{prec}$ and $\sigma' \geq \sigma'_{prec}$, respectively (Dassargues 1995).

The consolidation results can also be plotted on a ($\log \sigma'$, e) diagram (Figure 6.5a), where e is the void ratio. The "swelling index" C_s and the "compression index" C_c are used to describe the elastic stress-strain behavior ($\sigma' < \sigma'_{prec}$) and the plastic behavior ($\sigma' \geq \sigma'_{prec}$), respectively, leading to:

$$\begin{cases} e = -C_s \log \sigma' + Cst \\ e = -C_c \log \sigma' + Cst \end{cases} \text{and}$$

$$\begin{cases} de = -\dfrac{C_s}{2.3}\dfrac{d\sigma'}{\sigma'} & \sigma' < \sigma'_{prec} \\ de = -\dfrac{C_c}{2.3}\dfrac{d\sigma'}{\sigma'} & \sigma' \geq \sigma'_{prec} \end{cases} \tag{6.3}$$

where typical values for C_c range from 0.8 to 0.2 in clays and the values of C_s are approximately $C_c/3$.

Monte and Kritzen (1976) cited by Lewis and Schrefler (1987) have proposed a bilinear law linking $\ln K$ with the void ratio. On the basis of Equation 6.3 and Figure 6.5a, a similar law can be proposed and expressed in a ($\log K$, e) diagram as illustrated in Figure 6.5b. In this bilinear diagram, C_{K_1} and C_{K_2} describe the elastic and plastic rate, respectively, of the $\log K$ variation as a function of e according to:

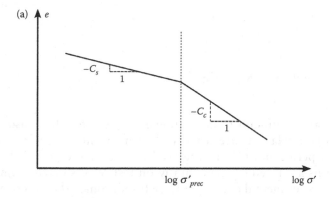

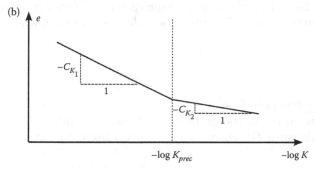

Figure 6.5 Schematic results of oedometer tests in a linearized (log σ', e) diagram (a), and the corresponding ($-$log K, e) diagram (b) showing the evolution of the hydraulic conductivity during a consolidation process (Dassargues 1998).

$$\begin{cases} e = C_{K_1} \log K + Cst \\ e = C_{K_2} \log K + Cst \end{cases} \quad \text{and}$$

$$\begin{cases} de = \dfrac{C_{K_1}}{2.3} \dfrac{dK}{K} & K > K_{prec} & \sigma' < \sigma'_{prec} \\[2mm] de = \dfrac{C_{K_2}}{2.3} \dfrac{dK}{K} & K \leq K_{prec} & \sigma' \geq \sigma'_{prec} \end{cases} \qquad (6.4)$$

where K_{prec} is the hydraulic conductivity corresponding to the preconsolidation effective stress σ'_{prec}. Combining Equations 6.3 and 6.4, gives:

$$\begin{cases} -\dfrac{C_s}{C_{K_1}} \dfrac{d\sigma'}{\sigma'} = \dfrac{dK}{K} & K > K_{prec} \\[2mm] -\dfrac{C_c}{C_{K_2}} \dfrac{d\sigma'}{\sigma'} = \dfrac{dK}{K} & K \leq K_{prec} \end{cases} \qquad (6.5)$$

This automatically leads to the equation:

$$K = C/\sigma'^{a} \tag{6.6}$$

where

$$\begin{cases} a = C_s/C_{K_1} & K > K_{prec} \\ a = C_c/C_{K_2} & K \leq K_{prec} \end{cases} \quad \text{and} \quad C = K_{prec}\left(\sigma'_{prec}\right)^{a}$$

However, in practice, it can be difficult to determine $a = C_c/C_{K_2}$ for the consolidating clays. Thus, many other relations are usually chosen to link the hydraulic conductivity K (or intrinsic permeability k) to the void ratio e or the porosity n. Most of the adopted relations are fitted experimentally on test results on the compressible medium. Once the experimental coefficients are fitted, some other physical or empirical parameters can be used, such as the average specific surface per volume unit (S_{sp}) in the Kozeny-Carman relation (Guéguen and Palciauskas 1994), which is:

$$k = \lambda \frac{n}{S_{sp}^2} \tag{6.7}$$

where λ is the experimentally fitted coefficient.

Another typical example is the use of the geotechnical "plasticity index" I_p in the relation of Nishida and Nakagawa (1969), which was generalized by Dassargues (1997) as follows:

$$K = e^{a \cdot e + b} \quad \text{where} \quad a = 2.3/\left(c\, I_p + d\right) \tag{6.8}$$

where b, c, and d are experimentally fitted.

6.4 Examples of sinking cities and famous case studies

Different authors have published reviews in the past (among others, Poland 1984, Galloway *et al.* 1999) and recently (Galloway and Burbey 2011, Gambolati and Teatini 2015) about the worldwide occurrence of anthropogenically induced land subsidence.

These reviews will not be reproduced here, but a few case histories that are representative of main anthropogenic land subsidence events in well-known key areas will be noted.

Due to fluid extraction, including the extraction of groundwater, gas, or oil by deep pumping, or water by groundwater drainage, settlements can be induced. The depth and the rate of the fluid extraction can vary from very shallow conditions to very deep gas and oil reservoirs (Gambolati and Teatini 2015).

Venice

The land subsidence in Venice is probably the most emblematic example of a groundwater pumping-induced effect. The Venetian lagoon lies above a thickness of approximately 1000 m of Quaternary sediments alternating fine sands and silts with compressible clays and peat layers originating from fluviatile-lagoonal and littoral environments (Gambolati *et al.* 1974). The upper 60 m are made of clay, sand, and peat lenses. A 10–12 m thick paleosoil, named "Caranto," is the only overconsolidated clay layer in this sedimentological sequence. More continuous bedding is found deeper than 60 m, with a series of seven confined aquifers before a depth of 330 m. Pumping wells in Marghera, Venice, and Lido generally tap different aquifers with fully screened equipment. Over time, the deepest aquifers have also been the most exploited, but accurate and detailed data are not available for each aquifer. Data on the history of the piezometric levels are rather limited (Gambolati 1972), and the available information is often mixed among different aquifers due to the depth-averaged conditions in the wells. The average maximum values of the piezometric decrease were 7.5 m for the 1931–1962 period, and 5.6 m for the 1961–1970 period (Gambolati 1972). In terms of subsidence, the maximum values are considered to be between 10 and 15 cm in 1974. Between 1973 and 1993, an additional 3 to 9 cm of subsidence was observed in the littoral zone of the lagoon (Carbognin *et al.* 1995). Since then, the land subsidence has officially been very limited; however, based on combined differential GPS and SAR data for the period between 2000 and 2010, it was recently determined that the city of Venice is still subsiding an average of approximately 1 to 2 mm a year (Bock *et al.* 2012). Overall, the assessed cumulative values of induced land subsidence range between 15 and 30 cm. Based on the predictive simulations of a project that is currently under investigation, seawater injections in the deep confined saline aquifers of Venice could create an elastic rebound of up to 25–30 cm over a 10-year period (Gambolati and Teatini 2015). However, predictive calculations of such rebounds are strongly dependent on uncertain assumptions linked to the choice of the actual boundary conditions influencing fluid pressures in the regional deep confined saline aquifers.

The Netherlands

A large part of the shallow geology of The Netherlands comprises recent peaty clay layers. In addition to this geological setup, land reclamation activities have occurred for centuries to fight against land loss, as one-third of the country lies below the mean sea level (Hoeksema 2007). The western part of the country is mostly composed of low-lying lagoon and estuary deposits with peat bogs and is affected by both natural consolidation processes and man-induced land subsidence. Drainage ditches and pumping during the "three stages in the history of land reclamation" (Hoeksema 2007) are the main causes of land subsidence increasing the need for coastal protections and further drainage. An average land subsidence of approximately 1 m per century is usually reported (Oude Essink *et al.* 2010). Processes involved in the subsidence are: (1) consolidation resulting from decreasing pore pressure and the corresponding increase in effective stress, (2) oxidation allowing the decay of the partially saturated peat, and (3) clay shrinkage due to drying.

Intensive drainage using windmills (Meijer 1996) and steam-powered, electrical, and diesel-powered pumps affect not only low-lying lands but also lagoons and lakes. The reclaimed areas are now "polders" located below the mean sea level, which are surrounded by dikes and have controlled groundwater levels. Piezometric levels must

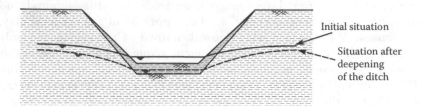

Figure 6.6 Drainage ditch in a polders area: the possible deepening of the ditch year after year can induce a lowering of the controlled piezometric head and consequently induces land subsidence.

be low enough to allow agricultural and urban activities but high enough that additional land subsidence is not produced. Thus, a difficult equilibrium must be found. In practice, achieving this equilibrium verges on a vicious circle of mitigation efforts resulting in increased risks. A simplified example can be schematized as follows (Figure 6.6): if a drainage ditch is annually cleaned with a backhoe, it is difficult to avoid a slight deepening of the ditch walls. This deepening induces the drainage of groundwater to a slightly lower level.

Bangkok

Sandy aquifers alternating with compressible clayey layers are found in the thick (>2000 m) stratigraphic sequence of recent clastic sediments in the "Lower Central Plain" of Bangkok. This is the result of the fluvial and marine deposits of the Chap Phraya River delta (Takaya and Thiramongkol 1982). The top layer (approximately 15–20 m thick) of clay is known to be highly compressible and is referred as to the "Bangkok Soft Clay" deposited in shallow marine conditions in the Holocene (Rau and Nutalaya 1983) and overlying a stiff clay layer and the first sandy aquifer. Layers of clay separate the different aquifers but leakages and even direct connections between aquifers exist due to the heterogeneity of the sediments in the braided sedimentological system. Despite well-known evidence of land subsidence, the total groundwater production increased from 700,000 m³/day in the seventies to more than 2 million m³/day in 2000 (Phien-wej *et al.* 2006). The most exploited aquifers are between 50 and 200 m deep, as brackish or saline groundwater is found below 200 m. Recharge to the aquifers is considered a slow process. Mainly occurring laterally, the recharge from the main rivers and surrounding fractured bedrock has an assessed value of approximately 1.6 million m³/day. Evidence of land subsidence is more obvious where many different and differential settlements are observed. This is particularly true in Bangkok and the following causes can be invoked: (a) the soft clay layer is highly heterogeneous; (b) building foundations (i.e., piles, slurry walls, and jet grouting) can reach various depths, so the compressible clay thickness differs under each type of foundation, thus inducing different total settlements (Figure 6.7). The largest cumulative subsidence from 1933 to 2015 is estimated to be approximately 2.30 m, and the highest recorded annual rate of subsidence was up to 120 mm in the beginning of the 1980s in the eastern zone of the city (Phien-wej *et al.* 2006).

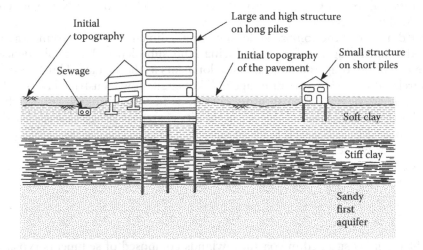

Figure 6.7 Different and differential settlements resulting from land subsidence induced by groundwater pumping in Bangkok: schematic situation showing different settlements for the large and the small building structures as a function of the depth reached by their foundations and the differential settlements of a small structure in contact with the larger one (Phien-wej *et al.* 2006).

Mexico City

The metropolitan area of Mexico City mainly lies on fine-grained and organically rich lacustrine sediments that were artificially drained approximately 200 years ago (Ortega-Guerrero *et al.* 1999). These Quaternary sediments can have a thickness of approximately 300 m in the center of the lacustrine plain and can possibly be interbedded with thin pyroclastic sand layers. Below, a regional alluvial and pyroclastic aquifer of approximately 200–300 m thickness has been the target of many pumping wells since the late nineteenth century. Large-scale and severe land subsidence was a consequence of the wells, as documented by Carillo (1947). Over the years, the subsidence in downtown Mexico has become so great (>9 m in some zones) that the old wells have progressively been replaced by new well fields in the suburbs. One typical example is the well field of the Chalco basin, which started in the 1980s and was documented by Ortiz-Zamora and Ortega-Guerrero (2010). There, the piezometric heads decreased at an approximate rate of 1.5 m/year and one of the most rapid land subsidence rates, 40 cm/year (Ortega-Guerrero *et al.* 1999), was occurring. The cumulative land subsidence reached approximately 13 m in 2006. Water pressure measurements in the thick compressible aquitard have clearly shown that the decrease in pressure induced by pumping is not yet detectable throughout the entire compressible unit (Ortiz-Zamora and Ortega-Guerrero 2010) as illustrated previously in Figure 6.1. A transient effect is induced and delayed land subsidence is still to be expected. On the basis of predictive transient groundwater flow and land subsidence modeling, Ortiz-Zamora and Ortega-Guerrero (2010) assessed that a maximum total subsidence of 19 m could be expected by 2020 in areas where the thickness of the lacustrine aquitard is at its maximum. Consequently, the risks of flooding and of New Chalco Lake growth will be increased.

The surface drainage of the compressible lacustrine sediments and the groundwater pumping in the underlying aquifer are inducing land subsidence. Land subsidence increases the risk of flooding and new lake developments at the surface. Thus, a kind of "vicious circle" starts, as further drainage of lakes or shallow lacustrine layers will trigger more land subsidence. We can observe that, as in other "sinking cities," the main well fields that induce land subsidence in the city center are progressively closed, but they are replaced by other well fields in suburban areas, creating the same type of consequences (or worse) depending on the local hydrogeological conditions. This clearly shows the lack of consideration of the hydrogeological conditions in the land-use planning process.

Shanghai

The city of Shanghai is situated in coastal lowlands composed of sediments typical of a transition from an estuarine to a fluvial environment (Baeteman 1994). The back and forth shifting of the estuary formed by the Yangtze River has left inter-bedded sediments (Baeteman 1989). Sandy layers are representative of the estuarine conditions and compressible clayey layers represent the subtidal and intertidal conditions. One of the clay layers is characteristic of a clay flood-basin and is known as the dark green stiff clay (DGSC) because it was over-consolidated and compacted by dewatering. However, it is absent in some zones due to Holocene erosion by the river avulsions in the flood plain. The center of the city has suffered human-induced subsidence resulting from groundwater pumping since the 1920s in the confined multi-aquifer system. Land subsidence was estimated to have a maximum value of 2.5 m in 1962 and a maximum yearly rate of 98 mm between 1956 and 1959 (Shi and Bao 1984). The variable thicknesses and properties of the compressible layers and their organic content greatly influence the observed subsidence rates (Dassargues *et al.* 1991). Since 1962, the recharge of the main exploited aquifers has led to a partial elastic rebound followed by a residual increasing of approximately 2–3 mm/year in the 1980s and 1990s. An average yearly rate of 16 mm/year has been reported (Zhang *et al.* 2007) for the 1990–present period but this is probably neglecting large local variations as groundwater pumping is decreased in the center and increased in the peripheral urban districts. During the same period, pumping yields from the shallow confined aquifers were also strongly restricted and those from deeper aquifers increased.

Between 1986 and 1989, an international Sino-Belgian research project was dedicated to the modeling of the 3D groundwater flow and the coupled 1D elastoplastic clay deformation. The main conceptual choices of this study are schematized in Figure 6.8. As a detailed example of the results, the calculated water pressure as a function of depth for the period from 1920 to 1965 is shown in Figure 6.9 for two 1D columns located in the central area of Shanghai. The calculated land subsidence from 1920 to 1991 is also shown (Figure 6.10) for the same 1D columns, integrating the effect of the recharge since 1962.

Indeed, an increasing number of detailed studies are being performed and recently, numerical simulations involving a 3D groundwater flow model and the 1D

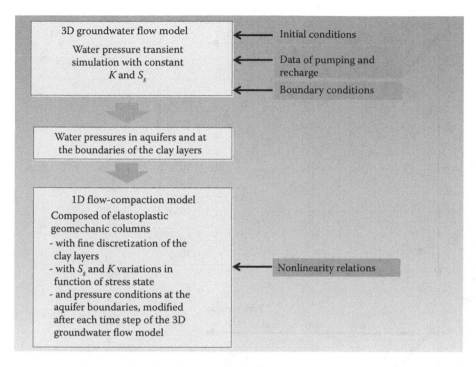

Figure 6.8 Conceptual schema of the coupled 3D groundwater flow model and the 1D flow-compaction model (Dassargues *et al.* 1993).

visco-elasto-plastic behavior of the compressible layer have been proposed by Wu *et al.* (2010). However, the area of land subsidence and the area where groundwater pumping occurs in the suburbs and newly developed districts of Shanghai continue to expand. The cumulative value of land subsidence continues to increase because of the strong transient water pressure propagation in clay layers and the corresponding delay in deformation.

6.5 New developments in measurements and remediation

Recently, interferometric synthetic aperture radar (SAR) methods have been used to measure the relative displacement of many radar reflector target points. After the necessary processing and corrections, SAR allows the detection of ground movement with a high measurement resolution (at the scale of a few millimeters). The greatest interest in SAR relates to its large-scale investigation possibilities. Ideally, it should be calibrated or conditioned with "reference measurements" from levelling or differential GPS surveys to provide "absolute values." In many areas of the world, SAR results have provided useful and sometimes surprising results on land subsidence resulting from groundwater pumping that were not detected previously by other surveying techniques (Gambolati and Teatini 2015).

SAR techniques are not very appropriate in farmland and forested areas with densely vegetated zones with very few reliable reflector points. In urban areas, another

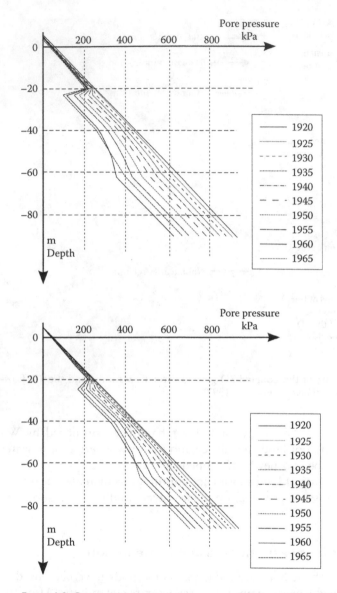

Figure 6.9 Computed water pressure as a function of depth for two 1D columns located in the central zone of Shanghai. Each 5-year curve shows the water pressure decrease in the layers and the computed delay and limited value in the very low permeability layers. The 1965 curve shows the effect of the increasing water pressure due to winter recharges greater than the summer pumping.

problem arises because of the different settlements of most of the reflector points on different building structures with various foundation depths (Figure 6.7).

In terms of remediation, there is no miracle solution. One can just rely on a partial rebound (uplift) corresponding to the elastic part of the clay behavior when water pressures are increased or restored in confined aquifers. In addition, the water used

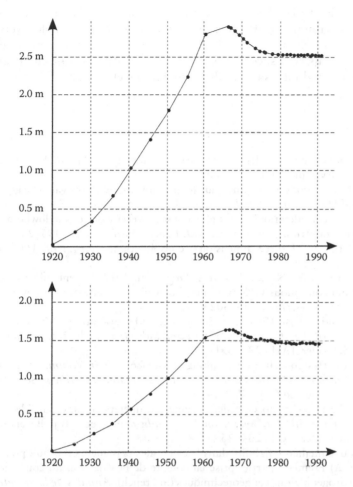

Figure 6.10 Calculated land subsidence from 1920 to 1991 in two columns (identical to those of Figure 6.9) located in the central zone of Shanghai. A delayed elastic rebound was computed due to recharge in the main aquifers. However, an additional and delayed subsidence of approximately 2–3 mm/year reappears (Dassargues *et al.* 1993).

to restore pressure in the aquifers should not induce any negative physicochemical effects on the compressibility of the sediments and their resistance to deformation. Oil and gas engineers know very well that seawater injection in depleted reservoirs can induce additional compaction (which has a positive effect in terms of oil/gas production) (De Gennaro *et al.* 2003). If an injection of seawater is foreseen in the deep confined saline aquifers of Venice, this would create uplift, according to the first simulations of the injection (Gambolati and Teatini 2015). Before starting such a delicate operation, the detailed physicochemical interactions between the sediments and the new injected (sea)water, which are not in chemical equilibrium, should be studied to determine possible adverse effect.

The methods used to mitigate land subsidence inevitably include groundwater pumping reductions most often associated with artificial recharge schemas to progressively restore water pressures in the depleted aquifers. Ultimately, the goal of many mitigation strategies is clearly to maintain an effective stress state in the compressible layers that is lower than the preconsolidation effective stress (the stress level experienced to date by the formation). This goal successfully guarantees that the geomechanical behavior of the compressible sediments is mainly limited to the elastic part.

References

Baeteman, C. 1989. The Upper Quaternary deposits of the Changjiang coastal plain, Shanghai area. Belgian Geological Survey (unpublished report).

Baeteman, C. 1994. Subsidence in coastal lowlands due to groundwater withdrawal: The geological approach. *Journal of Coastal Research* 12 (Special Issue): 61–75.

Baeteman, C. 2010. Geological considerations on the effect of sea-level rise on coastal lowlands, in particular in developing countries. *Bull. Séanc. Acad. R. Sci. Outre-Mer* 56: 195–207.

Biot, M.A. 1941. General theory of three-dimensional consolidation. *Journal of Applied Physics* 12: 155–164.

Bock, Y., Wdowinski, S., Ferretti, A., Novali, F. and A. Fumagalli. 2012. Recent subsidence of the Venice Lagoon from continuous GPS and interferometric synthetic aperture radar. *Geochemistry Geophysics Geosystems* 13: Q03023.

Carbognin, L., Marabini, F. and L. Tosi. 1995. Land subsidence and degradation of the Venice littoral zone, Italy. In *Proc. of the 5th Int. Symp. on Land Subsidence*, eds. B.J. Barends, F.J.J. Brouwer and F.H. Schröder, IAHS 234: 391–402.

Carillo, N. 1947. Influence of artesian wells in the sinking of Mexico City. *in Volumen Nabor Carrillo, Comision Impulsora y Coordinadora de la Investigacion Cientifica*, Sec. de Hacienda y Credito Publico, Mexico City, 47: 7–14.

Dassargues, A. 1995. On the necessity to consider varying parameters in land subsidence computations. In *Proc. of the 5th Int. Symp. on Land Subsidence*, eds. B.J. Barends, F.J.J. Brouwer and F.H. Schröder, IAHS 234: 259–268.

Dassargues, A. 1997. Vers une meilleure fiabilité dans le calcul des tassements dus aux pompages d'eau souterraine, A) Première partie: prise en compte de la variation au cours du temps des paramètres hydrogéologiques et géotechniques (in French). *Annales de la Société Géologique de Belgique*, 118(1995)(2): 95–115.

Dassargues, A. 1998. Prise en compte des variations de la perméabilité et du coefficient d'emmagasinement spécifique dans les simulations hydrogéologiques en milieux argileux saturés (in French). *Bull. Soc. Géol. France*, 169(5): 665–673.

Dassargues, A., Biver, P. and A. Monjoie. 1991. Geotechnical properties of the Quaternary sediments in Shanghai. *Engineering Geology* 31(1): 71–90.

Dassargues, A., Schroeder Ch. and X.L. Li. 1993. Applying the Lagamine model to compute land subsidence in Shanghai. *Bulletin of Engineering Geology (IAEG)* 47: 13–26.

De Gennaro, V., Delage, P., Cui, Y.-J., Schroeder, Ch. and F. Collin. 2003. Time-dependent behaviour of oil reservoir chalk: A multiphase approach. *Soils and Foundations* 43(4): 131–147.

Delage, P. and G. Lefebvre. 1984. Study of the structure of a sensitive Champlain clay and of its evolution during consolidation. *Canadian Geotechnical Journal* 21: 21–35.

Galloway, D.L. and T.J. Burbey. 2011. Review: Regional land subsidence accompanying groundwater extraction. *Hydrogeology Journal* 19: 1459–1486.

Galloway, D., Jones, D.R. and S.E. Ingebritsen (Eds.). 1999. *Land subsidence in the United States*. U.S. Geol. Surv. Circ. 1182, 177.

Gambolati, G. 1972. Estimate of subsidence in Venice using a one dimensional model of the subsoil. *IBM Journal of Research and Development* 6(2): 130–137.

Gambolati, G., Gatto, P. and R.A. Freeze. 1974. Mathematical simulation of the subsidence of Venice: 2. Results. *Water Resources Research* 10(3): 563–577.

Gambolati, G. and P. Teatini. 2015. Geomechanics and subsurface water withdrawal and injection. *Water Resources Research* 51: 3922–3955.

Gorelick, S.M. and C. Zheng. 2015. Global change and the groundwater management challenge. *Water Resources Research* 51: 3031–3051.

Gueguen, Y. and V. Palciauskas. 1994. *Introduction to the physics of rocks.* Princeton (NJ): Princeton University Press.

Hoeksema, R.J. 2007. Three stages in the history of land reclamation in The Netherlands. *Irrigation and Drainage* 56: S113–S126.

Lewis, R.W. and B.A. Schrefler. 1987. *The finite element method in the deformation and consolidation of porous media.* New York: Wiley & Sons.

Meijer H. 1996. *Water in, around and under the Netherlands.* IDG-Bulletin 1995/96. Utrecht: The Information and Documentation Centre for the Geography of the Netherlands.

Monte, J.L. and R.J. Kritzen. 1976. One -dimensional mathematical model for large-strain consolidation. *Geotechnique* 26(3): 495–510.

Nishida, Y. and S. Nakagawa. 1969. Water permeability and plastic index of soils. In *Proc. of the Int. Symp. on Land Subsidence* IAHS-UNESCO 89: 573–578.

Ortega-Guerrero, A., Rudolph, D.L. and J.A. Cherry. 1999. Analysis of long-term land subsidence near Mexico City: Field investigations and predictive modeling. *Water Resources Research* 35(11): 3327–3341.

Ortiz-Zamora, D. and A. Ortega-Guerrero. 2010. Evolution of long-term land subsidence near Mexico City: Review, field investigations, and predictive simulations. *Water Resources Research* 46: W01513.

Oude Essink, G.H.P., van Baaren, E.S. and P.G.B. de Louw. 2010. Effects of climate change on coastal groundwater systems: A modeling study in the Netherlands. *Water Resources Research*, 46: W00F04.

Phien-wej, N., Giao, P.H. and P. Nutalaya. 2006. Land subsidence in Bangkok, Thailand. *Engineering Geology* 82: 187–201.

Poland, J.F. 1984. *Guidebook to studies of land subsidence due to groundwater withdrawal, Studies and Reports in Hydrology.* Paris: IHP-UNESCO.

Poland, J.F. and G.H. Davis. 1969. Land subsidence due to withdrawal of fluids. *Reviews in Engineering Geology* 2: 187–270.

Rau, J.L. and P. Nutalaya. 1983. Geology of Bangkok clay. *Bulletin of the Geological Society of Malaysia* 16: 99–116.

Rieke, H.H. and G.V. Chilingarian. 1974. *Compaction of argillaceous sediments.* Amsterdam: Elsevier.

Takaya, Y. and N. Thiramongkol. 1982. *Chao Phraya delta of Thailand. Asian rice-land investigation. A description Atlas 1,* Center for Southeast Asian Studies, Kyoto University, Japan.

Terzaghi, K. 1943. *Theoretical soil mechanics.* London: Chapman and Hall.

Showstack, R. 2014. Scientists focus on land subsidence impacts on coastal and delta cities. *Eos* 95(19): 159.

Shi, L. and M. Bao. 1984. Case history n°9.2. Shanghai, China. In *Guidebook to studies of land subsidence due to groundwater withdrawal, Studies and Reports in Hydrology.* Paris: IHP-UNESCO, 155–160.

Wu, J., Shi, X., Ye, S., Xue, Y., Zhang, Y., Wei, Z. and Z. Fang. 2010. Numerical simulation of viscoelastoplastic land subsidence due to groundwater overdrafting in Shanghai, China. *Journal of Hydrologic Engineering.* 15(3): 223–236.

Zhang, Y., Xue, Y.Q., Wu, J.C., Ye, S.J. and Q.F. Li. 2007. Stress-strain measurements of deforming aquifer systems that underlie Shanghai, China. *Environmental & Engineering Geoscience* 13(3): 217–228.

Chapter 7

Introduction to groundwater quality and hydrochemistry

7.1 Introduction and units

Pure water is made of H_2O molecules. Natural waters are never pure, in fact, they are water-based or aqueous solutions. A study of the detailed water chemistry is beyond the scope of this chapter. The most important notions about groundwater hydrochemistry currently used by most hydrogeologists will be summarized to answer essential questions such as the following:

- How should the natural and human-influenced groundwater quality states be assessed?
- What parameters should be measured?
- How should a groundwater quality monitoring program be organized, and how should groundwater be sampled?
- How should chemical analysis results be interpreted?

The basic concepts of hydrochemistry are starting points for the study or evaluation of groundwater quality and contamination issues. Excellent books on this particular topic are available, including: Pankow (1991), Deutsch (1997), Appelo and Postma (2005), and Atteia (2005).

In a molecule of water, the spatial distribution of the electrical charges is not symmetric, inducing polar properties. This makes water an excellent *solvent* (i.e., with a high ability to dissolve chemically different liquids, solids, or gases) for many other ionic and polar substances. Recharge water flowing through the partially saturated zone of soils and shallow layers, and then, groundwater flowing in the saturated zone, chemically reacts with the gases, minerals, and organic compounds of the media along its entire path. The current composition of groundwater at a given place at a given time can be dependent on multiple factors, but the main factors are (1) the nature and chemical characteristics of the minerals and gases in the encountered rocks and (2) the transit time of groundwater as it determines if chemical equilibrium can be reached between the minerals and groundwater in the pores and fissures. Unfortunately, solutes have also been introduced in groundwater by human activities. Chemical processes in groundwater are complex and can be induced, boosted, or delayed by many factors, such as microorganism-based reactions. In some specific geological environments, natural groundwater (i.e., not affected by any human contamination) may not be suitable for drinking or, on the contrary, considered to have health benefits for some individuals. This high variability of groundwater chemical conditions results from the geological variability in rock compositions, the variability of rainfall composition, the variability of groundwater paths and residence times, and the variability of human activity-induced contaminants.

Phases and constituents

It is important to make the distinction between phases and constituents. In groundwater problems, a *phase* is a chemically defined part or region of the pore space that is physically separated from other phases by an interface (i.e., solid phase, aqueous phase, gas phase, and NAPL = nonaqueous liquid phase). In a REV of a porous and/or fractured medium, the quantification of each phase is expressed using volume ratios or mass to volume ratios (Figure 7.1 and Table 7.1). A *constituent* (here considered as equivalent to a compound, component, species, or substance) is any chemical substance that can be identified in a phase. Constituents of the solid phase are mostly minerals and possibly organic matter. Constituents of the aqueous phase or groundwater are solutes (or dissolved chemicals). The *concentration* of a constituent in a phase (in the broadest meaning) expressed as the constituent quantity per unit quantity of the considered phase. In fact, groundwater can be considered a *solution* (i.e., a homogeneous mixture composed of two or more constituents) where solutes are constituents dissolved in the pure water solvent. As the solvent is the major fraction of the solution, this latter is often considered to take on most of the characteristics of the solvent. For example, if the mass of the contained solutes is negligible, the density of the groundwater (solution) is considered equal to that of pure water (solvent).

Concentration units

The concentration of a solute in groundwater can be described in different ways (Table 7.2) depending on the use of these data. In chemical equilibrium calculations,

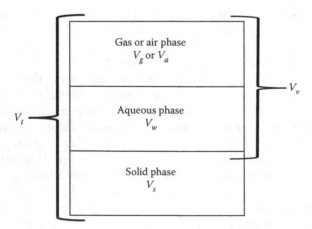

Figure 7.1 Concept model for the quantification of each phase using volume ratios or mass to volume ratios within a REV of a porous and/or fractured medium.

Table 7.1 Typical notations and definitions for quantifying the different phases in a REV

Definition	Units	Notation/Formula
Solid		
Solid mass	kg [M]	M_s
Effective solid volume	m³ [L³]	V_s
Volume fraction of solid	–	$\theta_s = V_s/V_t$
Total porosity	–	$n = V_v/V_t$
Density of the solid matrix	kg/m³ [ML³]	$\rho_s = M_s/V_s$
Bulk (apparent, dry) density	kg/m³ [ML³]	$\rho_b = M_s/V_t$
Water		
Water mass	kg [M]	M_w
Water volume	m³ [L³]	V_w
Water content (fraction)	–	$\theta = \theta_w = V_w/V_t$
Water saturation	–	$S_w = V_w/V_p$
Nonaqueous phase liquid = NAPL		
NAPL mass	kg [M]	M_{NAPL}
NAPL volume	m³ [L³]	V_{NAPL}
NAPL content (fraction)	–	$\theta_{NAPL} = V_{NAPL}/V_t$
NAPL saturation	–	$S_{NAPL} = V_{NAPL}/V_p$
Air (or gas)		
Air (gas) mass	kg [M]	M_a or M_g
Air (gas) volume	m³ [L³]	V_a or V_g
Air volume fraction	–	$\theta_a = V_a/V_t$
Air saturation	–	$S_a = V_a/V_p$

Subscripts *a, b, g, p, s, t, v,* and *w* are used for air, bulk, gas, pores, solid, total, void, and water, respectively.

the quantity of a dissolved constituent is expressed as the *molar concentration* (i.e., molarity), the number of moles of the constituent per liter of solution (mol/L):

$$C_{[i]} = n_i/V_w \tag{7.1}$$

where n_i is the number of moles of the *i*th constituent. This is logical as chemicals tend to react in direct proportion to the number of molecules in the solution. Sometimes, the number of moles is divided by the mass of the solution (i.e., solvent + constituents), which produces the *molal concentration* (i.e., molality) (mol/kg):

$$M_i = \frac{n_i}{M_w + \sum_i m_i} \cong \frac{n_i}{M_w} \tag{7.2}$$

where M_w is the mass of the groundwater/solvent. If the solution is highly diluted, one kg of solvent nearly corresponds to one liter of solution; in that case, the molality is considered equivalent to the molarity.

Table 7.2 Aqueous phase concentrations: names, definitions, units, remarks

Name	Definition	Units	Remark
Molarity	$C_{[i]} = \dfrac{n_i}{V_w}$ n_i = number of moles of the ith constituent	mol/L mmol/L $= 10^{-3}$ mol/L	Acceptable for equilibrium/ stoichiometric equations
Molality	$M_i = \dfrac{n_i}{M_w + \sum_i m_i} \sim \dfrac{n_i}{M_w}$	mol/kg mmol/kg $= 10^{-3}$ mol/kg	
Mass concentration	$C_i = \dfrac{m_i}{M_w + \sum_i m_i} \sim \dfrac{m_i}{M_w}$	– 1 ppm = 1 mg/kg 1 ppb = 1 μg/kg	
Volume concentration	$C_i^v = \dfrac{m_i}{V_w}$	mg/L μg/L	
Mole fraction	$x_i = \dfrac{n_i}{\sum_j n_j}$	–	$\sum_i x_i = 1$ in the mixture
Normality	$N_i = \dfrac{n_i(elec.charge)}{V_w}$ $N_i = C_{[i]} \ (elec.charge)$	eq/L meq/L	Acceptable for electrical balance calculations

Concentrations measured in water quality laboratories are usually expressed in the mass of a constituent per total mass of the solution (i.e., *mass concentration*):

$$C_i = \frac{m_i}{M_w + \sum_i m_i} \cong \frac{m_i}{M_w} \tag{7.3}$$

Mass concentration has no official unit (kg/kg), but in practice, it is usually expressed in mg/kg, equivalent to *ppm* (one part per million), or μg/kg, equivalent to *ppb* (one part per billion). If the density of the solution is considered to be unaffected by the solute content, the most common concentration unit is mass per unit volume (i.e., *volume concentration*), giving the mass of a constituent per total volume of the solution:

$$C_i^v = \frac{m_i}{V_w} \tag{7.4}$$

A volume concentration is commonly expressed in mg/L or μg/L [ML^{-3}]. However, mass concentrations are preferred to volume concentrations when the occurrence of saline groundwater is expected, inducing a *density effect* (i.e., a significant change of the groundwater solution density).

For electrochemical and, in particular, electrical balance or electroneutrality calculations, electrical charge units per liter (i.e., *normality*) are quantified for each ion in the solution. An equivalent is defined as mole of electrical charge (positive or negative). *Normality* is found by dividing the volume concentration by the molar mass (i.e., providing the molarity) and multiplying by the electrical charge of the concerned ion:

$$N_i = \frac{n_i \,(elec.charge)}{V_w} = C_{[i]}\,(elec.charge) \tag{7.5}$$

Therefore, equivalents per liter (eq/L) or, most often, milliequivalents per liter (meq/L) are used as the normality units.

In Table 7.2, the main concentration definitions are listed with mathematical definitions and units. For example, if the volume concentration of SO_4^{2-} is measured as 33.59 mg/L, the molar mass of S is 32.06 g/mol, and the molar mass of SO_4^{2-} is $32.06 + (4 \times 15.999) = 96.06$ g/mol. The molarity concentration of S is therefore $33.59 \times 10^{-3}/ 96.06 = 0.35$ mmol/L. The negative electrical charge is 2; thus, the normality concentration is 0.70 meq/L. A full example of unit conversions is provided in Table 7.3 using true chemical analysis results for a groundwater sample from a limestone aquifer.

Table 7.3 For a given groundwater chemical analysis expressed in volume concentrations, the following columns provide the molar concentration (mol/L) and normality concentrations (meq/L)

Constituent	Volume concentration (mg/L)	Molar mass (g/mol)	Molarity (mmol/L)	Normality (meq/L)	Comment/Remark
Ca^{2+}	103.29	40.078	2.58	5.15	
Mg^{2+}	26.01	24.305	1.07	2.14	
Na^+	50.36	22.99	2.19	2.19	
K^+	<0.5	39.098	<0.01	0.00	not detected
Fe^{3+}	0.07	55.845	1.25×10^{-3}	0.00	
Mn^{2+}	0.00	54.938	0.00	0.00	not detected
NH_4^+	<0.1	18.039	0.00	0.00	not detected
Cl^-	127.15	35.45	3.59	3.59	
SO_4^{2-}	33.59	96.06	0.35	0.70	
NO_2^-	<0.2	46.005	0.00	0.00	not detected
NO_3^-	28.06	62.004	0.45	0.45	
F^-	<0.2	18.998	<0.01	0.00	not detected
$H_2PO_4^-$	<0.2	96.985	0.00	0.00	not detected
CO_3^{2-}	3.91	60.008	0.065	0.13	
HCO_3^-	282.09	61.016	4.62	4.62	
CO_2 free	0.80	44.009	0.018	–	
SiO_2	4.89	60.065	0.081	–	
Tot. cations				9.48	expressing the
Tot. anions				9.49	electroneutrality

Physicochemical characteristics	Units	Value
Conductivity (25°C)	μS/cm	899
pH		8.40
pHs		7.19
Langelier Saturation Index (LSI)		1.21
Total Hardness (TH)	°f	36.5
Total Alkalinity (TAlk)	°f	23.8

The mass concentration of the ith constituent of the solid phase can also be defined when adsorption processes are involved (see Chapter 8):

$$C_{s_i} = \frac{m_i}{M_s + \sum_i m_i} \cong \frac{m_i}{M_s} = \frac{m_i}{\rho_b V_t} \tag{7.6}$$

where m_i is the mass of the ith constituent adsorbed on the solid matrix, M_s is the mass of the solid most often considered as the total mass of the solid and the adsorbed constituents, ρ_b is the bulk density, and V_t is the total volume of the REV.

A concentration can also be considered in the gas phase:

$$C_{g_i} = \frac{m_i}{V_g} \tag{7.7}$$

where m_i is the mass of the ith constituent in the gas phase and V_g is the gas volume in the REV of the porous medium. In this case, a partial pressure resulting from the molecules of the ith constituent occupying, alone, the total volume of gas in the REV can be defined:

$$p_i = x_i \, p_{gas\,tot} = \left(\frac{n_i}{n_{tot}}\right) p_{gas\,tot} \tag{7.8}$$

where x_i is the mole fraction (see Table 7.2) of the ith constituent in the gas phase (mixture) and $p_{gas\,tot}$ is the total gas pressure.

7.2 Natural solutes and main physicochemical characteristics of groundwater

Major and minor constituents

There is a large variety of solutes in groundwater. Inorganic solutes are classified as *major*, *minor*, and *trace* constituents. Conventionally, *major constituents* are considered to be those with a volume concentration higher than 5 mg/L in groundwater, *minor constituents* have concentrations between 0.1 and 5 mg/L, and the concentrations of all trace elements are lower than 0.1 mg/L. The classical list of the major constituents is given in Table 7.4. This list is often adapted to local specific geological conditions, inducing specific groundwater compositions resulting from strong contamination (e.g., nitrate). These constituents are systematically included in any groundwater quality survey and chemical analysis campaign. As primary solutes,

Table 7.4 Major and minor chemical constituents in groundwater

Major constituents ($C^v > 5$ mg/L)		Minor constituents ($0.1 < C^v < 5$ mg/L)	
Ca^{2+}	Calcium	$B^{2+/3+}$	Boron
Mg^{2+}	Magnesium	$Fe^{2+/3+}$	Iron
Na^+	Sodium	NH_4^+	Ammonium
HCO_3^-	Bicarbonate	K^+	Potassium
SO_4^{2-}	Sulfate	Sr^{2+}	Strontium
Cl^-	Chloride	Mn^{2+}	Manganese
CO_3^{2-}	Carbonate	NO_3^{-a}	Nitrate
Si or SiO_4^0	Silicon or silicon oxide	F^-	Fluoride

Note
a Natural concentrations of nitrate are always lower than 5 mg/L.

they directly influence most of the physicochemical characteristics of the considered groundwater. The minor constituents can be numerous but a classical short list of them is given in Table 7.4. They are frequently analyzed, particularly if their presence results from a specific geological context or expected anthropic influences. Trace elements can be various and numerous. Metallic trace elements such as As, Cd, Co, Cr, Cu, Hg, Ni, Pb, and Zn can reach high levels if the groundwater is contaminated which is also the case for elements such as Ba, Be, Li, Mn, Mo, Sb, Se, Te, Tl, and Ti.

Groundwater also contains organic molecules. In natural conditions, they are present in trace quantities and mostly composed of humic and fulvic acids from the decay of organic matters in near-surface layers. In a standard groundwater analysis, these specific organic molecules are not identified (Fitts 2002), but the total dissolved organic content is measured. Indeed, numerous man-made organic compounds have led to many severe contamination incidents, with the contaminants spreading through different processes such a multiphase fluid flow (see Chapters 8 and 9), solute transport, and vapor-phase flow. Depending on each case, many specific expected organic solute compounds are targeted in (usually) very expensive groundwater analyses.

TDS, electrical conductivity, and DOC

An indirect measurement of the total solute content of a groundwater sample is given by the *total dissolved solids* (*TDS*) content. After removing the suspended solids by filtration, complete evaporation of the water makes it possible to measure the mass of the dry residues. The *TDS* value in mg/L is defined as this mass of the residues divided by the initial volume of water. The only (small) source of error comes from the fact that some dissolved elements can also evaporate. Approximately 90% to 95% of the *TDS* is usually composed of the major elements.

Another proxy of the total dissolved ions is the water *electrical conductivity* (*EC*) or specific electrical conductance. Measured with a conductivity meter in a laboratory or in the field and expressed in μS/cm, it gives an approximation of the ionic (solute) content. This is a very quick and easy measure to conduct. As *EC* values are temperature dependent, measurements must be corrected. Usually, they are reported as if they were made at a reference temperature of 20°C or 25°C. The following linear relation can be used for diluted solution at 25°C:

$$EC_{25°} = \frac{EC_{T°}}{1 + 0.02(T° - 25)}$$

(7.9)

This relation is an empirical correction and for diluted and neutral solutions the linear temperature coefficient is approximately 2% per °C (less for acidic and more for basic waters). However, a conductivity measurement taken at the reference temperature will always be more accurate than a temperature-compensated value. In addition, there is a good correlation between measured and calculated EC values in the laboratory, but a poorer correlation when EC is measured in situ. The filtration and acidification of groundwater samples alter the measurement of EC (acceptable value differences are of the order of 10% to 20%).

When the groundwater composition is well known, a theoretical EC can also be calculated from the different molar ECs and concentrations.

An empirical relation is often cited between the TDS value (mg/L) and the EC value (in $\mu S/cm$) (Hem 1985):

$$TDS = f_c\, EC$$

(7.10)

where f_c is a conversion factor between 0.55 and 0.80. However, this factor relies on an assumed average mass of the conducting ions. A more accurate relation is the following experimental relation (at 25°C) (Appelo and Postma 2005):

$$\sum cations = \sum anions \text{ (meq / L)} \cong \frac{EC}{100} (\mu S / cm)$$

(7.11)

Typically, rainwater has a TDS value lower than 60 mg/L and an EC value between 2 and 100 $\mu S/cm$. Indeed, more variation is observed for groundwater with TDS values that are lower than 350 mg/L and EC values between 50 and 600 $\mu S/cm$. Seawater has a TDS of approximately 35,000 mg/L (i.e., 35 g of solutes per liter) typically corresponding to EC values of approximately 50,000 $\mu S/cm$. For brine or hypersaline water (i.e., water with a higher salinity than seawater), a TDS value up to 300,000 mg/l can be observed, with EC values on the order of 500,000 $\mu S/cm$. Therefore, EC can be used as a proxy for groundwater density (and thus salinity) estimations in coastal aquifers (Post 2012) in order to make measurement campaigns aiming to delineate the extension of seawater intrusions easier. Any electrical conductivity meter should be calibrated in solutions of known conductivities before its use in the field. The conductivity calibration range must be chosen with values similar to expected measured conductivities.

As mentioned above, the organic content of groundwater is usually identified globally in the form of the total *dissolved organic carbon* (DOC). DOC is a measure of the organic carbon available for possible oxidation in mg of carbon per liter (mgC/L). Analyses are performed by measuring the CO_2 generated by a strong oxidation process. In contamination problems, it can be extremely important to measure DOC, as it significantly influences the transport of metal constituents in groundwater because metals can form extremely strong complexes with dissolved organic matter, enhancing their solubility. A previous study has shown the complexity of the mobilization processes of the dissolved organic carbon in a pumped aquifer (Graham *et al.* 2015). In general, the concentration and flux of organic carbon in aquifers are largely influenced by recharge and abstraction, as well as (land use) surface and subsurface processes. Typical groundwater *DOC* concentrations are between 0.2 and 2 mgC/L and decrease with depth (Fitts 2002). Higher *DOC* levels can be expected near peat-rich wetlands and in aquifers in contact with fossil fuel deposits (Thurman 1985).

Electroneutrality or electrical charge balance

In a given sample of groundwater, the sum of the electrical charges of all the positive ions (cations) must balance the sum of all the negative ions (anions) to ensure an electrically neutral solution. Charges are expressed in meq/L. If only the major ions are considered to establish this balance, an error of a few percent can be accepted. Quantification of this charge balance error (*CBE*) provides a relatively simple way to determine if the chemical analysis results are globally reliable:

$$CBE = \frac{\left(\sum cations - \left|\sum anions\right|\right)}{\left(\sum cations + \left|\sum anions\right|\right)} 100 \tag{7.12}$$

For example, for the chemical values of the sample in Table 7.3, the *CBE* calculated from all the available results is lower than 0.06%. If only the major constituents (as listed in Table 7.4) were considered, a *CBE* of 2.3% would be found. The groundwater sample must be filtered prior to the analysis, as the charge balance must be calculated based on the dissolved constituents only. For very low mineralized waters, a large *CBE* can be found due to very small inaccuracies in the analysis taking a high relative importance.

pH and aqueous reactions

The hydrogen ion activity is usually given by the *pH*. The reactivity of hydrogen is very high, thus it is correct to replace the hydrogen concentration with the hydrogen activity. When a reaction is not yet in equilibrium, the equilibrium reaction is replaced by an inequality, and the concentrations of the constituents are replaced by activities. *pH* is defined as the negative logarithm of the hydrogen ion activity in an aqueous solution:

$$pH = -\log_{10} a_{[H^+]} \tag{7.13}$$

If the aqueous solution is highly diluted or considered to be in equilibrium $a_{[H^+]} \cong C_{[H^+]}$. It is far beyond the scope of this chapter to consider nonequilibrium reactions. For most of the following notions we will consider equilibrium conditions at a temperature of approximately 25°C and atmospheric pressure. As mentioned by Fetter (2001), small deviations (i.e., of a few atmospheres of pressure and temperatures between 10°C and 15°C) from these assumptions, will not lead to significant errors (Hem 1985). *pH* gives a measure of the availability of groundwater to supply protonic hydrogen ions to a base or to take up protons from an acid. The theoretical range is from 0 to 14 (with 7 being neutral, a *pH* less than 7 indicating acidity, and a *pH* higher than 7 indicating a basic solution). The *pH* values of most groundwater typically range between 6 and 9. For the reasons indicated above, pH is temperature dependent; thus, in situ measurements should be prioritized over laboratory measurements.

The concentrations or activities of H^+ are in fact the sum of several hydrated forms of H (i.e., H_3O^+ and $H_5O_2^+$) (Stumm and Morgan 1996). Any *pH*-meter should be calibrated before use, and the *pH* calibration range must be chosen in agreement with the values expected to be measured.

Carbonate system, pHs, Langelier saturation index, and hardness

As for other environmental compartments, the carbonate system (CO_2, HCO_3^-, CO_3^{2-}) especially in $CaCO_3$-rich lithologies, is by far the most important system in groundwater chemistry. For example, the effect of H_2CO_3 (dissolved CO_2) on the groundwater *pH* and the influence of solid $CaCO_3$ on the groundwater composition have many practical implications. The carbonate equilibria in groundwater in contact with CO_2 from the air are of considerable interest. The acid-base reactions between the carbonate constituents and the dissolution of CO_2 based on the form of H_2CO_3 in the groundwater can be listed as follow:

$$H_2O + CO_2 \Leftrightarrow H_2CO_3 \text{ (dissolution of } CO_2 \text{ in water)} \tag{7.14}$$

$$H_2CO_3 + H_2O \Leftrightarrow H^+ + HCO_3^- + H_2O \tag{7.15}$$

$$HCO_3^- + H_2O \Leftrightarrow H^+ + CO_3^{2-} + H_2O \tag{7.16}$$

The equilibrium constants of these reactions are known (and mentioned in many textbooks) for different temperatures. Therefore, with a pH measurement and the assumption of carbonate equilibrium, the relative fraction of the dissolved carbonate constituents can be calculated. The relative fractions (in log-scale) of the dissolved carbonate constituents as a function of the measured *pH* are given in Figure 7.2 for an open system (i.e., with the possible dissolution of CO_2 in groundwater) at 25°C. Open systems correspond to the hydrogeological conditions of the partially saturated zone and the upper zone of unconfined aquifers. If deeper or in confined conditions, the carbonate system can be considered closed without direct CO_2 dissolution possibilities. This means that globally, there is less dissolved CO_2 (or H_2CO_3 acid); consequently, the groundwater is less acidic. The *pH* is thus higher in confined carbonate aquifers than in unconfined conditions. For the chemical results of the sample in Table 7.3, the relative fractions of HCO_3^- and CO_3^{2-} are = 98.6% and 1.4%, respectively, and consistent with a *pH* of 8.4. By far, the dominant ion is bicarbonate (HCO_3^-), corresponding to *pH* values between 6.5 and 10.

There are a number of calcite-saturation parameters. All of them are temperature dependent; thus, it is important to measure them in situ.

The calcite-saturation *pH*, or *saturation pH* (*pHs*) is defined as the *pH* measured under conditions where groundwater should be in contact with pure $CaCO_3$ for an

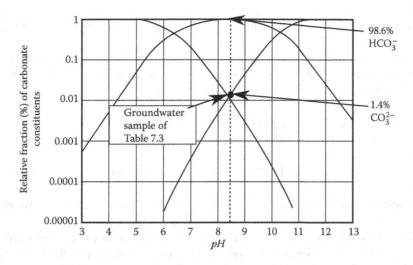

Figure 7.2 Relative fractions (in log-scale) of the dissolved carbonate constituents as a function of the measured *pH*.

infinite time (i.e., perfect equilibrium). The *Langelier saturation index* (*LSI*) is then defined very simply as:

$$LSI = pH - pHs \qquad (7.17)$$

From this definition, it can be easily deduced that:

- if $LSI < 0$, the groundwater is considered as "aggressive" because it is undersaturated in calcite and thus has a carbonate dissolution capacity;
- if $LSI > 0$, the groundwater is considered as "encrusting" because it is oversaturated in calcite and thus precipitates carbonates under the current conditions;
- if $LSI = 0$, the groundwater is considered as "inert" because it is in equilibrium (in terms of calcite saturation) under the current conditions.

A more general indicator of the groundwater mineralization is defined as the *total hardness* (*TH*): for this indicator, the sum of the molar concentrations of Ca^{2+} and Mg^{2+} (the two most prevalent divalent metal ions), $Fe^{2+/3+}$, $Al^{2+/3+}$, and $Mn^{2+/3+}$ can also be considered. In practice, water hardness could be expressed as a molar concentration, but it is not. Various local and odd units are used, such as German degrees (°dH), English or Clark degrees (°e or °Clark), French degrees (°f), and American degrees, that are equivalent to a mass concentration or ppm. The concept behind this indicator is to represent the equivalent mass of calcium carbonate ($CaCO_3$) (American, English, and French degrees) or calcium oxide (CaO) (German degrees) that, when dissolved in a unit volume of pure water, would result in the same total molar concentration of the above described ions. In detail,

- 1 ppm is usually considered 1 mg/L $CaCO_3$ for the American degree;
- 64.8 mg of $CaCO_3$ per Imperial gallon (4.55 liters), equivalent to 14.254 ppm, is the Clark degree (°Clark) or English degree (°e);

- 10 mg/L $CaCO_3$, equivalent to 10 ppm, is the French degree (°f);
- 10 mg/L CaO, equivalent to 17.848 ppm of $CaCO_3$, is the German degree (°dH).

Hardness unit conversion tables can be found in many books and on the internet. Groundwater can be considered as hard or soft water according to the following conventional classifications:

- Soft water has a value of 0–60 ppm (°f < 6, °e < 4.21 and °dH < 3.36)
- Moderately hard water has a value of 61–120 ppm (6 < °f < 1, 4.21 < °e < 8.42 and 3.36 < °dH < 6.72)
- Hard water has a value of 121–180 ppm (12 < °f < 18, 8.42 < °e < 12.63 and 6.72 < °dH < 10.09)
- Very hard water has a value above 181 ppm (°f > 18, °e > 12.63 and °dH > 10.09)

For example, for the chemical data in Table 7.3, the summed molar concentration of Ca^{2+}, Mg^{2+} and Fe^{3+} is 3.65 mmol/L, the molar mass of $CaCO_3$ is 100.0869 mg/mmol, and 3.65 mmol/L corresponds to 365.317 mg/L of $CaCO_3$ or a $TH = 36.53$°f; thus, it is very hard water. If more detail about the hardness is needed, a *permanent hardness* is defined considering only the hardness that persist after the water is boiled and calcium and magnesium precipitates form (i.e., sulfate and chloride remain). Indeed, the *temporary hardness* is defined with only calcium and magnesium bicarbonates. The temporary hardness is equal to the difference between the total hardness and permanent hardness.

Alkalinity

The *total alkalinity* (*TAlk*) is a parameter used to measure the buffering capacity or how acids can be neutralized in the water (i.e., the sum of all the dissolved constituents that can "consume" H^+ ions). Expressed in meq/L, it includes mainly carbonates and bicarbonates but also hydroxides and other constituents:

$$TAlk = m(HCO_3^-) + 2m(CO_3^{2-}) + \cdots + m(OH^-)$$
$$+ m(H_3SiO_4^-) + m(org.ions) - m(H^+) \tag{7.18}$$

In the laboratory, units are most often expressed in mg/L as[1] $CaCO_3$, similar to the different hardness units: $TAlk$ (meq/L) = $TAlk$ (mg/L) × 2 (meq/mmol of $CaCO_3$) × 1/100.0869 (mmol of $CaCO_3$/mg) so that $TAlk$ (mg/L) = 50.0434 × $TAlk$ (meq/L). For the chemical data in Table 7.3, $TAlk = 4.75$ meq/L is thus converted to 237.7 mg/L (or 23.8°f, as mentioned in Table 7.3).

Different alkalinity parameters can be used, such as *simple alkalinity* (*SAlk*), which is defined excluding the bicarbonates (i.e., including only strong bases):

$$SAlk = 2m(CO_3^{2-}) + \cdots + m(OH^-) + m(H_3SiO_4^-) + m(org.ions) - m(H^+) \tag{7.19}$$

---

[1] 'as' is used here because the alkalinity results actually from a mixture of ions.

Therefore:

$$TAlk - SAlk = m(HCO_3^-) \qquad (7.20)$$

Sometimes, a *carbonate alkalinity* (*CAlk*) is defined by taking into account only the carbonate and bicarbonate species (in many cases, the *CAlk* is nearly equal to the *TAlk*):

$$CAlk = m(HCO_3^-) + 2m(CO_3^{2-}) \qquad (7.21)$$

Rainfall has low or negative alkalinity. Subsequently, the alkalinity increases progressively along its underground path because of the dissolution of minerals from the porous/fractured medium.

Redox potential (Eh)

Another important parameter is the *redox potential* (*Eh*). Oxidation occurs when a chemical element loses electron(s), unlike reduction states when a chemical element gains electron(s). The transfer of electron(s) produces an electric current. In groundwater, each constituent has a *redox state* corresponding to an electrical potential. The most classical example is given by the dissolved ferrous and ferric ions Fe^{2+} and Fe^{3+}. Fe^{2+} and Fe^{3+} have the redox states +II and +III, respectively, as they can be reduced by gaining 2 and 3 electrons, respectively. The use of a measured redox potential can allow the calculation, from a total dissolved iron concentration, of the proportions of ferrous and ferric forms. This can be useful because many of the natural and anthropic dissolved ions in groundwater can occur in equilibrium at more than one redox state (e.g., Fe, Mn, N, S, O, Cr, Al, Co, Ni, As, and Se) (Deutsch 1997). However, the important assumption of redox equilibrium is not often met in natural groundwater conditions. It is therefore necessary, in each case, to choose the redox couples that are assumed to be reactive in the groundwater at the considered timescale (Deutsch 1997) and then, by obtaining the related *Eh* with the Nernst equation (described in any inorganic chemistry textbook), use the classical redox equilibrium equation. Without going into too much detail, it appears that due to redox reactions in several stages of nonequilibrium, it is difficult to establish clear relationships between the redox potential and redox capacity (i.e., oxidizing and reducing capacities in the aquifer). In most aquifer conditions, the oxidizing capacity results from the presence of "electrons acceptors": dissolved oxygen (O_2), sulfate (SO_4^{2-}), ferric iron (Fe^{3+}), manganese (Mn^{4+}), nitrate (NO_3^-), nitrites (NO_2^-), and even carbon dioxide (CO_2). These constituents then tend to be reduced. The reducing capacity in an aquifer results from the presence of 'electron donors': sulfur (S^{2-}), ferrous iron (Fe^{2+}), ammonium (NH_4^+), and hydrocarbons (CH_2O, and others). These hydrocarbons are often added by human induced contamination.

In principle, if the *pH* and *Eh* of groundwater are measured accurately, the stability of any mineral of the matrix in contact with the groundwater and any dissolved constituent can be determined. Producing *Eh-pH* diagrams allows the delineation of the stability domains for the different solid and dissolved forms of a species at its different oxidation levels. In practice, however, it is not easy to measure reliable in situ values of *Eh* in field conditions. A systematic bias logically occurs due to the oxidizing conditions as groundwater is in contact with the atmosphere in groundwater springs or wells. For electrical conductivity, Eh is usually reported as corrected values at 20°C or 25°C. Eh is temperature dependent, so measurements should be performed in situ. Oxidizing conditions are usually found in the shallowest parts of the unconfined aquifers or near the surface, and reducing conditions are often observed in deep confined aquifers. Highly oxidizing rainwater and infiltrating water are progressively reduced during their contact with the reducing constituents of the ground and underground (i.e., Eh decreases).

Biogeochemical redox reactions are very important for the natural or enhanced degradation of many organic contaminants in groundwater. In fact, biologically mediated redox reactions are often more rapid than abiotic reactions. In terms of the natural attenuation of organic contaminants, aerobic and anaerobic biodegradation and abiotic oxidation processes are invoked. Classical examples are BTEX biodegradation via aerobic respiration and the reductive dechlorination (halorespiration) of chlorinated solvents (Wiedemeier *et al.* 2007).

Dissociation, dissolution, and precipitation, rock weathering

Regarding the carbonate system and redox reactions, one of the simplest and most ubiquitous reaction is the dissociation of inorganic salts in groundwater. Classical examples are:

$$NaCl \Leftrightarrow Na^+ + Cl^- \tag{7.22}$$

$$MgSO_4 \Leftrightarrow Mg^{2+} + SO_4^{2-} \tag{7.23}$$

Groundwater is able to dissolve some minerals from the solid matrix of the medium. On the other hand, when highly mineralized groundwater undergoes changes (i.e., decreases) in pressure and temperature, minerals may precipitate. The expression of the dissolution/precipitation equilibrium is referred to as the *solubility product*. An example can be taken from the anhydrite mineral ($CaSO_4$). The equilibrium dissolution/precipitation reaction can be written as:

$$CaSO_4 \Leftrightarrow Ca^{2+} + SO_4^{2-} \tag{7.24}$$

with the solubility product equal to the equilibrium constant:

$$K_{Anhydrite\,solubility} = C_{[Ca^{2+}]}C_{[SO_4^{2-}]} \tag{7.25}$$

where the concentration $C_{[CaSO_4]} = 1$ because $CaSO_4$ is in the solid phase.

Values of the solubility products are known for different minerals and for different pressure and temperature conditions (Nordstrom *et al.* 1979, Morel 1983, Stumm and Morgan 1996, Deutsch 1997).

More generally, if the equilibrium is not reached, the *ion activity product* (*IAP*) can be expressed as:

$$IAP_{Anhydrite} = a_{[Ca^{2+}]}a_{[SO_4^{2-}]} \tag{7.26}$$

If $IAP < K_{Mineral\ solubility}$, then the groundwater is undersaturated for the considered mineral.

The *saturation index* (SI) is defined as:

$$SI = \log\left(\frac{IAP}{K_{Mineral\ solubility}}\right)\tag{7.27}$$

with $SI < 0$ indicating undersaturated groundwater, $SI = 0$ indicating groundwater in chemical equilibrium with the considered mineral in the matrix, and $SI > 0$ indicating oversaturated groundwater, which induces the precipitation of the considered mineral.

However, using the SI is not easy in practice because groundwater is a solution composed of numerous constituents that tend to be in equilibrium with a large variety of minerals (Deutsch 1997). The calculation of the saturation indices of the different minerals must account for (1) the respective and mutual influences of all the constituents in the solution and (2) the fact that the reactive mineral may already be a weathering product coating or cementing the solid grains of the rock matrix.

$T°$, pH, and Eh are useful indicators for the main chemical characteristics of the groundwater composition. Complexation processes linked to organic and inorganic species, the presence of different minerals that are the source of common ions, and ion shielding (i.e., ion binding by reciprocal electrostatic influences) exert influences on the actual constituent activities. It is far beyond the scope of this chapter to describe how to cope with these effects during the quantification of effective solubility. In practice, the PHREEQc code (Parkhurst 1995, Parkhurst and Appelo 1999), among others, allows speciation and saturation-index calculations, reversible and irreversible reactions, surface-complexation, and ion-exchange equilibria, kinetically controlled reactions, the mixing of solutions, and pressure and temperature changes to be addressed.

The most common minerals in groundwater come from the following reactions (Deutsch 1997):

$$CaCO_3 \Leftrightarrow Ca^{2+} + CO_3^{2-} \qquad IAP_{Calcite} = a_{[Ca^{2+}]}a_{[CO_3^{2-}]}$$

$$CaMg(CO_3)_2 \Leftrightarrow Ca^{2+} + Mg^{2+} + 2CO_3^{2-} \qquad IAP_{Dolomite} = a_{[Ca^{2+}]}a_{[Mg^{2+}]}\left(a_{[CO_3^{2-}]}\right)^2$$

$$CaSO_4 - 2H_2O \Leftrightarrow Ca^{2+} + SO_4^{2-} + 2H_2O \qquad IAP_{Gypsum} = a_{[Ca^{2+}]}a_{[SO_4^{2-}]}$$

$$Fe(OH)_3 + 3H^+ \Leftrightarrow Fe^{3+} + 3H_2O^- \qquad IAP_{Ferrihydrite} = \frac{a_{[Fe^{3+}]}}{\left(a_{[H^+]}\right)^3}$$

$$FeS_2 \Leftrightarrow Fe^{2+} + 2S^- \qquad IAP_{Pyrite} = a_{[Fe^{2+}]}\left(a_{[S^-]}\right)^2$$

Silicate minerals (i.e., most often including one or all of the following elements: Si, Al, and Fe) are only minor constituents in groundwater and result from the weathering of secondary minerals with a generally low solubility. For example, the weathering of a K-feldspar, boosted by the carbonic acid content of groundwater, produces kaolinite, bicarbonate, and silicic acid. The weathering of biotite produces kaolinite and ferrihydrite (Deutsch 1997). The quantity of these secondary minerals and their respective solubilities limit the solution concentration of their constituents: Fe^{3+} for ferrihydrite ($Fe(OH)_3$), excess silica for H_4SiO_4 (if not limited by the presence of another silicate mineral) and Al^{3+} for kaolinite ($Al_2Si_2O_5(OH)_4$).

Cation exchange capacity, sodium adsorption ratio

Ions can be sorbed on the solid grains of the rock matrix by two processes: *adsorption* referring to adherence of a constituent to the surface, and *absorption*, which suggests that the constituent is fully part of the solid (i.e., part of the chemical reactions with the solid constituents). In certain circumstances (e.g., electronic attractions and chemical energy), some ions are more sorbed than others. For example, cations are more sorbed than anions, and bivalent cations are generally more sorbed than monovalent cations. As summarized by Fitts (2002), the following selectivity sequences are generally observed (from the strong sorption to the weak sorption of cations): $Ca^{2+} > Mg^{2+}$ and $K^+ > Na^+ > Li^+$.

These sorption processes are highly dependent on the "selectivity" of the solid surfaces and, consequently, on the considered mineral. They are also highly dependent on the concentration levels reached by the groundwater composition.

The *cation exchange capacity* (CEC) is defined as the meq of cations that can be sorbed per unit of dry mass of the solid matrix. In practice, however, the CEC is not helpful because CECs are strictly reliable under the measured conditions. For any change of the groundwater composition, resulting from natural or human causes, the system tends to reach a new equilibrium but this can take hundreds of years. For example, the equilibrium of a Ca^{2+}- and HCO_3^--dominated groundwater composition (i.e., from calcite dissolution) is completely changed by a seawater intrusion or by Na^+-rich irrigation infiltration water, and the cation-exchange complex first becomes saturated with sodium (Fetter 2001). Next, the high concentrations of Na^+ (and K^+ and Mg^{2+} in seawater) are gradually attenuated by cation exchanges resulting from new infiltration of fresh rainwater. Since the quantity of Cl^- remains, the groundwater composition evolves from Na-Cl, Na-HCO_3, and Mg-HCO_3 types toward Ca-Cl_2 and Ca-HCO_3 types.

A specific way to evaluate the sodium presence in the groundwater composition is the *sodium adsorption ratio* (SAR) proposed by Richards (1954):

$$SAR = \frac{C_{[Na^+]}}{\sqrt{(C_{[Ca^{2+}]} + C_{[Mg^{2+}]})/2}}$$

(7.28)

High values of SAR indicate high concentrations of Na^+ and thus a high groundwater salinization danger.

7.3　Graphs, diagrams, and multivariate analysis of chemical groundwater compositions

Groundwater quality data sets include numerous variables measured and/or analyzed at many different points from different samples and at different times. Data sets are spatially distributed (i.e., not only horizontally but also as a function of depth) and time dependent. To address this difficulty, a series of conventional and specific plots or diagrams are proposed in order to allow a quick and easy interpretation of the data. Some of them show the whole mineralization content, while others are intended to show only the main hydrochemical facies of the groundwater.

Conventional bar and pie charts

Bar diagrams are used with a double-bar plot for each sample; conventionally, the left bar shows the cation concentrations (meq/L) and the right bar shows the anions. For each bar, the specific input of a given ion is visualized with a unique frame or color. The cumulative bars on each side directly show if an acceptable ionic balance has been reached with the results (Figure 7.3). As these plots can be drawn for each sample, they could be reproduced "in small" and located directly next to their sampling points in 2D detailed maps. This technique allows groundwater composition changes to be illustrated or highlighted across an area.

The principle is the same for pie diagrams that clearly show how cations should balance anions by representing the cations and anions in the two opposite parts of the pie. The ion proportions are translated in pie parts and the diameter of the circle is scaled proportionally to the total meq/L (Figure 7.4), allowing groundwater composition comparisons when the pie charts are shown in a 2D map.

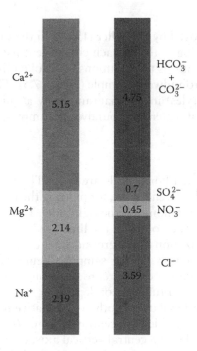

Figure 7.3 Bar chart for the groundwater composition sample shown in Table 7.3.

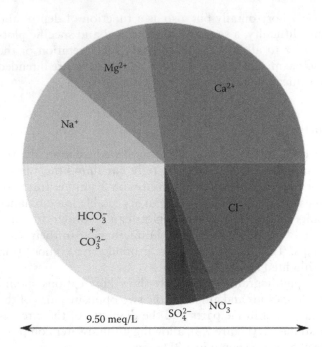

Figure 7.4 Pie chart for the groundwater composition sample shown in Table 7.3.

Semi-log Schoeller diagram

A fairly conventional semi-log diagram was proposed by Schoeller (1955). In the diagram, concentrations (meq/L) are represented in log scale for each constituent listed along the horizontal axis (from left to right, the cations to the anions). A broken line joins the successive concentrations for the same groundwater sample (Figure 7.5). The different broken lines (with different colors or styles) in the diagram allow a quick visual comparison of the chemical compositions of different groundwater samples.

Stiff diagrams

In Stiff diagrams (Stiff 1951), three (or four) parallel horizontals are used. The horizontals extend from both sides of a central vertical reference axis, for plotting the three (of four) main cations (on the left side) and the main anions (on the right side). The distances of ions from the central axis are scaled and the points on the lines are joined if they belong to the same sample; thus, direct visualization of the groundwater chemical facies is achieved by the obtained polygon (i.e., one polygon per sample) (Figure 7.6). These diagrams can be shown on a 2D horizontal map or can be reproduced on a vertical cross section (if different samples are analyzed as a function of depth).

In the latter case, when chemical results are available over depth, a cumulative diagram of the vertical profile can be used to show the added concentrations (meq/L) of the main cations and anions on the left and right sides of a central vertical axis, respectively. This can be very useful for visualizing vertical variations in ion concentrations in the partially and totally saturated zones in relation, for example, with the infiltration

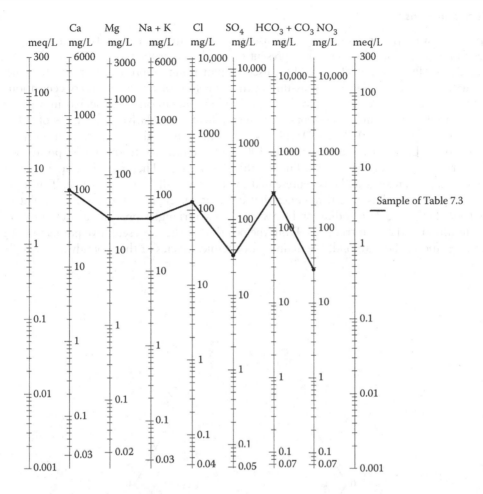

Figure 7.5 Schoeller semi-log diagram for the groundwater composition sample shown in Table 7.3.

water quality and changes in the depth of redox conditions. Among other processes, the NO_3 decreases with depth associated with the redox conditions or the NaCl increases with depth resulting from seawater intrusions can be clearly shown by these diagrams. However, they require a data set of chemical results corresponding to multilevel sampling procedures that are not easy to install and maintain (see Section 7.5).

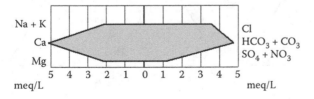

Figure 7.6 Stiff diagram showing the groundwater composition facies for the sample chemical results of Table 7.3.

Piper diagrams

Piper diagrams (Piper 1944) are very useful for visualizing results from many samples in a single plot. However, they present only relative proportions of the main ions. Conventionally, a lower left triangle is used to plot points representing the percentage of Ca^{2+}, Mg^{2+}, and $(Na^+ + K^+)$, assuming that the sum of these cation concentrations (meq/L) is 100%. The same procedure can be done for anions represented in a lower right triangle. The sample representative points indicate the relative percentages of SO_4^{2-}, $(Cl^- + NO_3^-)$, and $(HCO_3^- + CO_3^{2-})$ with respect to the sum of the concentrations of these anions (meq/L). Each corner represents 100% of a constituent, and the opposite side of the triangle represents the 0% line for this constituent. Therefore, in practice, each percentage, taken separately, is represented by a line parallel to its 0% line. The three percentage values thus give three lines that cross at the representative sample point in the triangle (Figure 7.7). A diamond between the two triangles provides a common field for the anion and cation triangles. From the position of the representative points in both triangles and in the diamond, the main hydrochemical facies of the groundwater sample

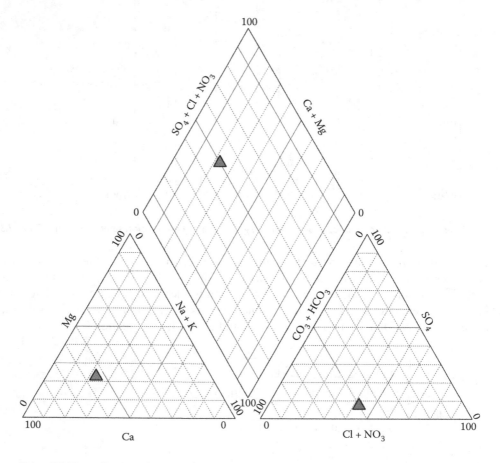

Figure 7.7 Piper diagram showing the groundwater composition facies for the sample chemical results of Table 7.3.

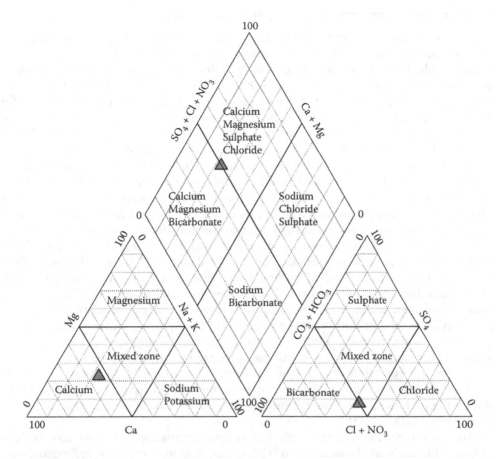

Figure 7.8 Piper diagram showing the main groundwater chemical types. The sample of Table 7.3 shows facies nearly balanced between the $Ca^{2+} - HCO_3^-$ type and the $Ca^{2+} - Cl^-$ type.

can be directly determined according to the classification shown in Figure 7.8. Among the different advantages of Piper diagrams, the most important are listed as follows:

- Results from many groundwater samples can be shown on the same diagram.
- If different samples are from the same monitoring point but were sampled at different times, it shows the possible time evolution of the groundwater chemical composition.
- If the sample results are from distributed points located in different parts of the same aquifer, the possible groundwater chemical trends or changes in the facies are shown, indicating possible influences of lithological changes, redox conditions, and/or confined/unconfined conditions.

Additionally, if different types of symbols are used for the representative points of the sample with different *TDS* values, the respective magnitudes of the considered concentrations are indicated. A background color scheme for the Piper plot can be proposed (Peeters 2013) so that spatial representations of these data can be later colored in the 2D map according to their location in the Piper diagram.

Principal component analysis and self-organizing maps

The regional and long-term monitoring of groundwater quality yields large and multivariate data sets. Multivariate statistical techniques are applied to summarize the available data, extracting the main useful information, and clustering groundwater types (Güler *et al.* 2002, Lambrakis *et al.* 2004, Love *et al.* 2004). This can be critical when preparing reports for decision-makers and formulating hypotheses for further investigations. As described in many reference books on statistics, *principal component analysis (PCA)* is a statistical procedure to reduce the dimensionality of the high dimensional original data sets. Linear combinations of the original variables are used to convert possibly numerous correlated variables into a limited set of uncorrelated variables (Davis 1986). Depending on the field of application, this technique has different names (e.g., eigenvalue decomposition of the covariance matrix in linear algebra). Applied to groundwater quality as diagnostic tool, *PCA* can supply the user with a clear picture (if not too many principal components are chosen).

Recently, artificial neural network techniques such as *self-organizing maps (SOMs)* have been used to explore groundwater quality data (e.g., Hong and Rosen 2001, Sanchez-Martos *et al.* 2002). Developed by Kohonen (1995), self-organizing map algorithms are neural networks designed to project multidimensional data onto a two-dimensional display in a topology-preserving way. This method allows complex and nonlinear relationships between variables to be captured, and it performs unsupervised classification and clustering. Based on the assumption that groundwater samples from locations that are close together are more likely to be related to each other, the incorporation of geographical coordinates in the SOMs analysis (Peeters *et al.* 2007) has provided an elegant way to account for the spatial correlation in the data set exploration. Including the three geographical dimensions, this technique has been named GEO3DSOM.

Other combined multivariate statistical analysis approaches, such as factor analysis (FA) and hierarchical cluster analysis (HCA), may help to interpret of hydrogeochemical processes and assess spatial patterns of groundwater types (Moeck *et al.* 2016).

7.4 Groundwater quality standards

General background and context

Water and, in particular, groundwater quality standards are generally defined as functions of their foreseen water uses. The aim is to protect these groundwater uses from possible contamination. The quality criteria are enacted by national agencies (e.g., EPA in the United States) from their own experiences, from world organizations, or from supranational directives (e.g., the EU Water Directives).

Different types of chemical elements could be considered during a groundwater quality assessment. Most often this assessment is performed as a function of the expected future uses of the water and thus quantifies how different constituents can possibly affect and limit these uses (Rentier *et al.* 2006). For example, the European Water Directive 2000/60/CE sets three general objectives concerning groundwater quality: (1) to prevent its deterioration, (2) to enhance and restore, as soon as possible, a "good" water status (i.e., defined as a function of the locally allowed uses of the produced groundwater), and (3) to reverse any significant and sustained contamination trend resulting from any pollutant. In the United States, the EPA and states define water quality standards (EPA 2016) with very similar aims.

Drinking water quality standards

The specific criteria for drinking water use are deduced from toxicological surveys on all possibly harmful constituents and under the basic assumption of a normal consumption of drinking water by every human. The World Health Organization (WHO) has published its criteria for "safe drinking water." However, there are no universally accepted international standards. Moreover, in many countries where standards exist, some are expressed only as guidelines rather than legal requirements. Two notable exceptions, among others, are the European Drinking Water Directive in the European Union and the Safe Drinking Water Act in the United States. The maximum concentrations for certain parameters are set, and groundwater monitoring must achieve these standards. The long lists of microbiologic parameters, organic compounds, and inorganic dissolved constituents that could potentially endanger human health are easily found in reference books, official national reports, and official online sites. These lists evolve continuously over time as new molecules and contaminants are produced by human and industrial activities. Among the organic pollutants, emerging organic contaminants (EOCs) are of particular concern (Jurado et al. 2012). They include pharmaceuticals, drugs of abuse, surfactants, and personal care products that are often detected in waste water. Therefore, they can, potentially, also be detected in groundwater. Many EOCs are endocrine disruptor compounds coming from a wide range of manufactured products and thought to have adverse developmental and reproductive effects in both humans and wildlife (Campbell et al. 2006). EOCs may be toxic and persistent, even at very low concentrations, and their eventual degradation products could even be more toxic than the parent products (Soares et al. 2008).

In the countries or states where legal requirements provide a list of drinking water standards, secondary standards are often published as guidelines to help decision makers manage the groundwater resource appropriately (e.g., the EPA secondary drinking water standards—NSDWR National Secondary Drinking Water Regulations).

In Table 7.5, some of the main inorganic constituents and physicochemical indicators are listed with values from the EU Drinking Water Directives, EPA Drinking Water Standards, and WHO guidelines.

Sometimes, surprising debates occurred about the prescription of necessary minimum groundwater contents for some constituents in order to fulfill groundwater quality requirements. For example, in France, there was a lively debate about the minimum content of fluoride (i.e., for dental disease prevention) in drinking water. In the end, logically, no minimum requirement was decided. It would have been very unusual to consider perfectly natural and uncontaminated groundwater as unfit for consumption because it lacked an ion desired for medical risk prevention purposes.

General groundwater quality problems are encountered as a result of, for example, salinization from seawater intrusions in coastal aquifers and heavy irrigation practices in semi-arid and arid zones. Additionally, in many countries, nitrate contaminations related to intensive agriculture and the use of artificial fertilizers are among the main quality issues. Pesticides and herbicides have, over time, been more frequently observed and measured. Worse, new molecules with no known toxicity are

Table 7.5 Quality standards for drinking water, according to the European Union (EU Water Directives), the United States (EPA Drinking Water Standards), and the World Health Organization (WHO guidelines)

Constituent Concentrations mg/L	European Union (EC Drinking Water Directives) MCLs[a]	United States (EPA Drinking Water Standards) MCLs[a]	WHO (World Health Organization) GVs[b]
Na^+	150	200	200
K^+	12	–	–
Ca^{2+}	–	200	500 (with Mg^{2+})
Mg^{2+}	50	125	500 (with Ca^{2+})
HCO_3^-	–	500	–
SO_4^{2-}	250	250	400
Cl^-	–	250	250
NO_3^-	50	20	44
F^-	1.5	4	1.5
Se^{2-}	0.010	0.050	0.040
Hg	0.001	0.002	0.006
Cd	0.005	0.005	0.003
Benzene	0.001	0.005	0.010
Cr	0.050	0.100	0.050
TDS	1500	500	1000
pH	6.5–8.5	–	6.5–8.5

Notes
a   MCLs = Maximum Contaminant Levels (mandatory)
b   GVs = Guide Values (target levels, not legally mandatory)

continuously developed. For reliability reasons and because of their large number of required observations, toxicology studies take years before they can be finalized.

Other classical groundwater quality issues are found in (old) industrial and mining zones with metal contamination including (among other metals) barium, nickel, selenium, cadmium, and cobalt. When groundwater rebound is observed after coal mining activities, high contents of arsenic and sulfate are typically observed with the other metals.

In addition, organic compounds such as BTEX compounds (i.e., benzene; toluene; ethylbenzene; and o-, m-, and p-xylenes; and others), hydrocarbons, and chlorinated solvents are associated with many industrial contamination events.

Last not but least, the microbiological contamination of groundwater is still a huge problem in many countries where protection areas around sources or pumping wells are poorly defined and where ineffective and decaying sewage systems and treatment plants are still deployed.

7.5 Groundwater sampling and monitoring strategies

When going on the field for groundwater sampling or in situ physicochemical parameter measurements, a series of pitfalls should be avoided. A field hydrogeologist should sample and measure in a way allowing an unbiased assessment of the groundwater quality within the aquifer (i.e., as if there was no drilling or groundwater contact with the atmosphere). In practice, this is far from easy due to changes in pressure, contact with air, the associated degassing, and changes in T°, Eh, and pH.

When sampling groundwater in a well, the mixing of groundwater coming from different depths and geological layers is dependent on the well equipment (unique or separate screens, gravel packs, clay stops, packers, and multilevel sampling probes). Thus, a "depth-averaged" sampling must be definitively distinguished from multilevel sampling procedures.

Depth-averaged versus multilevel sampling

Depth-averaged sampling corresponds to sampling groundwater that is screened across several aquifers or geological formations during a pumping of a well. Therefore, the sampled groundwater actually comes from any or all of the formations (Sanders 1998). The proportions in groundwater sample that come from each formation are very difficult to assess. In theory, they could be deduced from the respective hydraulic conductivities of each formation. In practice, however, these proportions are also highly dependent on the well equipment (among others, adequate choices and uniformity of the gravel pack characteristics, the screen characteristics, and the depth of the pump).

Multilevel sampling, on the other hand, is designed to obtain reliable and representative groundwater samples from determined depths. This can be critical when different successive aquifers or fractured rocks are concerned. The most reliable but expensive way of obtaining multilevel groundwater chemical data sets is to sample and measure in separate monitoring wells drilled at different depths with the adequate equipment, screens and gravel, and clay packs in the appropriate positions to enable the desired measurements and sampling.

For multilevel measurements and sampling inside a single well, different systems are now commercially available with multichannel tubing or multitubing (Einarson 2006). It remains difficult to use more than seven to eight different channels or narrow tubes within the same well unless the well diameter is expensively large. The main principle of multilevel measurements is to obtain (to the greatest possible extent) isolated zones in a single borehole and sample representative water from these zones. For both permanent and removable systems, modular systems are developed that can be adapted to each specific case. Individual channels or narrow tubes lead to monitoring ports (most often in PVC) at each desired zone. The operational system must be chosen and adapted as a function of the study objectives, the hydrogeological conditions, the contaminant type, and the drilling equipment used (Chapman *et al.* 2014). The narrow internal tubes or channels must be large enough to allow water level monitoring and sampling using a peristaltic pump (if the water table is not deeper than the net positive suction head of the pump, approximately 7.5 m in practice). Deeper, larger individual tubes are needed to accommodate submerged small centrifugal pumps of 1.5 inches or 3.81 cm in diameter (Figure 7.9) with a positive water displacement. One of the objectives is to obtain samples from isolated zones where the water does not mix with the groundwater coming from another depth through seepage along the borehole wall. The borehole can be backfilled using alternating layers of sand and bentonite pellets to create monitoring intervals between seals. This work is delicate and the accuracy is critical; a tagline for monitoring the settling progress, a

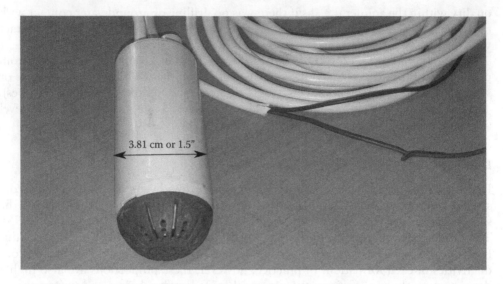

Figure 7.9 Submerged small centrifugal pumps (of 1.5 inches or 3.81 cm in diameter) that can be introduced in narrow tubes to sample groundwater in one port of a multilevel sampling system in a single well.

funnel, and a tremie should be used for a precision work (Chapman *et al.* 2014). Some prebuilt modular systems with inflatable packers (removable) or sand- and bentonite-filled packs (permanent) can be customized in advance and then introduced into the borehole. Otherwise, a large tube containing different narrow tubes or channels must be clamped against the bare borehole wall with the individual sampling ports directly facing the targeted zone to be measured and sample. The results of this operation are clearly uncertain and remain frequently unreliable.

For multilevel sampling in the partially saturated zone, inclined boreholes with multilevel systems should be preferred over vertical boreholes. This allows the avoidance of typical disturbances, such as preferential flow paths in the disturbed formation and along the vertical drilling and borehole equipment. These disturbances are even more important than the horizontal component of the groundwater flow and solute transport is very limited or nonexistent in the vadose zone. A sleeve is inserted in an uncased inclined borehole (Dahan *et al.* 2009) allowing the placement of vadose zone adapted monitoring or sampling ports that possibly include fracture samplers. This type of system is currently operational and can be applied to deep vadose zones ranging from 16 to 45 m depths (Fernandez De Vera *et al.* 2015).

Sampling procedure and representativity of the sampled groundwater

Whether groundwater is sampled in multilevel systems or not, a series of precautions should be taken to avoid biasing the chemical results of a groundwater quality campaign. Most problems arise because the water in the borehole no longer has the same chemical composition as the groundwater in the geological formation.

Physicochemical parameters may change rapidly with exposure to air, light, and temperature changes.

In situ measurements

In the field, when measuring temperature, pH, Eh, electrical conductivity (EC), dissolved gases (particularly dissolved oxygen), and possible turbidity, we must always consider where we are in the process of purging the well and whether the measured/sampled water in the well is representative of the groundwater in the aquifer formation. To improve reliability, some or all of these parameters should be monitored continuously during sampling operations and the purging of the well.

Filtration and acidification

If samples are taken for metal analyses (particularly the metals, iron and manganese), they should be filtered and acidified. The filtering option is often debated, but this is dependent on the aim of the study. In brief, if the groundwater quality must be assessed for its future use in contact with the atmosphere, acidification is not needed. Despite the precipitation of some hydroxide constituents after contact with the air, conditions in the sample would be similar to those in the distribution network. If in situ groundwater quality must be studied, filtration and acidification may be necessary to ensure that the main metal content remains as dissolved as possible.

Purging the well until reaching stabilized electrical conductivity

The purpose of purging the well before any sampling is to obtain samples of water that are chemically representative of the groundwater in the geological formation (Sanders 1998). Water in the borehole is most likely relatively stagnant in the well and has thus been in contact with air and equipment materials for a time. This water is no longer representative of the current aquifer conditions. After pumping starts, a purge of the well actually induces the feeding of the well by groundwater from the aquifer. The tempo that should be adopted for purging is also debatable. This tempo should be optimized as a function of the possible recharge of the well by the aquifer. Generally, three of four "well volumes" should be pumped before sampling. A more reliable way to determine the purge effect before sampling is to continuously measure the main physicochemical parameters, such as EC, $T°$, pH, and Eh. The assumption is made that these parameters will be stabilized (i.e., a kind of steady state will be reached) when they are mostly influenced by the formation water.

References

Appelo, C.A.J. and D. Postma. 2005. *Geochemistry, groundwater and pollution* (2nd Edition). Amsterdam: Balkema.

Atteia, O. 2005. *Chimie et pollutions des eaux souterraines (Chemistry and pollutions of groundwaters) (in French)*. Paris: Lavoisier.

Campbell, C.G., Borglin, S.E., Green, F.B., Grason, A., Wozei, E. and W.T. Stringfellow. 2006. Biologically directed environmental monitoring, fate, and transport of estrogenic endocrine disrupting compounds in water: A review. *Chemosphere* 65: 1265–1280.

Chapman, S., Parker, B., Cherry, J., Munn, J., Malenica, A., Ingleton, R., Jiang, Y., Padusenko, G. and J. Piersol. 2014. Hybrid Multilevel System for monitoring groundwater flow and agricultural impacts in fractured sedimentary bedrock. *Groundwater Monitoring & Remediation* 35(1): 55–67.

Dahan, O., Talby, R., Yechieli, Y., Adar, E. and Y. Enzel. 2009. In situ monitoring of water percolation and solute transport using a vadose zone monitoring system. *Vadose Zone Journal* 8(4): 916–925.

Davis, J.C. 1986. *Statistics and data analysis in geology*. New York: John Wiley & Sons.

Deutsch, W.J. 1997. *Groundwater geochemistry. Fundamentals and applications to contamination*. Boca Raton: CRC press.

Einarson, M.D. 2006. Multilevel ground-water monitoring. In *Practical handbook of environmental site characterization and ground-water monitoring* (2nd Edition), ed. D. Nielsen, chap. 11. Boca Raton: CRC Press, 808–848.

EPA. 2016. EPA Water Quality Standards Program (https://www.epa.gov/wqs-tech) and eCFR 40 Part 131, accessed in May 2018.

Fernandez de Vera, N., Dahan, O., Dassargues, A., Vanclooster, M., Nguyen, F. and S. Brouyère. 2015. Vadose zone characterisation at industrial contaminated sites. *CL:AIRE bulletin, AB* 7: 1–4.

Fetter, C.W. 2001. *Applied hydrogeology* (4th edition). Upper Saddle River (NJ): Pearson Education Limited.

Fitts, Ch. R. 2002. *Groundwater science*. London: Academic Press.

Graham, P.W., Baker, A. and M.S. Andersen. 2015. Dissolved organic carbon mobilisation in a groundwater system stressed by pumping. *Sci. Rep.* 5: 18487.

Güler, C., Thyne, G. D. and J.E. McCray. 2002. Evaluation of graphical and multivariate statistical methods for classification of water chemistry data. *Hydrogeology Journal* 10(4): 455–474.

Hem, J.D. 1985. *Study and interpretation of the chemical characteristics of natural water* (3rd Edition). USGS Water Supply Paper 2254, Alexandria (VA): US Government Printing Office.

Hong, Y. S. and M.R. Rosen. 2001. Intelligent characterisation and diagnosis of the groundwater quality in an urban fractured-rock aquifer using an artificial neural network. *Urban Water* 3(3): 193–204.

Jurado, A., Vàzquez-Suñé, E., Carrera, J., López de Alda, M., Pujades, E. and D. Barceló. 2012. Emerging organic contaminants in groundwater in Spain: A review of sources, recent occurrence and fate in a European context. *Science of the Total Environment* 440(1): 82–94.

Kohonen, T. 1995. *Self-organizing maps*. Berlin: Springer.

Lambrakis, N., Antonakos, A. and G. Panagopoulos. 2004. The use of multicomponent statistical analysis in hydrogeological environmental research. *Water Research* 38(7): 1862–1872.

Love, D., Hallbauer, D., Amos, A. and R. Hranova. 2004. Factor analysis as a tool in groundwater quality management: Two southern African case studies. *Phys. Chem. Earth* 29(15–18): 1135–1143.

Moeck, C., Radny, D., Borer, P., Rothardt, J., Auckenthaler, A., Berg, M. and M. Schirmer. 2016. Multicomponent statistical analysis to identify flow and transport processes in a highly-complex environment. *Journal of Hydrology* 542: 437–449.

Morel, F.M.M. 1983. *Principles of aquatic chemistry*. New-York: John Wiley.

Nordstrom, D.K., Plummer, L.N., Wigley, M.L., Wolery, T.J., Ball, J.W., Jenne, E.A., ... J. Thrailkill. 1979. A comparison of computerized chemical models for equilibrium calculations in aqueous systems. In *Chemical modelling in aqueous systems*, ed. E.A. Jenne. Washington, DC: American Chemical Society, 38: 857–892.

Pankow, J.F. 1991. *Aquatic chemistry concepts.* Chelsea (MI): Lewis Publishers.

Parkhurst, D.L. 1995. *User's Guide to PHREEQC, a computer model for speciation, reaction-path, advective-transport and inverse geochemical calculations.* U.S. Geological Survey Water-Resources Investigations Report 95-4227.

Parkhurst, D.L. and C.A.J. Appelo. 1999. *User's Guide to PHREEQC (version 2), a computer program for speciation, batch-reaction, one-dimensional transport and inverse geochemical calculations.* U.S. Geological Survey Water-Resources Investigations Report 99-4259.

Peeters, L. 2013. A background color scheme for Piper plots to spatially visualize hydrochemical patterns. *Groundwater* 52(1): 2–6.

Peeters, L., Bacao F., Lobo, V. and A. Dassargues. 2007. Exploratory data analysis and clustering of multivariate three-dimensional spatial hydrogeological data by means of GEO3DSOM, a variant of Kohonen's Self-Organizing Map. *Hydrology and Earth System Sciences* 11: 1309–1321.

Piper, A.M. 1944. A graphic procedure in the geochemical interpretation of water analysis. *AGU Transactions* 25: 914–923.

Post, V.E.A. 2012. Electrical conductivity as a proxy for groundwater density in coastal aquifers. *Groundwater* 50(5): 785–792.

Rentier, C., Delloye, F., Brouyère, S. and A. Dassargues. 2006. A framework for an optimised groundwater monitoring network and aggregated indicators. *Environmental Geology* 50(2): 194–201.

Richards, L.A.E. 1954. *Diagnosis and improvement of saline and alkali soil.* Washington: U.S. Department of Agriculture Handbook 60.

Sanchez-Martos, F., Aguilera, P.A., Garrido-Frenich, A., Torres, J.A. and A. Pulido-Bosch. 2002. Assessment of groundwater quality by means of self-organizing maps: Application in a semi-arid area. *Environmental Management* 30(5): 716–726.

Sanders, L.L. 1998. *A manual of field hydrogeology.* Upper Saddle River (NJ): Prentice Hall.

Schoeller, H. 1955. Géochimie des eaux souterraines (In French). *Revue de l'Institut du Pétrole* 10: 230–244.

Soares, A., Guieysse, B., Jefferson, B., Cartmell, E. and J.N. Lester. 2008. Nonylphenol in the environment: A critical review on occurrence, fate, toxicity and treatment in wastewaters. *Environment International* 34: 1033–1049.

Stiff Jr, H.A. 1951. The interpretation of chemical water analysis by means of patterns. *Journal of Petroleum Technology* 3: 15–17.

Stumm, W. and J.J. Morgan. 1996. *Aquatic chemistry* (3rd Edition). New York: Wiley and Sons.

Thurman, E.M. 1985. *Organic geochemistry of natural waters.* Dordrecht, The Netherlands: Martinus Nijhoff/Dr. W.Junk Publishers.

Wiedemeier, T.H., Rifai, H.S., Newell, Ch.J. and J.T. Wilson. 2007. *Natural attenuation of fuels and chlorinated solvents in the subsurface.* Hoboken (NJ): John Wiley & Sons.

Chapter 8

Contaminant transport, residence times, prevention, and remediation

8.1 Introduction

Many different kinds of contaminants can be introduced into the geological environment and their concentrations may vary significantly in space and time. It is often crucial to have a clear understanding about the current state and context of the contaminated site, the origin and the source of contamination, and what will happen in the future. In particular, if the groundwater environmental analysis is considered within a *Driver Pressure State Impact Response* (DPSIR) chain, we need to quantify each step with rigorous and process-based approaches.

Drivers represent all potential contaminating human activities but can also be due (in specific circumstances) to natural processes. *Pressures* describe the direct and actual effects of drivers (e.g., infiltration of a contaminant). These pressures influence the measurable *State* of a given system. *Impact* represents the measurable effect due to the considered pressure with regards to a reference state (e.g., measured aquifer contamination compared to its assessed "natural state"). Ideally, any *Response* should act on *Drivers* and not only on *Pressures* to be effective at mid- and long-term to improve the *State*.

In groundwater flow problems, quantification approaches (see Chapter 4) lead to calculation of groundwater fluxes and piezometric heads. In contamination problems, quantification of solute mass fluxes and concentrations is required, but additionally, groundwater contaminant velocities and travel times are most often needed. Whether it is to determine groundwater vulnerability, or to delineate the protection zones around pumping wells, or to determine the best remediation option for a contaminated aquifer, reliable residence and travel times in the geological medium (within the saturated and unsaturated zones) together with the mixing, the spreading, and attenuation of individual or mixtures of contaminants, must be quantified.

In this chapter, the fate of reactive and nonreactive dissolved contaminants (Section 8.2) are first described, followed by the fate of nonaqueous contaminants (Section 8.3). The different transport processes and their respective conceptual models are explained with their governing equations describing their effects in heterogeneous geological media. The aim is to tend toward a quantifiable understanding. However, despite recent important advances in the field, predictions of the fate of contaminants still generate extensive debate (Fiori and de Barros 2015). For years, several "hot" scientific topics have arisen with the following issues (among others): What are the main mechanisms and the best (i.e., the most effective and accurate) ways to represent them? How should data be obtained at an adequate scale required for predictions? How should we deal with complex processes such as channeling and retention that could be particularly important in highly heterogeneous geological formations?

At the same time, new high resolution and time lapse measurement techniques (not described in this book) allow, in some circumstances, to obtain 2D and 3D images of the actual heterogeneity with higher resolution than ever. This challenging topic is in a state of continual and rapid evolution with many research directions still to be discovered and tested.

8.2 Solute transport

In the following section, the main physical and chemical processes affecting the fate of solutes in groundwater are described. From the conceptualization needed for quantification purposes, and for each of the considered processes, an expression of the solute mass flux is derived. Finally, expressing the conservation of solute mass, solute transport equations will be considered to describe quantitatively solute transport in the saturated zone of geological porous media.

Diffusion

Molecular diffusion is due to the molecular Brownian motion resulting from thermal energy when the solute is at a temperature above absolute zero. Diffusion induces a mixing process at the pore scale, tending always to smooth out, or equilibrate solute concentrations. The result is a mass flux of solute from higher to lower concentration zones. The solute mass flux of diffusion f_m (expressed in kg/m²s), is described by a linear form of Fick's law:

$$f_m = -D_m \textbf{grad}\, C^v = -D_m \nabla C^v \tag{8.1}$$

$$f_m = -D_m \textbf{grad}(\rho C) = -D_m \nabla(\rho C) \tag{8.2}$$

where D_m is the *effective molecular diffusion coefficient* in m²/s [L^2T^{-1}], C^v is the *volume concentration* in kg/m³ (in practice often in mg/L or μg/L) [ML^{-3}] (see Chapter 7, Equation 7.4) giving the mass of the considered solute per total volume of the water solution (i.e., used when no significant change of the groundwater solution density is expected), C is the *mass concentration* in kg/kg (in practice often in mg/kg = *ppm* (one part per million) or μg/kg = *ppb* (one part per billion) [-] (see Chapter 7, Equation 7.3) giving the mass of a constituent per total mass of the solution, and ρ is the water solution density (see Chapter 4, Equation 4.5). Equation 8.2 is used in problems where a variable water solution density can be expected due to a high solute content so that the density is dependent on the solute concentration: $\rho(C)$.

At the macroscopic scale, in a geological porous medium, D_m will be less than the value of the *molecular diffusion coefficient in free water* d_m (ranging between 1.10^{-9} and 2.10^{-9} m²/s) due to the presence of solid matrix in the considered volume (REV), so that $D_m < d_m$ and:

$$D_m = \tau d_m \tag{8.3}$$

where τ is the tortuosity coefficient of the porous medium [-] expressing, at the macroscopic scale, how diffusion is affected by the microscopic or pore-scale heterogeneity of the medium. Some researchers (Fitts 2002, Bear and Cheng 2010, among others) consider that the tortuosity should be better expressed as a tensor to describe a possible anisotropy of the medium affecting the diffusion process. This can be particularly important in layered sediments and in fractured rocks. In this case, a tensor D_m is indeed used in Equations 8.1 and 8.2. Due to the fact that in most applications (especially in aquifers) diffusion can be considered as small with regards to other dispersive processes, practitioners most often assume that diffusion is an isotropic process in saturated porous media, with typical values of τ between 0.56 and 0.88 (Rausch *et al.* 2005). It is perhaps more important to note that D_m is temperature dependent, which can be described with:

$$\frac{D_{mT_1}}{D_{mT_2}} = \frac{T_1}{T_2} \frac{\mu_{T_2}}{\mu_{T_1}} \tag{8.4}$$

where μ_{T_1} and μ_{T_2} are the groundwater dynamic viscosity values respectively at temperatures T_1 and T_2. Applying this relation, one can note that D_m is multiplied by 2 when temperature increase from 5°C to 25°C (Rausch *et al.* 2005, see water viscosity values in Chapter 4, Table 4.4).

Molecular diffusion is independent of the flow direction but the effective diffusive flux occurs only through the saturated pores or fissures of the geological medium. Equations 8.1 and 8.2 are thus multiplied by the total porosity n to obtain:

$$f_m = -nD_m\nabla C^v = -n\tau d_m\nabla C^v \tag{8.5}$$

$$f_m = -nD_m\nabla(\rho C) = -n\tau d_m\nabla(\rho C) \tag{8.6}$$

Some authors consider the influence of the total porosity included in the tortuosity that is used to obtain the effective diffusion coefficient. Thus, the total porosity value does not always appear explicitly in equations describing the solute diffusion mass flux.

One of the most important characteristics of diffusion is the fact that it is totally independent of the groundwater flow direction. Any sharp gradient of solute concentration will be smoothed out with time (Figure 8.1). The relative importance of diffusion with regards to other solute transport processes is highly dependent on the groundwater flow conditions. The relative importance of diffusion increases as groundwater flow velocities and fluxes decrease. This process of solute transport becomes very important in low permeability media, for example when studying local waste containment or regional confining conditions. Diffusion can also become of critical importance for security assessment at a "geological time scale" of radioactive waste disposal in clayey layers (Huysmans and Dassargues 2005, 2007).

Advection

Advection is the more evident process of solute transport as it describes the mass of solute transported by groundwater flow due to an existing piezometric head gradient. If the conditions are such that temperature or density gradients can also affect

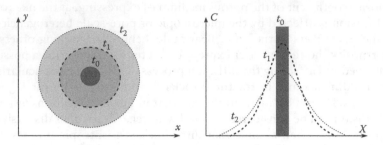

Figure 8.1 Effects of molecular diffusion as a function of time from a high concentration point-source contaminant in a saturated zone without groundwater motion.

groundwater flow, then the term convection is used instead of advection (Rausch *et al.* 2005; see Chapters 10 and 11). If groundwater is flowing, the dissolved substances in the groundwater will be transported in the same direction. Averaged (as usual) over a representative volume of porous medium (REV), the advective solute mass flux of f_a (expressed in kg/m^{2}s) is written as:

$$f_a = qC^v \tag{8.7}$$

$$f_a = q\rho C \tag{8.8}$$

where q is the groundwater Darcy flux as expressed in Equation 4.28 (see Chapter 4) in m/s [LT^{-1}]. As stated previously, C^v and C are respectively the *volume concentration* and *mass concentration* (i.e., the latter being used when a variable water solution density can be expected). Applying Equation 4.28 in Equations 8.7 and 8.8, we obtain:

$$f_a = -K \cdot \nabla h C^v = -K \cdot \nabla h (\rho C) \tag{8.9}$$

For practical purposes, it is often important to determine at which velocity this migration of dissolved species occurs, i.e., which groundwater velocity should be considered for advection? As discussed in Chapter 4 (Equations 4.21, 4.22, and 4.23), different velocities can be calculated depending on the considered porosity. The term "Darcy velocity" is to be banished definitively herein because it induces much confusion. The porosity to be chosen corresponds to the actual mobile water portion in the saturated REV of the considered aquifer. It is named *effective transport porosity* or *mobile water porosity for transport* (n_m) in Equation 4.23. The velocity vector for advective transport (or transport velocity) can thus be written as:

$$v_a = q/n_m \tag{8.10}$$

Therefore, when the solute velocity in groundwater is the main object of study, it appears that the useful porosity for solute transport is this *effective transport porosity* (Box 8.1). This is confirmed when tracer tests are performed, and where effective porosity must be distinguished from the effective porosity that most often describes effective drainage porosity. In fact, it is logical to point out that the porosity available to transmit solute concentrations (i.e., corresponding to the most mobile part of the water occupying the pores, fissures, and fractures) is not automatically the same as that corresponding to mobile water for drainage. This observation is still surprisingly little mentioned (or not at all) in the traditional texts on hydrogeology. It is, however, a well-known and accepted observation by hydrogeologists, especially by those dealing with aquifers affected by multiple porosities (Worthington 2015). In this book, the distinction will be made systematically between the effective drainage porosity and the effective transport porosity which is typically less (see Chapter 4, Section 4.4, Equations 4.22 and 4.23). The two hydrogeological parameters/properties of the saturated geological medium that are needed to calculate advective transport are thus the hydraulic conductivity and the effective transport porosity.

Pure advection as shown in Figure 8.2 does not exist in reality. Solute spreading, or dispersion, is always observed as velocity variations occur at different scales in the geological media. Thus, even in the absence of diffusion, variations in direction and magnitude of the pore water velocity at a scale smaller than that used for averaging the groundwater velocity (<REV) will disperse the solute.

Box 8.1   About effective transport porosity

Effective transport porosity is in reality very dependent on what we define as mobile and immobile groundwater in the representative volume of geological medium. For many contamination problems and associated risk assessments, it is very important to predict the first arrival time of solute contaminant at a "target." This is the case, for example, for protection zone delineation around a pumping well, for security assessment of waste disposal, for assessment of risk created by industrial contamination hazards, and for groundwater vulnerability mapping. To be conservative (i.e., on the security side) with regards to most practical applications where solute transport is calculated, one should choose to define the effective porosity relevant to solute advection as a relatively small part of the total porosity. This is especially true and important if the geological medium is known as a "dual" porosity medium or is highly heterogeneous at the micro- and macroscales. If the Darcy flux (i.e., specific discharge, see Chapter 4, Section 4.4) would be divided by the total porosity n, this would produce a kind of averaged velocity between the mobile and immobile groundwater in the medium which can be highly misleading (i.e., leading to underestimated velocities). However, some can argue that this could be partially and artificially compensated or corrected by considering larger values of the longitudinal dispersivity.

Dispersion

Solute *mechanical dispersion* accounts for the mixing and spreading of a solute that occurs along and across the main flow direction (Delleur 2000) due to aquifer

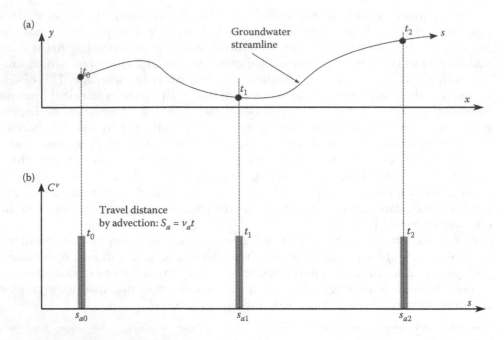

Figure 8.2 Concept of pure advection of solute. A given concentration at time t_0 is transported along the groundwater streamline until time t_1, and then until t_2, ... (a). If a uniform groundwater velocity field is assumed, the advective transport distance during a time t can be calculated by $s_a = v_a t$, so that successive advective positions $s_{a_0}, s_{a_1}, s_{a_2}$ ... can be calculated (b).

heterogeneities at the pore scale (<REV) (i.e., *microdispersion*) and at larger scale (macroscale) (>REV) (i.e., *macrodispersion*). Trying to be selective with terms that are commonly used to describe dispersion effects, Kitadinis (1994) specified that *mixing* due to microdispersion induces a dilution of the solute in a larger volume than the volume in which concentrations (due to advection) are considered, while *spreading* due to macrodispersion affects the plume shape. Indeed, if a more global scale concentration is considered (i.e., at a larger scale than the REV and in the long term), one could consider that spreading due to geological heterogeneities also contributes to the dilution of the solute by making the spatial distribution of the solute plume more complex.

Pore-scale velocity variations in magnitude and direction are due to different mechanical effects (Figure 8.3): (a) different pathways of different lengths within the REV, (b) a nearly zero velocity in contact with the solid grains and a parabolic distribution within different size pores, and (c) flow velocities which depends on the pore size distribution. *Mechanical dispersion* is the net result of these effects considered at the scale of the representative elementary volume (REV). The multiscale heterogeneity of the geological media also influences the flow velocities at larger scales.

The actual flow velocity field is highly complex and variable in direction and intensity. *Macrodispersion* describes the longitudinal and transverse spreading with

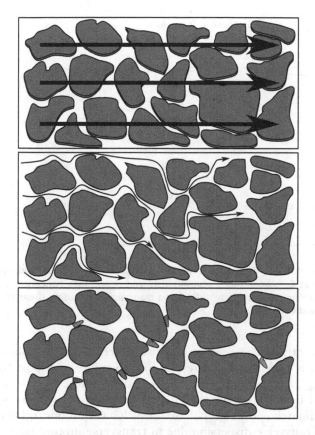

Figure 8.3 Within a representative elementary volume (REV), pore-scale variations of water velocities induce solute *mechanical dispersion* that is not described by simple advection considered at the REV scale (above). The contributing causes of dispersion are different pathways, velocity variations depending on pore size distribution, and velocity profiles within each pore.

respect to a general flow direction. It is observed that the longitudinal spreading is greater than the transverse spreading, but the primary direction of groundwater flow can also be highly variable (spatially and as a function of time) so that the actual direction of this longitudinal component of dispersion can change. If diffusion is added to mechanical dispersion, the total mixing and spreading of solute is described as the *hydrodynamic (macro)dispersion*.

As a first illustration of dispersion, let us consider conceptually a Dirac injection (i.e., an instantaneous or slug injection) of solute at a concentration C_{max} in a 2D horizontal uniform saturated flow field with an averaged velocity v_a (Figure 8.4). With time, the solute plume becomes wider and less concentrated and tends to show a greater longitudinal (along the uniform flow direction) than transverse spreading.

A second example could be considered with a 1D column experiment and a continuous injection of solute under steady saturated groundwater flow. If one records the concentration as a function of time at a given point located downgradient (i.e., at the base of the column), the obtained *breakthrough curve* (Figure 8.5) clearly shows

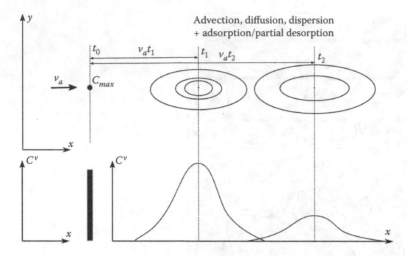

Figure 8.4 Conceptual example of solute dispersion in a 2D horizontal uniform flow field. An initial slug of solute at C_{max} becomes a continuously wider and less concentrated plume over time showing different longitudinal and transverse spreading (Bear and Verruijt 1987, Bear and Cheng 2010).

a progressive increase of concentration instead of an abrupt arrival of the solute. The 0.5 relative concentration is observed as corresponding to solute transport by advection only at a time $t = l_{column}/v_a$ with l_{column} the length of the column.

Results from many lab and field tests have shown that longitudinal dispersion can be considered as approximately proportional to the velocity (among others, Schwartz and Zhang 2003). Transverse dispersion due to transverse mixing occurs in both directions orthogonal to the groundwater flow direction. Transverse dispersion can be particularly important in continuous contamination source cases, and also for transport of reactive compounds especially at the plume's fringes (Chiogna *et al.* 2011).

Contrarily to diffusion, dispersion is dependent on water velocities. In terms of effects, both processes contribute to the spreading of solute around a theoretical upscaled (i.e., at the REV scale) advective position. For practical reasons, and although it is becoming more debatable (see among others, Hadley and Newell 2014, Cushman and O'Malley 2015), a Fickian approach is most often conceptually adopted to describe in practice (macro)dispersion transport in geological media. Linked to this strong conceptual choice, the solute mass flux of mechanical dispersion f_d expressed in kg/m²s, can also be described (i.e., as in diffusion) by a linear form of Fick's law:

$$f_d = -n_m D \cdot grad \, C^v = -n_m D \cdot \nabla C^v \tag{8.11}$$

$$f_d = -n_m D \cdot grad(\rho C) = -n_m D \cdot \nabla(\rho C) \tag{8.12}$$

where D is the dispersion tensor whose components are expressed in m²/s [L²T⁻¹] and n_m is the effective transport porosity [-]. As previously defined, C^v and C are respectively the *volume concentration* and *mass concentration* (i.e., the latter being used

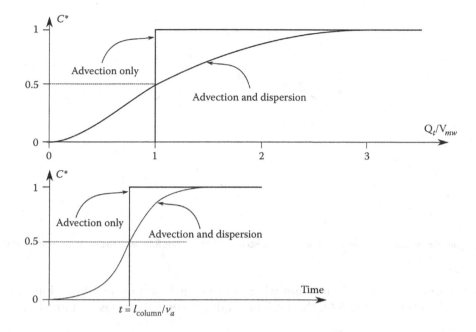

Figure 8.5 Solute breakthrough curve as obtained in a 1D column of homogeneous porous medium for a continuous injection of solute under steady-state saturated ground-water flow. Relative concentrations are used where $C* = (C - C_0)/(C_{max} - C_0)$ with $0 < C* < 1$. For such standardized column tests, it is usual to express time in dimensionless units, multiplying the time (t) by the ratio between the total discharge (Q) and the mobile water pore volume of the column ($V_{mw} = n_m V_t$): Qt/V_{mw} (also referred to as "pore volumes" or PV's) (Bear and Cheng 2010, Pinder and Celia 2006).

when a variable water solution density can be expected). The components of D are dependent on the velocity components and dispersivities characterizing the geological medium at the considered scale. In 3D, the 9 components of this tensor can be calculated as a function of the velocity components and the two main *dispersivities* which are: a_L and a_T respectively the *longitudinal* and *transverse dispersivities* expressed in m [L] (Scheidegger 1961, Bear 1972):

$$D_{ij} = (a_L - a_L)\frac{v_{a_i} v_{a_j}}{|v_a|} + a_T |v_a| \delta_{ij} \tag{8.13}$$

where indicial notations are used and i and $j = 1, 2$ or 3 respectively, δ_{ij} is the Kronecker function with $\delta_{ij} = 1$ if $i = j$ and $\delta_{ij} = 0$ if $i \neq j$, $|v_a| = \sqrt{(v_{a_1}^2 + v_{a_2}^2 + v_{a_3}^2)}$. Burnett and Frind (1987) (cited by Schwartz and Zhang 2003, Bear and Cheng 2010, among many others) also developed this equation but as a function of three main dispersivities, differentiating between a_{TH}, and a_{TV}, respectively the transverse horizontal and transverse vertical dispersivities. Also, Lichtner *et al.* (2002) introduced a 4-component form, where the longitudinal and transverse dispersivities vary depending on the flow

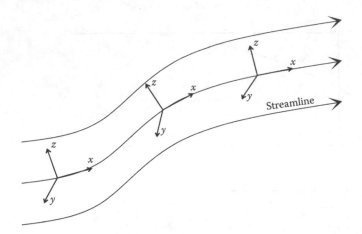

Figure 8.6 Isotropic medium and reference system with the x-axis following the local flow direction.

angle. For an isotropic 3D medium where the x-axis is always oriented along the velocity direction (Figure 8.6), the mechanical dispersion tensor D is simplified as:

$$D = \begin{bmatrix} D_{xx} & & \\ & D_{yy} & \\ & & D_{zz} \end{bmatrix} \tag{8.14}$$

and $D_{xx} = a_L v_{a_x}$, $D_{yy} = a_T v_{a_x}$, $D_{zz} = a_T v_{a_x}$, with $v_{a_y} = v_{a_z} = 0$.

When measured at the lab scale, a_L is representative of pore-scale dispersion leading most often to values ranging between 10^{-4} and 10^{-2} m. When applied to the field scale, where larger-scale heterogeneities exist at a scale greater than the volume chosen as representative, the Fickian dispersion concept leads to longitudinal macrodispersivities a_L greater by at least two orders of magnitude (for example, Zech *et al.* 2015, Zech *et al.* 2016). In this case, a_L was found as scale-dependent but limited by an apparent asymptotic value. Extensive literature exists on this topic to try to exhibit and deduce upscaling relations for dispersivities (among others, Beims 1983, Arya *et al.* 1988, Neuman 1990, Gelhar *et al.* 1992, Xu and Eckstein 1995). The main difficulty lies in the reliability of reported values from multiple field tests performed under various actual hydrogeological conditions. One can remark that if large heterogeneities are present at the aquifer scale, macrodispersion should be time-dependent but this time-dependency may be relaxed when the correlation scale of macrodispersion has a finite value or, in other words, if it has reached its asymptotic value (Cornaton and Perrochet 2006).

A recent rigorous reevaluation of the previous results by Zech *et al.* (2015) has led to the conclusion that macrodispersivities are "formation-specific." Calculating solute transport should be based on realistic characterization of the local aquifer and especially the spatial distribution of hydraulic conductivities. A local upscaling law should be considered as site or formation specific (Zech *et al.* 2015). Field-scale macrodispersion could also be strongly affected by unresolved transient flow variations (Kinzelbach and Ackerer 1986).

A dimensionless *Peclet number* is used to provide a ratio between the rates of solute transport by advection and dispersion (Bear and Cheng 2010). This *Peclet number* (Box 8.2) is often used to characterize the relative importance of these two transport processes. When advection dominates, mechanical dispersion is far more important than diffusion thus diffusion can be neglected. When advection is weak, diffusion can become more important than mechanical dispersion.

Box 8.2   Peclet number for solute transport

A dimensionless Peclet number was initially defined by:

$$Pe = \frac{v_a l}{D_m} = \frac{q l}{n_m D_m}$$

where $v_a = q/n_m$ is the velocity, D_m is the effective diffusion coefficient, and l is a "characteristic" length of the problem (Freeze and Cherry 1979). Unfortunately, different Peclet number definitions exist and, for a given particular case, they lead to diverse values (Huysmans and Dassargues 2005). In particular, there are significant differences in the definition of the "characteristic" length such as the characteristic length of the pores (Bear and Verruijt 1987), hydraulic radius of the saturated pore space (Bear and Cheng 2010), $\sqrt{k}$ with k the intrinsic permeability (de Marsily 1986), average grain size, the correlation scale assuming a Gaussian correlation structure of the medium heterogeneity, or the distance from the injection source (lab and field tracer tests). Note that not all of the possible differences in the l definitions are due to the scale of study. In very low permeability environments (e.g., as in plastic clays or claystones in the context of radioactive waste storage), it could be very useful to assess if transport by advection and dispersion are to be neglected compared to diffusion. It is often considered that if $Pe \ll 1$, diffusion dominates. To this end, and with this objective in mind, different definitions of the Pe number were tested by Huysmans and Dassargues (2005). Generalizing this approach, for analytical interpretation of tracer tests, Sauty (1980) defined a *Peclet number* as the ratio between advection and dispersion as follows:

$$Pe = \frac{v_a L}{D}$$

where L is the distance from the point of injection to the point of measurement of the tracer test, and D is the hydrodynamic dispersion coefficient (including diffusion).

An important point to observe is that dispersion is an anisotropic process. If a porous medium is anisotropic, the anisotropy of the medium is added to the anisotropy of the mechanical dispersion and the dispersion tensor D becomes a fourth-order tensor with 81 components:

$$D_{ij} = \sum_{k=1}^{3}\sum_{m=1}^{3} a_{ijkm} \frac{v_{a_k} v_{a_m}}{|v_a|} \tag{8.15}$$

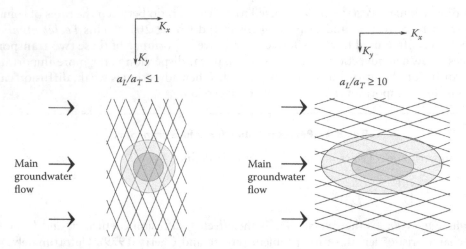

Figure 8.7 Influence of the medium anisotropy in K (e.g., represented by the combined effects of two preferential fracture directions in a fractured medium assimilated to an equivalent anisotropic porous medium, see Chapter 4, Section 4.1) on the a_L/a_T ratio using a "classical" Fickian concept applied to an assumed isotropic porous medium.

where i, j are indices of the principal directions of dispersion anisotropy, and k, m are the indices of the principal directions of the permeability (i.e., anisotropy of the medium). In practice however, it is far more convenient to return to Equation 8.14 assuming the medium as isotropic, and consequently including all anisotropic effect in the ratio between a_L and a_T. In real isotropic media, $1 < a_L/a_T < 10$ is usually considered at the pore scale (Bijeljic and Blunt 2007) as well as at the field scale (Pickens and Grisak 1981). Other extreme values of this a_L/a_T ratio could be clear indicators of the medium anisotropy in K as shown, for example, in Figure 8.7.

One of the main reasons to adopt Fick's law to describe the dispersive mass flux was the possibility to obtain a similar expression as used for diffusion. Dispersion and diffusion are physically different processes but they both produce mixing and spreading of the solute. As mentioned previously, the combined effect of dispersion and diffusion is referred to as *hydrodynamic dispersion*. The solute mass flux of hydrodynamic dispersion f_h (expressed in kg/m²s), can be written as:

$$f_h = -n_m(\mathbf{D_h}) \cdot grad C^v = -n_m(\mathbf{D_h}) \cdot \nabla C^v \tag{8.16}$$

$$f_h = -n_m(\mathbf{D_h}) \cdot grad(\rho C) = -n_m(\mathbf{D_h}) \cdot \nabla(\rho C) \tag{8.17}$$

where $\mathbf{D_h} = \mathbf{D} + \mathbf{D_m} = \mathbf{D} + D_m\mathbf{I}$ with $\mathbf{D_m}$ the effective diffusion coefficient tensor ($\mathbf{D_m}$) or the scalar (D_m) multiplied by the identity tensor $\mathbf{I}$. Note that in this equation, n_m has replaced n for the diffusion component which implies that diffusion would be neglected in the immobile water of the saturated porous medium (this assumption will be discussed later in the section about immobile water effect and "matrix diffusion").

The spreading of a solute as shown in Figures 8.4 and 8.5 is actually a combination of advection and hydrodynamic dispersion. The corresponding solute mass balance equation can be written in 3D for the mobile water as follows:

$$n_m \frac{\partial(\rho C)}{\partial t} = -div(f_a + f_b) = -\nabla \cdot (f_a + f_b) \tag{8.18}$$

If ρ can be considered constant over time and throughout the domain:

$$n_m \frac{\partial C^v}{\partial t} = -div(f_a + f_b) = -\nabla \cdot (f_a + f_b) \tag{8.19}$$

Taking the f_a and f_b values respectively from Equations 8.7 and 8.16, Equation 8.19 becomes:

$$n_m \frac{\partial C^v}{\partial t} = -div(qC^v - n_m(D_b) \cdot grad\, C^v) = -\nabla \cdot (qC^v - n_m(D_b) \cdot \nabla C^v) \tag{8.20}$$

This advection-dispersion equation (ADE) can be written in 1D along the x-axis in the direction of the assumed uniform groundwater flow velocity and dividing all terms by n_m:

$$\frac{\partial C^v}{\partial t} = -v_a \frac{\partial C^v}{\partial x} + a_L v_a \frac{\partial^2 C^v}{\partial x^2} \tag{8.21}$$

For an instantaneous injection of solute at time $t = 0$ and $x = 0$ in a uniform flow field along x-axis, as shown in Figure 8.4, this Equation 8.21 can be solved to find the concentration as a function of the coordinate x and time t. The boundary conditions of such a 1D problem are:

$$\lim_{x \to \pm\infty} C^v(x,t) = 0 \tag{8.22}$$

$$C^v(0,0) = C_0^v \tag{8.23}$$

This differential equation can then be solved analytically (Ogata and Banks 1961) approximated by:

$$C''(x,t) = \frac{C_0^v}{2} erfc\left[\frac{x - v_a t}{\sqrt{4a_L v_a}}\right] \tag{8.24}$$

This approximation is acceptable when the solute has moved sufficiently far from the injection point and the "advective displacement" is greater than the longitudinal dispersivity: $v_a t \gg a_L$ or when $Pe \gg 1$. Other analytical solutions are presented by Fetter (1993) to obtain solute concentrations at a given distance as a function of time for this 1D problem with different boundary conditions.

Equation 8.24 can also be written as:

$$C^v(x,t) = \frac{C_0^v}{\sqrt{4\pi a_L v_a}} exp\left[-\frac{(x - v_a t)^2}{4a_L v_a t}\right] \tag{8.25}$$

which is known as the Gaussian integral centered on the advective position at the time t. In statistics, a normal (or Gaussian) distribution is often assumed to describe a continuous probability distribution. If the mean is μ and the standard deviation σ, the probability distribution function (PDF) of the assumed normal distribution is:

$$PDF = \frac{1}{\sqrt{2\pi\sigma^2}} exp\left[-\frac{(x - \mu)^2}{2\sigma^2}\right] \tag{8.26}$$

So, it appears that adopting a Fickian concept to describe dispersion is equivalent to making an analogy between solute dispersion in a porous medium and Gaussian statistics about the position of solute particles around a mean advective position (i.e., $\mu \leftrightarrow v_a t$) and with a dispersion statistic assimilated to solute dispersion (i.e., $\sigma^2 \leftrightarrow 2a_L v_a t$). As shown in Figure 8.4, at time t_0 the distribution is located at $x = 0$. At later times, the distribution has been advected downstream but the dispersion is centrally located around a mean advective position. As the concentration distribution can be assimilated to a PDF, one can calculate the (mathematical) moments of these concentration distributions (Box 8.3; Dagan 1989).

This conceptual model leading to a Gaussian description of the dispersion has led to many studies about quantitative approaches relating macrodispersivity values to geostatistical properties of hydraulic conductivity (i.e., in practice ln K, as hydraulic conductivity is often considered as a log-normal property, see Chapter 12, Section 12.6). In this way, using a correlation length λ (i.e., the distance beyond which pairs of values are no longer correlated, the range of a semi-variogram, see Chapter 12, Section 12.5), the following theoretical relation is often proposed (Dagan 1982, Gelhar and Axness 1983, Dagan 1989):

$$a_L = \sigma_{lnK}^2 \lambda \tag{8.27}$$

where σ_{lnK}^2 is the variance of the ln K property.

Probabilistic approaches to solve the advection-dispersion transport equation (ADE) have also led to using numerical random walk particle tracking methods (See Chapter 13; Box 8.4).

Fundamental progress in this field has been made since the 1980s, demonstrating that actual dispersion and especially macrodispersion can rarely be adequately described by a Fickian law. Without entering too much in detail, some authors have pointed out that if we could reproduce the actual detailed groundwater velocity field in the porous medium we would need only advection and diffusion processes. This is linked to the observation that macrodispersion requires an unrealistically long travel distance or travel time before so-called Fickian conditions are reached, which are probably never reached in many natural heterogeneous systems (Sudicky et al. 1983). As summarized by Molz (2015), the advection-diffusion concept has been illustrated

---

**Box 8.3   Moments analysis for a pulse of solute
mass (Dagan 1989, Zech et al. 2015)**

For assumed Gaussian spatial concentration distributions, the zeroth moment (M_0) is the total probability (i.e., one), the first moment (M_1) is the mean, the second central moment (M_2) is the variance, the third central moment is the skewness (measuring asymmetry), and the fourth central moment (with normalization and shift) is the kurtosis (measuring heavy- or light-tailed curve) relative to a *normal distribution*:

$$M_0 = \int_{-\infty}^{+\infty} n_m C^v(x,t)dx \quad M_1 = \int_{-\infty}^{+\infty} n_m x C^v(x,t)dx \quad M_2 = \int_{-\infty}^{+\infty} n_m x^2 C^v(x,t)dx$$

M_0 corresponds to the total mass of injected solute. The global motion of the solute plume is described by the central first moment, with a mean location of the solute distribution given by $\mu = M_1/M_0$. The longitudinal spreading of the plume is described by the second central moment:

$$\sigma^2 = \frac{M_2}{M_1} - \mu^2 = \frac{\int_{-\infty}^{+\infty} n_m(x-\mu)^2 C^v(x,t)dx}{M_0}$$

If the mean location is μ, the transport velocity can be found by: $v_a = d\mu/dt$ and:

$$v_a = \frac{1}{M_0}\frac{d}{dt}\int_{-\infty}^{+\infty} x C^v(x,t)dx$$

The longitudinal spreading being σ^2, the longitudinal dispersivity can be deduced by:

$$a_L = \frac{D_L}{v_a} = \frac{1}{2v_a}\frac{d\sigma^2}{dt} \quad \text{and} \quad a_L = \frac{1}{2v_a M_0}\frac{d}{dt}\int_{-\infty}^{+\infty} n_m(x-\mu)^2 C^v(x,t)dx$$

Spatial moment analysis assumes that the plume is sufficiently large to justify ergodicity of the moments (i.e., the plume size is large enough with respect to the considered heterogeneity scale). The 1D analysis can be extended in 2D and 3D and similar definitions can be applied to characterize transverse spreading. If a breakthrough curve (BTC) is measured at a point, a similar approach can be used giving the mean arrival time $t_{mean} = x/v_a$ and dispersivity $a_L = (1/2v_a^2)(d\sigma_t^2/dx)$ where σ_t^2 is the second central *temporal moment*. In practice, the longitudinal dispersivity is deduced even if the plume distribution is not Gaussian (Zech et al. 2015).

(for example, Gillham *et al.* 1984) but measurements and computational issues have led to more development of the stochastic macrodispersion approach as proposed by Gelhar and Axness (1983). Nowadays, with new high resolution measurement techniques, combined with multiple point geostatistics (for example, Huysmans and Dassargues 2009, Renard and Allard 2013) and/or facies driven stochastic analysis, it is becoming more common to show that small-scale hydraulic conductivity fields (i.e., decreasing drastically the REV size) associated with adequately described connectivity (Russo 2015) can greatly improve results. Under these conditions, even conventional advection-dispersion approaches can be used for capturing complex plume-scale transport (Dogan *et al.* 2014).

Detailed experimental characterizations and images at the pore scale, and (more challenging) at the field scale should be performed to adequately assess the complex interaction between diffusion, advection, and solute reactions (Dentz *et al.* 2011).

Box 8.4   Random walk approaches for advective-dispersive solute transport

An assumed equivalence between the Langevin (1908) equation (i.e., describing here the advective motion of noninteracting particles and an uncorrelated random velocity being the noise) and the advection-dispersion solute transport equation (Kinzelbach 1988) is assumed. The random process is defined in time and space: the particle motion occurs with variable spatial increment ΔL (transition length) and a variable time increment $\Delta \tau$ (transition time). The fundamental properties of transport are governed by the asymptotic behavior of the coupled space-time probability distribution function (PDF) $p(\Delta L, \Delta \tau)$ (Berkowitz *et al.* 2002). As mentioned by Russian *et al.* (2016), "*the heterogeneous medium or flow properties are mapped onto the coupled distribution of transition length and time which renders an ensemble average transport picture.*" The term asymptotic behavior is used to refer to respectively long-distance or long-time behavior.

This technique is known as the continuous time random walk (CTRW) method (among others, Dentz and Berkowitz 2003, Frippiat and Holeyman 2008). However, classical random walk particle tracking methods are still often implemented with discrete time steps (discretization of time in given time steps) and variable spatial increments that depend on the velocity field and the random noise (see Chapter 13).

CTRW methods have allowed simulating anomalous (non-Fickian) transport behavior in heterogeneous media at different scales (among others, Berkowitz *et al.* 2006, Le Borgne *et al.* 2008). Solute transport can thus be considered "spatially anomalous" and/or "temporally anomalous." This topic is currently in rapid development and new techniques similar to, or improving the CTRW approach are emerging (e.g., the time domain random walk TDRW where the transition times are exponentially distributed [Russian *et al.* 2016]).

This issue is also linked to the problem of knowing which part of the saturated porosity is occupied by mobile water, on which the transport velocity is based (Payne *et al.* 2008, Hadley and Newell, 2014). In particular, in rock aquifers where fracture porosity is usually <1%, very rapid transport velocities are observed (Worthington 2015). The immobile water contributes to solute transport through the "matrix diffusion" process (see next paragraphs). In aquifers, advection combined with dispersion and matrix diffusion will determine first solute arrival times with much less dilution than expected when applying the macrodispersion Fickian concept (Molz 2015).

Adsorption-desorption

All physical, chemical, and biological processes leading to the adhesion of groundwater solutes to the surface of the solid matrix or grains are described under the term *adsorption*. In some circumstances, the processes can be (partially or totally) reversible and *desorption* is observed. Adsorption-desorption processes (often qualified as *adsorption-desorption reactions*) can be very important in groundwater (in both the partially saturated and saturated zone) as they influence solute transport. In this section, the reference condition will be in saturated geological (porous) media, but indeed all described processes can also be considered active in partially saturated conditions. As observed at many contaminated sites, and especially for inorganic contaminants, adsorption-desorption can often be one of the most important solute transport processes to be taken into account. Adsorption is more important in organic matter and clay rich sediments. The main effect of adsorption-desorption processes on the fate of solute contaminant in groundwater is a retardation and possibly a decrease of the observed concentrations in the breakthrough curve (Figure 8.8).

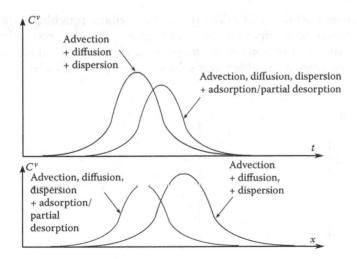

Figure 8.8 Effect of adsorption-desorption processes on a breakthrough curve of a solute as measured at a given distance from the injection point (above), and on a plume measured in the direction of the (assumed uniform) groundwater velocity (below).

When considering solute transport in groundwater, the term adsorption-desorption usually includes many processes such as physical sorption due to van der Waals forces, chemical sorption due to covalent bonding, electrostatic attraction, precipitation-dissolution, and oxidation-reduction. As part of geochemistry and hydrochemistry, all those processes are described in many specialized books. If we do not want to consider the complexity of surface complexation models (SCMs), adsorption processes are usually described by empirical relations whose parameters are fitted to experimental lab or field measurements and based on strong assumptions such as equilibrium and isothermal conditions (Davis and Kent 1990). Indeed, this is only a conceptual way of describing retardation, neglecting the impact of variable chemical conditions, and not explicitly relating the adsorption reactions to thermodynamic data.

Interactions with solid surfaces is also the main "filtration" process associated with decay of pathogens in saturated porous and fissured media. Filtration is favored by the fact that pathogens are larger than solutes. Since outbreaks of human disease can be induced by a very small exposure, particular attention should be given to pathogen transport in preferential flowpaths and the associated leading edges of the channeling plume that is affected by adsorption and decay (Hunt and Johnson 2017).

Linear adsorption isotherm

The *partitioning* between the mass of solute adsorbed on the solids and the mass remaining in the mobile groundwater in the REV is assumed linear under isothermal and equilibrium conditions. If the mass concentration adsorbed on the solid is expressed as a mass of the considered species per unit mass of solid $\bar{C}$ (-), a linear empirical adsorption isotherm partitioning relation is written as:

$$\bar{C} = K_d \rho C \quad \text{or} \quad \bar{C} = K_d C^v \tag{8.28}$$

where K_d is the *partitioning coefficient* in m³/kg [L³M⁻¹] describing roughly, at the REV scale, the global affinity of a solute for the solid (Figure 8.9a). As previously, C^v and C are respectively the *volume concentration* and *mass concentration* (i.e., the latter being used where and when a variable water solution density can be expected).

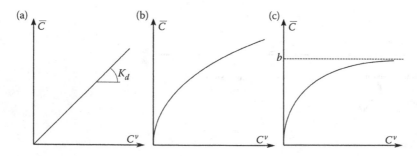

Figure 8.9 The different simplified adsorption-desorption equilibrium and isotherm relations (a) K_d linear relation, (b) Freundlich relation, and (c) Langmuir relation (Goldberg *et al.* 2007).

When the contribution of adsorption-desorption is taken into account in the solute mass balance equation, the coefficient K_d will contribute to the retardation factor R (See Equation 8.46). The use of this assumed constant K_d partitioning coefficient (to be fitted on experimental data) leads to strong limitations (Goldberg et al. 2007): (a) it cannot represent the contributions of different processes to solute retardation, (b) there is no maximum adsorption limit, and (c) its application is problematic for inorganic solutes when variations of pH, alkalinity, or complexing ligand concentrations are expected. Thus K_d does not adequately describe adsorption-desorption processes in geological media where large variations of lithology and hydrochemistry are expected. Another bias comes from the fact that K_d values are often deduced from lab tests where conditions are relatively homogeneous and controlled. In sediments, higher K_d values are found for clay-rich formations. But K_d values can still reflect that cations are adsorbed more than anions, bivalent cations are adsorbed more than monovalent ones, etc. For adsorption of organic species (from NAPLs) that are hydrophobic, a partition coefficient K_{ow} is defined as the octanol—water partition coefficient (-) (see Section 8.3).

Freundlich adsorption isotherm

A nonlinear equilibrium isotherm relation was suggested by Freundlich (1926) (Figure 8.9b):

$$\bar{C} = K_{Fr}\rho C^m \quad \text{or} \quad \bar{C} = K_{Fr}C^{v^m} \tag{8.29}$$

where K_{Fr} and m are empirical coefficients to be fitted on experimental data. If m is found equal to 1, this isotherm relation becomes linear and identical to the partition coefficient approach (i.e., $K_{Fr} = K_d$). The disadvantages of this method are generally similar to those of the previous one.

Langmuir adsorption isotherm

Still assuming equilibrium conditions, but introducing an upper limit to the available solid surface area for adsorption, Langmuir (1918) proposed the following adsorption isotherm (Figure 8.9c):

$$\bar{C} = \frac{K_{La}b\rho C}{1 + K_{La}\rho C} \quad \text{or} \quad \bar{C} = \frac{K_{La}bC^v}{1 + K_{La}C^v} \tag{8.30}$$

where K_{La} is an empirical constant, b is the maximum capacity of the solid matrix in the REV. The equations are to be fitted on experimental data (i.e., nonlinear regression). The advantage of this Langmuir relation lies in the description of the decrease in sorption capacity (Figure 8.9c) as the solid matrix surface becomes partially saturated with the adsorbed species (Goldberg et al. 2007).

The literature describing these equilibrium relations is broad and other formulations have been proposed but are less used in practice.

Kinetic (nonequilibrium) adsorption isotherms

When adsorption is considered as operating slowly with regards to the solute velocity, kinetic adsorption relations are preferred. For example, the simplest nonequilibrium relation is written for an irreversible adsorption as:

$$\frac{\partial \bar{C}}{\partial t} = k_1 \rho C \quad \text{or} \quad \frac{\partial \bar{C}}{\partial t} = k_1 C^v \tag{8.31}$$

where k_1 is a constant coefficient to be fitted on experimental adsorption data.

If the process can be reversible, the simplest first-order isotherm can take the form:

$$\frac{\partial \bar{C}}{\partial t} = k_1 \rho C - k_2 \bar{C} \quad \text{or} \quad \frac{\partial \bar{C}}{\partial t} = k_1 C^v - k_2 \bar{C} \tag{8.32}$$

where k_2 is a constant to describe desorption to be determined experimentally. Note that at equilibrium (i.e., $\partial \bar{C}/\partial t = 0$), Equation 8.32 is reduced to the linear adsorption relation with $K_d = k_1/k_2$.

To compare with the Freundlich approach, the nonequilibrium relation of Equation 8.32 can be generalized as:

$$\frac{\partial \bar{C}}{\partial t} = k_1 \rho C^n - k_2 \bar{C} \quad \text{or} \quad \frac{\partial \bar{C}}{\partial t} = k_1 C^{v^n} - k_2 \bar{C} \tag{8.33}$$

where n is an empirical exponent to be found on the basis of experimental data. At equilibrium, Equation 8.33 is reduced to the Freundlich adsorption relation with $K_{Fr} = k_1/k_2$.

A combination of two relations is also possible to describe an initial rapid reversible adsorption of the solute followed by a slow and less reversible adsorption. The first stage is described by an equilibrium reversible relation while the second by a first-order reversible kinetic relation (Cameron and Klute 1977).

At this stage, the adsorption-desorption term to be used in the solute mass balance equation can be expressed by $-\rho_b \mathcal{R}_{w,s}$ (in kg/m³s) where $\mathcal{R}_{w,s}$ is the rate of mass flux in kg/(kg s) from water to the solid, ρ_b is the dry bulk density of the medium ($\rho_b = (1 - n)\rho_s$) with ρ_s being the solid matrix density, the negative sign comes from the fact that the mass flux is leaving the groundwater.

Decay/degradation

Some solutes may be subject to decay or degradation due to various physical or biochemical processes. Decay is here understood to be the result of various possible processes or reactions degrading the considered solute (parent species) into one or several daughter product(s). If the daughter solutes are to be taken into account then transport of multiple reactive solutes must be considered. If only the mass balance of the initial solute is considered, decay is handled as a sink (i.e., disappearance) of mass of the considered solute species in the REV. The main result of decay processes on the fate of a solute contaminant in groundwater is a general attenuation of the concentrations in the breakthrough curve (Figure 8.10). One describes here the general possible

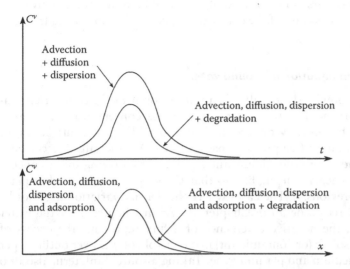

Figure 8.10 Effect of decay/degradation on a breakthrough curve of a solute as measured at a given distance from the injection point (above), and on a plume measured in the direction of the (assumed uniform) groundwater velocity (below).

effect on solute concentrations by several tens of reactions or partial reactions, under equilibrium or nonequilibrium conditions, or under locally heterogeneous equilibrium conditions, depending on the nature of the medium and of the groundwater flow velocity. At this stage, it is not the aim to consider the full complexity of geochemical reactions which are highly dependent on local thermodynamic data. The simplest law describing decay which is linearly proportional to the current solute concentration can be expressed as:

$$\frac{\partial(\rho C)}{\partial t} = -\lambda \rho C \quad \text{or} \quad \frac{\partial C^v}{\partial t} = -\lambda C^v \tag{8.34}$$

where λ is a first-order (linear) decay constant in s^{-1} [T^{-1}]. This law is often used for radioactive decay where λ takes the value:

$$\lambda = \ln \frac{2}{t_{1/2}} \tag{8.35}$$

where $t_{1/2}$ is the half-life or the time it takes for half of the substance to decay. Sometimes, for reactions causing decay of a solute, a more general power law is considered:

$$\frac{\partial(\rho C)}{\partial t} = -\Gamma \rho C^n \quad \text{or} \quad \frac{\partial C^v}{\partial t} = -\Gamma C^{v^n} \tag{8.36}$$

where n is an exponent depending on the reaction stoichiometry, Γ is a decay constant to be fitted on experimental data.

Adopting the linear degradation law, the degradation term to be used in the solute mass balance equation can be expressed by $-n\lambda \rho C$ (kg/m^{3}s) where the total porosity n

is used since degradation occurs in mobile as well as immobile water, the negative sign comes from the fact that a mass flux of the considered species is disappearing from the groundwater solution.

Solute mass conservation equation in groundwater

The processes described previously are all taken into account in the conceptual diagram in Figure 8.11. For mobile groundwater, the solute conservation or balance equation can be written for an REV or a control volume adding the contribution of the different solute mass fluxes (see previous paragraphs). Additionally, we consider that the density of groundwater is not significantly changed by varying solute concentrations (i.e., density changes are negligible) so that C^v (kg/m³) can be used in place of $\rho(C)C$ (as explained previously). The cases where the density variation may not be neglected will be considered in detail in Chapter 10. As in Chapter 4 for the water mass balance equations, the negative divergence of a flux represents the excess of inflow of the considered solute (or contaminant) in a control volume over outflow, per unit volume of porous medium and per unit time. Taking a source/sink term also into account, the solute mass balance equation is written:

$$\frac{\partial}{\partial t}(n_m C^v) = -\nabla \cdot f_a - \nabla \cdot f_d - \nabla \cdot f_m - \rho_b \mathcal{R}_{w,s} - n\lambda C^v + M^v \tag{8.37}$$

where M^v is an external solute mass source (if >0) or sink (if <0) term in kg/m³s [ML⁻³T⁻¹]. Note that all terms of this equation are expressed in kg/m³s [ML⁻³T⁻¹], a mass of solute per volume of REV and per time. Introducing in Equation 8.37 the values of f_a, f_d, and f_m from, respectively, Equations 8.7, 8.11, and 8.5:

$$\frac{\partial}{\partial t}(n_m C^v) = -\nabla \cdot (q C^v) + \nabla \cdot n_m (D \cdot \nabla C^v) + \nabla \cdot n(D_m \nabla C^v) - \rho_b \mathcal{R}_{w,s} \tag{8.38}$$
$$- n\lambda C^v + M^v$$

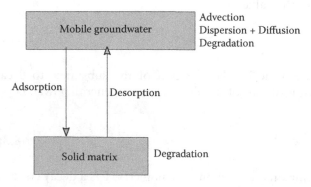

Figure 8.11 Conceptual diagram showing the main solute transport processes taken into account in the solute mass balance Equations 8.39 and 8.40, respectively written for the mobile groundwater and the solid matrix.

As proposed by Rausch *et al.* (2005), for the sake of simplicity, let's assume that the total porosity n is reduced to the effective transport porosity n_m in the diffusive mass flux and for the degradation. With this hypothesis, diffusion and degradation are limited to the mobile water (see next paragraph for more details). Equation 8.38 then becomes:

$$\frac{\partial}{\partial t}(n_m C^v) = -\nabla \cdot n_m(v_a C^v) + \nabla \cdot n_m(D_b \cdot \nabla C^v) - \rho_b \mathcal{R}_{w,s} - n_m \lambda C^v + M^v \tag{8.39}$$

Due to adsorption-desorption processes and degradation, a solute mass balance on the solid matrix is thus considered as:

$$\frac{\partial}{\partial t}(\rho_b \bar{C}) = \rho_b \mathcal{R}_{w,s} - \lambda \rho_b \bar{C} \tag{8.40}$$

Assuming a linear adsorption-desorption K_d relation (Equation 8.28), Equation 8.40 becomes:

$$\frac{\partial}{\partial t}(\rho_b K_d C^v) = \rho_b \mathcal{R}_{w,s} - \lambda \rho_b K_d C^v \tag{8.41}$$

and

$$\rho_b \mathcal{R}_{w,s} = \lambda \rho_b K_d C^v + \frac{\partial}{\partial t}(\rho_b K_d C^v) \tag{8.42}$$

Entering Equation 8.42 in Equation 8.39 gives:

$$\frac{\partial}{\partial t}(n_m C^v) + \frac{\partial}{\partial t}(\rho_b K_d C^v) = -\nabla \cdot n_m(v_a C^v) + \nabla \cdot n_m(D_b \cdot \nabla C^v)$$
$$- \lambda \rho_b K_d C^v - n_m \lambda C^v + M^v \tag{8.43}$$

Then dividing Equation 8.43 by the assumed constant effective transport porosity n_m gives:

$$\left(1 + \frac{\rho_b}{n_m} K_d\right)\frac{\partial C^v}{\partial t} = -\nabla \cdot (v_a C^v) + \nabla \cdot (D_b \cdot \nabla C^v) - \left(1 + \frac{\rho_b}{n_m} K_d\right)\lambda C^v + \frac{M^v}{n_m} \tag{8.44}$$

The *retardation factor* is defined as the factor multiplying $\partial C^v/\partial t$:

$$R = \left(1 + \frac{\rho_b}{n_m} K_d\right) \tag{8.45}$$

Indeed, the expression of Equation 8.45 is linked to the assumed linear adsorption relation (i.e., using K_d). If, for some reasons (e.g., groundwater flow in low permeability layers), the groundwater velocity was calculated using the total porosity instead of the effective mobile water porosity, a similar definition of the retardation factor is found using n in place of n_m. Then, this replacement of n_m by n should also take place and in all relevant terms of Equation 8.39. This is justified as in low velocity flow fields, no distinction would be possible between immobile and mobile water in the saturated pore space. For transport calculation in low permeability porous media the retardation factor is thus:

$$R = \left(1 + \frac{\rho_b}{n} K_d\right) \tag{8.46}$$

If more elaborated adsorption-desorption models are adopted, other expressions of the retardation factor must be used. Examples are given in Fetter (1993) for the Freundlich and Langmuir relations. Entering the *retardation factor* definition in Equation 8.44, the advection-dispersion equation becomes:

$$R\frac{\partial C^v}{\partial t} = -\nabla \cdot (v_a C^v) + \nabla \cdot (D_b \cdot \nabla C^v) - R\lambda C^v + \frac{M^v}{n_m} \tag{8.47}$$

Equation 8.47 assumes that degradation occurs in both the mobile mass phase as well as the sorbed phase (Figure 8.11). This applies, for example to a radioactive species. On the contrary, for some organic contaminants, the sorbed phase may not decay (e.g., due to a limited accessibility for the microbes), thus one would have a slightly different form for this equation. If Equation 8.47 is divided by R, one can observe that the migration of any sorbing solute depends on the ratio v_a/R which is indeed slower than v_a. The source/sink term has until now been represented by M^v in kg/m³s [$ML^{-3}T^{-1}$]. In fact, two kinds of sources/sinks can be distinguished: (a) sources/sinks of solute mass linked to a groundwater flow rate exchanged with the external world with regards to the domain for which the balance equation is expressed; (b) sources/sinks of solute mass not associated with any flow rate (i.e., resulting from chemical reactions and immobile water effects, see next paragraphs). At this stage, if we neglect sources/sinks associated with chemical reactions and immobile water effects, the sources/sinks term is written as:

$$M^v = q_s C_s^v \tag{8.48}$$

where q_s is the volumetric flow rate per unit volume of porous medium (s^{-1})[T^{-1}] flowing into ($q_s > 0$) or flowing out from ($q_s < 0$) the control volume, and C_s^v is the associated concentration (in kg/m³) [ML^{-3}]. Note that if $q_s < 0$, automatically $C_s^v = C^v$ because any groundwater flowing out is assigned the local concentration in the control volume.

Expansion of the advective term in Equation 8.47 gives:

$$\nabla \cdot (v_a C^v) = C^v \nabla \cdot v_a + v_a \cdot \nabla C^v \tag{8.49}$$

If the fluid (groundwater) is assumed incompressible, the only internal change in v_a is due to internal sources or sinks so that:

$$C^v \nabla \cdot v_a = C^v q_s \tag{8.50}$$

Introducing this value in Equation 8.49 and combining Equations 8.49, 8.48, and 8.47, the solute mass balance equation is written as:

$$R\frac{\partial C^v}{\partial t} = -\boldsymbol{v}_a \cdot \nabla C^v + \nabla \cdot (\boldsymbol{D}_h \cdot \nabla C^v) - R\lambda C^v - \frac{q_s}{n_m}(C^v - C_s^v) \tag{8.51}$$

Indeed, the groundwater flow field must be known before solving this equation. Note that if the source/sink concentration C_s^v is equal to the local concentration C^v, the last term of Equation 8.51 is equal to zero meaning that it does not affect concentrations (Rausch *et al.* 2005). This is usually the case for sinks corresponding to an outflowing flow rate with an associated local concentration. On the other hand, if a highly concentrated water is injected ($C_s^v > C^v$), this source will produce increasing concentrations in the considered volume. Finally, to be complete, if an injection of water occurs with a lower concentration than the local one ($C_s^v < C^v$), a dilution of local solute concentrations is obtained.

Now, if we want to take into account sources/sinks of solute mass that are not associated with any flow rate but resulting from an immobile water effect and/or from chemical reactions, additional terms would need to be expressed in which would appear on the right-hand side of this equation.

Immobile water effect/matrix diffusion

As mentioned previously in the discussion about solute advection velocity, relatively immobile water may occupy a significant part of the saturated pores. Immobile water generally corresponds to subzones of very low permeability (e.g., clay and loam lenses in sedimentary deposits) or dead-end pores or microfissures where groundwater can be considered as quasi-immobile. In this case, a dual-porosity concept, with micropores in the solid porous matrix and macropores corresponding to fractures, fissures, or bedding, is considered (Gerke and van Genuchten 1993). If applied at larger scales, subzones of low and high hydraulic conductivity also form a dual permeability medium. In this conceptual model, a fraction of the groundwater is much less mobile compared to highly mobile water in preferential pathways or fissures (Figure 8.12). Diffusion may therefore occur between mobile

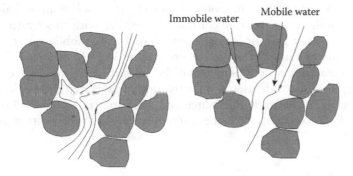

Figure 8.12 Schematic microscopic view of highly mobile water in fissures or macropores and quasi-immobile water in intergranular pores of a dual-porosity medium.

water and immobile water as a result of solute concentration differences (i.e., between the solute concentration in the mobile water (C_m) and the solute concentration in the immobile water (C_{im})). When a contamination occurs, for example in a fissured medium, the concentration (C_m) first rises in the very mobile groundwater in the fissures. This mobile water will be in contact with quasi-immobile water at a concentration (C_{im}) in the matrix pores. Assuming $C_m > C_{im}$, the concentrations difference induces diffusion from the mobile into the immobile water, and a slow increase of C_{im} is observed. When the peak concentration has passed, C_m decreases and may become lower than C_{im}, inducing back diffusion from the immobile into the mobile water. In a breakthrough curve (BTC), the net result is a longer tailing of the curve (Figure 8.13).

Detected first as a result of tracer tests in fissured and porous rocks, this effect of heterogeneity on solute behavior at the REV scale has since been observed in almost all geological media. Even in high hydraulic conductivity formations, as for example alluvial aquifers, heterogeneous conditions are often observed due to sand, silt, loam, and peat that can be mixed with gravels, appearing as small low permeability subzones or lenses. A conceptual model including preferential flowpaths with highly mobile groundwater and subzones with quasi-immobile water is often a realistic way of describing the actual hydrogeological conditions.

A conceptual approach distinguishing two apparent water phases (i.e., mobile and immobile water) is usually adopted to represent this *matrix diffusion* or *immobile water effect* (Coats and Smith 1964, Skopp and Warrick 1974, Rao *et al.* 1982, Bear and Verruijt 1987). In this case, a linear exchange relationship is expressed:

$$f_m^{im} = \alpha_d^m(\rho C - \rho_{im}C_{im}) \quad \text{or} \quad f_m^{im} = \alpha_d^m(C^v - C_{im}^v) \tag{8.52}$$

where f_m^{im} is the diffusive solute mass flux (kg/m^{3}s) [ML^{-3}T^{-1}] from mobile to immobile water (or vice-versa), C_{im} is the mass concentration in the immobile water and ρ_m the corresponding density of immobile water, α_d^m is defined as the matrix diffusion coefficient or immobile water diffusion coefficient (s^{-1}) [T^{-1}], similar to a diffusion coefficient (m^2/s) divided by a surface area (m^2) intended to describe the mobile water—immobile water contact area. This empirical coefficient is most often determined experimentally from lab and field tracer tests results. In practice, in solute transport models which account for the immobile water effect, a solute mass balance equation in the immobile water must be considered together with this transfer Equation (8.52) to obtain the net gain or loss of solute mass in the mobile water at each time step (see dual-domain mass transfer in Chapters 12 and 13). In other words, we are facing a set of interacting equations that will be simplified by making reasonable assumptions. If we neglect diffusion within the immobile water and neglect adsorption-desorption between the immobile water and solid matrix (Figure 8.14), a simplified solute mass balance equation in the immobile water can be written as follows:

$$\frac{\partial}{\partial t}(n_{im}C_{im}^v) = f_m^{im} - \lambda n_{im}C_{im}^v \tag{8.53}$$

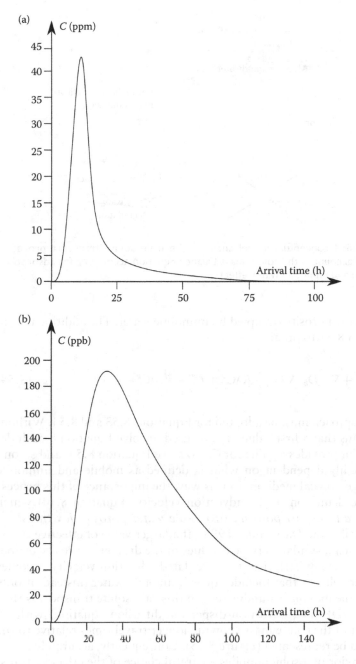

Figure 8.13 Measured BTCs in the pumping well for tracer tests performed under radially convergent conditions, in a fissured and slightly karstified chalk (a), and in fluvial sediments (b), showing a clear tailing of the curve (Hallet and Dassargues 1998, Haerens *et al.* 1999).

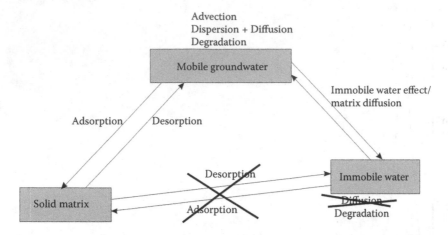

Figure 8.14 The simplified conceptual model showing the main solute transport processes taken into account in the solute mass balance equation accounting for adsorption-desorption and immobile water effects.

where n_{im} is the part of the porosity occupied by immobile water. The additional term f_m^{im} is added to Equation 8.51 to obtain:

$$R\frac{\partial C^v}{\partial t} = -v_a \cdot \nabla C^v + \nabla \cdot (D_h \cdot \nabla C^v) - R\lambda C^v + f_m^{im} - \frac{q_s}{n_m}(C^v - C_s^v) \qquad (8.54)$$

A coupled solution approach must be adopted for Equations 8.53 and 8.54. Without too much detail, it means that a first value of C^v is used to solve Equation 8.53 independently, which in turn, provides a value for C_{im}^v to solve Equation 8.54, and so on.

Note that we are highly dependent on what is defined as mobile and immobile water in the saturated geological medium. In this way, the importance of this process is closely linked to the definition of the advection velocity (Equation 8.10) which depends on the *effective transport porosity* or *mobile water porosity for transport* (n_m) (Equation 4.23; Hallet and Dassargues 1998). If a larger value of effective transport porosity is chosen (e.g., similar to typical values of the drainage porosity or specific yield in an unconfined aquifer), then the calculated advection velocity becomes an REV-averaged lower velocity that includes quasi immobile water (instead of only being representative of the moving groundwater). In this case solute transport is then only described by the ADE (i.e., advection-dispersion-diffusion equation) with no tailing (i.e., a large part of the solute mass showing longer travel times relative to the median in a BTC) could be represented (Figure 8.15). Consequently, accurately representing the tailing behavior of a solute implies a careful choice of the effective transport porosity in combination with dispersivities and the immobile water diffusion coefficient (α_d^m). If α_d^m is too high (i.e., the exchange between the mobile and immobile waters becomes increasingly and unrealistically fast), results will be more and more similar to the case where the chosen porosity approaches the total porosity (Zheng and Wang 1999). On the other hand, if α_d^m is nearly zero, results become similar to the

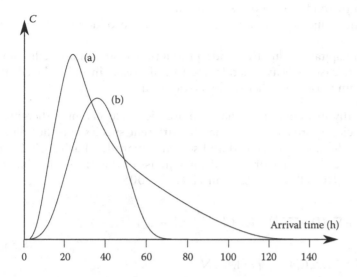

Figure 8.15 In order to represent the tailing behavior of a measured BTC (curve a), a relatively small effective transport porosity must be chosen (i.e., for a fast advection) combined with dispersion and the immobile water effect. If advection is calculated with a higher porosity, including some or all of the immobile water, a classical ADE is used (curve b) which will produce almost no tailing.

case where the chosen porosity approaches the mobile water porosity (sensu stricto) which may unrealistically neglect diffusion effects with immobile water. Zheng and Wang (1999) also argued that introducing an immobile water effect may account for implicitly introducing a component of dispersion (i.e., at least microdispersion).

Reactive solute transport

An additional chemical reaction term could be added to Equations 8.51 or 8.54, to include the effect of any biochemical or chemical reactions on solute contaminant transport. Indeed, a number of reactions are already included under adsorption-desorption and degradation processes, but various other reactions involving complexation, dissociation, and redox reactions need to be included with more complex higher-order relationships.

In this case, the solute transport problem becomes a multispecies reactive solute transport problem. Multicomponent reactive transport is particularly important for realistically describing and predicting behavior of contaminant plumes but it is also widely used to assess the best remediation techniques.

A few classical examples can be cited to show the importance of these multispecies reactions:

a. redox reactions leading to the oxidation of pyrite may contribute to denitrification processes in chalk aquifers but may also contribute, in other environments, to acid drainage for example in mine tailings
b. reductive dechlorination of chlorinated organic contaminants

   c. biodegradation by oxidation of organic contaminants

   d. calcite dissolution due to microbiological/microbial production of carbon dioxide

   e. heavy metal precipitation due to sulfide production from sulfate reduction or, conversely, release of heavy metals due to a decrease in pH induced by acidification from pyrite (or other sulfide) oxidation

Biotic (microbiologically mediated) reactions and abiotic reactions must therefore be coupled. A multispecies approach means that N_s different species specific solute mass balance equations should be developed and solved in parallel. Each includes a "reaction term" expressing the source or sink of solute mass of the ith species due to reactions with the others (Kinzelbach 1992, Rausch *et al.* 2005):

$$R_i \frac{\partial C_i^v}{\partial t} = -v_{a_i} \nabla C_i^v + \nabla \cdot (D_b \cdot \nabla C_i^v) - R_i \lambda_i C_i^v - \frac{q_s}{\theta_i}(C_i^v - C_{s_i}^v)$$

$$+ \frac{1}{\theta_i} \sum_{j=1}^{N_s} S_{ij}(C_1^v, \ldots, C_n^v) \qquad i = 1, \ldots, N_s \tag{8.55}$$

where S_{ij} is the source/sink term representing the effect of reactions (kg/m³s)[ML⁻³T⁻¹], and θ_i is the groundwater specific volume fraction of the REV where the species i is located. There are as many equations as species being considered in the reaction system: N_s which are coupled through the $S_{ij}(C_1^v, \ldots, C_n^v)$ terms. If all reactions occur in the water phase, θ_i are all equal to n_m and the components of v_{a_i} are all equal to v_a (i.e., the advection velocity), which is defined as a *homogeneous reaction system*. On the contrary, if a part of the involved species is on the solid matrix or in the immobile water, the reaction system is defined as heterogeneous, v_{a_i} and D_b being equal to zero for the species in those immobile phases. Equation 8.55 is thus used for multispecies reactive transport in mobile groundwater (i.e., only a single moving phase) taking into account immobile water and solid matrix (i.e., two immobile phases). Multiphase flow and transport are beyond the scope of this book except for what is given in Chapter 9, Section 9.4.

As summarized by Rausch et al. (2005), some reactions can be considered as *equilibrium reactions* if they are reversible and if their reaction rate is high (faster than transport). Chemical equilibrium equations can then be used to calculate the S_{ij} components. This is the case, for example, for dissociation reactions and some precipitation/dissolution reactions (e.g., carbonates in groundwater). All other reactions are considered as *kinetic reactions* (i.e., not having time to reach equilibrium under the assumed transport conditions). A reaction rate is defined for the appearance and disappearance of each species:

$$\frac{dC_i^v}{dt} = \pm \kappa_i (C_i^v)^{r_i} \tag{8.56}$$

where κ_i is the reaction rate coefficient (+ or − sign indicates appearance or disappearance) of species i and r_i is the reaction order with respect to the ith species. If r_i is zero, it means that the reaction rate does not depend on this species. In practice, it is often assumed that the reaction evolution can be described by first-order reactions ($r_i = 1$) in groundwater systems (Fetter 1993). In a sequential reaction system, the total reaction rate of the ith species is:

$$\frac{dC_i^v}{dt} = \nu_i \kappa_{i-1}(C_{i-1}^v)^{r_{i-1}} - \kappa_i(C_i^v)^{r_i} \tag{8.57}$$

where ν_i is the stoichiometric coefficient of the reaction producing species i (as daughter species) from the species $(i - 1)$ (parent species), $-\kappa_i(C_i^v)^{r_i}$ accounts for the disappearance of the ith species (now a parent species) in favor of the $(i + 1)$th species (daughter species of ith species). Taking the theoretical example (modified from Sun et al. 1999, Diersch 2014) of Figure 8.16 where a reaction network is $A \to B \to C_1, C_2$ and $C_2 \to D$, with all species considered as solutes in the mobile water (i.e., a homogeneous reaction), the reaction rates are respectively:

$$\frac{dC_A^v}{dt} = -\kappa_A C_A^v \tag{8.58}$$

$$\frac{dC_B^v}{dt} = \nu_B \kappa_A C_A^v - \kappa_B C_B^v \tag{8.59}$$

$$\frac{dC_{C_1}^v}{dt} = \nu_{C_1} \kappa_B C_B^v - \kappa_{C_1} C_{C_1}^v \tag{8.60}$$

$$\frac{dC_{C_2}^v}{dt} = \nu_{C_2} \kappa_B C_B^v - \kappa_{C_2} C_{C_2}^v \tag{8.61}$$

$$\frac{dC_D^v}{dt} = \nu_D \kappa_{C_2} C_{C_2}^v - \kappa_D C_D^v \tag{8.62}$$

where $\nu_B, \nu_{C_1}, \nu_{C_2}$ and ν_D are stoichiometric coefficients for reactions inducing appearance of B, C_1, C_2, and D respectively. These reaction rates can be used in the five solute transport equations written on the basis of Equation 8.55: one written for each species using respectively Equations 8.58 through 8.62.

Unfortunately, most of the microbiologically enhanced reactions are not well described with first-order and zero-order reactions. Reactions rates could be rate-limited and sometimes they could even be inhibited by the presence of a specific species. Many examples can be found in reactions that are favored either under anaerobic or in aerobic conditions. One of these is the sequential dehalogenation of chlorinated

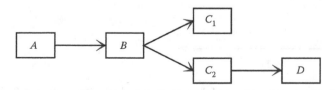

Figure 8.16 Sequential and parallel reaction network (Sun *et al.* 1999, Diersch 2014).

solvents (PCE) by anaerobic bacteria which is inhibited under aerobic conditions. Therefore, when contamination occurs, initial aerobic conditions favor aerobic *mineralization* (i.e., transformation into inorganic products) of TCE. When oxygen concentrations decrease to below 0.5 mg/L and nitrate concentrations to below 1 mg/L (according to Wiedemeier *et al.* 1999), anaerobic sequential reductive dechlorination starts. A single plume may include zones where aerobic and anaerobic conditions are encountered. Aerobic conditions are most often encountered in the furthest downgradient parts of the plume (i.e., fringes) while anaerobic conditions mainly prevail in the center of the plume or near the source zone. As Monod recognized that the growth rate of the useful microbial population could be restricted by some species (as dissolved solutes or sorbed on the solid), an empirical Monod model describes a limited reaction rate (Figure 8.17):

$$\frac{dC_i^v}{dt} = -\frac{S_{Mo}C_i^v}{\kappa_{Mo} + C_i^v} \tag{8.63}$$

where S_{Mo} is the Monod maximum reaction rate and κ_{Mo} is the Monod reaction constant. One can observe (Figure 8.17) that for high concentrations (i.e., with regards to κ_{Mo}), a zero-order reaction is found with a fixed maximum reaction rate $(-S_{Mo})$, and for low concentrations, the reaction rate can be approximated by a first order reaction with $\kappa_i = S_{Mo}/\kappa_{Mo}$. Multiple Monod models can be needed to describe limitations of the reaction rate due to multiple species. In some circumstances, a model describing the inhibition of the considered reaction could be needed.

Many complex models representing various microbially-enhanced chemical reactions in groundwater have been developed. They are very useful for an adequate description and prediction of moving contaminant plumes in saturated and partially saturated zones. However, details of these multispecies reactions and interactions, within the different geological contexts, fall beyond the scope of this

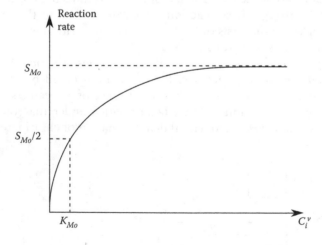

Figure 8.17 Evolution curve of the reaction rate following a Monod relation (Equation 8.63) (Diersch 2014).

book. More useful details dedicated specifically to multicomponent reactive solute transport in geological media can be found in the following references (among others): Appelo and Postma 1993, Knox *et al.* 1993, Parkhurst 1995, Waddill and Widdowson 1998, Deutsch 1997, Wiedemeier *et al.* 1998, Zheng and Wang 1999, Toride *et al.* 1999, Prommer 2002, Prommer *et al.* 2003, Bear and Cheng 2010, Diersch 2014.

8.3 NAPL contaminant transport

As organic compounds became very useful over the last century in many aspects of day-to-day life, their appearance as groundwater contaminants has grown. While some cases of groundwater contamination by organics have been accidental, others were not. Between the 1940s and 1990s, in particular, much groundwater contamination from NAPL products occurred simply by total ignorance of the threats they may induce in the environment. Many contaminated sites resulted from a direct product dump and infiltration into the soil surface. Once the pollutant had infiltrated, it was considered gone, or no longer a concern for most of the industrial and military users of the time.

Many organic contaminants which form a separate liquid phase migrate through the subsurface first under the form of a separate NAPL (i.e., nonaqueous phase liquid, see Section 7.1). Further subsurface migration of these contaminants depends on their physical properties in combination with the heterogeneity of the geological medium. Depending on their solubility, for example, they can partially or completely dissolve in groundwater and can then migrate as do other dissolved solutes transported by flowing groundwater. Depending on their hydrophobicity, organic contaminants can also be retained by the solid matrix, and according to their vapor pressure, they can volatilize into the air phase within the partially saturated zone. Indeed, depending on their density, they can move in the saturated zone as a DNAPL or LNAPL. A DNAPL (Dense Nonaqueous Phase Liquid) is denser than water and tends to sink in the partially saturated zone, as well as in the saturated zone. An LNAPL (Light Nonaqueous Phase Liquid) is lighter than water and tends to float on the water table.

NAPL solubility in groundwater

NAPL solubility in groundwater is defined as the maximum concentration $[ML^{-3}]$ of the NAPL component as a result of its dissolution at equilibrium in groundwater:

$$S^o_{NAPL\,in\,gw} = C^v_{NAPL,eq\,in\,gw} \tag{8.64}$$

This solubility is higher for higher temperatures. Except for MTBE and DCM (see Box 8.5), organic compounds have relatively low solubilities due to their nonpolar behavior. However, most often, NAPL contaminants are mixtures of different compounds. For example, a gasoline can be a variable mixture of benzene, toluene, ethylbenzene, m-, p-, and o-xylenes with MTBE as an additive (Box 8.5). Each individual compound of the complete NAPL (i.e., NAPL gasoline) will dissolve into water at different rates and solubilities. The actual dissolved concentration of each compound will be less than what would be allowed by its respective "pure" solubility (i.e., if 100% of the NAPL was composed of this compound). For a mixed-component NAPL in groundwater, under ideal and equilibrium conditions, the dissolved concentration C_i^v [ML^{-3}] of the ith compound is expressed by Raoult's law:

$$C_i^v = X_i S_i^o \tag{8.65}$$

where X_i is the mole fraction of compound i (number of moles of the compound i divided by the total number of moles in the substance), S_i^o is the "pure" solubility of the compound i (for a single-component NAPL or if 100% of the NAPL was composed of this compound).

Box 8.5   Most common organic contaminants (among hundreds of other substances)

- *Benzene, toluene, ethylbenzene, m-, p-, and o-xylenes* as main constituents of petroleum-fuels (gasoline, jet fuel…), they are often mentioned under the general term of *BTEX*.
- *Benso(a)pyrene* is a polycyclic aromatic hydrocarbon (*PAH*) resulting from incomplete combustion, a constituent of coal-tar, coal-tar and wood-tar creosotes, motor oils, and gasoline.
- *Polychlorinated biphenyls* (*PCBs*) are organic chlorine compounds widely used as dielectric and coolant fluids in electrical applications, in hydraulic fluids, in lubrication oils, plastics, paints, inks, adhesives, and sealants.
- *Tetrachloroethylene (PCE), trichloroethylene (TCE), 1,1,1 trichloroethane (1,1,1-TCA), and 1,2-dichlorethane (1,2-DCA)* with intermediate compositions are solvents for metal degreasing, dry cleaning, and paints.
- *Dichloromethane (DCM)* or *methylene chloride* is also a degreasing solvent, a paint stripper, also used in aerosols and foams.
- *Methyl-tert butyl ether (MTBE)*, mainly used as an additive for boosting the octane rate of gasoline and improving combustion, was introduced extensively after lead was banned.

Source: Fitts 2002.

Taking the example of a gasoline in equilibrium contact with groundwater in a porous medium, the individual dissolved concentrations can be calculated (as shown in Table 8.1) following these different steps (Fitts 2002):

- Assess an average molecular weight taking into account all individual compounds in the gasoline (NAPL mixture).
- Calculate the molecular weight of the ith compound from its chemical formula.
- Calculate X_i from the known proportion in weight of the ith compound in the gasoline (NAPL mixture).
- Use Equation 8.65 to find C_i^v [ML^{-3}].

Table 8.1 Individual equilibrium dissolved concentrations (C_i^v in mg/L) and pure solubility for different BTEX compounds of a given gasoline mixture (with a global averaged molecular weight of 101 g/mole)

Compound	Chemical formula	% in weight in mixture	X_i	C_i^v	S_i^o	Molecular weight
Benzene	C_6H_6	3	0.039	69.04	1780	78.12
Toluene	C_7H_8	20	0.219	113.99	520	92.15
Ethylbenzene	C_8H_{10}	2.5	0.024	4.28	180	106.18
m-, p-xylenes	C_8H_{10}	12	0.114	18.26–21.69	160–190	106.18
o-xylenes	C_8H_{10}	3	0.029	4.99	175	106.18
MTBE	$(CH_3)_3COCH_3$	9	0.103	4639	45,000	88.17

Sources: Fitts 2002, Cline *et al.* 1991.

NAPL affinity for solids and groundwater

The mobility of an organic contaminant under the form of a NAPL in groundwater is influenced by its "affinity" for soil particles or matrix solids. The ratio between the NAPL solubility in octanol and its solubility in water is used to provide a relative affinity:

$$K_{ow} = \frac{S_{NAPL\,in\,gw}^o}{S_{NAPL\,in\,octanol}^o} \tag{8.66}$$

where K_{ow} is the octanol-water partition coefficient (-) measuring hydrophobicity. Organic contaminants characterized by a high K_{ow} are generally retained on the solid matrix and thus quasi-immobile in the porous medium except if the NAPL pressure in the pores is sufficiently high, for example, in the case of a thick layer of NAPL.

If the natural organic content of the geological medium is not negligible, the nonpolar organic contaminant can sorb onto the natural organic fraction of the solid matrix (for more details about sorption see Section 8.2). An organic carbon-water partition coefficient can be defined as:

$$K_d = f_{oc}K_{oc} = f_{oc}\left(\frac{\bar{C}_{NAPL}}{S_{NAPL\,in\,gw}^o}\right) \tag{8.67}$$

where f_{oc} is the organic matter fraction (i.e., the quantity of natural organic matter in the REV matrix) [ML^{-3}], K_{oc} is the ratio of the adsorbed NAPL onto the matrix ($\bar{C}_{NAPL}$) on the dissolved NAPL in groundwater (taken equal to its solubility in groundwater ($S_{NAPL\,in\,gw}^o$) in the REV [ML^{-3}]).

Vapor mobility of NAPL organic contaminants

In the partially saturated zone, the NAPL vapor pressure ($p_{NAPL\,vapor}$) determines the equilibrium pressure of the gas phase in contact with the liquid NAPL phase. If a NAPL has a high vapor pressure, it is highly volatile and belonging to the VOCs (volatile organic compounds) family of contaminants.

A NAPL dissolved in groundwater in the partially saturated zone will be volatilized and mixed with air depending on its vapor pressure using the Henry's law:

$$p_{NAPL\,vapor} = HS^o_{NAPL\,in\,gw} \tag{8.68}$$

where $S^o_{NAPL\,in\,gw}$ is the solubility of the NAPL in the groundwater, H is Henry's constant (in SI units it should be in Pa m³/mol, but is often expressed in atm m³/mol), $p_{NAPL\,vapor}$ is the partial vapor pressure of the NAPL mixture. Similar to the solubility, the vapor pressure of an individual compound in a NAPL mixture is lower than what would be expected for the corresponding "pure" vapor pressure (i.e., if the NAPL substance was made of 100% of this compound). Again, applying Raoult's law, under ideal equilibrium conditions, the vapor pressure of the mixture is expressed as a function of the mole fraction of the compound i, and the "pure" vapor pressure of the compound i (if 100% of the NAPL was composed of this compound):

$$p_{NAPL\,vapor} = X_i\, p_{vapor\,of\,pure\,compound\,i} \tag{8.69}$$

NAPL mobility in unsaturated and saturated zones

NAPL mobility as a separate liquid phase indeed depends on its density as well as its viscosity. As mentioned previously, a DNAPL tends to sink in the unsaturated and saturated zones, while an LNAPL tends to float on the water table. For both liquids, as the NAPL thickness (or column height) increases, the influx of NAPL will also increase. The behavior of a NAPL in partially saturated and saturated media is in fact a multiphase flow problem (see Section 9.4). Infiltrating from a surface source, for example, the migration of a NAPL from one pore to another is highly dependent on the micro- and macroheterogeneities of the medium. As the migrate, NAPLs always leave behind small quantities mostly in the larger size pores because NAPLs are less wetting than water with respect to the solid matrix (Chapter 9, Section 9.4). The residual blobs can potentially act as sources for further volatilization (in the partially saturated zone) and dissolution inducing additional mobile contamination. In Figure 8.18, different situations are described for LNAPL and for DNAPL showing the most frequent specificities of those contaminations.

LNAPL

Moving generally vertically (depending on local heterogeneities) in the partially saturated zone, an LNAPL forms a floating lens above the water table or at the top of the capillary fringe. This floating phase will move in the direction of the water table gradient. If the gradient is low and the input of the LNAPL is large, the thickness of the floating phase can increase, possibly creating a slight depression of the water table interface (Figure 8.18). In unconfined conditions and with large water table fluctuations, an LNAPL could be smeared out over a large thickness inducing a complex spatial distribution of the NAPL including small entrapment zones. Note that problems can be encountered when assessing the actual thickness of the floating phase in the geological medium from measurements in monitoring wells. First, comparing measured values of the piezometric head without LNAPL and with floating LNAPL,

(a)

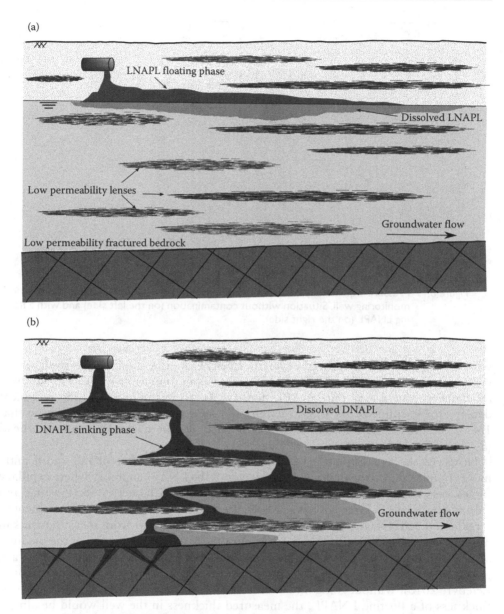

(b)

Figure 8.18 General migration behavior of an LNAPL (a), and a DNAPL (b) from a buried source in the partially saturated and saturated zones of a heterogeneous geological medium.

the pressure equilibrium induces a relation between the total thickness of the floating LNAPL (Δh_{LNAPL}) and the part of this thickness located above the natural (no NAPL) piezometric head (Figure 8.19):

$$\Delta h_{LNAPL} = \left(h_{LNAPL} - h\right)\frac{\rho_w}{\rho_w - \rho_{LNAPL}} \tag{8.70}$$

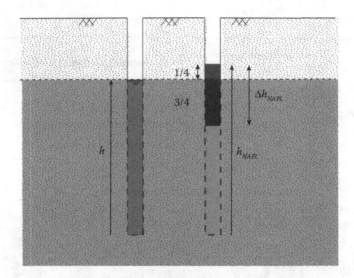

Figure 8.19 Pressure equilibrium between groundwater and a typical gasoline LNAPL in a monitoring well. Situation without contamination (on the left side) and with a floating LNAPL (on the right side).

where h_{LNAPL} is the measured level of LNAPL in the monitoring well, h the piezometric head in the well if there was no contamination, ρ_w the groundwater density, and ρ_{LNAPL} the LNAPL density. As an example, assuming a typical density for gasoline of 0.75 kg/m³, Equation 8.70 can be used to determine that 1/4 of the floating gasoline thickness will be above the natural piezometric head ($h_{gasoline} = h + \Delta h_{gasoline}/4$).

However, the actual situation in the geological medium is different and significantly more complex due to the multiphase character of the LNAPL mobility where capillary forces, relative wettability, pore apertures, and shapes (Chapter 9, Sections 9.2 and 9.4) all play a role. A large base of scientific literature is dedicated to the assessment of a realistic LNAPL floating thickness (for volume assessment) from measurements in monitoring wells. One must take into account the capillary properties of the medium as well as the interface properties between the LNAPL and the other phases (i.e., air, groundwater, solids) in terms of density and interfacial tension (Charbeneau 2000, Lefebvre 2010). Without going into details, it is observed (Figure 8.20) that for a large thickness of a floating LNAPL, the measured thickness in the well would be almost the same as in the medium. On the other hand, a very small thickness of free LNAPL implies that the thickness of the LNAPL in the well should be at least equivalent to the entry-pressure (see Chapter 9) of the LNAPL in the considered geological medium for a LNAPL-water system.

For remediation purposes, priority must be given to free-product recovery from aquifers using, for example, skimmer-, single-, and dual-pump well systems (Figure 8.21). Indeed, as long as a NAPL is present in the pores or fissures of the medium, it will continue to dissolve into the passing groundwater (i.e., the groundwater flow conditions bring new groundwater in contact with the NAPL, so that dissolution is continuous).

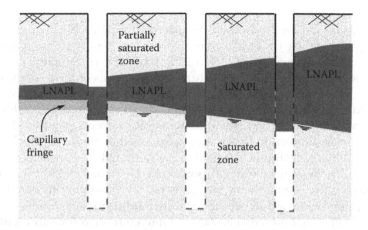

Figure 8.20 The ratio between Δh_{LNAPL} measured in the well and the actual Δh_{LNAPL} in the medium decreases with increasing thickness of the Δh_{LNAPL} (Lefebvre and Boutin 2000).

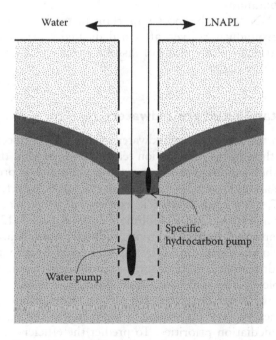

Figure 8.21 Example of a dual-pump system: drawdown of the water table is induced and optimized by groundwater pumping with (hopefully) low dissolved concentrations of LNAPL. The free product recovery is performed by a second (specific and smaller) pump. Continuous or intermittent regulation of both pumping rates is in practice very delicate as are also the initial setup, operation, and maintenance (EPA 1996).

DNAPL

Due to their higher density and lower viscosity compared to water, DNAPLs generally migrate vertically (depending on local heterogeneities) in the unsaturated and saturated zones (Figure 8.18). Migration of DNAPLs is actually very complex, more influenced by the permeability and porosity structures and contrasts of the geological medium. For example, water-saturated clays or loam-rich lenses form effective barriers for the downward migration of DNAPLs. DNAPL lateral migration will follow any slight dip of the layers or will accumulate until an overflow occurs (Figure 8.18). Migration can thus occur in a direction that is more controlled by the dip of the low permeability interfaces than the hydraulic gradient in the saturated zone, which often complicates the actual distribution of residual pools and blobs of DNAPL. These complex residual phases form contaminant source zones by dissolving in contact with "new" flowing groundwater following the local hydraulic gradient conditions. In fractured rocks, DNAPLs migration can be rapid in interconnected fractures with large apertures, while limited (due to capillary pressures) in fracture networks characterized by small apertures (see Chapter 9, Section 9.2). Remediation of DNAPL contamination is difficult and sometimes considered nearly impossible. The pure product remaining over relatively low permeability layers can be pumped using horizontal or vertical wells but it is unlikely to obtain efficient results. Other methods of remediation can be bioremediation, reactive barriers, or pump and treat operations, but these address the dissolved phase of the contaminant.

In the case of either LNAPLs or DNAPLs, the most important factor is to limit further dissolution in the moving groundwater. Even after a dedicated removal of the "pure" NAPL, the risk of generating a solute plume of dissolved compounds can remain important for many years (Figure 8.22).

8.4 In situ remediation of contaminated groundwater

Although undoubtedly one of the hottest topics in groundwater science and engineering, a comprehensive description of the various remediation techniques is beyond the scope of this book. A pragmatic synthesis about groundwater decontamination is proposed here. For more detailed information, the reader may consult the following references (among others): Suthersan 1997, Brady et al. 1998, Wiedemeier et al. 1999, Naftz et al. 2002, Suthersan 2002, Alvarez and Illman 2006, Cullimore 2008, Russel 2012.

The general purpose of remediation is indeed to treat polluted groundwater by removing or at least reducing the contamination. Many factors make this activity specific to each considered site. Hydrogeological conditions, even though they may appear similar, will never be completely the same and small differences (not only with respect to physicochemical groundwater properties but also in terms of microbiological conditions) may sometimes be very important to characterize the main active processes and to establish remediation priorities. To predict the efficiency of a particular remediation method, a clear understanding and modeling of pollutant transport is generally advised. However, local contaminant plumes often have characteristic lengths smaller than the horizontal integral scale of the hydraulic conductivity heterogeneity. Macrodispersion will thus often predict smoothed large plumes while, in reality, plumes to be remediated will show evidence of channeling and complex

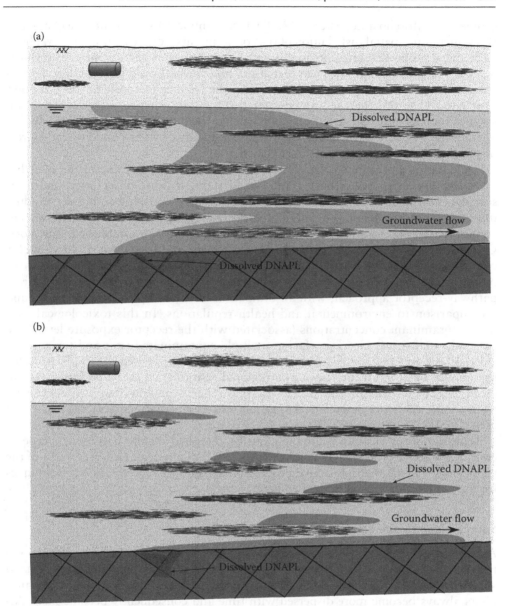

Figure 8.22 For initial conditions as depicted in Figure 8.18(b) and even after complete removal of the accessible "pure" DNAPL, different plumes of dissolved DNAPL are migrating with the groundwater flow system (a). After a few (or more) years, and despite possible remediation or stimulated/natural attenuation, groundwater is still contaminated by different residual plumes fed by dissolution and diffusion from the remaining DNAPL blobs and pools trapped in the heterogeneities and fractures of the geological medium (Heron et al. 2016).

geometries (Moreno and Paster 2017). Each contaminated site has unique characteristics, including spatial distribution of the contaminants, source zone longevity, physicochemical properties, and potentially complex associations of contaminants forming reactive "cocktails." Finally, local regulatory and societal conditions with respect to environmental problems may be specific from one country to another. This last point associated with the current groundwater quality state (due to many contaminant sources such as in urban areas) may in particular lead to less ambitious remediation operations (Fitts 2002) including only a partial removal of the contaminant source, or plume monitoring instead of active remediation.

Remediation operations often share common characteristics due to the fact that the contaminant source is buried and out of sight, making observation and measurements difficult and expensive. Uncertainty lies in identifying and understanding interacting processes, but also more fundamentally, what is the actual distribution of contaminant concentrations and mass fluxes with regards to established indicators. Concerning this last point, classical risk assessment and management concepts for contaminated sites are usually based on a univocal relationship between a source of pollution and a potentially exposed receptor, commonly referred to as the source-pathway-receptor approach, with an evaluation of the receptor exposure level and a comparison to environmental and health regulations. In this toxicological context, contaminant concentrations (associated with the receptor exposure levels) are the main indicators. However, facing multiple contaminated sites and for ecological risk assessments at the regional scale, flux-based approaches as indicators for the quality of groundwater (seen as a regional resource) could be preferred (Jamin et al. 2012, Horneman et al. 2017). Mass discharge or flow-weighted concentrations should therefore be preferred as groundwater remediation metrics (Hadley and Newell 2012, Einarson 2017).

Remediation or cleanup techniques can be classified on the basis of the type of considered active processes: biological, chemical, and physical. However, most of the techniques include multiple active processes. Off-site (ex-situ) treatment techniques for excavated soils are not addressed here.

Source cleanup and containment-stabilization

As mentioned in Section 8.3 (for NAPL contaminants), the priority is given to the removal of the contaminant source. It is indeed the first priority for any contaminated site. One of the general principles in environmental contamination is that contaminants always become more dispersed with time and consequently become more difficult to cleanup.

For NAPL sources, free-product recovery is crucial to avoid further dissolution in groundwater. LNAPL free-product sources can be removed from aquifers with skimmer-, single-, and dual-pump well systems (Figure 8.21). These systems are technically complex and must be adapted to particular conditions in boreholes. New technical improvements are continuously being proposed for increasing cleanup efficiency. For DNAPLs, the pure product remaining above relatively low permeability layers should be pumped using horizontal or vertical wells but results show that it is often not very efficient. In heterogeneous formations, DNAPLs are usually highly disseminated under the form of residual blobs or pools of free product. DNAPL

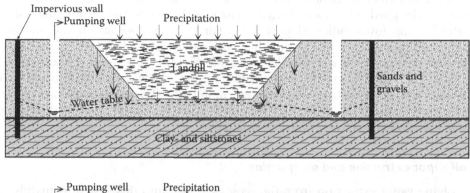

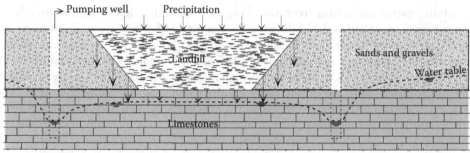

Figure 8.23 Hydraulic engineered containment option for a contaminant source from an old landfill (Monjoie *et al.* 1992).

sources are thus recognized as being nearly impossible to completely cleanup. If excavation is not feasible, the only effective ways of recovering DNAPLs are by vapor extraction or dissolved products remediation by bioremediation, reactive barriers, or pump and treat operations.

More generally, excavation of highly contaminated soils, if feasible, is a common practice allowing surface off-site or on-site treatment of the contaminated materials. An alternative to excavation is source containment or isolation by engineered barriers combined with, or using, modified local hydrogeological conditions. Synthetic membranes, slurry walls, steel sheet piles, grout injection, bentonite, and compacted clays can be used in various ways to isolate the source zone. The actual effectiveness of this type of operation is often related to the size of the area to be isolated. Remediation of many old leaking landfills is often started by such containment or by applying isolation operations (Figure 8.23). In some cases, the contaminant source zone can be stabilized using cement-based binders. The contaminated soil or waste is encompassed into a hardened engineering material (Bates and Hills 2015).

Pump and treat systems

One of the oldest and still most widely used remediation techniques consists of pumping of contaminated groundwater, treatment at the surface by biochemical processes and reinjection. This operation serves two purposes: removal of a part of the contaminant and control of the solute plume. Indeed, pumping and reinjection, when properly designed, allows an effective control of the plume and avoids contaminated water flowing downgradient (Figure 8.23). For each case, optimization of the chosen pump and treat scheme must be studied with care by considering the actual capture zone of each pumping well (and/or the combined effects of several pumping wells) in detailed models where the heterogeneity of the geological medium is duly taken into account. As for other remediation techniques, performance assessment must be performed with accuracy as this assessment is an important issue with respect to remedial action objectives (Truex *et al.* 2017).

Soil vapor extraction and air sparging

Involving vapor extraction from the geological medium, this system is mainly used for removal of volatilized NAPL mixed with air in the unsaturated zone. As mentioned in Section 8.3, vapor pressure can be calculated using Henry's and Raoult's laws. Vapor extraction is particularly efficient for the highly volatile family of contaminants known as VOCs (volatile organic compounds). Wells that are screened in the unsaturated zone are connected to a depressurizing or vacuum system. Vapor flow rates should be estimated to assess the efficiency of such a system and must take into account the permeability pattern of the heterogeneous medium, its varying water saturation, and contaminant gas densities. Critical factors include the limited volatilization of some contaminants and the low efficiency in low permeability media.

Air sparging consists of injecting air into the saturated zone of an aquifer. Along the pathways of air bubbles toward the unsaturated zone, contaminant stripping occurs by volatilization of contaminant molecules. A vapor/vacuum extraction system is then needed to recover the contaminated air. The same limitations apply as for vapor extraction in addition to the difficulties in predicting the actual efficiency of the air stripping in the saturated zone.

Bioremediation

In general terms, bioremediation refers to any remediation process involving microbial populations as biocatalysts for contaminant degradation reactions. The topic has become very broad in recent years and covers both natural attenuation and nutrients enhanced processes. Entire books have been dedicated to the application of various natural or enhanced bioremediation techniques to a variety of contaminants and in different hydrogeological contexts (among others, Brady *et al.* 1998, Wiedemeier *et al.* 1999, Suthersan 2002, Alvarez and Illman 2006, Cullimore 2008). Microorganisms and especially bacteria in soils and groundwater may be responsible for natural attenuation. After having gained a clear understanding of the natural active processes and their limiting factors at a given site (i.e., environmental conditions not appropriate for the desired microbial activity), enhanced bioremediation can be chosen by adding appropriate nutrients that help enhance the growth of useful microbial populations. Applied to the degradation of organic contaminants, biodegradation leads to mineralization of elements such as carbon, nitrogen, phosphorus, sulfur, carbon dioxide, or other inorganic components.

Biodegradation by microorganisms can occur under aerobic and anaerobic conditions and in many different types of hydrogeological contexts. Bacteria and fungi can degrade many synthetic compounds and probably every natural product (Suthersan 1997). In practice, bioremediation is used mainly for cleanup of hydrocarbon and chlorinated organics. Aerobic and anaerobic microbially catalyzed reactions degrade organic compounds by dechlorination, hydrolysis, cleavage, oxidation, reduction, dehydrohalogenation, and substitution (Norris *et al.* 1993). If the contaminated site is well characterized and monitored, the efficiency of the bioremediation techniques can be predicted during ongoing operations and calculated afterward by using multireactive transport models taking the considered processes into account. The obtained degradation rate of each component depends on many factors including the microbial populations, the actual mix of organic contaminants, the relative abundance of oxygen and other nutrients, as well as pH and temperature conditions (Fitts 2002). Bioventing and biosparging techniques combine oxygen and nutrient addition with a vapor extraction system.

Reactive barriers and zones

Since favorable conditions for biodegradation or abiotic chemical degradation do not exist everywhere, engineered reactive zones, barriers, or walls can be placed across the path of the plume (Fitts 2002, Naftz *et al.* 2002). Efficient and optimized conditions are prepared in these barriers to obtain the desired degradation processes. The barriers must be large enough and their efficiency must be maintained at mid- and long-term periods considering the predicted spatial and compositional evolution of the plume. Reactive barriers are most often to be combined to induced changes in the groundwater flow and contaminant transport conditions by containment, or by drainage and pumping operations. In this approach, the plume could be "channelized" in a given zone using "funnel and gate" geometries (Starr and Cherry 1994). Bioscreens, biological fences, or by simply creating bioreactive zones by injection wells located in the contaminant plume are all techniques that have been extensively developed in recent years. In all cases, these semi-passive modes of remediation require a detailed knowledge of the local groundwater flow conditions.

Phytoremediation

Phytoremediation is a plant-based bioremediation method. Some natural or transgenic plants are able to concentrate toxic elements from the soil or groundwater. Another advantage is that this bioaccumulation is found in the above-ground part of the plants, which are then harvested for removal by incineration or industrial recycling. Toxic heavy metals and organic pollutants are the major targets for phytoremediation. New biological and engineering strategies are developed for optimizing and improving phytoremediation. The approach is particularly effective for soil cleaning and very shallow groundwater conditions where plant roots can reach the contaminated water. Plant growth can also, in many cases, be significantly affected by the toxicity of the contaminated soil.

8.5 Tracer tests

Aquifer heterogeneity is the source of many uncertainties for the simulation of groundwater flow and contaminant transport. Moreover, in fractured aquifers, fracture frequency and interconnection combined with matrix porosity greatly influence this heterogeneity. With tracer tests, one can obtain directly, and at the field scale, a set of transport parameters including the effect of small-scale heterogeneities. If several tracer tests are carried out in the same aquifer, a large dataset becomes available to constrain further solute transport modeling. Tracer tests with injection of artificial tracers are addressed here. The use of environmental tracers and isotopes will be described in Section 8.7. The actual injection of a chosen artificial tracer is normally well known and quantified which is less the case for natural environmental tracers. On the other hand, the disadvantages of using artificial tracers are mainly: (a) chemicals must be introduced into groundwater which may contaminate the system and (b) the timescale of the tracer test is quite limited in relation to the timescale of some transport processes (e.g., such as diffusion).

Initially, tracer tests were used mainly in karstic systems. In this context, the main aim was to establish connections between preferential infiltration points (i.e., sinkholes) and the resurgence springs. Eventually, the travel times between the recharge zone and the spring are recorded and interpreted qualitatively. Tracer tests performed under these conditions were not always considered useful by hydrogeologists since they were often applied under inappropriate conditions to allow quantitative interpretation. Dilution of the tracer in groundwater has led to many "failures" in terms of tracer detection. Better preparation and operational designs together with the continuous improvement of analytical detection techniques have greatly improved the situation. Tracer tests carried out with care and organization may provide quite useful information. Today, the typical objective of tracer tests is the identification, at the appropriate scale of interest, of the main solute and fate processes and therefore the resulting parametrization (values to be given to the transport properties). They are also more and more used to obtain information on natural or enhanced biodegradation processes, and possibly also on the volume of NAPLs. Quantitative results are obtained in the form of breakthrough curves in one or more monitoring or pumping well. Specialized books and dedicated publications are available describing artificial tracers and their properties, including details of an implementation strategy for quantitative interpretations (among others, Käss 1998, Ptak et al. 2004, Divine and McDonnell 2005, Leibundgut et al. 2009, Maliva 2016).

Generally, preliminary numerical modeling is useful to design and dimension the tracer test. During the test, breakthrough curves are obtained after tracer injection. Measured normalized tracer concentrations (i.e., tracer concentration divided by the total mass of injected tracer) are represented as a function of time. After the test, inverse modeling procedures can be applied on the breakthrough curve data to obtain solute transport properties, such as effective transport porosity, longitudinal, and transversal dispersivities, and possibly retardation factors and immobile water diffusion coefficients. The resulting transport parameter values are valid at the applied scale of the tracer test. Results are nonunique (see Chapter 12) which may become more complex if the flow field was not previously well simulated or calibrated by adapting the hydraulic conductivity field to allow a better fit between observed and simulated results. Simulating tracer tests involves all the classical issues linked to solute transport modeling. The user should be very careful to avoid numerical dispersion

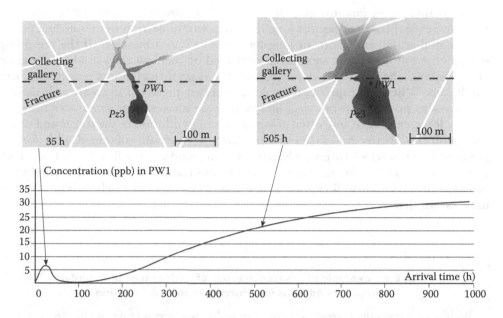

Figure 8.24 Typical tracer test breakthrough curve in a dual porosity chalk showing a fast first peak due to rapid advection in the fractures and a slowly increasing concentration due to matrix diffusion and slow advection-dispersion in the porous chalk matrix (Dassargues *et al.* 2011).

and oscillations due to inadequate choices of grid and time step sizes (see details in Chapter 13).

Multiple peaks in breakthrough curves can be normally[1] interpreted as reflecting different pathways for the tracer in a heterogeneous aquifer (Dassargues *et al.* 2011). Dual porosity and dual permeability models could be used to reproduce multipeak BTCs (Figure 8.24) and BTCs which clearly show matrix diffusion (i.e., immobile water effect) and higher macrodispersion values due to the high degree of advection velocity variations.

A measure of the total tracer recovery (i.e., especially in a pumping well) always provides important information and is an additional constraint to be honored afterward by the inverse modeling interpretation. A lower recovery than expected indicates that unforeseen losses have occurred due to the heterogeneous nature of the actual advection velocity field and/or that strong adsorption and/or degradation is occurring.

Tracer tests configurations

Various tracer tests can be performed depending on the objectives. The relative complexity of the test depends on the following characteristics (Maliva 2016): whether a

---

[1] If the injection and sampling procedures of the tracer test were performed under reliable conditions.

test is qualitative or quantitative, under natural or forced gradient conditions, with reactive or ideal tracers, short or long distances from injection to monitoring points (i.e., spatial scale), and short- or long-term results (i.e., timescale). If the prevailing piezometric gradient is not significantly affected by the flow rates used for injection and sampling, the tracer test can be considered as performed under natural gradient conditions. Where appropriate, these conditions are preferred to create as little perturbation as possible. In practice, however, these *natural gradient tracer tests* are usually long, expensive, and often affected by a high dilution of the tracer and detection problems. Better and faster tracer recovery is usually obtained with *forced-gradient tracer tests* where groundwater pumping modifies the flow field. These tests can be most suitable for situations where solute transport toward a groundwater production well (e.g., well capture zones and protection zone delineation Box 8.6) must be studied.

Box 8.6   Example of a systematic use of multitracer tests under pumping conditions for protection zone delineation

Wellhead protection areas around groundwater production wells are often defined in terms of time-of-travel zones, based on isochrones of a first contaminant arrival at the pumping well. For example, in Wallonia (Belgium), 1-day and 50-days isochrone contours correspond in the regulation to respectively two different levels of protection (i.e., in terms of authorized activities within the delineated areas). The actual shape and extent of protection zones are strongly dependent on the heterogeneity of the aquifer. The defined methodology for accurately delineating the protection zones involves the following steps (Derouane and Dassargues 1998, Hallet *et al.* 2000): (a) a preliminary regional hydrogeological study, (b) a local geophysical survey (e.g., electrical resistivity tomography ERT) for lithological changes and fractures detection, (c) drilling of four to five monitoring wells (with optimized locations) around the production well, (d) pumping tests in each available well and piezometric surveys, (e) a preliminary model calibrated on pumping tests data for dimensioning tracer tests, (f) multitracer tests performed between each observation well and the production well under maximum production conditions, (g) calibration of the flow and transport model using the measured breakthrough curves, (h) following a sensitivity analysis, simulations of contaminant injections from different locations, and (i) delineation of the protection zones on the basis of the computed times respecting the local regulations. This complete methodology is entirely applied in a deterministic framework. It was proposed to place this methodology in a stochastic framework where all available hydrogeological and geophysical data contribute to a coconditional stochastic model (Rentier and Dassargues 2002, Dassargues 2006). However, in this particular case, local decision-makers preferred to maintain a deterministic approach that seemed more understandable to everyone.

Forced gradient and especially radially converging flow conditions to the pumping well are known to provide the most realistic effective transport porosities for the pumping conditions prevailing during the tracer test (Box 8.7), but make it difficult to deduce a transverse dispersivity and can bias the longitudinal dispersivity (underestimation of a_L as shown by Ptak *et al.* 2004). The most classical situations are either a two-well configuration with an injection well and a pumping well, or multiple wells around a single pumping well (Derouane and Dassargues 1998; Box 8.6). For a better control of the tracer plume, some transect planes with monitoring wells can be placed orthogonally to the predicted path of the tracer between the injection and pumping wells (Figure 8.25). Whether injection and pumping are performed in fully-screened wells or between packers, dual monitoring wells (i.e., allowing separate measurements in the lower and upper part of the aquifer) are often preferred in a stratified aquifer to obtain more detail on the vertical distribution of the plume.

Single well tracer tests with injection and possible pumping (push-pull tracer test) in the same well have the advantage or requiring only one well but the disadvantage

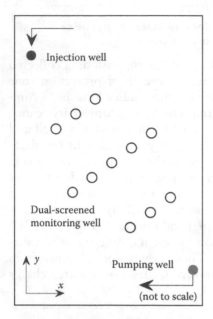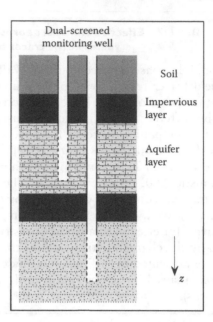

Figure 8.25 A two-well injection-pumping configuration for a tracer test with three transect panels of dual-screened monitoring wells (Wildemeersch *et al.* 2014).

of a relatively small investigation volume. Very rigorous measurements of the injected and recovered tracer quantities must be recorded over time. If a preexisting natural gradient exists the flow conditions can be very different (i.e., not purely radially convergent or divergent) during the injection and pumping periods. Interpretation provides a local value of the dispersion and possibly local information on biogeochemical processes, but must be based on a previous detailed knowledge of the hydraulic conductivity field.

In *single well dilution tests*, there is no pumping and the decrease in concentrations observed in the well depends on the dilution occurring in the well due to the influence of the local flow field (West and Odling 2007). This influence is not so easy to account for as the presence of the well equipment (i.e., the screen and gravel pack) can distort the natural flow field in a way that is difficult to anticipate. By varying the injection setup, many variants of this test have been developed, including instantaneous or continuous injections, and point or depth-averaged injections in the well. These methods are often used in contaminated sites as there is no need for pumping contaminated groundwater. The *point dilution method* (PDM) (Drost *et al.* 1968, Hall 1993) consists of monitoring concentrations over time after injection of a tracer. It can be applied conventionally in an open well, to identify qualitatively the more permeable layers of a stratified formation. Many variants are possible. Among others, as reported by Maliva (2016), the method proposed by Brainerd and Robbins (2004) for an open well in fractured rocks provides a qualitative view on the most highly fractured zones. Their approach involves injection of a tracer at the bottom of a well, where optimized limited pumping is performed to obtain steady-state flow conditions toward the well and then measuring diluted concentrations along a vertical profile. Modifying the classical PDM, Brouyère *et al.* (2007) proposed a finite volume point dilution method (FVPDM) generalizing the single-well point dilution method to the case of finite volumes of tracer and water flush. Given a continuous injection of a tracer at a very low rate, an analysis of the temporal evolution of the concentrations in the injection well allows identifying the well-aquifer interactions. With a detailed description of these interactions (Brouyère *et al.* 2007), the (possibly varying) local groundwater fluxes can be measured. For example, this approach allows monitoring Darcy fluxes in groundwater in relation to changes in hydrogeological conditions and groundwater—surface water interactions.

Tracers, injection, and sampling operations

A first review of the expected characteristics of an "ideal" tracer was provided by Davis et al (1980). Logically, it includes detectability aspects (by lab and field analytical devices), nontoxicity aspects, low adsorption and degradation, low background concentrations, and operational costs (tracer, security, and analysis costs). The actual choice of tracers is made on the basis of test-specific circumstances and objectives. Often, a tracer is expected to behave similarly to groundwater, while in other circumstances, a dedicated tracer is used to mimic a specific contaminant behavior in the geological environment. An overview of tracers and their main characteristics is given below on the basis of a recent review by Maliva (2016).

For *ionic tracers*, anions are preferred to cations as they are less adsorbed and less reactive with clay minerals. Chloride and bromide are used when their background concentrations are very low. For ionic tracers the background concentration is particularly relevant. In groundwater where background concentrations are higher than a few mg/L (for example for Cl^- or NO_3^-), one can easily imagine the large quantity of injected Cl^- or NO_3^- needed for resolving a BTC given the small induced differences in measured concentrations. Chloride concentrations must also stay low for avoiding any density effect on the groundwater flow and transport conditions (see Chapter 10). Indeed, some results have shown that anions are actually less conservative than thought (Korom and Seaman 2012). Even bromide, known as one of the most conservative anions, is adsorbed onto Fe and Al oxides and kaolinite (Goldberg and Kabengi 2010). If the aquifer is characterized by high background concentrations for some ions, an "inverse" or "negative" tracer test could be invoked with injection of deionized water. However, a large quantity of water would need to be injected and a quantitative interpretation would be difficult.

Multiple *fluorescent dyes* are also commonly used in tracer tests, with each dye having its characteristic fluorescence peak wavelength (Maliva 2016), so that multiple dyes can be used together in multitracer tests. They can be detected by using lab spectrofluorometers or field fluorometers. A large body of literature exists on their use and properties (among others, Käss 1998, Kasnavia *et al.* 1999, Sabatini 2000, Leibundgut *et al.* 2009). To summarize the precautions needed for use, one can point out that the detectability of a given dye can be dependent on the general groundwater background fluorescence which can be temperature and pH dependent.

Losses can occur, for example, by adsorption and absorption on organic matter and clay particles (especially for rhodamine-WT). Samples must not be exposed to daylight. Beside other fluorescent tracers such as uranine, eosine, pyranine, and sodium naphtionate, *fluorescein* $(C_{20}H_{12}O_5)$ is probably the most used tracer in the world. Usual detection levels achieved by lab spectrofluorometers are the following: 0.01 ppb for fluorescein and uranine, 0.02 ppb for rhodamines, and 0.05 ppb for eosin and naphtionates.

Other specific tracers such as particles can also be used. *Microspheres* are traditionally used for imitating the suspended matter movement in fissured and karstified aquifers. New developments are expected using *nanoparticles* for some specific applications as, for example, zero-valent iron nanoparticles which play a role in the dehalogenation of chlorinated hydrocarbons, or inert nanoparticle tracers are preferred to ionic tracers for their less diffusive behavior. DNA sequence-based tracers have been tried by Sabir *et al.* (1999), which can be detected at very low concentration but are costly.

Partitioning tracers which partially dissolve into a hydrocarbon or a NAPL (e.g., they have some specific affinity for DNAPLs) are used in combination with ideal tracers in order to assess the remaining NAPL trapped in the geological medium (Istok *et al.* 2002). The presence of an organic phase increases the apparent partitioning tracer travel time by dissolution or partitioning into the phase. This delay is used to estimate the saturation distribution of the organic phase.

Some researchers are using *dissolved gas tracers* (i.e., noble gases such as helium and neon) because they are easy to transport on-site in pressurized containers (Maliva 2016). In this case, since loss of tracer gas must be avoided, the sampling procedure is more complex and expensive (Sanford *et al.* 1996).

Heat has been used as a tracer (see Chapter 11) since the 1960s (Anderson 2005), mainly for river-aquifer exchange flux assessment, where it is also possibly combined with other hydrochemical and isotope measurements. Fiber-optic distributed-temperature-sensing (DTS) (Selker *et al.* 2006) enables high spatial resolution measurements that can be useful to constrain inverse problems.

Rigorous procedures for *injection and sampling* are very important to obtain reliable tracer test results. A thorough review on the influence of the injection conditions on the measured BTCs for a two-well tracer test was provided by Brouyère *et al.* (2005). The nonnegligible duration of injection must be taken into account in such tests. The actual well equipment and/or well conditions (i.e., well-bore, screens, clay and gravel packs, clogging) must be known as they possibly trap the tracer in the injection well. The injected water, forcing the tracer into the formation, must be optimized: a sufficiently large volume is desired with respect to the borehole volume but as small a volume as possible is desired to avoid modifying the local gradient. Depending on the objectives, injection can be performed (a) at the top of the well with a volume of additional water, (b) by a small diameter tube through a packer, (c) by perforated small diameter tubes, or (d) with a uniform spatial distribution of the tracer concentration by recirculation (Maliva, 2016).

Sampling procedures and techniques have also evolved significantly over the past few years. Indeed, when a BTC is analyzed, one must know how the sample was obtained. Is the sample from a pumping well with full screens inducing "depth-averaged" conditions? Or is it a BTC from a point sampling device between packers (or between bentonite packs) or from another type of multilevel sampling device? Depending on the tracer test configuration and objectives, passive samplers can be placed in combination with sequential automatic samplers and/or down-hole devices as, for example, fluorometers. Indeed, passive samplers provide only a cumulative value of absorbed tracer mass, while automatic samplers followed by lab analysis provide very accurate tracer concentrations at given times, and down-hole fluorometers allow more frequent or nearly continuous measurements of a given fluorescent tracer. Combined with the sampling procedure, hydrogeophysical measurements can also be performed especially with tracers having an influence on the electrical resistivity (Hermans *et al.* 2012). Time-lapse hydrogeophysics applied to tracer experiments (Hermans *et al.* 2015) and to the characterization of preferential pathways at contaminated sites is a rapidly developing topic. As mentioned previously (see Dispersion paragraph in Section 8.2), these high-resolution measurement techniques are very useful to constrain inverse modeling procedures.

8.6 Transport and residence times

It has always been a dream for hydrogeologists to be in a position to determine exactly the "age" of a drop of water. Indeed, many important groundwater management decisions could be based on knowing locations of "young" and/or "old" groundwater. For example, old groundwater in an aquifer could indicate a relatively low current recharge rate (e.g., with regards to pumped groundwater). In contamination problems, the time required for a plume to reach a pumping well or any "target" zone to be protected is crucial. The relative importance of reactive transport, adsorption, and degradation processes influencing solute transport could be highly dependent on the time spent by contaminated water in the partially saturated and saturated zones in contact with the mineral heterogeneity of the geological medium. Determining legacy contamination sources (e.g., agricultural practices, industrial activities) which have led to contamination, is another application where groundwater age can be very useful.

If we follow a drop of water, age is an evolving variable with time, meaning the difference between an initial moment and the determination moment. With this concept, age is linked to an individual entity not mixing with anything else during the whole journey. In short, this important assumption means that age of groundwater is an idealized concept, or at least a misleading term, as it neglects mixing.

Piston-flow groundwater age: an idealized concept

Conceptually, if we imagine an infinitely small water parcel, unaffected by mixing processes, the idealized age between an initial moment and the present moment can be defined (Suckow 2014). It is known as the "piston-flow" concept of age or *piston flow age* (Bethke and Johnson 2008). In the piston flow model, only advection is considered. This concept is clear and easy to understand but there is no way to determine such an age experimentally because the assumptions are not in agreement with reality.

Groundwater age: A misleading term

The reality

In reality, a groundwater sample, technically considered as corresponding to a point in a regional groundwater flow system, represents a population of water molecules of different ages (Suckow 2014) (i.e., having resided in the groundwater system for varying lengths of time). Any measured age would be an averaged value of an unknown age distribution in the sample. What kind of age distribution? It would be useful to know but we have no way to determine it. The mixture of different groundwater origins is due to the complex heterogeneous conditions combined with varying boundary conditions in space and in time. Even the H_2O molecules exchange their H atoms among each other within picoseconds (Suckow 2014).

The measured/interpreted reality using dating techniques

Using dating methods, based on isotope tracer analysis (Section 8.7), *tracer-based groundwater ages* are assessed for groundwater samples.

First, each dating technique assumes the sample was from an infinitesimal volume of water in a closed system like in the idealized piston flow model. The conversion of a tracer concentration to an *idealized age* is done using assumed asymptotic empirical degradation laws (see Section 8.7) for the considered tracer. However, very often, this conversion is not unique due to combined effects of the actual input of the tracer in the system and infiltration conditions (Suckow 2014). Therefore, to decrease uncertainty, it is useful to use several isotope or environmental tracers together (with different time ranges; Section 8.7). Normally each tracer is known as an *ideal tracer* for water (i.e., transported in a similar way as water), but in reality, slight differences can exist for input (i.e., infiltration) conditions, and for particular diffusion, retardation, and degradation conditions.

Second, for taking into account the effect of mixing in the sample, the term *apparent tracer-based groundwater age* (e.g., apparent ^{14}C groundwater age) is used to describe each tracer-based groundwater age that is found. The term *apparent* reminds the user that piston flow is assumed in place of an actual but unknown mixing. Bethke and Johnson (2008) have shown that any apparent radioactive tracer-based age is biased toward younger values when compared to the mean age of the mixture.

The statistical approach for calculating ages: Mean age and residence time

For calculation purposes, many authors have suggested to regard groundwater age in a control volume (or in an REV) as a statistical, or probabilistic distribution (among others, Goode 1996, Cornaton and Perrochet 2006, Molson and Frind 2012) rather than as an average single value. A link can be made with the notion of travel time probabilities studied by many authors in groundwater solute transport (for example, Jury 1982, Dagan 1989). In practice, the notion of tracer travel time is used to describe the time spent by the tracer to travel between two locations in the aquifer. For example, this time-of-travel can be important for delineation of wellhead protection areas

(Frind *et al.* 2002) and well intrinsic vulnerability (Frind *et al.* 2006). If water is considered as entering into the system through infiltration at the land surface, a zero age is given. Then, the groundwater molecule is aged continuously at the rate of one unit per considered time unit. Groundwater age should be viewed in a probabilistic framework rather than absolute values at each point in space and time (Cornaton 2004). In doing so, the approach honors mixing processes that have a significant impact on the mean transit time distribution under heterogeneous conditions. The concepts of *mean groundwater age* $A(x, y, z)$ considered at any point of the groundwater system and its complementary *mean life expectancy* $E(x, y, z)$ were defined in seminal important papers by Goode (1996) followed by Cornaton and Perrochet (2006) and Molson and Frind (2012). The *total residence time* (TRT) between the infiltration point at the surface and the outflow point is the sum of the *mean groundwater age A* (also named *mean residence time MRT*) at any point and the remaining *life expectancy E* of water in the considered system: $TRT = A + E$ (Figure 8.26). A grows at a rate of 1 time unit per time unit in the forward flow direction and E grows at the same rate in the backward flow direction.

If it can be assumed that the mean age of mixed groundwater is a mass-weighted average (Box 8.8), then there is an analogy between mean age and conservative solute concentration that can be exploited to derive a governing transport equation for mean age A (Goode 1996). A given mass of water with a mean age A is thus assumed to be characterized by an "age mass." This concept is used to describe the distribution of age due to the effect of the different transport processes. By analogy with the transport of solute where C (i.e., the *mass concentration* in kg/kg) is used as the main variable in equations but the transported quantity is the solute mass that is $\rho C V$ (V being the REV or control volume in m³), here, the analogous transported quantity is the age mass $\rho A V$ (in kg s) (Bethke and Johnson 2008).

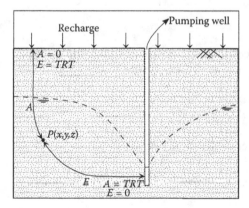 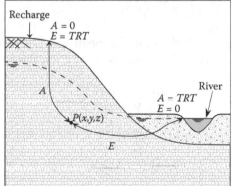

Figure 8.26 2D vertical conceptual models showing the notions of mean groundwater age (A) and life expectancy (E) at any point P(x, y, z). At the inlet boundary (i.e., where the recharge is considered as entering the system) A = 0 and E = TRT, while at the outlet boundary (i.e., pumping well or draining river) A = TRT and E = 0 (Molson and Frind 2012).

For steady-state flow conditions, at the aquifer scale, time-independent solute transport parameters are considered as time-independent (or having reached their asymptotic values) and the derived transport equation for mean age A can be written as (Goode 1996, Cornaton and Perrochet 2006):

$$-v_a \cdot \nabla A + \nabla \cdot (D_h \cdot \nabla A) + 1 = 0 \tag{8.71}$$

where D_h is the hydrodynamic macrodispersion tensor and v_a is the advection velocity vector.

Box 8.8 Analogy between mean concentrations and mean age of a groundwater mixture

Neither C (i.e., the *mass concentration* in kg/kg) nor A (i.e., the *mean age of water* in s) are additive variables. If we consider a mixture of two samples of concentrations C_1 and C_2, the mean concentration C of the mixture will be found by the following weighted average:

$$C = \frac{\rho_1 C_1 V_1 + \rho_2 C_2 V_2}{\rho_1 V_1 + \rho_2 V_2}$$

where the subscripts 1 and 2 distinguish the samples. If the density variations with concentration can be neglected, it becomes a volume-weighted average:

$$C = \frac{C_1 V_1 + C_2 V_2}{V_1 + V_2}$$

and if the volume of each sample is identical ($V_1 = V_2$):

$$C = \frac{C_1 + C_2}{2}$$

It is the same for mean age A, if we consider a mixture of two samples respectively of mean ages A_1 and A_2, the mean age of the mixture will be found by:

$$A = \frac{\rho_1 A_1 V_1 + \rho_2 A_2 V_2}{\rho_1 V_1 + \rho_2 V_2}$$

If no change of the water density is considered, it becomes:

$$A = \frac{A_1 V_1 + A_2 V_2}{V_1 + V_2}$$

and if the volume of each sample is identical ($V_1 = V_2$):

$$A = \frac{A_1 + A_2}{2}$$

This equation is inspired by Equation 8.51 where adsorption, degradation, and chemical reactions are not considered (as they do not affect the water) and where the +1 source term expresses the increasing mean (steady-state) age at the rate of 1 time unit per time unit in the forward flow direction. As shown in Figure 8.26, $A = 0$ at the inlet boundary and $A = TRT$ at the outlet.

The mean *life expectancy* E at a given point P (Figure 8.26) is the mean time remaining before reaching the outlet taking all mixing effects into account. As shown on Figure 8.26, it can be calculated easily in a backward flow direction (i.e., with the opposite sign of the advective term). The equation for the mean (steady-state) life expectancy (E) is written:

$$v_a \cdot \nabla E + \nabla \cdot (D_h \cdot \nabla E) + 1 = 0 \tag{8.72}$$

Logically, E grows in the backward flow direction by 1 time unit per time unit. Mean age and mean life expectancy are continuously generated during groundwater flow due to the +1 source term in Equations 8.71 and 8.72. It has also been demonstrated (Cornaton and Perrochet 2006) that logically, "longitudinal dispersion significantly affects the aging process while lateral dispersion rejuvenates the system due to transverse mixing." Still, considering the backward flow field away from pumping well, a capture probability approach can be used to determine time-of-travel zones of a water supply well (Molson and Frind 2012). Even if the problem is considered under steady-state flow conditions, for computational purposes, it is often more pragmatic to use a time-stepping procedure to calculate the capture probability, so that the capture probability $P(x, y, z, t_{max})$ (Figure 8.26) is found by solving the equation:

$$v_a \cdot \nabla P + \nabla \cdot (D_h \cdot \nabla P) = \frac{\partial P}{\partial t} \tag{8.73}$$

where $P(x, y, z, t_{max})$ is the probability that water or an ideal tracer would be captured by the pumping well during the time period $t < t_{max}$. Note that this time derivative term will be equal to 0 at steady-state. This term allows a time-stepping solution until steady-state is reached. There is no source/sink term. The boundary condition at the pumping well is $P = 1$ (i.e., 100% probability of capture), then we track the P plume, like a contaminant plume, in the upgradient direction. Equations 8.73 describes the upgradient growing plume of capture probability from the pumping well (i.e., where $P = 1$).

Other methods have also been developed, with ρA as the conserved quantity, using volume averaged temporal moment equations (for example, Varni and Carrera 1998). Theoretically, as the age distribution at a given outlet is not known, an infinite number of moments would be needed but Harvey and Gorelick (1995) have shown that the first five temporal moments of a BTC allow determining the needed information on the age distribution with reasonable accuracy.

Similar to Equations 8.71 and 8.72, transient age distributions have been derived (Engdahl *et al.* 2016), but these have been, until now, less often used in practice (even if more realistic then assuming steady-state flow) (Engdahl 2017).

These recent developments, first linked to age of water, have drifted toward an even greater interest when applied in contaminant transport applications. Residence times in subsurface hydrological systems have become a hot topic (de Dreuzy and Ginn 2016).

Adding reactive transport, adsorption, and degradation terms to Equation 8.71 allows to generalize its application for specific solute contaminants (Ginn 1999). This adaptation also allows more reasonable comparisons and calibrations with results from isotopes and environmental tracer dating methods (see Section 8.7) (see, among many others, Solder *et al.* 2016, Batlle-Aguilar *et al.* 2017, Cook *et al.* 2017), and is opening wide avenues for applications in contaminant hydrogeology.

8.7 Isotopes and environmental tracer interpretations

Isotopes and environmental tracers in groundwater include stable and radioactive isotopes and any chemicals that are observed in groundwater due to natural or anthropogenic processes. Knowing the spatial distribution of concentrations in groundwater provides a better understanding of the studied system. Added insight value from analysis of these tracers may include contaminant origins, pathways, and processes affecting groundwater from its recharge until its outlet, often with the added possibility to asses a mean apparent age of groundwater samples. More specifically, it can be particularly useful to determine the spatial distribution of infiltration, interactions between surface water and groundwater, how long groundwater remains in the system, and which transport processes are the most effective along the groundwater pathway. Isotopes and environmental tracer interpretations are not the only dating techniques. As mentioned in Section 8.6, ages that are obtained using radioactive tracers are actually apparent ages for a groundwater mixture over the sample volume. Some authors (for example, Suckow 2014) have even suggested avoiding the conversion from tracer concentrations to ages as this conversion is a nonunique modeling process. However, for qualitative interpretation and to help develop a hydrogeological conceptual model, dating provides useful information.

Many textbooks and review publications are dedicated to this topic, including the reference books of Clark and Fritz (1997), Clark (2015), and multiple author books like those edited by Kendall and McDonnell (1998), Aggarwal *et al.* (2005) for the International Atomic Energy Agency (IAEA), Kazemi et al. (2006), and Leibundgut *et al.* (2009). Only a very brief summary will be given herein for this broad and interdisciplinary topic. Advances in new analytical measurement techniques are inducing continuous progress in this field with improved detection limits and compound specific analysis of isotopes. As mentioned previously, many applications of those isotope techniques have been published in the specialized literature, with different combinations of stable and radioactive isotopes (among the most recent, Solder *et al.* 2016, Batlle-Aguilar *et al.* 2017, Cook *et al.* 2017).

Note that a general rule of good practice must always be respected for a reliable interpretation: spatial and temporal signatures of tracer isotope must be significantly greater than the sampling and analytical error ranges.

Stable isotopes

Natural stable and radioactive isotopes of H, O, C, N, and S are currently used in a variety of applications (Table 8.2). For each groundwater sample, on the basis of the analysis results, an *isotope ratio* R_i between the less abundant isotope species N_i and its corresponding abundant species N is systematically expressed as:

Table 8.2 Most frequently used isotopes of H, O, C, N, and S, their mean natural abundance (in ppm), and molecules which are most frequently chosen for their detection in groundwater.

Light and abundant isotope	Heavy isotope in groundwater	Mean natural abundance (10^{-6})	Molecules most frequently chosen for analysis
Hydrogen = ^{1}H	Deuterium ^{2}H = D	200	H_2O
	Tritium ^{3}H = T (Radioactive isotope)	10^{-11} (1/2 life = 12.3 yr)	H_2O
Oxygen = ^{16}O	Oxygen-18 ^{18}O	2,000	$H_2O, CO_2, HCO_3^-,$ $CaCO_3, SO_4^{2-}, CaSO_4$
Carbon = ^{12}C	Carbon-13 ^{13}C	10,000	CO_2, CH_4
	Carbon-14 ^{14}C (Radioactive isotope)	10^{-6} (1/2 life = 5,730 yr)	$CO_2, HCO_3^-, CaCO_3$
Nitrogen = ^{14}N	Nitrogen-15 ^{15}N	4,000	NO_3^-
Sulfur = ^{32}S	Sulfur-34 ^{34}S	40,000	$SO_4^{2-}, H_2S, CaSO_4$

$$R_i = \frac{N_i}{N} \tag{8.74}$$

Stable isotope contents are then given as differences in isotope ratios between the groundwater sample (R_{gw}) and a recognized standard (R_{st}). This difference, very small, is represented by a delta (δ) and is usually expressed in parts per thousand (‰):

$$\delta_{gw_i} = \left[\frac{R_{gw_i} - R_{st_i}}{R_{st_i}}\right]1000 = \left[\frac{R_{gw_i}}{R_{st_i}} - 1\right]1000 \tag{8.75}$$

where i denotes the isotope species. For water the most commonly used standard value is the so-called SMOW (standard mean ocean water) value published by the IAEA. For the most common groundwater isotope ratios ^{18}O/^{16}O and ^{2}H/H (which are ideal tracers as intrinsic component of the water molecule):

$$R_{SMOW_{18O/16O}} = 2005.2 \pm 0.45 \; 10^{-6}$$

and

$$R_{SMOW_{2H/1H}} = 155.76 \pm 0.05 \; 10^{-6}$$

A positive δ indicates an enrichment of ^{18}O or ^{2}H compared to the recognized standard (SMOW) and a depletion of heavier isotopes is shown by negative δ values.

$\delta^{18}O$ and δD ratios

^{18}O and ^{2}H (Deuterium also noted D) are selectively partitioned at each step of the hydrological cycle (Clark 2015): primary evaporation over the oceans, condensation and precipitation as rainfall, evapotranspiration at the land surface and in soils before recharge, and groundwater evaporation and runoff as surface water back to the seas. Fractionation at each phase change occurs according to mass differences:

heavier isotopes form stronger chemical bonds, and thus have a greater tendency to stay bonded in a solid rather than in a liquid, and a tendency to remain in a liquid rather than in a gas. As water evaporates, the remaining liquid water is therefore enriched in heavier isotopes (^{2}H, ^{3}H, ^{18}O) while, on the other hand, the water vapor is depleted in these isotopes. Fractionation is higher at lower temperatures compared to higher temperatures (i.e., the equilibrium fractionation is a decreasing function of temperature; Leibundgut *et al.* 2009). The isotopic signature of precipitation at a given location thus depends on the temperature of condensation and on the degree of "rainout" (Rayleigh rain-out effect) in the air, defined as the ratio between already condensed water from vapor and the initial amount of water vapor. A series of macroscale effects are then observed on the isotope ratios in precipitation:

- A latitude effect, with an increasing degree of "rain-out" with increasing latitude, global maps of ^{18}O/^{16}O in precipitations have been published by IAEA (Yurtsever and Gat 1981), about $-0.6‰$ per degree of latitude for ^{18}O for coastal and continental stations in Europe and the United States, and $-0.2‰$ per degree of latitude in the colder Antarctic continent.
- An altitude effect, due to mixed temperature and pressure effects, topographically uplifted air masses are precipitating water with less heavier isotopes, -0.15 to $-0.5‰$ per 100 m for ^{18}O, and -1.5 to $-4‰$ per 100 m for ^{2}H or D. This pattern is weaker in interior continental mountain chains, and not observed for snow.
- A continental effect, also referred to a "distance-from-coast" effect (Figure 8.27), isotope ratios decrease in precipitation with increasing distance from the coast, but this trend varies considerably from area to area and from season to season.
- An amount of precipitated water effect, due to the "rain-out" effect as fractionation depends on the air humidity.

The isotopic composition of rainwater can therefore be highly variable since they are a function of the evaporation and condensation processes, which are themselves very dependent on temperature, pressure, and other climatic conditions. Isotopically heavier precipitation occurs as rain (typically $\sim-3‰$–$0‰$ for ^{18}O), while snow is dramatically lighter ($\sim-20‰$, ^{18}O). Local sampling and isotope analysis of precipitation waters are always needed for the further interpretation of groundwater isotope ratios in terms of mixing and origins.

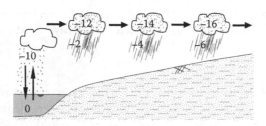

Figure 8.27 Continental effect on the ^{18}O isotope ratios of meteoric waters showing a relation $\delta^{18}O_{Precipitation} \approx \delta^{18}O_{Vapor} + 10‰$.

On the basis of many measurements from the "Global Network of Isotopes in Precipitation" (GNIP) established by IAEA and WMO, Craig (1961) observed that there is a strong and general correlation between ^{18}O and D in global freshwaters (i.e., on the average global scale). The regression line of this correlation (with r^2 better than 0.95) gives the *"global meteoric water line"* (GMWL), defined by (Figure 8.28):

$$\delta D = 8\delta^{18}O + 10‰ \tag{8.76}$$

This is a global regression line which represents the alignments of changes in isotopic compositions of ^{18}O and D with precipitation. Similarly, due to the combined effects described above, the relationship between $\delta^{18}O$ and δD in rainfall can be plotted for a specific area on a local meteoric line (LMWL). The relationship can also be specific for some periods (such as seasons). As mentioned before, this LMWL is needed for any further interpretation of groundwater isotope ratios in terms of mixing and origins.

Groundwater analysis results not lying on this line imply the groundwater has undergone processes (i.e., often named "secondary processes" in the isotope community) which have modified this correlation. In a ($\delta^{18}O$, δD) diagram, the effect of each process is reflected by a line segment joining the point of the initial composition and the point representing the composition of the water having undergone the considered process (Figure 8.28). Without entering into too much detail about isotope fractionation, one can summarize the following influences on the $\delta^{18}O$ and δD ratios (Figure 8.28):

- Evaporation makes the remaining surface water or groundwater enriched in both heavy isotopes. In the ($\delta^{18}O$, δD) diagram, the evaporating groundwater line is below the local MWL having a slope of 2–3 while the evaporating surface water line has a slope of 4–6.

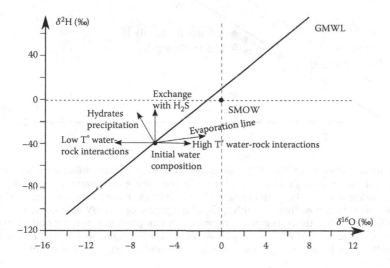

Figure 8.28 Relation between the $\delta^{18}O$ and δD values, showing the GMWL and deviation from this line for different processes that water undergoes (Gascoyne and Kotzer 1995).

- During water-rock interaction, especially with calcite and silicates. δD is not affected since H is rarely present in the minerals composition. However, $\delta^{18}O$ is modified and is highly dependent on temperature so that this "isotopic thermometer" is used for geothermal groundwater studies. At high temperature the positive $\delta^{18}O$ can be significant, while at low temperatures ($<100\ °C$) water-rock interactions induce only a slight decrease of $\delta^{18}O$.
- Precipitation of hydrated minerals increases δD and decreases $\delta^{18}O$ since many of these minerals (clay, gypsum) are enriched in ^{18}O and depleted in ^2H.
- Isotopic exchange with NAPLs and gas species such as H_2S does not change $\delta^{18}O$ but increases the presence of ^2H in groundwater, thus increases δD.

An illustration of some of these effects is given in Figure 8.29 with, on one hand, groundwaters showing typical ($\delta^{18}O$, δD) signatures of evaporated remaining waters and, on the other hand, groundwaters showing depleted $\delta^{18}O$ due to a low temperature exchange with CO_2 from a magmatic origin.

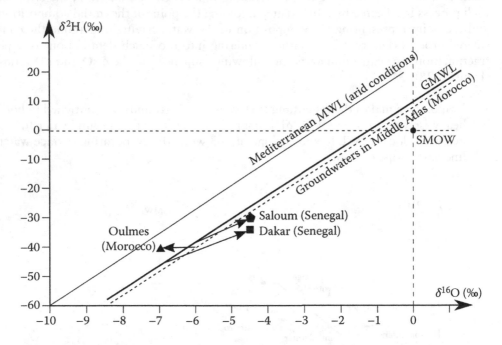

Figure 8.29 Relation between ($\delta^{18}O$, δD) values for the evaporation influenced Saloum groundwaters (Ndeye *et al.* 2017) and Quaternary groundwaters near Dakar (Madioune *et al.* 2014), contrasting with the Oulmes groundwaters influenced by low temperature exchanges with CO_2 of magmatic origin (Wildemeersch *et al.* 2010). For these three regions, precipitation compositions correspond or are very close to the GMWL showing an Atlantic Ocean origin. Note that a regional Middle Atlas (Morocco) groundwater line is found slightly below the GMWL.

$\delta^{34}S$ and $\delta^{18}O$ ratios in sulfate

Other stable isotopes are also used. Isotopic compositions of dissolved sulfate (relation $\delta^{34}S$ - $\delta^{18}O$) and nitrate (relation $\delta^{15}N$ - $\delta^{18}O$) in surface water and groundwater, for example, are useful to identify their sources and active biogeochemical processes. Sulfate ion (SO_4) (i.e., dissolved SO_4 must first be converted to pure $BaSO_4$) is used to determine $\delta^{34}S$ and $\delta^{18}O$. For both, however, standards are not very well defined (Mayer *et al.* 1995). For other sulfur compounds in groundwater (e.g., H_2S), it is essential to prevent their oxidation in SO_4 during sampling. Besides natural origins of dissolved sulfate from the atmosphere, pedosphere, or lithosphere, anthropogenic sulfate sources are now found in most aquifers. Summarized information about the relation $\delta^{34}S$ - $\delta^{18}O$ measured in various contexts is shown in Figure 8.30. In evaporitic rocks, the $\delta^{34}S$ composition is $8 < \delta^{34}S < 35‰$ (depending on the rock-formation ages) with $7 < \delta^{18}O < 20‰$. Atmospheric sulfate depositions have $-10 < \delta^{34}S < 22‰$ with $6 < \delta^{18}O < 16‰$, and have very wide windows. Sulfate produced by various oxidation processes from many different sulfides, also has a wide range, showing generally lower values for both ratios and negative $\delta^{18}O$ values (Figure 8.30). On the other hand, ocean water sulfate is characterized by $\delta^{18}O \approx 9.5‰$ and $\delta^{34}S = 21‰$, and ratios for sulfate from sulfur oxidation of magmatic and volcanic rocks are near $0‰$. Anthropogenic sources of sulfate have quite diverse characteristics depending on the actual product origins and/or objectives

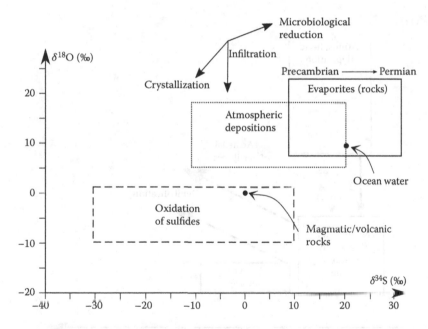

Figure 8.30 Combinations of $\delta^{34}S$ and $\delta^{18}O$ ratios for different sulfate origins. Above: most likely evolution (from any initial SO_4 composition) due to crystallization, infiltration, and microbiological reduction (Mayer 2005).

(e.g., water additive, fertilizers, leakage from landfills). When water infiltrates in the partially saturated zone, $\delta^{34}S$ remains unchanged while $\delta^{18}O$ is depleted due to the fact that SO_4 may react with a series of compounds. Evaporation increases the SO_4 concentrations without changing the ratios. Crystallization of sulfate induces a slight decrease of $\delta^{34}S$ and $\delta^{18}O$ along a line with a slope of about 2 in the remaining groundwater. Dissolution of evaporites does not change the isotope ratios of sulfate in groundwater. Microbiological reduction of sulfate involves enrichment of the remaining sulfate in both $\delta^{34}S$ and $\delta^{18}O$ (Mayer 2005).

$\delta^{15}N$ and $\delta^{18}O$ ratios in nitrate

Isotopic compositions of dissolved nitrate ($\delta^{15}N$ - $\delta^{18}O$ signatures) can be very useful to trace its origin and to identify associated processes. Different reference values for $\delta^{15}N$ are available in the specialized literature. Indeed, each anthropogenic nitrate source has its own $\delta^{15}N$ signature (e.g., sewage, manure, fertilizers), but determining respective contributions from many sources remains challenging, especially since various nitrogen transformations in the biosphere and pedosphere may significantly affect isotope ratios. Summarized information about the measured $\delta^{15}N$ - $\delta^{18}O$ signatures in various contexts is shown in Figure 8.31. Nitrification in soils typically produces nitrate $\delta^{18}O$ values between 0 and 15‰. Microbial denitrification increases

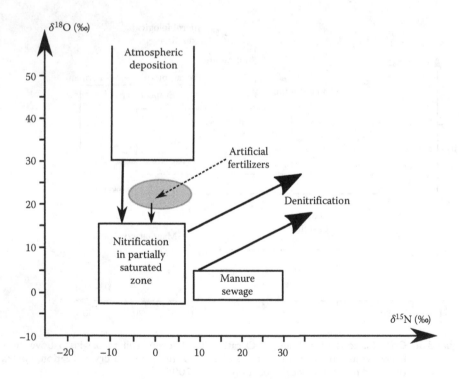

Figure 8.31 Combinations of $\delta^{15}N$ and $\delta^{18}O$ ratios for different nitrate origins. Arrows show the trend (from any initial NO_3 composition) due to denitrification (Mayer 2005).

both ratios $\delta^{15}N$ and $\delta^{18}O$ in the remaining nitrate (Mariotti *et al.* 1981). During infiltration in the partially saturated zone, nitrate from an atmospheric origin and from fertilizers may undergo various reactions generally inducing lower values of $\delta^{18}O$ (Figure 8.31). It is particularly difficult to assess origins of NO_3 without different measurement campaigns which would allow spatial and temporal trends in $\delta^{15}N$ and $\delta^{18}O$ ratios to be determined (Mayer 2005). Combined approaches involving other isotope ratios like $\delta^{11}B$ can also be recommended to highlight the origin of contamination. For example, urban effluent contamination has isotopic signatures with a low $\delta^{11}B$ ratio, whereas agricultural contamination (i.e., due to fertilizers) has a significantly higher boron isotope ratio. Another well-known signature is observed when denitrification occurs simultaneously with sulfide (e.g., pyrite) oxidation: a low $\delta^{34}S$ ratio, similar to the sulfide ratio with an inorganic origin of sulfate, is observed together with increasing $\delta^{15}N$ and $\delta^{18}O$ ratios with decreasing nitrate concentration (Aravena and Robertson 1998).

^{4}He, ^{20}Ne, SF$_6$, CFCs

Various sources of ^{4}He in groundwater can be defined including the atmosphere and productions from rocks and sediments. Taking into account errors due to sampling, ^{4}He increases with time due to subsurface production. However, ^{4}He dating methods that could theoretically be applicable in a wide range of ages (Figure 8.32), are limited by uncertainties about local production rates in the geological medium (Kazemi *et al.* 2006).

The only source of ^{20}Ne in groundwater is the equilibration of infiltrated water with the atmosphere. Isotopic compositions of Ne (^{20}Ne), combined with other noble gas compositions (Ar, Kr, Xe) are used to characterize recharge salinities (Mazor 1991). If Ne losses during sampling are taken into account, measured depletion with regards to the atmospheric composition may indicate increasing groundwater salinities during recharge through the partially saturated zone.

SF_6 is an inorganic gas used initially as a dielectric gas in electrical applications, but is also used in medical fluids and construction glass. Its use is now highly regulated in most countries as it is an extremely potent greenhouse gas. Time-series of

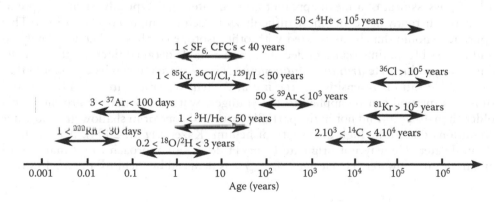

Figure 8.32 Useful time intervals of the different isotopes for groundwater dating (Suckow 2014).

atmospheric concentrations are well known and for a dating range over 40 years, no other source needs to be considered (Busenberg and Plummer 2000). Similarly, recent abrupt changes in the atmospheric composition of *CFCs* (chlorofluorocarbons) completely mask minor natural sources. Interpretation approaches, as for tritium, address the question if recharge of groundwater occurred before or after introduction of these gases into the atmosphere. Additional details can often be deduced from the use of the known input function. However, due to their high diffusivity in the partially saturated zone, these tracers equilibrate with groundwater at the water table, therefore results do not represent percolation through the unsaturated zone.

Radioactive isotopes

Tritium (^3H)

Tritium (^{3}H) is a radioactive isotope of hydrogen with a half-life of 12.43 years, disintegrating to stable ^{3}He. It is an excellent tracer as part of the water molecule. Natural atmospheric content of ^{3}H is due to interactions of cosmic radiation with nitrogen and deuterium. Between 1952 and 1980, tritium of artificial origin was released in the atmosphere through tests of thermonuclear bombs. The concentrations are usually expressed in "tritium units" (TU) defined as:

$$1\,TU = \frac{^3\mathrm{H}}{^1\mathrm{H}}10^{-18} = 0.118\,\mathrm{Bq/L} \tag{8.77}$$

In present-day studies, an absence of tritium shows that groundwater would correspond to relatively "old water" infiltrated before the nuclear test period. On the other hand, if tritium is detected the interpretation could be more complex, not excluding an "old water" component mixed with other more recent waters, which is difficult to distinguish from very recently infiltrated waters. The recent temporal evolution (since 1950) of *reference tritium concentrations* in precipitation have been published by IAEA showing a peak in 1963 and strong latitude differences. The local reference time evolution must be used to interpret groundwater tritium results. As for other radioactive tracers having a half-life, the most important use of tritium was traditionally the assessment of a mean apparent groundwater age. Typically, such interpretations are currently more and more difficult with such low input concentrations. This approach should thus be associated with other isotope or physicochemical analysis, and if possible on time scales of decades, for a reliable detailed interpretation. If the daughter product ^{3}He (*tritium-helium method*) is measured, it can be considered that the ^{3}H/^{3}He ratio is insensitive to the initial content of tritium in groundwater and thus this method can be applied to the youngest waters, but not for groundwaters older than 40 years and not in the partially saturated zone or in shallow groundwater conditions due to the high diffusivity of helium (Kazemi *et al.* 2006). Although all groundwater ^{3}He may not originate from ^{3}H decay, an approach to estimate mean apparent groundwater ^{3}H/^{3}He age ($\tau_{^3\mathrm{H}/^3\mathrm{He}}$) was deduced by (Schlosser *et al.* 1988):

$$\tau_{^3\mathrm{H}/^3\mathrm{He}} = \frac{1}{\lambda}\ln\left(\frac{C_{^3\mathrm{He}\,gw} - C_{^3\mathrm{He}\,atm}}{C_{^3\mathrm{H}\,gw}} + 1\right) \tag{8.78}$$

where λ is the linear degradation coefficient ($\lambda = \ln(2)/\tau_{\text{half-life}}$), $C_{^3\text{H}\,gw}$ is the measured tritium concentration in groundwater, and $C_{^3\text{He}\,gw}$ and $C_{^3\text{He}\,atm}$ are the ^3He concentrations, respectively, in groundwater and derived from the atmosphere.

^{14}C Carbon-14

Radiocarbon (^{14}C) with a half-life of 5730 years, is one of the most popular isotopes used for dating in archeology, recent geology, and hydrogeology. ^{14}C enters the groundwater composition as dissolved inorganic carbon by soil respiration of CO_2. A mean apparent ^{14}C groundwater age ($\tau_{^{14}C}$) is assessed by:

$$\tau_{^{14}C} = -\frac{1}{\lambda} \ln\left(\frac{A_{^{14}C\,gw}}{A_{^{14}C\,atm}}\right) \tag{8.79}$$

where λ is the linear degradation coefficient ($\lambda = \ln(2)/\tau_{\text{half-life}}$), and $A_{^{14}C\,gw}$ and $A_{^{14}C\,atm}$ are the ^{14}C activities (i.e., number of disintegrations per second) respectively in carbon from groundwater and in carbon in the infiltration water in equilibrium with the atmosphere; more details on the estimation of this last value can be found in Clark and Fritz (1997). This dating can be biased leading to erroneous old groundwater ages due to "dead" carbon in the medium (from carbonates or organic matter) which deplete the global measured ^{14}C activity (Fontes and Garnier 1979). A detailed understanding of the ongoing reactive processes is needed to interpret the measurements.

Biodegradation of organic pollutant can be detected by compound-specific carbon isotope analysis (CSIA) (Hunkeler *et al.* 2008, Batlle-Aguilar *et al.* 2009).

^{36}Cl, ^{40}Ar, ^{81}Kr and ^{85}Kr

Other radioactive environmental tracers can be used for groundwater dating. ^{36}Cl, ^{40}Ar, and ^{85}Kr can be useful as they are known as nonreactive tracers in groundwater systems. Attempts have also been made with dissolved ^{26}Al, ^{10}Be, ^{32}Si, ^{129}I, and ^{137}Cs, all being very rare isotopes originating from nuclear processes. Despite the doubts of Létolle and Olive (2004) about their added value, combined radioactive isotope studies can provide new constraints for further calibrating groundwater flow and transport models. Isotope hydrology is progressing very rapidly with many newly-developed techniques becoming useful. ^{36}Cl with a half-life of 3.01×10^5 years is naturally produced in the atmosphere by cosmic interactions with ^{40}Ar and ^{36}Ar. It can be used similarly to tritium but for dating very old groundwaters (Figure 8.32). However, the thermonuclear released ^{36}Cl and ^{129}I (with a half-life of 1.7×10^7 years) have made it possible to use the $^{36}\text{Cl}/\text{Cl}$ and $^{129}\text{I}/\text{I}$ ratios to detect qualitatively very recent infiltration waters. More quantitatively, for old ground water dating, a reduced production of ^{36}Cl occurs by thermal neutron activation of ^{35}Cl. A "secular" ratio $^{36}\text{Cl}/\text{Cl}$ can be predicted from the U and Th concentrations and from the bulk chemistry of the aquifer rock (Bentley *et al.* 1986) after which only decay affects the ratio.

Argon has two stable isotopes ^{36}Ar and ^{40}Ar with a constant ratio in the atmosphere. So for infiltrating water and groundwater, the ratio increases due to the dissolution of ^{40}Ar produced from the decay of ^{40}K of the rock minerals (Gascoyne and

Kotzer 1995). For dating purposes, ^{39}Ar with a half-life of 269 years could be very useful to determine mean apparent groundwater ages between 50 and 1000 years. It is almost the only available isotope allowing dating in this range of time (Figure 8.32). In deep groundwater systems, the production of ^{39}Ar is possible by dissolution of minerals with possible ^{39}K - ^{39}Ar ongoing reactions. If anthropogenic ^{39}Ar was also detected, it may be used to date shallow recent groundwaters.

^{81}Kr, with a half-life of 2.1×10^5 years, is in the same time dating range as ^{36}Cl (Figure 8.32). It can be used for dating very old groundwaters assuming that the production in the geological medium (from spontaneous ^{235}U fission) is known from the mineral content. On the other hand, ^{85}Kr, with a half-life of 10.8 years, adds to this natural origin (i.e., spontaneous ^{235}U fission) the anthropogenic signal in the atmosphere from the reprocessing of spent nuclear fuel rods (Gascoyne and Kotzer 1995). The yearly mean ^{85}Kr concentrations in the atmosphere are continuously increasing and values are published. Different combined ratios can also be useful, as for example, the ^{3}H/^{85}Kr ratio that increases with groundwater residence time.

For very short periods of time, ^{82}Br and ^{51}Cr (half-lives respectively of 36 hours and 27.8 days) can be suitable for use in tracer tests of limited duration. However, for these short periods, typically interesting for groundwater-surface water interactions, ^{222}Rn and ^{37}Ar are preferred.

^{222}Rn and ^{37}Ar for short-time dating ranges

Radon -222 is a radiogenic noble gas produced from the decay of thorium-uranium series isotopes present in most rocks and sediments. This is a particularity because its half-life of 3.82 days is very short but it is continuously being regenerated in the subsurface since thorium and uranium are common radioactive elements in geological layers and have very long half-lives. Surface waters have very low ^{222}Rn contents as it is rather quickly lost to the atmosphere. For the theoretical case of surface water suddenly saturating a volume of alluvial sediment, it would take about 20 days to reach a stabilized equilibrium content in ^{222}Rn (Bourke *et al.* 2014). This is the result of equilibrium between production (from the aquifer sediment matrix) and decay. It is thus theoretically an excellent tool to characterize groundwater-surface water interactions, assessment of the water residence time in hyporheic zones, and quantification of interacting fluxes. However, for losing streams, ^{222}Rn profiles in the streambed remain difficult to interpret (Bourke *et al.* 2014) due to the heterogeneity of the sediment which affects the production rates. For draining or gaining streams, it is sometimes difficult to distinguish clearly between the effective groundwater discharge and the more diffuse water exchanges between the stream and its hyporheic zones (Cook *et al.* 2006). As for other similar approaches, it is only the combination of different methods that will help to constraint the problem and allow more reliable assessment of those interactions. The ^{222}Rn balance in groundwater can be described (Hoehn *et al.* 1992) by:

$$A_{222_{Rn\,meas}} = A_{222_{Rn\,eq}}(1 - e^{-\lambda_{222_{Rn}}\tau}) \tag{8.80}$$

where $A_{222_{Rn\,meas}}$ and $A_{222_{Rn\,eq}}$ are respectively the measured and equilibrium activity concentrations of ^{222}Rn, $\lambda_{222_{Rn}}$ is the degradation constant equal to 0.182 (in days^{-1}), and τ is the time (in days) or ^{222}Rn-based residence time $\tau_{222\,\text{Rn}}$ (i.e., apparent age). This last term can thus be obtained by:

$$\tau_{222\,\text{Rn}} = \left(\lambda_{222\,\text{Rn}}\right)^{-1} \ln\left(1 - \frac{A_{222\,\text{Rn}\,meas}}{A_{222\,\text{Rn}\,eq}}\right) \tag{8.81}$$

Interpretation becomes problematic for ^{222}Rn residence times longer than 12 days ($\sim$3 half-lives) as measured and equilibrium ^{222}Rn activity concentrations converge toward very similar values (e.g., Cecil and Green 2000, Schilling 2017, Schilling et al. 2017).

Schilling et al. (2017) proposed to complement ^{222}Rn measurements by ^{37}Ar to close the residence time gap characterization between a few days and a few tens of days (Figure 8.32).

Argon -37 with a half-life of 35.1 days is a very rare radioactive tracer that is now measurable at natural levels (Loosli and Purtschert 2005). Recent progress in low level detection and water sample reduction has increased field application for groundwater dating. Without entering in details, the production of ^{37}Ar in the subsurface (mainly due to Ca activation by cosmic rays) can be considered as reaching a peak between 2 and 4 m depth and then decreases with depth (more details can be found in Riedmann and Purtschert 2011). Thus, the ^{37}Ar production-decay equilibrium activity concentration depth profile must be known to use this technique for further assessment of groundwater residence times (especially for recently infiltrated waters). Similar to ^{222}Rn, but including activity concentration dependency with depth (i.e., decreases with depth), ^{37}Ar-based residence time $\tau_{37\,\text{Ar}}$ (i.e., apparent age) can be obtained (Cecil and Green 2000) by:

$$A_{37\,\text{Ar}\,meas}(d) = A_{37\,\text{Ar}\,eq}(d)(1 - e^{-\lambda_{2237\,\text{Ar}}\tau}) \tag{8.82}$$

where $A_{37\,\text{Ar}\,meas}(d)$ and $A_{37\,\text{Ar}\,eq}(d)$ are respectively the measured and equilibrium activity concentrations of ^{37}Ar at the depth d, $\lambda_{37\,\text{Ar}}$ is the degradation constant equal to 0.0197 (in days^{-1}), and τ is the time (in days) or ^{37}Ar-based residence time $\tau_{37\,\text{Ar}}$ (i.e., apparent age). This last term can thus be obtained by:

$$\tau_{37\,\text{Ar}} = (\lambda_{37\,\text{Ar}})^{-1} \ln\left(1 - \frac{A_{37\,\text{Ar}\,meas}(d)}{A_{37\,\text{Ar}\,eq}(d)}\right) \tag{8.83}$$

The reasoning is similar to that as previously given, considering a zero starting point at the surface (atmospheric conditions) and an increase in ^{37}Ar concentration over 100 days ($\sim$3 half-lives) in the first few meters. A negative exponential evolution of the activity concentration is added to account for the ^{37}Ar decrease as a function of depth (Schilling 2017, Schilling et al. 2017).

8.8 Vulnerability and protection of groundwater

Facing groundwater contamination threats from anthropogenic and natural origins, prevention is always more effective then remediation. Sustainable management of groundwater requires a regulation framework combining prevention and protection of the groundwater resources in general, as well as protecting the groundwater receptors (i.e., natural springs or production wells). Delineation of protection zones and capture zones of drinking water wells, and a rigorous control of the land use where groundwater is more vulnerable to possible contamination, are necessary and complementary actions which ensure effective management. In most countries, legal frameworks have been established together with guidelines of methods and tools for both aspects:

a. establishing protection zones around groundwater pumping wells or springs
b. determining groundwater vulnerability or groundwater sensitivity to pollution for assessing the natural protective capacity or the sensitivity of groundwater to contamination (for example, Albinet and Margat 1970, Tripet *et al.* 1997, Troiano *et al.* 1999, Focazio *et al.* 2008, Arthur *et al.* 2012)

Protection zones

In protection zones, also named wellhead protection areas, human activities are strictly regulated or restricted in order to minimize the risk of contamination. Protection zones are thus prevention-oriented. Most protection zones are defined in terms of time-of-travel (see Section 8.6, Box 8.6; US-EPA 1993, Lallemand-Barres and Roux 1999, Thomsen and Thorling 2003, Molson and Frind 2012) of an ideal tracer to the considered production well or source. By estimating pollutant travel times to the abstraction point, they are implicitly based on the minimum time needed for urgent remediation operations if pollution occurs in the area. In this sense, these protective measures are only partial (i.e., focused only on protection of groundwater production points) and only provide short- to mid-term protection of groundwater quality. If the calculated solute travel times are only considered in the saturated zone, they do not provide any insight into the natural protection and attenuation capacity of the soil and partially saturated zone with regards to the various contamination issues. In summary, protection zones protect production wells against new point-source contaminants but do not address diffuse pollution issues.

The first approaches provided *circle-like protection zones*. Homogeneity and isotropy of the aquifer were assumed and only advective transport was considered under saturated conditions. Even if local hydraulic conductivities were accurately assessed (e.g., from pumping tests in the production well), this was not the case for effective transport porosity. This problem has resulted in many underestimated circular protection perimeters due to underestimated velocities as they were calculated with values closer to the total porosity (see Box 8.7 for realistic effective transport porosity values).

Facing the heterogeneous nature of aquifers, and considering advection + macrodispersion, deterministic and stochastic approaches were then developed to improve the reliability of time-of-travel protection zones. *Stochastic protection zones* methods were developed mainly on the basis of estimating macrodispersivity from the (log) hydraulic conductivity variance (Equation 8.27) and solving the coupled space-time probability distribution function (PDF) (Box 8.4). In the best cases, these techniques inferred the heterogeneous aquifer properties mainly from the local measured hydraulic conductivity values (among others, Varljen and Shafer 1991, Tiedeman and

Gorelick 1993, Evers and Lerner 1998, van Leewen *et al.* 1998, Vassolo *et al.* 1998, Guadagnini and Franzetti 1999, Kunstmann and Kinzelbach 2000, Feyen *et al.* 2001).

On the other hand, *deterministic protection zone* approaches were developed based on extensive additional field work including hydrogeophysical measurements, drilling observations, tracer injection-wells, and multitracer tests (see Box 8.6). ADE-based calculations were calibrated on the tracer tests before prediction of travel times from any point source to the production well (Derouane and Dassargues 1998). A deterministic delineation of protection zones is the final result. If Fickian dispersion is considered, theoretically a (infinitesimally) small first arrival of tracer arrives at the production well at a (infinitesimally) short time after injection. In practice, a relative threshold concentration ($C_{tracer}/M_{injected\ tracer}$) may be chosen in convention with the regulators and the water companies for defining what is meant by a significant "first arrival" of tracer. Attempts have been made to move these field-based deterministic protection zones in a stochastic framework (McKenna and Poeter 1995, Rentier and Dassargues 2002, Rentier *et al.* 2002) by using coconditional stochastic simulations (Dassargues *et al.* 2006), but unfortunately regulators and decision-makers are not always very receptive to complex calculations of uncertainties.

As mentioned previously, most of the existing techniques address solute travel times under saturated conditions only. All processes occurring in the partially saturated zone are therefore ignored and a (possibly strong) conservative delineation of protection zones is produced. An elegant way to include calculation of the travel time in the unsaturated and saturated zones is developed considering *well vulnerability* (Frind *et al.* 2006, Graf 2016) in terms of transit time from ground surface to the production well. If this last concept is adopted with process-based calculations, protection zones can be delineated corresponding to certain times of solute travel to the production well.

Vulnerability and sensitivity mapping

The notion of aquifer vulnerability to contaminations was first used in the seventies (Albinet and Margat 1970). In environmental sciences, the term vulnerability is the "degree to which human and environmental systems are likely to experience harm due to a perturbation or stress" (Füssel 2007). Unfortunately, but logically, within each application domain, "vulnerability" has been interpreted and conceptualized in very different ways. For groundwater, the fundamental question is: will groundwater be easily contaminated by any contamination hazard at the surface? As shown by the reviews on groundwater vulnerability mapping techniques (among others, Vrba and Zaporozec 1994, Frind *et al.* 2006, Popescu *et al.* 2008) conceptualization of the above mentioned definition is quite diverse.

Further details are also often introduced, since the definition of *intrinsic vulnerability* depends only on the inherent physical, geological, and hydrogeological characteristics that control the contamination due to a given input of a conservative ideal solute contaminant. On the other hand, *specific vulnerability* includes the biogeochemical behavior of a specific contaminant in the geological medium. The probability of *hazard* (i.e., mass input of contaminant) which can also be contaminant specific, is combined with specific or intrinsic vulnerability to assess the *risk* (Figure 8.33).

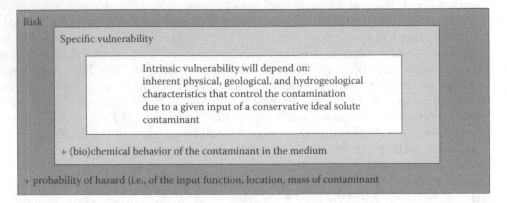

Risk

Specific vulnerability

Intrinsic vulnerability will depend on:
inherent physical, geological, and hydrogeological
characteristics that control the contamination
due to a given input of a conservative ideal solute
contaminant

+ (bio)chemical behavior of the contaminant in the medium

+ probability of hazard (i.e., of the input function, location, mass of contaminant

Figure 8.33 Intrinsic vulnerability becomes contaminant specific if the specific reactive behavior of a contaminant is taken into account. Risk assessment is evaluated by combining the probability of hazard occurrence with the vulnerability of the system.

One can also distinguish *groundwater resource vulnerability* (GRV) from *groundwater source vulnerability* (GSV). The latter considers that the main target is the receptor (or pumping well), while the former takes the entire saturated zone as the "target" (Figure 8.34). GRV assessments and mapping are thus based on processes occurring between the contamination release at the land surface and the saturated zone of the aquifer. For GSV assessment, additional transport processes in the saturated zone toward the pumping well or other receptors must be considered.

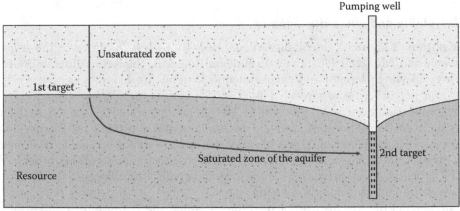

Unsaturated zone = pathway for 'groundwater resource vulnerability' (GRV)

Unsaturated zone + saturated zone = pathway for 'groundwater source vulnerability' (GSV)

Figure 8.34 Groundwater resource vulnerability (GRV) is based on the pathway from the contamination release at the land surface to the saturated zone of the aquifer. For groundwater source vulnerability (GSV), transport processes in the saturated zone must also be considered toward the pumping well (Goldscheider 2005).

Popular methods have been developed on the basis of multicriteria decision analysis which are easily developed on Geographical Information Systems (GIS). These approaches consider various physical factors such as depth to groundwater, hydraulic conductivity, soil type, infiltration, topography, unsaturated zone lithology, and other properties or variables that may increase or decrease groundwater vulnerability to contamination. This has resulted in a wide proliferation of methods (Gogu and Dassargues 2000), among others: DRASTIC (Aller *et al.* 1987), AVI (Van Stempvoort *et al.* 1993), SINTACS (Civita, 1994), GOD (Robins *et al.* 1994), a so-called German method (von Hoyer and Söfner 1998), REKS (Malik and Svasta 1998), a so-called Hungarian system (Madl-Szonyi and Fule 1998), a so-called Irish approach (Daly and Drew 1999), EPIK (Doerfliger *et al.* 1999), PI (Goldscheider *et al.* 2000), and PaPRIKa (Kavouri *et al.* 2011). While each individual factor has its own physical significance in these methods, the same does not apply for the empirical combination obtained by the weighting system in order to calculate a final index. Moreover, a multicriteria analysis requires that the criteria would be independent, which is not necessarily the case for most methods. The resulting vulnerability maps are relatively easy to implement and usually require little data. However, the empirical character of the final vulnerability index prevents any serious validation. Interpretation is difficult because results obtained by different methods on a same area can be dramatically different as shown by Gogu *et al.* (2003).

Process-based methods have been developed to address the following issues: what are the possible processes that can influence the impact of a contaminant, and how these processes can be quantified (conceptually simplified in different ways) for a reliable assessment (Brouyère *et al.* 2001, Perrin *et al.* 2004, Focazio *et al.* 2008, Popescu *et al.* 2008). In line with the DPSIR concept (as described in Section 8.1), the challenging question remains how to quantify the "impact" of a hazard (i.e., the "pressure," a contaminant release) from the changing groundwater "state." Two approaches can be proposed:

- Identify quantifiable key criteria describing contaminant transport processes that may occur from the surface to the target (i.e., the saturated zone for GRV, the production well or source for GSV), to proceed with a quantitative estimate of each process, and then according to conventional agreements, to determine a relative importance (weighting) for building a global vulnerability index.
- Develop an aquifer sensitivity model to describe how "pressure" on the groundwater "state" induces an "impact."

Process-based criteria methods

The contaminant impact at the "target" of any impulse or Dirac-type source at the land surface can be monitored as a breakthrough curve (BTC) of the contaminant. The impact at the target can thus be evaluated according to three criteria (Brouyère *et al.* 2001): the first arrival time, the pollution duration, and the maximum relative concentration (Figure 8.35). Considering only advection-dispersion of an ideal tracer (i.e., neglecting any adsorption and decay that can be highly contaminant-specific),

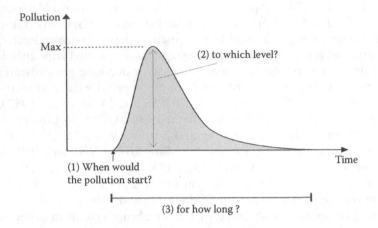

Figure 8.35 Breakthrough curve (BTC) of a contaminant at the target. Three criteria can be adopted to evaluate the potential harm at the target: the first arrival time (for a given threshold concentration), the pollution duration (above this given threshold concentration), and the relative maximum concentration (normalized to the injected mass) (Brouyère *et al.* 2001).

these three criteria can be assessed from deterministic ADE transport calculations (Popescu *et al.* 2008) which can be preceded by calculation of the surface pathway and spatial distribution of the contaminated water before infiltration. An intrinsic vulnerability index can then be built in agreement with the local priorities of decision-makers, conventionally attributing index values and weighting to each of the above defined criteria. If a specific vulnerability is needed, the same approach can be used taking into account the contaminant specific reactions, including adsorption and degradation, in the transport equations. This method has the advantage of clearly separating the physically-based analysis from the conventional criteria induced by local societal, environmental, or political priorities (Dassargues *et al.* 2009). Another option is to only use the groundwater age or the mean residence time (see Section 8.6) as an integrated indicator of vulnerability (Graf 2016). Indeed, different kinds of complexity can be adopted to calculate this contaminant travel time from the land surface to the groundwater resource (or receptor): from simple piston-flow to detailed (specific) reactive solute transport models. This last approach could be used to introduce uncertainty in vulnerability mapping under the form of a PDF (probability distribution function) as explained in Section 8.6 about using the statistic approach to calculate groundwater ages.

Sensitivity methods

A sensitivity of the groundwater resource (for GRV) or of the groundwater receptor or (pumping) well (for GSV) can be defined according to (Beaujean *et al.* 2013):

$$S \equiv S_{i,j} = \frac{\partial I_j}{\partial P_i} \tag{8.84}$$

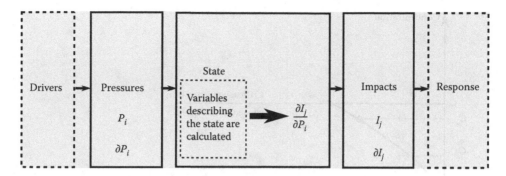

Figure 8.36 Sensitivity analysis within the DPSIR concept.

where $S_{i,j}$ is the sensitivity of the jth impact (I_j) to a change in the jth pressure (P_i). Fully in line with the DPSIR concept (Figure 8.36), as $S_{i,j}$ increases, the groundwater state becomes more sensitive. This general sensitivity does not take into account the margin available between the present state of the groundwater resource/receptor and the critical state from where it should be considered as effectively damaged. To deduce a vulnerability from the sensitivity as defined above, Luers *et al.* (2003) therefore proposed integrating a ratio (ST/ST_0) that reflects the "distance" between the current state (ST) and its "damaged state" (ST_0), so that:

$$V = \frac{S}{ST/ST_0} \tag{8.85}$$

if the "damaged" state is a critical minimum value (e.g., a lowest piezometric head for a vulnerability assessment linked to groundwater quantity), and inversely:

$$V = S.\left(\frac{ST}{ST_0}\right) \tag{8.86}$$

if the "damaged" state is a critical maximum value (e.g., a highest concentration for a vulnerability assessment linked to groundwater quality).

Accordingly, vulnerability of the groundwater resource (GRV) or of the groundwater source or (pumping) well (GSV) can be calculated. As an example, we assume that the groundwater quality state of an aquifer is assessed on the basis of a contaminant concentration (C_{gw}). The pressure is a recharge with infiltrating water at a concentration C_R. The impact on the state of the aquifer can be quantified by the concentration sensitivity (S_C):

$$S_C = \frac{\partial C_{gw}}{\partial C_R} \tag{8.87}$$

The vulnerability of the aquifer is calculated with respect to a given critical maximum concentration (C_0) corresponding to a damaged state (Figure 8.37):

$$V_C = S_C\left(\frac{C_{gw}}{C_0}\right) = \frac{\partial C_{gw}}{\partial C_R}\frac{C_{gw}}{C_0} \tag{8.88}$$

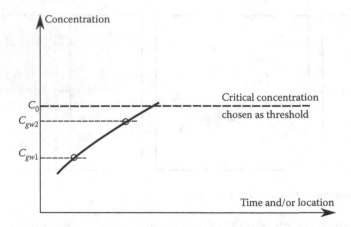

Figure 8.37 Even in the improbable case where the sensitivity to contaminated infiltrating water would be the same, the vulnerability for two different aquifer states will be different according to the "distance" to the critical concentration chosen as the threshold value. In this case, the vulnerability of the aquifer at State 2 is higher than at State 1.

For the case of Figure 8.37 (i.e., assuming $S_{C_1} \approx S_{C_2}$), the vulnerability of the aquifer at State 2 is higher than at State 1 since the distance to the critical concentration is lower.

$$V_{C_1} = S_{C_1} \frac{C_{gw1}}{C_0} < S_{C_2} \frac{C_{gw2}}{C_0} = V_{C_2} \qquad (8.89)$$

This kind of approach allows calculating the groundwater sensitivity/vulnerability to any stress factor by physically-based indicators. This allows addressing both GRV and GSV problems with regards to both groundwater quality and quantity issues. In summary, three steps are needed: (1) defining the physically-based indicators (i.e., groundwater state variables) that are able to reflect/quantify the sensitivity of impact changes to stress factors, (2) calculation of sensitivity coefficient using adequately calibrated numerical models, and (3) definition of a threshold that enables calculation of vulnerability coefficients. Step 3 depends on site-specific conditions and consists of a maximum/minimum acceptable value for the aquifer state (e.g., the lowest acceptable groundwater level for vulnerability of groundwater quantity, and highest acceptable groundwater concentration for vulnerability of groundwater quality).

In practice, and depending on the situation, sensitivity calculation methods can be performed by the perturbation method, the sensitivity equation method, or the adjoint operator method (Beaujean *et al.* 2013). The latter one is more convenient for sensitivity calculations at a single point with respect to a simultaneous change in each other location or with respect to a changing spatially-distributed change (GSV), while the two former methods are most convenient for sensitivity calculated at different locations of the domain with respect to a change in one point (GRV). More details on those methods can be found in Sykes *et al.* (1985), Sun and Yeh (1990), Jyrkama and Sykes (2006), and Hill and Tiedeman (2007).

References

Aggarwal, P.K., Gat, J.R. and K.F.O Froehlich (eds). 2005. *Isotopes in the water cycle: past, present and future of a developing science*. IAEA, Dordrecht: Springer.

Albinet, M. and J. Margat. 1970. Cartographie de la vulnérabilité à la pollution des nappes d'eau souterraines. (Mapping aquifer vulnerability to pollution) in French. *Bulletin BRGM* 2(3-4): 13–22.

Aller, L., Bennett, T., Lehr, J., Petty, R. and G. Hackett. 1987. *DRASTIC: A standardized system for evaluating ground water pollution potential using hydrogeologic settings*. EPA-600/2-87-035. Ada (OK): National Water Well Association.

Alvarez P.J.J. and W.A. Illman. 2006. *Bioremediation and natural attenuation: Process fundamentals and mathematical models*. Hoboken (NJ): John Wiley & Sons.

Anderson, M.P. 2005. Heat as a ground water tracer. *Ground Water* 43(6): 951–968.

Appelo, C.A.J. and D. Postma. 1993. *Geochemistry, groundwater and pollution*. Rotterdam: A.A. Balkema, Rotterdam.

Aravena, R. and W.D. Robertson. 1998. Use of multiple isotope tracers to evaluate denitrification in ground water: study of nitrate from a large-flux septic system plume. *Ground Water* 36: 975–982.

Arthur, J.D., Wood, H.A.R., Baker, A.E., Cichon, J.R. and G.L. Raines. 2007. Development and implementation of a bayesian-based aquifer vulnerability assessment in Florida. *Natural Resources Research* 16(2): 93–107.

Arya, A., Hewett, T.A., Larson, R.G. and L.W. Lake. 1988. Dispersion and reservoir heterogeneity. *SPE Reservoir Engineering* 3(01): 139–148.

Bates, E. and C. Hills. 2015. *Stabilization and solidification of contaminated soil and waste: A manual of practice*. Report for US EPA. Hyggemedia: www.hyggemedia.com.

Batlle-Aguilar, J., Banks, E.W., Batelaan, O., Kipfer, R., Brennwald, M.S. and P.G. Cook. 2017. Groundwater residence time and aquifer recharge in multilayered, semi-confined and faulted aquifer systems using environmental tracers. *Journal of Hydrology* 546: 150–165.

Batlle-Aguilar, J., Brouyère, S., Dassargues, A., Morasch, B., Hunkeler, D., Hohener, P., Diels, L., Vanbroekhoeven, K., Seuntjens, P. and H. Halen. 2009. Benzene dispersion and natural attenuation in an alluvial aquifer with strong interactions with surface water. *Journal of Hydrology* 361: 305–317.

Bear, J. 1972. *Dynamics of fluids in porous media*. New York: American Elsevier.

Bear, J. and A.H.D. Cheng. 2010. *Modeling groundwater flow and contaminant transport*. Dordrecht: Springer.

Bear, J. and A. Verruijt. 1987. *Modeling groundwater flow and pollution*. Dordrecht: Reidel Publishing Company.

Beaujean, J., Lemieux, J.M., Gardin, N., Dassargues, A., Therrien, R. and S. Brouyère. 2013. Physically-based groundwater vulnerability assessment using sensitivity analysis methods. *Groundwater* 52(6): 864–874.

Beims, U. 1983. Planung, durchführung und auswertung von gütepumpversuchen (in German). *Zeitschrift für Angewandte Geologie* 29(10): 482–490.

Bentley, H.W., Phillips, F.M., Davis, S.N., Habermehl, M.A., Airey, P.L., Calf, G.E., Elmore, D., Gove, H.E. and T. Torgersen. 1986. Chlorine 36 dating of very old groundwater, 1, The Great Artesian Basin, Australia. *Water Resources Research* 22: 1991–2001.

Berkowitz, B., Cortis, A., Dentz, M. and H. Scher. 2006. Modeling non-Fickian transport in geological formations as a continuous time random walk. *Review Geophysical* 44: RG2003.

Berkowitz, B., Klafter, J., Metzler, R. and H. Scher. 2002. Physical pictures of transport in heterogeneous media: advection-dispersion, random-walk, and fractional derivative formulations. *Water Resources Research* 38(10): 1191.

Bethke, C.M. and T.M. Johnson. 2008. Groundwater age and groundwater age dating. *Annual Review of Earth and Planetary Sciences* 36: 121–152.

Bijeljic, B. and M.J. Blunt. 2007. Pore-scale modelling of transverse dispersion in porous media. *Water Resources Research* 43: W12S11.

Bourke, S.A., Cook, P.G., Shanafield, M., Dogramaci, S. and J.F. Clark. 2014. Characterisation of hyporheic exchange in a losing stream using radon-222. *Journal of Hydrology* 519: 94–105.

Brady, P.V., Brady, M.V. and D.J. Borns. 1998. *Natural attenuation: CERCLA, RBCA's and the future of environmental remediation*. Boca Raton: CRC Press.

Brainerd, R.J. and G.A. Robbins. 2004. A tracer dilution method for fracture characterization of bedrock wells. *Ground Water* 42: 774–780.

Briers, P., Dollé, F., Orban, P., Piront, L. and S. Brouyère. 2017. *Etablissement des valeurs représentatives par type d'aquifère des paramètres hydrogéologiques intervenant dans l'Evaluation des risques pour les eaux souterraines en application du Décret du 5 décembre 2008 relatif à la Gestion des Sols (in French)*. University of Liège - GEOLYS final report. SPW/DGO3, Walloon Region (Belgium).

Brouyère, S., Batlle-Aguilar, J., Goderniaux, P. and A. Dassargues. 2007. A new tracer technique for monitoring groundwater fluxes: The Finite Volume Point Dilution Method. *Journal of Contaminant Hydrology* 95: 121–140.

Brouyère, S., Carabin, G. and A. Dassargues. 2005. Influence of injection conditions on field tracer experiments. *Ground Water* 43(3): 389–400.

Brouyère, S., Jeannin, P.Y., Dassargues, A., Golscheider, N., Popescu, I.C., Sauter, M., Vadillo, I., and F. Zwahlen. 2001. Evaluation and validation of vulnerability concepts using a physically based approach. In *Proc. of the 7th Conf. on Limestone Hydrology and Fissured Media*, eds. J. Mudry and F. Zwahlen. Besançon, France: Université de Franche-Comté, Mémoire 13: 67–72.

Burnett, R.D. and E.O. Frind. 1987. Simulation of contaminant transport in three dimensions: 1. The alternating direction Galerkin technique. *Water Resources Research* 23(4): 683–694.

Busenberg, E. and L.N. Plummer. 2000. Dating young groundwater with sulfur hexafluoride: natural and anthropogenic sources of sulfur hexafluoride. *Water Resources Research* 36(10): 3011–3030.

Cameron, D.R. and A. Klute. 1977. Convective–dispersive solute transport with a combined equilibrium and kinetic adsorption model. *Water Resources Research* 13: 183–188.

Cecil, L.D. and J.R. Green. 2000. Radon-222. In *Environmental tracers in subsurface hydrology*, eds. P.G. Cook and A.L. Herczeg. New York: Springer, 31–77.

Charbeneau, R.J. 2000. *Groundwater hydraulics and pollutant transport*. Upper Saddle River (NJ): Prentice Hall.

Chiogna, G., Cirpka, O.A., Grathwohl, P. and M. Rolle. 2011. Relevance of local compound-specific transverse dispersion for conservative and reactive mixing in heterogeneous porous media. *Water Resources Research* 47: W07540.

Civita, M. 1994. *Le carte della vulnerabilità degli acquiferi all'inquinamento. Teoria e practica* (in Italian). Pitagora: 13, Bologna.

Clark, I.D. 2015. *Groundwater geochemistry and isotopes*. Boca Raton: CRC Press Taylor & Francis Group.

Clark, I.D. and P. Fritz. 1997. *Environmental isotopes in hydrogeology*. Boca Raton: Lewis Publishers CRC Press LLC.

Cline, P.V., Delfino, J.J. and P.S.C. Rao. 1991. Partitioning of aromatic constituents into water from gasoline and other complex solvent mixtures. *Environmental Science and Technology* 25(5): 914–920.

Coats, K.H. and B.D. Smith. 1964. Dead-end pore volume and dispersion in porous media. *Society of Petroleum Engineers Journal* 4(1): 73–84.

Cook, P., Dogramaci, S., McCallum, J. and J. Hedley. 2017. Groundwater age, mixing and flow rates in the vicinity of large open pit mines, Pilbara region, northwestern Australia. *Hydrogeology Journal* 25: 39–53.

Cook, P.G., Lamontagne, S., Berhane, D. and J.F. Clark. 2006. Quantifying groundwater discharge to Cockburn River, southeastern Australia, using dissolved gas tracers Rn-222 and SF6. *Water Resources Research* 42(10): W10411.

Cornaton, F.J. 2004. *Deterministic models of groundwater age, life expectancy and transit time distributions in advective-dispersive systems*. PhD diss., University of Neuchâtel, Switzerland.

Cornaton, F.J. and P. Perrochet. 2006. Groundwater age, life expectancy and transit time distributions in advective–dispersive systems: 2. Reservoir theory for sub-drainage basins. *Advances in Water Resources* 29: 1292–1305.

Craig, H. 1961. Isotopic variations in meteoric waters. *Science* 133:1702–1703.

Cullimore, D.R. 2008. *Practical manual of Groundwater Microbiology*. Boca Raton: CRC Press Taylor & Francis.

Cushman, J.H. and D. O'Malley. 2015. Fickian dispersion is anomalous. *Journal of Hydrology* 531: 161–167.

Dagan, G. 1982. Stochastic modeling of groundwater flow by unconditional and conditional probabilities, 2, The solute transport. *Water Resources Research* 18(4): 835–848.

Dagan, G. 1989. *Flow and transport in porous formations*. New York: Springer.

Daly, D. and D. Drew. 1999. Irish methodologies for karst aquifer protection. In *Hydrogeology and engineering geology of sinkholes and karst*, ed. B. Beck. Rotterdam: Balkema, 267–272.

Dassargues, A. 2006. Geo-electrical data fusion by stochastic co-conditioning simulations for delineating groundwater protection zones. In *Proc. CMWR XVI-Computational Methods for Water Resources*, Copenhagen (Denmark). http://hdl.handle.net/2268/3348

Dassargues, A., Goderniaux, P., Daoudi, M. and Ph. Orban. 2011. Measured and computed solute transport behaviour in the saturated zone of a fractured and slightly karstified chalk aquifer. In *Proc. H2Karst, 9th Conference on Limestone Hydrogeology*, eds. C. Bertrand, N. Carry, J. Mudry, M. Pronk and F. Zwahlen. Besançon: Université de Franche-Comté. 111–114.

Dassargues, A., Popescu, I.C., Beaujean, J., Lemieux, J.M. and S. Brouyère. 2009. Reframing groundwater vulnerability assessment for a better understanding between decision makers and hydrogeologists. In *The Role of Hydrology in Water Resources Management (Proc. of IAHS - IHP2008)*, eds. H.J. Liebscher, R. Clarke, J. Rodda, G. Schultz, A. Schumann, L. Ubertini and G. Young. Capri: IAHS Press, Publ. 327: 278–284.

Dassargues, A., Rentier, C. and M. Huysmans. 2006. Reducing the uncertainty of hydrogeological parameters by co-conditional simulations: lessons from practical applications in aquifers and in low permeability layers. In *Calibration and Reliability in Groundwater Modelling: From Uncertainty to Decision Making (Proc. of ModelCARE'2005)*, The Hague, IAHS Publ. 304: 3–9.

Davis, J.A. and D.B. Kent. 1990. Surface complexation modeling in aqueous geochemistry. *Review Mineral* 23: 177–260.

Davis, S.N., Thompson, G.M., Bentley, H.W. and G. Stiles. 1980. Ground-water tracers – A short review. *Ground Water* 18: 14–23.

de Dreuzy, J.R. and T.R. Ginn. 2016. Residence times in subsurface hydrological systems, introduction to the Special Issue. *Journal of Hydrology* 543: 1–6.

Delleur, J.W. 2000. Elementary groundwater flow and transport processes. In *The Handbook of Groundwater Engineering*, ed. J.W. Delleur, Chapter 2. Boca Raton: CRC Press and Heidelberg: Springer-Verlag, 190–230.

de Marsily, G. 1986. *Quantitative hydrogeology: groundwater hydrology for engineers*. San Diego: Academic Press.

Dentz, M. and B. Berkowitz. 2003. Transport behavior of a passive solute in continuous time random walks and multirate mass transfer. *Water Resources Research* 39(5): 1111.

Dentz, M., Le Borgne, T., Englert, A. and B. Bijeljic. 2011. Mixing, spreading and reaction in heterogeneous media: a brief review. *Journal of Contaminant Hydrology* 120-121:1–17.

Derouane, J. and A. Dassargues. 1998. Delineation of groundwater protection zones based on tracer tests and transport modelling in alluvial sediments. *Environmental Geology* 36(1-2): 27–36.

Deutsch, W.J. 1997. *Groundwater geochemistry: Fundamentals and applications to contamination.* Boca Raton: Lewis Publishers CRC Press LLC.

Diersch, H-J.G. 2014. *Feflow – Finite element modeling of flow, mass and heat transport in porous and fractured media.* Heidelberg: Springer.

Divine, C. E. and J.J. McDonnell. 2005. The future of applied tracers in hydrogeology. *Hydrogeology Journal* 13: 255–258.

Doerfliger, N., Jeannin, P.Y. and F. Zwahlen. 1999. Water vulnerability assessment in karst environments: a new method of defining protection areas using a multi-attribute approach and GIS tools (EPIK method). *Environmental Geology* 39(2): 165–176.

Dogan, M., Van Dam, R.L., Liu, G., Meerschaert, M.M., Butler Jr., J.J., Bohling, G.C., Benson, D.A. and D.W. Hyndman. 2014. Predicting flow and transport in highly heterogeneous alluvial aquifers. *Geophysical Research Letters* 21: 7560–7565.

Drost, W., Klotz, D., Arnd, K., Heribet, M., Neumaier, F. and W. Rauert. 1968. Point dilution methods of investigation ground water flow by means of radioisotopes. *Water Resources Research* 4(1): 125–146.

Einarson, M. 2017. Spatially averaged, flow-weighted concentrations – A more relevant regulatory metric for groundwater cleanup. *Groundwater Monitoring & Remediation* 37: 11–14.

Engdahl, N.B. 2017. Transient effects on confined groundwater age distributions: Considering the necessity of time-dependent simulations. *Water Resources Research* 53: 7332–7348.

Engdahl, N.B., McCallum, J.L. and A. Massoudieh. 2016. Transient age distributions in subsurface hydrologic systems. *Journal of Hydrology* 543: 88–100.

EPA. 1996. *How to effectively recover free product at leaking underground storage tank sites - A guide for state regulators.* EPA 510-R-96-001. Washington: US Environmental Protection Agency.

Evers, S. and D.N. Lerner. 1998. How uncertain is our estimate of a wellhead protection zone? *Ground Water* 36(1): 49–57.

Fetter, C.W. 1993. *Contaminant Hydrogeology.* New York: Macmillan Publishing Company.

Feyen, L., Beven, K.J., De Smedt, F. and J. Freer. 2001. Stochastic capture zone delineation within the generalized likelihood uncertainty estimation methodology: conditioning on head observations. *Water Resources Research* 37(3): 625–638.

Fiori, A. and F.P.J. de Barros. 2015. Groundwater flow and transport in aquifers: insight from modeling and characterization at the field scale (editorial). *Journal of Hydrology* 531(1): 1.

Fitts, Ch. R. 2002. *Groundwater science.* London: Academic Press.

Focazio, M.J., Reilly, T.E., Rupert, M.G. and D.R. Helsel. 2008. *Assessing groundwater vulnerability to contamination: Providing scientifically defensible information for decisions makers.* Denver: U.S. Geological Survey Circular 1224.

Fontes, J.C. and J.M. Garnier. 1979. Determination of initial ^{14}C activity of total dissolved carbon: a review of existing models and a new approach. *Water Resources Research* 15: 399–413.

Freeze, R.A. and J.A. Cherry. 1979. *Groundwater.* Upper Saddle River (NJ): Prentice Hall.

Freundlich, H. 1926. *Colloid and capillary chemistry.* London: Methuen.

Frind, E.O., Mohammed, D. and J.W. Molson. 2002. Delineation of three-dimensional capture zones in a complex aquifer. *Ground Water* 40(6): 586–598.

Frind, E.O., Molson, J.W. and D.L. Rudolph. 2006. Well vulnerability: a quantitative approach for source water protection. *Ground Water* 44(5): 732–742.

Frippiat, Ch.C. and A.E. Holeyman. 2008. A comparative review of upscaling methods for solute transport in heterogeneous porous media. *Journal of Hydrology* 362: 150–176.

Füssel, H.-M. 2007. Vulnerability: A generally applicable conceptual framework for climate change research. *Global Environmental Change* 17: 155–167.

Gascoyne, M. and T. Kotzer. 1995. *Isotopic methods in hydrogeology and their application to the underground research laboratory.* Manitoba. Chalk River Laboratories, A.E.C.L. - 11370, Chalk River.

Gelhar, L.W. and C.L. Axness. 1983. Three-dimensional stochastic analysis of macro-dispersion in aquifers. *Water Resources Research* 19(1): 161–180.

Gelhar, L.W., Welty, C. and K.R. Rehfeldt. 1992. A critical review of data on field-scale dispersion in aquifers. *Water Resources Research* 28(7): 1955–1974.

Gerke, H.H. and M.Th. van Genuchten. 1993. A dual-porosity model for simulating the preferential movement of water and solutes in structured porous media. *Water Resources Research* 29(2): 305–319.

Gillham, R.W., Sudicky, E.A., Cherry, J.A. and E.O. Frind. 1984. An advection–diffusion concept for solute transport in heterogeneous unconsolidated geologic deposits. *Water Resources Research* 20(3): 369–374.

Ginn, T.R. 1999. On the distribution of multicomponent mixtures over generalized exposure time in subsurface flow and reactive transport: foundations, and formulations for groundwater age, chemical heterogeneity, and biodegradation. *Water Resources Research* 35(5): 1395–1407.

Gogu, R. and A. Dassargues. 2000. Current and future trends in groundwater vulnerability assessment using overlay and index methods. *Environmental Geology* 39(6): 549–559.

Gogu, R., Hallet, V. and A. Dassargues. 2003. Comparison between aquifer vulnerability assessment techniques. Application to the Néblon river basin (Belgium). *Environmental Geology* 44(8): 881–892.

Goldberg, S., Criscenti, L.J., Turner, D.R., Davis, J.A. and K.J. Cantrell. 2007. Adsorption-desorption processes in subsurface reactive transport modelling. *Vadose Zone Journal* 6: 407–435.

Goldberg, S. and N.J. Kabengi. 2010. Bromide adsorption by reference minerals and soils. *Vadose Zone Journal* 9(3): 780–786.

Goldscheider, N. 2005. Karst groundwater vulnerability mapping - application of a new method in the Swabian Alb, Germany. *Hydrogeology Journal* 13(4): 555–564.

Goldscheider, N., Klute, M., Sturm, S. and H. Hötzl. 2000. The PI method – a GIS-based approach to mapping groundwater vulnerability with special consideration of karst aquifers. *Z Angew Geol Hannover* 46(3): 157–166.

Goode, D.J. 1996. Direct simulation of groundwater age. *Water Resources Research* 32: 289–296.

Graf, T. 2016. *Physically-based assessment of intrinsic groundwater resource vulnerability.* PhD diss., University Laval, Canada.

Guadagnini, A. and S. Franzetti. 1999. Time-related capture zones for contaminants in randomly heterogeneous formations. *Ground Water* 37(2): 253–260.

Hadley, P.W. and C.J. Newell. 2012. Groundwater Remediation: The Next 30 Years. *Ground Water* 50: 669–678.

Hadley, P.W. and C.J. Newell. 2014. The new potential for understanding groundwater contaminant transport. *Groundwater* 52(2): 174–186.

Haerens, B., Brouyère, S. and A. Dassargues. 1999. Detailed calibration of a deterministic transport model on multi-tracer tests: analysis and comparison with semi-analytical solutions. In *ModelCARE'99 Pre-published Proc.*, eds. F. Stauffer, W. Kinzelbach, K. Kovar and E. Hoehn. 1: 319–324, Zurich: ETH.

Hall, S.H. 1993. Single well tracer tests in aquifer characterization. *Ground Water Monitoring and Remediation* 13(2): 118–124.

Hallet, V. and A. Dassargues. 1998. Effective porosity values used in calibrated transport simulations in a fissured and slightly karstified chalk aquifer. In *Groundwater Quality 1998*, eds. M. Herbert and K. Kovar. Tubingen: Tübinger Geowissenschaftliche Arbeiten (TGA), C36: 124–126.

Hallet, V., Nzali, T., Rentier, C. and A. Dassargues. 2000. Location of protection zones along production galleries: an example of methodology. In *Proc. of TraM'2000 Conference on Tracers and Modelling in Hydrogeology*, ed. A. Dassargues. IAHS Publication 262: 141–148, Wallingford: IAHS Press.

Harvey, C.F and S.M. Gorelick. 1995. Temporal moment generating equations: modeling transport and mass transfer in heterogeneous aquifers. *Water Resources Research* 31: 1895–1912.

Hermans, T., Vandenbohede, A., Lebbe, L., Martin, R., Kemna, A., Beaujean, J. and F. Nguyen. 2012. Imaging artificial salt water infiltration using electrical resistivity tomography constrained by geostatistical data. *Journal of Hydrology* 438-439: 168–180.

Hermans, T., Wildemeersch, S., Jamin, P., Orban, P., Brouyère, S., Dassargues, A. and F. Nguyen. 2015. Quantitative temperature monitoring of a heat tracing experiment using cross-borehole ERT. *Geothermics* 53: 14–26.

Heron, G., Bierschenk, J., Swift, R., Watson, R. and M. Kominek. 2016. Thermal DNAPL source zone treatment impact on a CVOC plume. *Groundwater Monitoring Research* 36: 26–37.

Hill, M.C. and C.R. Tiedeman. 2007. *Effective groundwater model calibration with analysis of data, sensitivities, predictions, and uncertainty*. Hoboken (NJ): John Wiley & Sons.

Hoehn, E., Von Gunten, H.R., Stauffer, F. and T. Dracos. 1992. Radon-222 as a groundwater tracer. A Laboratory Study. *Environmental Science and Technology* 26: 734–738.

Horneman, A., Divine, C., Sandtangelo-Dreiling, T., Lloyd, S., Anderson, H., Smith, M.B. and J. McCray. 2017. The case for flux-based remedial performance monitoring programs. *Groundwater Monitoring and Remediation* 37(3): 16–18.

Hunkeler, D., Meckenstock, R.U., Sherwood Lollar, B., Schmidt T.C. and J.T. Wilson. 2008. *A guide for assessing biodegradation and source identification of organic ground water contaminants using compound specific isotope analysis (CSIA)*. Ada (Oklahoma): U.S. Environmental Protection Agency, EPA 600/R-08/148.

Hunt, R.J. and W.P. Johnson. 2017. Pathogen transport in groundwater systems: contrasts with traditional solute transport. *Hydrogeology Journal* 25: 921–930.

Huysmans, M. and A. Dassargues. 2005. Review of the use of Peclet numbers to determine the relative importance of advection and diffusion in low permeability environments. *Hydrogeology Journal* 13(5-6): 895–904.

Huysmans, M. and A. Dassargues. 2007. Equivalent diffusion coefficient and equivalent diffusion accessible porosity of a stratified porous medium. *Transport in Porous Media* 66: 421–438.

Huysmans, M. and A. Dassargues. 2009. Application of multiple-point geostatistics on modelling groundwater flow and transport in a cross-bedded aquifer. *Hydrogeology Journal* 17(8): 1901–1911.

Istok, J.D., Field, J.A., Schroth, M.H., Davis, B.M. and V. Dwarakanath. 2002. Single-well "push-pull" partitioning tracer test for NAPL detection in the subsurface. *Environmental Science and Technology* 36: 2708–2716.

Jamin, P., Dolle, F., Chisala, B., Orban, Ph., Popescu, I.C., Hérivaux, C., Dassargues, A. and S. Brouyère. 2012. Regional flux-based risk assessment approach for multiple contaminated sites on groundwater bodies. *Journal of Contaminant Hydrology* 127(1-4): 65–75.

Jury, W.A. 1982. Simulation of solute transport using a transfer function model. *Water Resources Research* 18(2): 363–368.

Jyrkama, M.I. and J.F. Sykes. 2006. Sensitivity and uncertainty analysis of the recharge boundary condition. *Water Resources Research* 42(1): W01404.

Kasnavia, T., Vu, D. and D.A. Sabatini. 1999. Fluorescent dye and media properties affecting sorption and tracer selection. *Ground Water* 37(3): 376–381.

Käss, W. 1998. *Tracing technique in geohydrology*. Rotterdam: Balkema, Boca Raton/ CRC Press Taylor & Francis.

Kavouri, K., Plagnes, V., Tremoulet, J., Dörfliger, N., Rejiba, F. and P. Marchet. 2011. PaPRIKa: a method for estimating karst resource and source vulnerability - application to the Ouysse karst system (southwest France). *Hydrogeology Journal* 19: 339–353.

Kazemi, G.A., Lehr, J.H. and P. Perrochet. 2006. *Groundwater Age*. Hoboken (NJ): Wiley-Interscience.

Kendall, C. and J.J. McDonnel (eds). 1998. *Isotope tracers in catchment hydrology*. Amsterdam: Elsevier.

Kinzelbach, W. 1988. The random walk method in pollutant transport simulation. In *Groundwater flow and quality modelling*, eds. E. Custodio, A. Gurgui and J.P.L. Ferreira. Netherlands: Springer, 227–245.

Kinzelbach, W. 1992. *Numerische methoden zur modellierung des transports von schadstoffen im grundwasser (in German)*. Schriftenreihe GWF Wasser, Abwasser, Bd. 21, -2 Aufl., Munchen: Oldenbourg.

Kinzelbach, W. and Ph. Ackerer. 1986. Modélisation de la propagation d'un contaminant dans un champ d'écoulement transitoire (in French). *Hydrogéologie* 2: 197–205.

Kitadinis, P.K. 1994. The concept of the dilution index. *Water Resources Research* 30(7): 2011–2026.

Knox, R.C., Sabatini, D.A. and L.W. Canter. 1993. *Subsurface transport and fate processes*. Boca Raton: Lewis Publishers CRC press LLC.

Korom, S.F. and J.C. Seaman. 2012. When "conservative" anionic tracers aren't. *Ground Water* 50: 820–824.

Kunstmann, H. and W. Kinzelbach. 2000. Computation of stochastic wellhead protection zones by combining the first-order second-moment method and Kolmogorov backward equation analysis. *Journal of Hydrology* 237: 127–146.

Lallemand-Barres, A. and J.C. Roux. 1999. *Périmètres de protection des captages d'eau souterraine destinée à la consommation humaine (in French)*. Manuel et méthodes 33, BRGM.

Langevin, P. 1908. Sur la théorie du mouvement brownien (in French) [On the Theory of Brownian Motion]. *Comptes rendus de l'Académie des Sciences (Paris)* 146: 530–533.

Langmuir, D. 1918. The adsorption of gases on plane surfaces of glass, mica, and platinum. *Journal of the American Chemical Society* 40: 1361–1403.

Le Borgne, T., Dentz, M. and J. Carrera. 2008. A Lagrangian statistical model for transport in highly heterogeneous velocity fields. *Physical Review Letters* 101: 090601.

Lefebvre R. 2010. *Ecoulement multiphasique en milieu poreux (in French)*. Course notes: 280-306. INRS Eau Terre Environnement, Québec, Canada.

Lefebvre, R. and A. Boutin. 2000. Evaluation of free LNAPL volume and productibility in soils. In *Proc. 1st Joint IAH-CNC and CGS Groundwater Specialty Conference, 53rd Canadian Geotechnical Conference*, Oct. 15–18, 2000, Montreal, Canada, 143–150.

Leibundgut, C., Maloszewski, P. and C. Küll. 2009. *Tracers in Hydrology*, 432. West Sussex, UK: John Wiley & Sons Ltd.

Létolle, R. and P. Olive. 2004. A short history of isotopes in hydrology. In *The Basis of civilization – Water science?*. Proc. of UNESCO/IAHS/IWHA Symp. Roma, IAHS Publ. 286: 49–66.

Lichtner, P.C., Kelkar, S., and B. Robinson. 2002. New form of dispersion tensor for axisymmetric porous media with implementation in particle tracking. *Water Resources Research* 38(8): 1146.

Loosli, H.H. and R. Purtschert. 2005. Rare Gases. In *Isotopes in the water cycle: past, present and future of a developing science*. eds. P.K. Aggarwal, J.R. Gat and K.F.O Froehlich, IAEA, Chapter 7, 91–96. Dordrecht: Springer.

Luers, A.L., Lobel, D.B., Sklar, L.S., Addams, C.L. and P.A. Matson. 2003. A method for quantifying vulnerability, applied to the agricultural system of the Yaqui Valley, Mexico. *Global Environmental Change* 13(4): 255–267.

Madioune, D.H., Faye, S., Orban, P., Brouyère, S., Dassargues, A., Mudry, J., Stumpp, C. and P. Maloszewski. 2014. Application of isotopic tracers as a tool for understanding hydrodynamic behavior of the highly exploited Diass aquifer system (Senegal). *Journal of Hydrology* 511: 443–459.

Madl-Szonyi, J. and L. Fule. 1998. Groundwater vulnerability assessment of the SW Trans-Danubian central range. *Environmental Geology* 35: 9–17.

Malik, P. and J. Svasta. 1998. *Groundwater vulnerability maps for the areas with karst-fissure and fissure aquifers (in Slovak)*. Slovak Republic: Arch. Geol. Surv.

Maliva, R.G. 2016. *Aquifer characterization techniques, Schlumberger Methods in Water Resources Evaluation Series No. 4*. New York, Berlin: Springer.

Mariotti, A., Germon, J.C., Hubert, P., Kaiser, P., Letolle, R., Tardieux, A. and P. Tardieux. 1981. Experimental determination of nitrogen kinetic isotope fractionation: some principles; illustration for the denitrification and nitrification processes. *Plant and Soil* 62: 413–430.

Mayer, B. 2005. Assessing sources and transformations of sulphate and nitrate in the hydrosphere using isotope techniques. In *Isotopes in the water cycle: past, present and future of a developing science*, eds. P.K. Aggarwal, J.R. Gat and K.F.O. Froehlich, IAEA, Chapter 6, 67–89. Dordrecht: Springer.

Mayer, B., Feger, K.H., Giesemann, A. and H.-J. Jäger. 1995. Interpretation of sulphur cycling in two catchments in the Black Forest (Germany) using stable sulphur and oxygen isotope data. *Biogeochemistry* 30: 31–58.

Mazor, E. 1991. *Applied chemical and isotopic groundwater hydrology*. Buckingham: Open University Press.

McKenna, S.A. and E.P. Poeter. 1995. Field example of data fusion in site characterization. *Water Resources Research* 31(12): 3229–3240.

Molson, J.W. and E.O. Frind. 2012. On the use of mean groundwater age, life expectancy and capture probability for defining aquifer vulnerability and time-of-travel zones for source water protection. *Journal of Contaminant Hydrology* 127: 76–87.

Molz, F. 2015. Advection, dispersion, and confusion. *Groundwater* 53(3): 348–353.

Monjoie, A., Rigo, J.M. and C. Polo-Chiapolini. 1992. *Vade-mecum pour la réalisation des systèmes d'étanchéité-drainage artificiels pour les sites d'enfouissement technique en Wallonie (in French)*. Liège, Belgium: University of Liège FSA.

Moreno, Z. and A. Paster. 2017. Prediction of remediation of a heterogeneous aquifer: A case study. *Groundwater* 55(3): 428–439.

Naftz, D.L., Morrison, S.J., Davis, J.A. and Ch.C. Fuller (eds). 2002. *Handbook of Groundwater remediation using permeable reactive barriers*. London: Academic Press Elsevier Science (USA).

Ndeye, M.D., Orban, P., Otten, J., Stumpp, C., Faye, S. and A. Dassargues. 2017. Hydrochemical and isotopic evidence of groundwater salinization evolution in the area of Saloum (Senegal). *Journal of Hydrology Regional Studies* 9: 163–182.

Neuman, S. 1990. Universal scaling of hydraulic conductivities and dispersivities in geologic media. *Water Resources Research* 26(8): 1749–1758.

Norris, R.D., Hinchee, R.E., Brown, R., McCarty, P.L., Semprini, L., Wilson, J.T., Kampbell, D.H., Reinhard, M., Bouwer, E.J., Borden, R.C, Vogel, T.M., Thomas, J.H., Ward, C.H. and J.E. Matthews. 1993. *Handbook of Bioremediation*. Boca Raton: Lewis Publishers.

Ogata, A. and R.B. Banks. 1961. *A solution of the differential equation of longitudinal dispersion in porous media.* USGS Professional Paper 411-I. Washington: US Government Printing Office.

Parkhurst, D.L. 1995. *User's guide to PHREEQC - A computer program for speciation, reaction-path, advective-transport, and inverse geochemical calculations.* Technical Report 4227, U.S. Geological Survey Water-Resources Investigations Report.

Payne, F.C., Quinnan, A. and S.T. Potter. 2008. *Remediation hydraulics.* Boca Raton: CRC Press/ Taylor & Francis.

Perrin, J., Pochon, A., Jeannin, P.-Y. and F. Zwahlen. 2004. Vulnerability assessment in karstic areas: validation by field experiments. *Environmental Geology* 46(2): 237–245.

Pickens, J.F., and G.E. Grisak. 1981. Scale-dependent dispersion in a stratified granular aquifer. *Water Resources Research* 17(4): 1191–1211.

Pinder, G.F. and M.A. Celia. 2006. *Subsurface hydrology.* Hoboken (NJ): John Wiley & Sons.

Popescu, I.C., Gardin, N., Brouyère, S. and A. Dassargues. 2008. Groundwater vulnerability assessment using physically based modelling: from challenges to pragmatic solutions. In *Calibration and Reliability in Groundwater Modelling: Credibility in Modelling,* (eds) J.C. Refsgaard, K. Kovar, E. Haarder and E. Nygaard. IAHS Publ. 320: 83–88. Wallingford: IAHS Press.

Prommer, H. 2002. *PHT3D - A reactive multicomponent transport model for saturated porous media.* Technical report, Contaminated Land Assessment and Remediation Research Centre, The University of Edinburgh.

Prommer, H., Barry, D.A. and C. Zheng. 2003. MODFLOW/MT3DMS-based reactive multi-component transport modelling. *Ground Water* 42(2): 247–257.

Ptak, T., Peipenbrink, M. and E. Martac. 2004. Tracer tests for the investigation of heterogeneous porous media and stochastic modelling of flow and transport – a review of some recent developments. *Journal of Hydrology* 294: 122–163.

Rao, P.S.C., Jessup, R.E. and T.M. Addiscot. 1982. Experimental and theoretical aspects of solute diffusion in spherical and nonspherical aggregates. *Soil Science* 133(6): 342–349.

Rausch, R., Schäfer, W., Therrien, R. and Chr. Wagner. 2005. *Solute transport modelling – An introduction to models and solution strategies.* Berlin-Stuttgart: Gebr.Borntraeger Verlagsbuchhandlung Science Publishers.

Renard, Ph. and D. Allard. 2013. Connectivity metrics for subsurface flow and transport. *Advances in Water Resources* 51: 168–196.

Rentier, C., Brouyère, S. and A. Dassargues. 2002. Integrating geophysical and tracer test data for accurate solute transport modelling in heterogeneous porous media. In *Groundwater Quality: Natural and Enhanced Restoration of Groundwater Pollution, Proc. of GQ'2001.* Sheffield, IAHS Publ. 275: 3–10.

Rentier, C. and A. Dassargues. 2002. Deterministic and stochastic modelling for protection zone delineation. In *Calibration and Reliability in Groundwater Modelling: A Few Steps Closer to Reality (Proc. of ModelCARE'2002).* IAHS Publ. 277: 489–497.

Riedmann, R.A. and R. Purtschert. 2011. Natural [37]Ar concentrations in soil air: implications for monitoring underground nuclear explosions. *Environmental Science and Technology* 45: 8656–8664

Robins, N., Adams, B., Foster, S. and R. Palmer. 1994. Groundwater vulnerability mapping: the British perspective. *Hydrogéologie* 3: 35–42.

Russel, D.L. 2012. *Remediation manual for contaminated sites.* Boca Raton: CRC Press Taylor and Francis.

Russian, A., Dentz, M. and P. Gouze. 2016. Time domain random walks for hydrodynamic transport in heterogeneous media. *Water Resources Research* 52: 3309–3323.

Russo, D. 2015. On the effect of connectivity on solute transport in spatially heterogeneous combined unsaturated-saturated flow systems. *Water Resources Research* 51: 3525–3542.

Sabatini, D.A. 2000. Sorption and intraparticle diffusion of fluorescent dyes with consolidated aquifer media. *Ground Water* 38(5): 651–656.

Sabir, I.H., Torgersen, J., Haldorsen, S. and P. Aleström. 1999. DNA tracers with information capacity and high detection sensitivity tested in groundwater studies. *Hydrogeology Journal* 7(3): 264–272.

Sanford, W.E., Shropshire, R.G. and D.K. Solomon. 1996. Dissolved gas tracers in groundwater: Simplified injection sampling and analysis. *Water Resources Research* 23: 1635–1642.

Sauty, J.P. 1980. An analyse of hydrodispersive transfer in aquifers. *Water Resources Research* 16(1): 145–158.

Scheidegger, A. 1961. General theory of dispersion in porous media. *Journal of Geophysical Research* 66(10): 3273–3278.

Schilling, O. 2017. *Advances in characterizing surface water - groundwater interactions: combining unconventional data with complex, fully-integrated models.* PhD diss., CHYN University of Neuchâtel, Switzerland.

Schilling, O.S., Gerber, C., Partington, D.J., Purtschert, R., Brennwald, M.S., Kipfer, R., Hunkeler, D. and P. Brunner. 2017. Advancing physically-based flow simulations of alluvial systems through atmospheric noble gases and the novel 37Ar tracer method. *Water Resources Research* 53: 10,465–10,490.

Schlosser, P., Stute, M., Sonntag, C. and K.O. Münnich. 1988. Tritium/3-Helium dating of shallow groundwater. *Earth and Planetary Science Letters* 89(3-4): 353–362.

Schwartz, F.W. and H. Zhang. 2003. *Fundamentals of Ground Water.* New York: Wiley.

Selker, J.S., Thévanez, L., Huwald, H., Mallet, A., Luxemburg, W., Van de Giesen, N., Stejskal, M., Zeman, J., Westhoff, M. and M.B. Parlange. 2006. Distributed fiber-optic temperature sensing for hydrologic systems. *Water Resources Research* 42: W12202.

Skopp, J. and A. Warrick. 1974. A two-phase model for the miscible displacement of the reactive solutes in soils. *Soil Science Society of America Proceedings* 38(4): 545–550.

Solder, J.E., Stolp, B.J., Heiweil, V.M. and D.D. Susong. 2016. Characterization of mean transit time at large springs in the Upper Colorado River Basin, USA: a tool for assessing groundwater discharge vulnerability. *Hydrogeology Journal* 24: 2017–2033.

Starr, R.C. and J.A. Cherry. 1994. In situ remediation of contaminated ground water: the funnel and gate system. *Ground Water* 32(3): 465–476.

Suckow, A. 2014. The age of groundwater – Definitions, models and why we do not need this term. *Applied Geochemistry* 50: 222–230.

Sudicky, E.A., Cherry, J.A. and E.O. Frind. 1983. Migration of contaminants in groundwater at a landfill: A case study: 4. A natural gradient dispersion test. *Journal of Hydrology* 63(1-2): 81–108.

Sun, Y., Petersen, J., Clement, T. and R. Skeen. 1999. Development of analytical solutions for multispecies transport with serial and parallel reactions. *Water Resources Research* 35(1): 185–190.

Sun, N.-Z. and W.W.-G. Yeh. 1990. Coupled inverse problems in groundwater modeling 1. Sensitivity analysis and parameter identification. *Water Resources Research* 26(10): 2507–2525.

Suthersan, S.S. 1997. *Remediation engineering, Design Concepts.* Boca Raton: CRC Press.

Suthersan, S.S. 2002. *Natural and enhanced remediation systems.* Boca Raton: CRC Press.

Sykes, J.F., J.L. Wilson and R.W. Andrews. 1985. Sensitivity analysis for steady state groundwater flow using adjoint operators. *Water Resources Research* 21(3): 359–371.

Thomsen, R., and L. Thorling. 2003. Use of protection zones and land management restore contaminated groundwater in Denmark. *Eos Transaction* 84(7): 63–65.

Tiedeman, C. and S.M. Gorelick. 1993. Analysis of uncertainty in optimal groundwater contaminant capture design. *Water Resources Research* 29(7): 2139–2153.

Toride, N., Leij, F. and M. van Genuchten. 1999. *The CXTFIT code for estimating transport parameters from laboratory or field tracer experiments.* Version 2.1. Technical report 137, Riverside: U.S. Salinity Laboratory.

Tripet, J.-P., Doerfliger, N. and F. Zwahlen. 1997. Vulnerability mapping in karst areas and its uses in Switzerland. *Hydrogéologie* 3: 51–57.

Troiano, J., Spurlock, F. and J. Marade. 1999. Update of the Californian Vulnerability Sail Analysis for Movement of Pesticides to Ground Water. Sacramento (CA): Environmental Protection Agency (EPA) - Department of Pesticide Regulation (DPR).

Truex, M., Johnson, C., Macbeth, T., Becker, D., Lynch, K., Giaudrone, D., Frantz, A. and H. Lee. 2017. Performance assessment of Pump-and-Treat Systems. *Groundwater Monitoring and Remediation* 37(3): 28–44.

U.S. EPA (Environmental Protection Agency). 1993. *Guidelines for delineation of wellhead protection areas.* EPA-440/5-93-001.

van Leewen, M., te Stroet, C.B.M., Butler, A.P. and J.A. Tompkins. 1998. Stochastic determination of well capture zones. *Water Resources Research* 34(9): 2215–2223.

Van Stempvoort, D., Evert, L. and L. Wassenaar. 1993. Aquifer vulnerability index: a GIS compatible method for groundwater vulnerability mapping. *Canadian Water Resources Journal* 18: 25–37.

Varljen, M.D. and J.M. Shafer. 1991. Assessment of uncertainty in time-related capture zones using conditional simulation of hydraulic conductivity. *Ground Water* 29(5): 737–748.

Varni, M, and J. Carrera. 1998. Simulation of groundwater age distributions. *Water Resources Research* 34(12): 3271–3281.

Vassolo, S., Kinzelbach, W. and W. Schäfer. 1998. Determination of a well head protection zone by stochastic inverse modelling. *Journal of Hydrology* 206: 268–280.

von Hoyer, M. and B. Söfner. 1998. *Groundwater vulnerability mapping in carbonate (karst) areas of Germany.* Hannover: Fed Inst Geosci. Nat. Resources. Archiv 117854.

Vrba, J. and A. Zaporozec. 1994. *Guidebook on mapping groundwater vulnerability.* IAH International Contribution for Hydrogeology. Series: Hydrogeology of selected karst regions. 13. Hannover: Heise.

Waddill, D.W. and M.A. Widdowson. 1998. *SEAM3D, A numerical model for three-dimensional solute transport and sequential electron acceptor-based bioremediation in groundwater.* Technical report, Blacksburg: Virginia Tech.

West, L.J. and N.E. Oldling. 2007. Characteristics of a multilayer aquifer using open well dilution tests. *Ground Water* 45: 74–84.

Wiedemeier, T.H., Rifai, H.S., Newell, C.J. and J.T. Wilson. 1999. *Natural attenuation of fuels and chlorinated solvents in the subsurface.* New York: John Wiley & Sons.

Wiedemeier, T., Swanson, M., Moutoux, D., Kinzie Gordon, E., Wilson, J., Wilson, B., Kampbell, D., Haas, P., Miller, R., Hansen, J. and F. Chapelle. 1998. *Technical protocol for evaluating natural attenuation of chlorinated solvents in ground water.* Technical report EPA/600/R-98/128, U.S. Environmental Protection Agency.

Wildemeersch, S., Jamin, P., Orban, Ph., Hermans, T., Klepikova, M., Nguyen, F., Brouyère, S. and A. Dassargues. 2014. Coupling heat and chemical tracer experiments for estimating heat transfer parameters in shallow alluvial aquifers. *Journal of Contaminant Hydrology* 169: 90–99.

Wildemeersch, S., Orban, P., Ruthy, I., Grière, O., Olive, P., El Youbi, A. and A. Dassargues. 2010. Towards a better understanding of the Oulmès hydrogeological system (Mid-Atlas, Morocco). *Environmental Earth Science* 60(8): 1753–1769.

Worthington, S.R.H. 2015. Diagnostic tests for conceptualizing transport in bedrock aquifers. *Journal of Hydrology* 529: 365–372.

Xu, M. and Y. Eckstein. 1995. Use of weighted least-squares method in evaluation of the relationship between dispersivity and field scale. *Ground Water* 33(6): 905–908.

Yurtsever, Y. and J.R. Gat. 1981. Atmospheric waters. In *Stable Isotope Hydrology*, eds. J.R. Gat and R. Gonfiantini. IAEA Tech. Rep. 210: 103–142. Vienna: IAEA.

Zech, A., Attinger, S., Cvetkovic, V., Dagan, G., Dietrich, P., Fiori, A., Rubin, Y. and G. Teutsch. 2015. Is unique scaling of aquifer macrodispersivity supported by field data? *Water Resources Research* 51: 7662–7679.

Zech, A., Attinger, S., Cvetkovic, V., Dagan, G., Dietrich, P., Fiori, A., Rubin, Y. and G. Teutsch. 2016. Reply to comment by S. Neuman on 'Is unique scaling of aquifer macrodispersivity supported by field data?'. *Water Resources Research* 52: 4203–4205.

Zheng, Ch and P.P. Wang. 1999. *MT3DMS A modular three-dimensional multispecies transport model for simulation of advection, dispersion and chemical reactions of contaminants in groundwater systems*. Documentation and user's guide. Technical report, Contract Report SERDP-99-1, Vicksburg (MS): U.S. Army Engineer Research and Development Center.

Chapter 9

Groundwater flow and transport under partially saturated conditions

9.1 Introduction

The flow and transport processes in the *vadose zone* are particularly important because this zone is considered to play a major role in many ecohydrological and hydrogeological problems. This is the central part of what is now referred to as the "critical zone" (i.e., including the pedosphere, the vadose, and the saturated zone) influencing most of the energy and mass exchanges that are necessary for biological and chemical production, degradation, and recycling. It also plays a critical role in water storages. In addition, the complex processes in the vadose zone occur over a large range of spatial and temporal scales. Simplifications are often needed to calculate and quantify mass or energy balances, as well as spatial and temporal behaviors.

A detailed physical and mathematical description of all these processes is beyond the scope of this short chapter, particularly because entire books are devoted to this topic. Here, only a practical synthesis will be proposed using physical principles and equations with the dual purpose of understanding how the main parameters/properties vary and, preparing readers for the quantification and further modeling of this complex zone.

The groundwater flow in the partially saturated zone or the vadose zone can be considered as a particular case of a two-phase flow, with air and water in the pores. This is usually further simplified by noting that, at usual temperatures, the air mobility (i.e., resulting from its lower viscosity) is far greater than the water mobility. Thus, any change in pressure will be transmitted much faster in the air phase than in the water phase, so the air phase is often considered perfectly continuous in the pore space and at atmospheric pressure (Szymkiewicz 2013). All other classic assumptions are taken: the dissolution of air in water and the evaporation of water are neglected, water density is dependent only on pressure (i.e., water isothermal and homogeneous conditions), matrix compressibility is neglected, and Darcy's law is applicable in each phase.

9.2 Capillary pressures

Resulting from the different molecular cohesion within each phase, the interface between two phases is characterized by a *surface tension*. Compared to air and most contaminant nonaqueous phase liquids (NAPLs), water is clearly preferentially attracted to the solid surface; thus, water can be considered as the *wetting phase*.

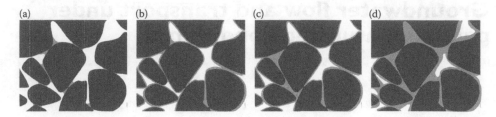

Figure 9.1 Microscopic schematic view of the spatial distribution of water for a nearly dry to a nearly saturated porous geological medium: (a) water only in the smallest pores or corners, (b) water film around each solid grain, (c) partially localized flow, and (d) continuous partially saturated flow.

The difference between the air pressure (p_a) and the water pressure (p), resulting from this surface tension, is named *capillary pressure* (p_c):

$$p_a - p = p_c \tag{9.1}$$

In the narrow pores, the capillary pressure increases strongly, as classically shown with the capillary rise of water in a small diameter tube. In this case, p_c is proportional to $1/r_c$, where r_c is the tube radius. A geological porous medium is far more complex than any assemblage of capillary tubes, but observations confirm that capillary water is preferentially trapped in the narrower pores or corners between grains (Figure 9.1a). With the increasing global water content, a kind of thin, continuous film of water is formed around each solid grain (Figure 9.1b). Therefore, more water creates possible local water flow through the merged pendular rings bound to solid grains (Figure 9.1c), and the continuity of this partial flow is increasingly more effective with increasing water content (Figure 9.1d) until the total saturation of the porous medium is reached.

If we return to the usual scale of consideration (i.e., Darcy's law scale, see Section 4.4), one uses the water content $(\theta = V_w/V_t)$ (Equation 4.13) and the water saturation (S_w) to describe the partially saturated conditions in the REV:

$$S_w = \frac{\theta}{n} = \frac{V_w}{V_t} \frac{V_t}{V_p} = \frac{V_w}{V_p} \tag{9.2}$$

where V_p, V_w and V_t are the pores, water and total volumes, respectively, and n is the total porosity (see Chapter 4). At this REV scale, the capillary pressure is assumed to be a quantifiable function of the saturation (or of the water content). We must note that many authors use different terms, such as *suction* or *matrix potentials*, to represent this average capillary pressure (expressed in water pressure head, in m) at the REV scale that accurately describes accurately the physical water pressure in any microscopic pore zone within the REV. As mentioned previously, the air phase is assumed to be perfectly continuous and at the atmospheric pressure, so for a given REV, the capillary pressure is expressed by:

$$p_c(S_w) = p_c(\theta) = -p \tag{9.3}$$

Using the Bernoulli equation (Equation 4.16), this can also be written as a function of the pressure head. The *capillary head* or *suction head* (h_c) is defined as:

$$\psi = h_c = \frac{p_c}{\rho g} = -h_p = -\frac{p}{\rho g} = z - h \tag{9.4}$$

where h_p is the pressure head.

Capillary pressure is considered to be a nonnegative value, but when the continuity between the saturated and partially saturated medium is considered in further equations, the main variable will be the water pressure (or the pressure head) that therefore takes negative values in the vadose zone. In other words, as the water content decreases, the capillary pressure increases, and the pressure head becomes more negative (Schwartz and Zhang 2003). This negative water pressure could be interpreted as the pressure that would be needed to extract water from the partially saturated medium. This negative pressure is typically measured using *tensiometers* generally consisting of a water-filled tube with a porous ceramic cup at one end. A partial vacuum can be created in the tube by a hand pump to subtract the dissolved oxygen. The water in the tube is pulled out by the pressure conditions in the partially saturated medium, increasing the vacuum inside the tube measured by the pressure sensor. After a while, the water pressure in the tensiometer is considered to be in equilibrium with the soil water pressure.

For a given geological medium (i.e., lithology), it is important to obtain information about the *water retention curves* (i.e., equivalent to *capillary pressure (or head)—water (or moisture) content* curves) (Figure 9.2) or, alternatively, the *capillary pressure (or head)—saturation* curves.

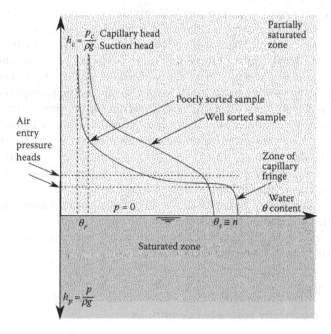

Figure 9.2 Typical schematic retention curves (i.e., capillary pressure head curve) for poorly and well-sorted porous media.

Typical curves are shown in Figure 9.2 for well sorted and poorly sorted porous media. For saturated conditions, the maximum water content is reached and considered equal to the total porosity $\theta = \theta_s = n$ and $S = S_{max} = 1$, neglecting the entrapped air. As the suction first increases, the water content remains near the full saturation. To obtain a significant desaturation, an increase in the suction above a threshold value is necessary. This pressure threshold is named *air-entry pressure* head or bubbling pressure, and it depends on the pore size. In a vertical profile of the porous medium, these conditions occur in the *capillary fringe* (see Chapter 1). Then, effective desaturation corresponds to the drainage of the medium under increasing suction conditions (i.e., decreasing water pressure), and this drainage is more rapid for large pores than small pores. For greater suction values, the curve derivative tends to the infinite (i.e., vertical asymptotic line). Therefore, an additional increase of the suction does not significantly change the *residual water content* (θ_r), also known as *field capacity*, corresponding to the *residual saturation* (S_r).

For practical purposes, the water content and the water saturation are often normalized with respect to the variation range in the medium:

$$\Theta = \frac{\theta - \theta_r}{\theta_s - \theta_r} \tag{9.5}$$

$$S_e = \frac{S - S_r}{1 - S_r} \tag{9.6}$$

where Θ is defined as the (normalized) effective water content, and S_e is defined as the (normalized) effective saturation. Retention curves obtained for each characterized porous medium express $\Theta(h_p)$ or $S_e(h_p)$.

Theoretically, these curves are used to determine, at any given time, the amount of water retained in the partially saturated medium for a given pressure or head. However, in practice, they are strongly affected by hysteresis and thus the direction and history of drainage and wetting. These processes are induced by (Bear and Verruijt 1987): "ink-bottle" effects resulting from the various shapes of the pores; "raindrop" effects resulting from a larger contact angle for the advancing (compared to the receding) water-air interface on a solid surface; air entrapment, particularly during rewetting; and various geomechanical effects (i.e., consolidation, swelling, and shrinkage). Many studies have been reported about hysteresis processes in air-water, oil-water, and gas-oil systems in each of the specialized bodies of literature. However, for evident simplifying reasons, only an average curve (sometimes only the drainage curve) is most often considered in practice (Bear and Cheng 2010).

Another problem comes from the heterogeneity of the partially saturated zone and the *saturation jumps* that this heterogeneity can induce in the corresponding vertical profile (Figure 9.3). These create many difficulties (see Section 9.3) because jumps in the hydraulic parameters values are consequently induced.

9.3 Partially saturated flow

Hydraulic conductivity under partially saturated conditions

For a given REV of a partially saturated porous medium, hydraulic conductivity varies with the water content and thus as a function of the water pressure head. The curves

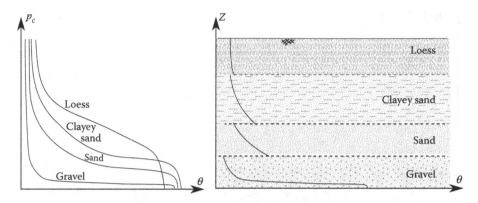

Figure 9.3 Capillary pressure discontinuities in a 4-layer heterogeneous profile showing the equilibrium discontinuous water content conditions (Bear and Cheng 2010).

describing hydraulic conductivity values under partially saturated conditions are closely dependent on the grain-size distribution and lithology of the medium. Logically, for a decreasing saturation (i.e., for lower water pressure and greater capillary pressure), a rapid decrease in the value is generally observed as shown in Figure 9.4 for coarser materials. If a fine-grained medium is above a coarse-grained layer in a partially saturated vertical profile, this can cause some *capillary barrier effects* (Yeh *et al.* 2015). For a given water pressure maintained low enough (i.e., corresponding to capillary pressure conditions above a "threshold"), a medium made of finer grains can have a larger hydraulic conductivity than a medium made of coarser material (Figure 9.4). The downward unsaturated flow occurring in the fine-grained material is blocked by the lower conductivity conditions in the material with coarser pores. This effect is commonly used in civil engineering designs for waste containment to prevent water infiltration by sandwiching sediments between earth liners of contrasting lithologies.

Usually, the partially saturated hydraulic conductivity is normalized by its saturated value; thus, the relative hydraulic conductivity is written as:

$$K_r(h_p) = \frac{K(h_p)}{K_s} \tag{9.7}$$

where K_s is the saturated hydraulic conductivity of the porous medium. K_r is often called the scalar relative permeability coefficient, and it varies from a value of 1 for the full saturation of the pores by water to a value of 0 when the water phase is considered immobilized (i.e., for a saturation $S \leq S_r$) [-]. This concept is generalized in multiphase problems for i fluid phases sharing a pore space, with the relative permeability of the medium for the ith fluid phase:

$$k_{r_i} = \frac{k_i}{k} \tag{9.8}$$

where k_i and k are the effective permeability for the ith phase and the intrinsic permeability respectively, and both are expressed in m² [L²]. Each k_{r_i} value ranges

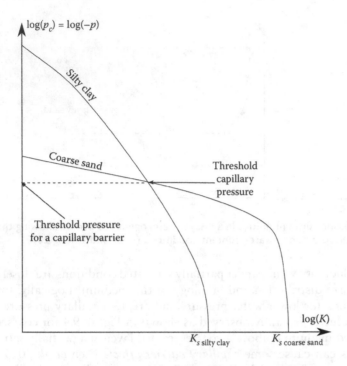

Figure 9.4 Schematic evolution of the partially saturated hydraulic conductivity (log scale) as a function of the capillary pressure (log scale) for a silty clay and a coarse sand. These two curves are crossing at a "threshold capillary pressure": for greater capillary pressure, the hydraulic conductivity of the silty clay is higher than that of the coarse sand.

between 0 and 1 depending on the respective saturation of each fluid phase, and $\sum_{i=1}^{n} k_{r_i} = 1$.

Many different models have been published in the specialized literature for calculating and modeling partially saturated flow (among others, Gardner 1958, Campbell 1974, Fredlung and Xing 1994). Because of its flexibility and relative simplicity, the most widely used model is the van Genuchten (1980) model based on the Mualem model (1976):

$$K_r(h_p) = \frac{\left(1-\left(\alpha|h_p|\right)^{n_{vg}-1}\left[1+\left(\alpha|h_p|\right)^{n_{vg}}\right]^{-m_{vg}}\right)^2}{\left[1+\left(\alpha|h_p|\right)^{n_{vg}}\right]^{-m_{vg}/2}} \tag{9.9}$$

$$\Theta(h_p) = \left[1+\left(\alpha|h_p|\right)^{n_{vg}}\right]^{-m_{vg}} \tag{9.10}$$

where $\alpha[L^{-1}]$, and the dimensionless exponents n_{vg} and m_{vg} are specific parameters that must be adjusted for the geological medium such that $m_{vg} = 1 - 1/n_{vg}$. The n_{vg}

parameter allows the slope of the hydraulic conductivity curve to be fitted, and α^{-1} is related to the air-entry pressure head. Note that Equation 9.10 can also be written as:

$$\theta(h_p) = (\theta_s - \theta_r)\left[1 + \left(\alpha|h_p|\right)^{n_{vg}}\right]^{-m_{vg}} + \theta_r \tag{9.11}$$

Many discussions have been published about the validity of the van Genuchten model, particularly for a low moisture content. For example, the model tends to an infinite pressure head for a residual water content that is not consistent with the reality. As it became the most commonly used model, some authors published detailed (statistical) studies providing the average values of α and n_{vg} with the θ_r, θ_s, and K_s values for different media corresponding to the USDA or simpler classifications (among others, Rawls et al. 1982, Carsel and Parrish 1988, Khaleel and Freeman 1995). These values can be considered as first estimates if no site-specific data are available.

Darcy-Buckingham law under partially saturated conditions

As adapted by Buckingham (1907), Darcy's law in a 3D partially saturated zone can be written as:

$$q = -K(\theta) \cdot [\nabla(h_p) + \nabla(z)] = -\frac{K_r(\theta)k}{\mu} \cdot [\nabla p + \rho g \nabla z] \tag{9.12}$$

where $K_r(\theta)$ is the scalar relative hydraulic conductivity (Equation 9.7) multiplying k the intrinsic permeability tensor and $\rho g/\mu$ to obtain $K(\theta)$, the tensor of partially saturated hydraulic conductivity. With the variation of the water content as a function of the capillary pressure, the variation of the hydraulic conductivity may preferably be expressed as a function of the water pressure head or of the water pressure:

$$q = -K(h_p) \cdot [\nabla(h_p) + \nabla(z)] = -\frac{K_r(p)k}{\mu} \cdot [\nabla p + \rho g \nabla z] \tag{9.13}$$

Generalized storage coefficient under partially saturated conditions or moisture capacity

In the partially saturated zone, the storage coefficient represents the water storage change (i.e., stored or drained volume of water) per unit volume of the medium and per unit of the pressure head change (m^{-1}) $[M^{-1}]$ at a given pressure head:

$$C(h_p) = \frac{\partial \theta}{\partial h_p} \tag{9.14}$$

The concept is similar to the specific storage coefficient (S_s) in the saturated zone (see Section 4.10); however, the involved physical processes are quite different. If a problem involves both partially saturated and saturated conditions, the generalized storage coefficient $C(h_p)$ can easily include the influence of S_s (i.e., the water storage change in the saturated zone linked to the compressibility is most often considered negligible in comparison to the water storage change from the drainage or wetting of the unsaturated porous medium).

Richards equation for flow under partially saturated conditions

Under transient conditions, for a variably saturated heterogeneous and anisotropic porous medium, the 3D flow equation can be written by expressing the mass (of the water) conservation as (Celia *et al.* 1990):

$$\nabla \cdot \rho[K(h_p) \cdot \nabla h_p + K(h_p) \cdot \nabla z] + \rho q' = \rho C(h_p) \frac{\partial h_p}{\partial t} \tag{9.15}$$

The different terms of this equation are expressed in kg/(m³s) [ML⁻³T⁻¹]. This equation is expressed as a function of the water pressure head, but it can also be formulated in a mixed way as a function of the water content and the pressure head (Richards 1931) as:

$$\nabla \cdot \rho[K(\theta) \cdot \nabla h_p + K(\theta) \cdot \nabla z] + \rho q' = \rho \frac{\partial \theta}{\partial t} \tag{9.16}$$

If the volume conservation is expressed, each term is divided by ρ (considered constant) and expressed in s⁻¹ [T⁻¹].

These equations may also be formulated as a function of the capillary head or suction head $\psi = h_c = -h_p$ (Equation 9.4). Equations 9.15 and 9.16 (with a constant ρ) may then be written as:

$$\nabla \cdot [-K(\psi) \cdot \nabla \psi + K(\psi) \cdot \nabla z] + q' = -C(\psi) \frac{\partial \psi}{\partial t} \tag{9.17}$$

and

$$\nabla \cdot [-K(\theta) \cdot \nabla \psi + K(\theta) \cdot \nabla z] + q' = \frac{\partial \theta}{\partial t} \tag{9.18}$$

Regrouping the hydraulic properties (hydraulic conductivity and storage coefficient) in a *diffusivity* tensor $D(\theta) = K(\theta)/C(\theta)$ [L²T⁻¹], Equation 9.16 becomes (Brouyère 2002):

$$\nabla \cdot [D(\theta) \cdot \nabla \theta + K(\theta) \cdot \nabla z] + q' = \frac{\partial \theta}{\partial t} \tag{9.19}$$

For practical and computational reasons, this equation is often reduced to a 1D-vertical equation:

$$\frac{\partial}{\partial z}\left[D(\theta)\frac{\partial \theta}{\partial z} + K(\theta)\right] + q' = \frac{\partial \theta}{\partial t} \tag{9.20}$$

9.4 Contamination and transport under partially saturated conditions

Contamination and transport processes in the partially saturated zone are usually complex processes that can be assessed by classifying them into two categories: (1) solute transport within the water phase in the variably saturated medium, and (2) multiphase flow where the NAPL contaminant(s) act as separate phase(s) relative to the water. Indeed, for NAPL contaminations, the relative solubility of the contaminant in the water may induce the involvement of both types of processes (see Section 8.3).

Solute transport processes in the variably saturated zone are usually described following the same conceptual principles as those under saturated conditions (see Chapter 8). However, the groundwater velocity is expected to be less constant and less uniform than it is in the saturated zone, increasing the spreading of the solute (Leij and van Genuchten 2002). Quantifying hydrodynamic dispersion under unsaturated conditions is more complex than it is in the saturated zone (Toride *et al.* 2003). The water content, through its influence on the water velocity and the shape of the saturated subzones at the pore scale, plays a critical role, but this role is not easy to characterize. Conceptually, this influence can be characterized by considering the dispersivities to be θ dependent. At the pore scale, it has been shown (Raoof and Hassanizadeh 2013) that the relation between the (longitudinal) dispersivity and the saturation (α_L, S_w) is not monotonic, with a maximum dispersivity observed for a critical saturation corresponding to the main slope changes in a relative permeability-saturation diagram (K_r, S_w).

Other influences of the saturation or water content on the adsorption processes are also logically observed. If the larger pores or macropores are occupied by air, part of the matrix solid is not available or accessible to groundwater and solute adsorption. Thus, the K_d values (as defined in Section 8.2) are lower and are dependent on the saturation (Bear and Cheng 2010). The clear occurrence of immobile water zones or subzones within the REV of partially saturated porous media increases the influence of *immobile water effects* or *matrix diffusion* processes (see Section 8.2). Thus, it is generally recognized in the specialized literature that two apparent fluid phases, immobile and mobile water, should be considered to account for very slow local velocities compared to the average REV velocities (among others, Gerke and van Genuchten 1993, Padilla *et al.* 1999, Toride *et al.* 2003). As described previously (see Section 8.2), this leads to two mass balance equations: one for mobile and the other for immobile water. Transport equations can be written with the water content relation $\theta = \theta_m + \theta_{im}$ replacing the porosity relation $n = n_m + n_{im}$. Equation 8.38 can now be written as:

$$\frac{\partial}{\partial t}(\theta_m C^v) = -\nabla \cdot (q C^v) + \nabla \cdot \theta_m(D \cdot \nabla C^v) + \nabla \cdot \theta(D_m \nabla C^v) - \rho_b R_{w,s}$$
$$- \theta \lambda C^v + M^v \tag{9.21}$$

For mobile water, the assumption that diffusion and degradation are limited to the mobile water content θ_m can be taken. Then, similar to the case in Section 8.2, the use of the retardation factor to describe adsorption-desorption, the development of the source/sink term and the addition of the exchange term for the solute mass flux of matrix diffusion from the mobile to immobile water (or vice-versa) (f_m^{im}) gives:

$$R\frac{\partial C^v}{\partial t} = -v_a \cdot \nabla C^v + \nabla \cdot (D_h \cdot \nabla C^v) - R\lambda C^v + f_m^{im} - \frac{q_s}{\theta_m}(C^v - C_s^v) \tag{9.22}$$

where all terms are in kg/m³s [ML^{-3}T^{-1}]. This equation is similar to Equation 8.54, except that θ_m replaces n_m. Now, for immobile water, Equation 8.53 becomes (i.e., accounting for the diffusion and degradation in immobile water):

$$\frac{\partial}{\partial t}(\theta_{im}C_{im}^v) = f_m^{im} - \lambda\theta_{im}C_{im}^v \tag{9.23}$$

where $f_m^{im} = \alpha_d^m(C^v - C_{im}^v)$ (Equation 8.52).

These Equations 9.22 and 9.23 may now be written for a 1D vertical solute transport in a partially saturated zone as:

$$R\frac{\partial C^v}{\partial t} = -\frac{q_z}{\theta_m}\frac{\partial C^v}{\partial z} + \frac{\partial}{\partial z}\left(\theta_m D_{h_m}(\theta_m)\frac{\partial C^v}{\partial z}\right) - R\lambda C^v - \alpha_d^m(C^v - C_{im}^v) - \frac{q_s}{\theta_m}(C^v - C_s^v)$$

$$\tag{9.24}$$

$$\frac{\partial}{\partial t}(\theta_{im}C_{im}^v) = \alpha_d^m(C^v - C_{im}^v) - \lambda\theta_{im}C_{im}^v \tag{9.25}$$

where the subscripts m and im refer to the mobile and immobile water, respectively.

NAPL contamination and multiphase flow

Here, the discussion will be limited to the case of a NAPL contamination, so that three phases are occupying the pore volume: air, water, and NAPL. The same principles (in terms of the capillary pressures and wetting phases) as those described above (Section 9.2) are applicable to this three-phase system: a wetting phase, an intermediate wetting phase, and a nonwetting phase. In most situations, water is the wetting fluid, and NAPL and air are considered as the nonwetting fluids (Figure 9.5). The saturations of these three phases, describing their relative abundance in the pore volume, are linked by the relation:

$$S_w + S_a + S_{NAPL} = 1 \tag{9.26}$$

As previously described (Section 8.3), the movements of the NAPLs are highly dependent on their densities. A DNAPL (Dense Nonaqueous Phase Liquid) is denser than water and tends to sink in the saturated porous medium and the partially saturated zone. An LNAPL (Lighter Nonaqueous Phase Liquid) is lighter than water and tends to float above the saturated zone. Most hydrocarbon fuels can be considered LNAPLs, while most chlorinated hydrocarbons are DNAPLs (Fitts 2002). Except when a NAPL is only present as isolated blobs at very low saturation, this phase is considered (like the other phases) a continuum through the multiple pores/fractures

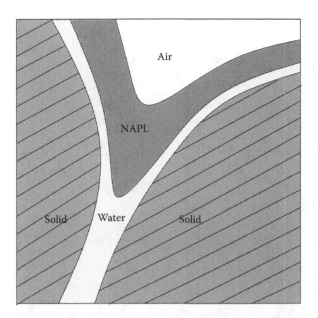

Figure 9.5 Schematic example of the spatial distribution of the water, NAPL, and air phases at the pore scale (Bear and Cheng 2010).

of the medium. The prediction of the different fluid flows becomes challenging (Payne *et al.* 2008), and simplifying approaches (beyond the scope of this book) are often inspired by the work of Buckley and Leverett (1942) as proposed by Pinder and Celia (2006). The multiphase flow of water and NAPL in the partially saturated zone can be classically described by Darcy's law written for variable density fluids and using relative permeabilities:

$$q_i = -\frac{K_{r_i}k}{\mu} \cdot [\nabla p_i + \rho_i g \nabla z] \tag{9.27}$$

where the *i*th fluid phase is water or NAPL.

Figures 9.6 and 9.7 give examples of saturation vertical profiles for water-LNAPL-air and water-DNAPL-air systems, respectively.

In contamination problems, the NAPL pools directly exposed to the air phase are more likely to produce vapor-phase contaminant plumes within the partially saturated zone (Petri *et al.* 2015). On the other hand, if the NAPL is limited to isolated blobs trapped in water-saturated zones, it can be considered mostly occluded from the air phase. Although most NAPLs have a low solubility in groundwater, the dissolved part of a NAPL contamination event will be transported as a solute (see Section 8.3), and dissolved NAPL plumes usually induce many problems resulting from the high toxicity of NAPL at very low concentrations in the groundwater. As mentioned in Section 8.3, if the solubility (i.e., the concentration of a substance in water that is in equilibrium with the pure phase of this substance) of a given NAPL in the groundwater is known for the current conditions (pressure and temperature), the actual quantity of

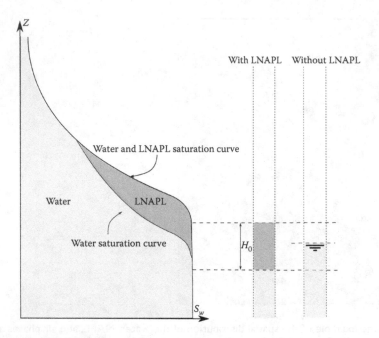

Figure 9.6 Saturation vertical profile and indicative measured piezometric heads in an observation well in conditions with and without the contamination of a LNAPL.

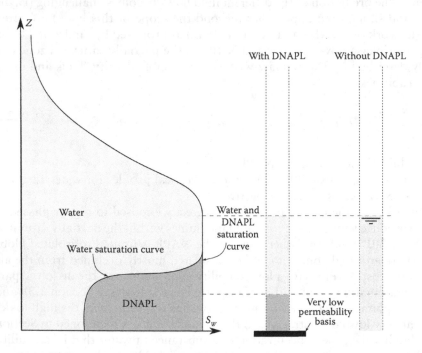

Figure 9.7 Saturation vertical profile and indicative measured piezometric heads in an observation well under DNAPL contamination conditions.

the dissolved NAPL in the given REV of a partially saturated porous medium can be estimated using Raoult's law (see Equation 8.2):

$$S_i^o = X_i S^o \tag{9.28}$$

where S^o is the solubility of a substance, i is partially present in the pore volume of the medium [ML^{-3}], and X_i is the mole fraction of the substance i (number of moles of the substance i divided by the total number of moles).

References

Bear, J. and A.H.D. Cheng. 2010. *Modeling groundwater flow and contaminant transport.* Dordrecht: Springer.

Bear, J. and A. Verruijt. 1987. *Modeling groundwater flow and pollution.* Dordrecht: Reidel Publishing Company.

Brouyère, S. 2002. Etude et modélisation du transport et du piégeage de solutés en milieu variablement saturé (in French). *PhD diss.*, University of Liège.

Buckingham, E. 1907. *Studies on the movement of soil moisture.* Bull. 38, Washington, DC: USDA Bureau of Soils.

Buckley S.E. and M.C. Leverett. 1942. Mechanism of fluid displacements in sands. *Transactions AIME* 146: 107.

Campbell, G.S. 1974. A simple method for determining unsaturated conductivity from moisture retention data. *Soil Science* 117(6): 311–314.

Carsel, R.F. and R.S. Parrish. 1988. Developing joint probability distributions of soil water retention characteristics. *Water Resources Research* 24: 755–769.

Celia, M.A., Bouloutas, E.T. and R.L. Zarba. 1990. A general mass-conservative numerical solution for the unsaturated flow equation. *Water Resources Research* 26(7): 1483–1496.

Fitts, Ch. R. 2002. *Groundwater science.* London: Academic Press.

Fredlung, D.G. and A. Xing. 1994. Equations of the soil-water characteristic curve. *Canadian Geotechnical Journal* 31: 521–532.

Gardner, W.R. 1958. Some steady state solutions of unsaturated moisture flow equations with application to evaporation from a water table. *Soil Science* 85: 228–232.

Gerke, H.H. and M.Th. van Genuchten. 1993. A dual-porosity model for simulating the preferential movement of water and solutes in structured porous media. *Water Resources Research* 29: 305–319.

Khaleel, R. and E.J. Freeman. 1995. *Variability and scaling of hydraulic properties for 200 area soils, Hanford Site.* WHC-EP-0883, Richland, WA: Westinghouse Hanfort Co.

Leij, F.J. and M.Th. van Genuchten. 2002. Solute transport. In *Soil physics companion,* ed. A.W. Warrick. Boca Raton: CRC Press, 189–248.

Mualem, Y. 1976. A new model for predicting the hydraulic conductivity of unsaturated porous media. *Water Resources Research* 12: 513–522.

Padilla, I.Y., T.C. Jim Yeh and M.H. Conklin. 1999. The effect of water content on solute transport in unsaturated porous media. *Water Resources Research* 35: 3303–3313.

Payne, F.C., Quinnan, A. and S.T. Potter. 2008. *Remediation hydraulics.* Boca Raton: CRC Press/ Taylor & Francis.

Petri, B.G., Fucík, R., Illangasekare, T.H., Smits, K.M., Christ, J.A., Sakaki, T. and C.C. Sauck. 2015. Effect of NAPL source morphology on mass transfer in the vadose zone. *Groundwater* 53(5): 685–698.

Pinder G.F. and M.A. Celia. 2006. *Subsurface hydrology.* Hoboken, New Jersey: Wiley & Sons.

Raoof, A. and S. M. Hassanizadeh. 2013. Saturation-dependent solute dispersivity in porous media: Pore-scale processes. *Water Resources Research* 49: 1943–1951.

Rawls, W.J., Brakensiek, D.L. and K.E. Saxton. 1982. Estimating soil water properties. *Trans. Am. Soc. of Agricultural Engineers* 25(5): 1316–1320.

Richards, L.A. 1931. Capillary conduction of liquids through porous mediums. *Physics* 1(5): 318–333.

Schwartz, F.W. and H. Zhang. 2003. *Fundamentals of Ground Water.* New York: Wiley.

Szymkiewicz, A. 2013. *Modelling water flow in unsaturated porous media accounting for non linear permeability and material heterogeneity.* Berlin-Heidelberg: Springer-Verlag.

Toride, N., Inoue, M. and F.J. Leij. 2003. Hydrodynamic dispersion in an unsaturated dune sand. *Soil Science Society of America Journal* 67(3): 703–712.

van Genuchten, M. Th. 1980. A closed-form solution for predicting the conductivity of unsaturated soils. *Soil Science Society of America Journal* 44: 892–898.

Yeh, T.C. J., Khaleel, R. and K.C. Carroll. 2015. *Flow through heterogeneous geologic media.* New York: Cambridge University Press.

Chapter 10

Salinization and density dependent groundwater flow and transport

10.1 Salinization processes

The salinization of groundwater is probably one of the most common but difficult problems to solve. Salinization occurs under different types of natural and anthropogenic conditions. Rainwater compositions may include some sea-salt sprays. The dissolution of evaporitic rocks and minerals along the groundwater path in the partially and fully saturated zones contributes a nonnegligible salt content. Evapotranspiration processes potentially lead to higher salt concentrations in the remaining water. Sediments deposited under marine conditions can also maintain a relatively high salinity content in the groundwater before they are fully washed out by incoming freshwater. Finally, unavoidable seawater intrusions in coastal aquifers produce significant groundwater salinization.

Evaporite dissolution

The weathering and dissolution of minerals determine the natural groundwater mineral composition. As shown by many authors (among others, Appelo and Postma 2005), the geology and the water travel time underground can have very important influences on the predominant ions in the groundwater. As mentioned in Chapter 7, the *TDS* (i.e., Total Dissolved Content) gives an indirect and global measurement of the total solute content of a groundwater (see Section 7.2). Evaporitic rock formations and their dissolution by groundwater are sources of enrichment in chlorides, sulfates, and carbonates. In general, calcite and dolomite are more common than gypsum, which is more common than halite, which is more common than potassium and magnesium salts. In each case, the specific geological formations condition the groundwater quality. Logically, the dissolution of KCl and $MgCl_2$ is easier than that of $NaCl$ (halite), which is easier than that of $CaSO_4$ (anhydrite) and $CaSO_4$-$2H_2O$ (gypsum), which is easier than that of $CaCO_3$, which is easier than that of $CaMg(CO_3)_2$ (dolomite).

Usually, the salinity of groundwater is defined including all types of salts. While seawater salinity is highly dominated by $NaCl$ (i.e., the general salt components of seawater are: Cl^- 55%, Na^+ 30.6%, SO_4^{2-} 7.7%, Mg^{2+} 3.7%, Ca^{2+} 1.2%, K^+ 1.1%, and others 0.7%), groundwater salinity is more site-specific as a result of dissolution influenced by the local lithologies and past geological events (e.g., sea transgression/regression and, salt-specific tectonics).

Mixing between (relatively) freshwater and a saltier water can also occur naturally based on hydrogeological conditions. This may lead, by cation exchanges, to different but identifiable groundwater compositions. For example, if groundwater with a composition dominated by Ca^{2+} and HCO_3^- (i.e., from calcite dissolution) is contaminated by seawater (where Na^+ and Cl^- are the dominant ions), an exchange of cations occurs as Na^+ is adsorbed and Ca^{2+} is released in the groundwater, with Cl^- remaining as the main anion. In this case, the seawater intrusion in the freshwater produces $CaCl_2$-rich groundwater. In contrast, when freshwater mixes with NaCl-rich saltwater, then Ca^{2+} is adsorbed on the sediments, and the Na^+ is released, thereby producing $NaHCO_3$-rich groundwater (Appelo and Postma 2005).

Evapotranspiration

Globally, continental groundwater always contains a nonnegligible salt content resulting from mineral weathering and the dust and sea-sprays in rainwater. Water that evaporates or is consumed by plants has very low salt contents, which automatically leads to higher salt concentrations in the remaining water. Evapotranspiration processes that occur with varying intensity according to the degree of aridity induce increases in salt concentrations. Sometimes this leads to precipitation and significant accumulations of salt. Groundwater salinization by evaporation can occur for any water in the first few meters below the land surface. In a warm and arid environment, the process is particularly exacerbated when the water table (i.e., partially saturated—saturated limit) is less than 5-meters deep. Consequently, in arid zones, land use practices have a determinant influence on land and groundwater salinity. It is striking and alarming that global maps of irrigation and groundwater salinity are very well correlated. Increases in groundwater salinity due to irrigation are the result of the combined effects of the following chain of processes (Figure 10.1 and Box 10.1):

- Almost all irrigation waters contain dissolved salts (in the best cases a few mg/L of salt content), and the evapotranspiration of this water leaves salt precipitates at the land surface inducing a soil salinity increase.
- Salts left in the soil decrease agricultural productivity.
- Salts are leached out of the plant root zone by the application of additional water, so the infiltration of saltwater is artificially boosted toward the groundwater and the saturated zone.
- In the worst cases, waterlogging (i.e., the saturation of soil with water) may induce specific anaerobic conditions and a loss of the drainage capacity of the soil, thus promoting further evaporation and increasing salt concentrations.
- The underlying groundwater is enriched in salts by saltwater infiltration; additionally, the boosted infiltration may locally induce rising groundwater piezometric heads, which in turn promotes evaporation by capillary action if the aquifer is shallow.
- If further irrigation must occur during a period when surface water is not available, groundwater pumping provides the irrigation water with a higher salt content than it previously had.

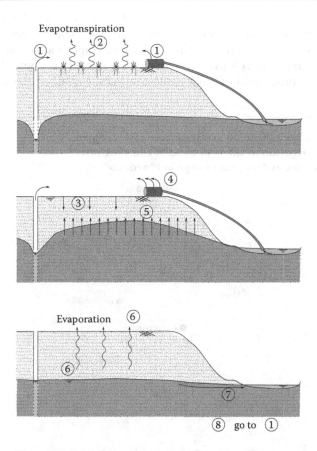

Figure 10.1 Schematic description of the main salinization processes induced by irrigation in arid zones. (1) irrigation with surface water or groundwater, (2) evapotranspiration at the land surface, in the soil and root zone, (3) leaching out the soil salinity by infiltration toward groundwater, (4) if the groundwater salt content is known to be high, surface water is preferred, (5) possible waterlogging at the surface and rising of groundwater piezometric heads, (6) easier groundwater evaporation from the partially saturated zone and the shallower saturated zone, (7) groundwater drainage by the river network, and (8) the cycle returns to (1) for the next growing season.

- If the groundwater is drained by the local river network, the river water acquires higher concentrations of salts than it previously had.
- The river water used for further irrigation downstream will have a higher initial salt background than the irrigation water upstream.

More generally, a local imbalance in the water budget induced by excesses of irrigation water is often the cause of a season-after-season progressive rising of the groundwater piezometric heads. Consequently, more direct groundwater evapotranspiration occurs, and new groundwater seepage discharge zones may appear in the landscape favoring soil and groundwater salinization. The clearing of trees can also favor dryland salinity as the deep rooting of trees are replaced by the very shallow rooting of agricultural crops. The water consumption of deep roots actually stabilizes the depth of the piezometric levels. In cleared zones, piezometric variations can be more important than in uncleared zones, demonstrating greater responses to rainfall events and inducing more evaporation periods (Peck and Williamson 1987). Combined with irrigation, the excess water contributes to the rising of the water table and, consequently, to more salinization (as explained above).

Box 10.1   Increasing salinity in groundwater and streams resulting from irrigation in the arid zone of the southern flank of the Atlas Mountain range in Morocco

The valleys of the Drâa and Ziz wadis on the southern flank of the central Anti-Atlas Mountains (Morocco) represent a particularly serious salinization problem resulting from irrigation. The Drâa and Ziz wadis flow southward reaching their southernmost points adjacent to the Sahara foreland (Klose 2012). Irrigation is needed for basic cereal crops, and in particular, the date palm trees grown in a series of oases located along the main wadis. Both the surface water of the wadis (i.e., released from dams located upstream) and the groundwater are used for irrigation. Water from the southern flank of the high Atlas Mountain range is collected by the El Mansour Eddahbi dam in Ouarzazate (the upper Drâa) and the Hassan Addakhil dam near Errachidia (the upper Ziz) (see map). Water from these dams has a nonnegligible salt content from evaporite dissolution in the southern foothills of the Atlas range, as well as from evaporation in the dam reservoirs. Irrigation occurs using water released from the dams in the wadis and is organized in the oases through complex networks of traditional canals named "seguias." If the released water is not sufficient or occurs at an inappropriate time for the crops, local groundwater pumping takes place in the alluvial aquifers corresponding to the different palm oases using thousands of shallow wells. The chain of processes described in Figure 10.1 is triggered, resulting in a vicious circle of intensifying salinization. One of the unfortunate consequences of this is that the salinity of the stream (i.e., the wadi when flowing) is increased downward. This in turn accentuates the same phenomena in the downstream oasis and so on. Unfortunately, the final result is a complete salinization of the soils and waters in the most downstream oases (i.e., Mhamid in the lower Drâa valley and Ouzina in the lower Ziz valley), which forces inhabitants to leave their land. One solution to this problem would be to organize drip irrigation but it would be very expensive to install.

Contamination by seawater

Groundwater contamination by seawater occurs in different ways. Indeed, the natural or artificial flooding of coastal zones by seawater during storms (Terry and Falkland 2010), tsunamis (Violette *et al.* 2009), or human activities have resulted in massive contamination of the underlying fresh groundwater. Infiltration and mixing processes combined with the density difference between saltwater and freshwater and cation exchange reactions are unfortunately leading to significant and long-lasting deterioration of the groundwater quality. However, it is more common for groundwater contamination by seawater to occur by underground seawater intrusions. This is clearly a global issue naturally induced by the density difference between seawater and freshwater (see Section 10.2). Seawater intrusions can be aggravated by human activities and are predisposed to be exacerbated by the rising sea levels associated with changing climate. As seawater has a higher mineral content than freshwater, seawater is denser. If an average seawater salinity of approximately 3.5% is considered (i.e., 35 g/L with 10.76 g Na^+, 0.387 g K^+, 1.294 g Mg^{2+}, 0.413 g Ca^{2+}, 19.353 g Cl^-, 0.142 g HCO_3^-, and 2.712 g SO_4^{2-}), the average density of seawater is $1.025 \cdot 10^3$ kg/m^3 while the average freshwater density is 1.10^3 kg/m^3 (at 4°C) (see Table 4.2). As a result, the pressure under a seawater column (p_{sw}) is higher than that under a freshwater column (p) of the same height (Figure 10.2). In other words, seawater has as a higher piezometric equivalent freshwater head than freshwater, and this difference increases with the total height of the column and, consequently, with the considered depth. Seawater can move into coastal aquifers.

However, in the coastal zone, a fresh-groundwater flux most often flows from inland regions toward the coast because of the natural piezometric gradient in that direction. This piezometric gradient allows a fresh-groundwater discharge toward the sea in the relatively shallow zone, while seawater, due to the density difference, increases with depth, which pushes the seawater inland beneath the fresh groundwater in a wedge shape (Figure 10.3). The inland extent of the seawater wedge is limited by the fresh-groundwater gradient in the opposite direction as the result of infiltration in higher land elevations. Any local groundwater drainage or pumping tends to exacerbate the salinization by the seawater intrusion and decreases the pressure (or "equivalent freshwater head"; Carabin and Dassargues 1999). In addition, tidal effects can be expected to create a perched local seawater intrusion zone in the upper part of the beach or coastal zone (Figure 10.3; Vandenbohede and Lebbe 2006, 2007, Evans and Wilson 2016).

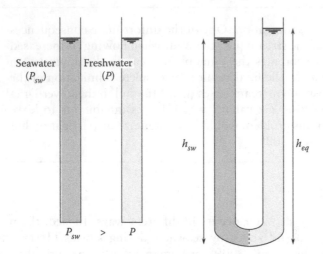

Figure 10.2 The pressure under a seawater column (p_{sw}) is higher than the pressure (p) under a freshwater column of the same height. A higher freshwater column is needed (h_{eq}) to produce a pressure equilibrium with the seawater column (h_{sw}).

The remediation of a seawater intrusion is difficult and costly, which is why many "critical seawater management questions remain unresolved" (Werner *et al.* 2013). Remediation mainly involves three types of measures: artificial recharge to boost the inland fresh-groundwater flux and create "hydraulic barriers" (among others, Pool and Carrera 2010, Reichard *et al.* 2010), land reclamation to push the underground

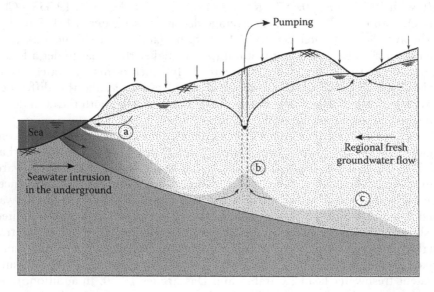

Figure 10.3 Schematic 2D vertical description of a seawater intrusion increasing with depth in a wedge shape and influence of (a) tidal effects, (b) pumping, and (c) draining river on groundwater salinization.

mixing zone between sea- and freshwater seaward, or building a physical barrier to decrease the hydraulic conductivity in the lower zone of the coastal aquifer (Luyun *et al.* 2011, Abdoulhalik *et al.* 2017). Seawater intrusion processes are understood and described in many groundwater books and publications giving rise to the popular concept of a seawater sharp interface that is described in Section 10.2. The sharp interface theory can be considered as an (over)simplified concept because in reality, seawater and fresh groundwater are mixed by diffusion, dispersion, and associated cation exchange reactions.

Field measurements and sampling

Particular precautions must be taken for field measurements where and when groundwater salinization is suspected or expected to occur. For example, piezometric head measurements must be accompanied by simultaneous electrical conductivity (*EC*) measurements (in $\mu S/cm$), as they give an approximation of the ionic content (see Section 7.2). This *EC* value allows the calculation of the groundwater density in the monitoring well. As mentioned previously (Chapter 4), the spatial salinity distribution influences the pressure field (Yoon *et al.* 2017). A measured piezometric or hydraulic head in salt or brackish groundwater is written as (Equation 4.16):

$$h_{sw} = z + \frac{p_{sw}}{\rho_{sw} g} \tag{10.1}$$

where h_{sw} is the saltwater piezometric head (m) [L] measured in the field, and ρ_{sw} and p_{sw} are the saltwater density and the salt water pressure at the measured point in the aquifer (Well 2 in Figure 10.4).

In order to compare hydraulic heads in a zone where salinity influence the temporal and spatial distribution of groundwater density, a choice must be made to work with pressure or with the equivalent freshwater piezometric heads (shortened as "equivalent freshwater head"; see Section 10.3). The equivalent freshwater head (h_{eq}) of the measured saltwater head (h_{sw}) is defined:

$$h_{eq} = z + \frac{p_{sw}}{\rho g} \tag{10.2}$$

where ρ is the fresh groundwater density. Entering the value of p_{sw} from Equation 10.1 into Equation 10.2 gives:

$$h_{eq} = z + \frac{\rho_{sw}}{\rho} (h_{sw} - z) \tag{10.3}$$

Thus, the correction for the equivalent freshwater heads can be calculated from the ratio (ρ_{sw}/ρ).

An example of such a calculation is given in Box 10.2. If a piezometric head is measured in a monitoring well screened over a large thickness of the aquifer, the

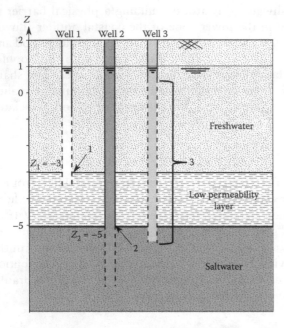

Figure 10.4 To compare head measurements in well 1 screened in the freshwater shallow aquifer, and in well 2 screened in the deep saline aquifer, equivalent freshwater heads must be calculated according to Equation 10.3. If an observation well 3 is unfortunately screened along the whole depth, one should consider density to evolve in the water column from freshwater in z_1 to saltwater in z_2, and an equivalent freshwater head should be calculated using Equation 10.4 (see Box 10.2).

measured head is a "depth-averaged" value that may be the global result of a vertical density driven gradient in the water column within the well between the altitudes z_1 and z_2 (Figure 10.4). In this case, the equivalent freshwater head (h_{eq}) of the measured (brackish) water head is calculated:

$$h_{eq} = z_2 + \frac{1}{\rho} \int_{z_1}^{z_2} \rho(z) dz + (h_{measured} - z_1) \tag{10.4}$$

where z_1 is the altitude with the minimum groundwater density and from which the salinity is increasing and z_2 is the lowest altitude with the maximum groundwater density (Figure 10.4). An example of this calculation is given in Box 10.2. Practically, it is indeed more accurate to use pressure sensors to measure water pressure p_{sw} directly at a point. This measurement is ideally associated with EC measurements (e.g., using a classical Eh-pH-T probe), allowing the assessment of the groundwater salinity.

For sampling purposes, no particular measures must be adopted except, for interpretation purposes, it is advisable to account for possible groundwater density stratification in the sampled water column.

Box 10.2   Calculation of equivalent freshwater heads

According to the case illustrated in Figure 10.4, for well 1, which is located in the freshwater upper unconfined aquifer:

$$h_1 = z_1 + \frac{p_1}{\rho g} = 1$$

For well 2, the measured saltwater head is $h_{2_{sw}} = z_2 + (p_{2_{sw}}/\rho_{sw}g) = 1$ and the equivalent freshwater head is calculated using Equation 10.3 (with $\rho_{sw} = 1025$ and $\rho = 1000$ kg/m^3):

$$h_{2_{eq}} = z_2 + \frac{\rho_{sw}}{\rho}(h_{2_{sw}} - z_2) = -5 + \frac{1025}{1000}(1+5) = 1.15$$

This clearly shows that, in this case, the groundwater flow through the low conductivity unit is directed from the bottom aquifer toward the upper aquifer. Let us assume that, as is often the case, this vertical groundwater flow is negligible compared to the horizontal freshwater flow in the upper aquifer. If a measured head $h_3 = 1$ is available in the fully screened well 3, and the density evolution is assumed to be the same in the water well column as the geological layers, an equivalent freshwater head can be calculated using Equation 10.4:

$$h_{3_{eq}} = z_2 + \frac{1}{\rho}\int_{z_1}^{z_2} \rho(z)dz + (h_3 - z_1).$$

If we take a mean value of ρ in the aquitard:

$$h_{3_{eq}} = -5 + \frac{1012.5}{1000}2 + (1+3) = 1.025$$

If the function $\rho(z)$ is more finely described for further integration (e.g., $\rho(z) = az^2 + b$), then the calculated equivalent freshwater head is:

$$h_{3_{eq}} = -5 + \frac{2022.92}{1000} + (1+3) = 1.023$$

Seawater intrusions involve complex processes that may affect the thickness of freshwater-seawater mixing zones associated with the spatial heterogeneity of each specific case. Many field techniques could be combined to retrieve all the required information that leads to understanding the processes and imaging their effects in a quantitative way: groundwater levels, hydrogeochemical surveys, environmental and isotope tracers, geophysical data combined with geological data, and hydrological stress factors (Werner et al. 2013).

10.2 Saltwater–freshwater interface concept

Considering an interface of equal pressure between the freshwater and saltwater (Figure 10.2), a sharp interface concept was developed starting with the approximation of Ghyben-Herzberg (Badon-Ghijben 1888, Herzberg 1901). This interface delineates an abrupt idealized contact between the seawater intrusion wedge and the fresh groundwater (Figure 10.5). Indeed, this simplifying concept neglects the true mixing zone that exists when any saltwater contaminates fresh groundwater. The Ghyben-Herzberg approximation tends to determine the shape and position of this sharp interface by assuming steady-state and mostly horizontal groundwater flow conditions (Dupuit assumption, see Chapter 4, Section 4.6). The free surface and the seawater wedge are considered to be flow lines so that equipotential lines (see Section 4.6) can be considered (in an isotropic medium) to be oriented at right angles to them (de Marsily 1986). Currently, the main utility of this concept is to provide a rough approximation and a nice illustration of the interface position as a function of the local piezometric levels in the coastal zone. To maintain simplicity, if the two columns illustrated in Figure 10.2 are connected, an equilibrium pressure interface $p_{sw} = p$ is found by (Figure 10.5):

$$p_{sw} = \rho_{sw} g h_{sw} = \rho g h_{eq} = \rho g (h_{sw} + h) = p \tag{10.5}$$

and

$$h_{sw} = \frac{\rho}{(\rho_{sw} - \rho)} h \tag{10.6}$$

where h is the thickness of the freshwater zone above sea level (most often taken as the reference level), h_{sw} is the seawater head and, consequently, also the depth to the interface from the reference level, with $h_{eq} = h_{sw} + h$. For example, if the considered

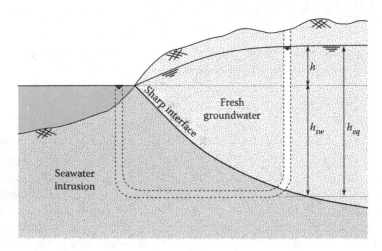

Figure 10.5 The sharp interface concept of Ghyben-Herzberg describing a pressure equilibrium between seawater and freshwater columns (Oude Essink 2001).

seawater has a density of $\rho_{sw} = 1025$ kg/m^3 and freshwater with $\rho = 1000$ kg/m^3, Equation 10.6 gives $h_{sw} = 40\ h$. Therefore, a freshwater drawdown of 1 m resulting from a pumping well will induce upconing (Reilly and Goodman 1987) of the seawater wedge by approximately 40 m toward the well.

In fact, this interface concept is not correct. Many authors have discussed various corrections to the shape and position of the interface (Strack 1976, 1984) if some of the main assumptions were relaxed in order to be more realistic. One of the most important corrections is that the interface should be shifted lower to allow an outlet for the freshwater seeping out to the sea (among others, Koussis *et al.* 2015), as described schematically in Figure 10.3. Other proposed corrections concern the shape and position of the interface for a confined or semi-confined coastal aquifer (Sikkema and van Dam 1982, Bakker 2006), for layered heterogeneous and stratified conditions (Dagan and Zeitoun 1998, Lu *et al.* 2013, Strack and Ausk 2015), for varying fresh groundwater flow from inland regions (Abdollahi-Nasab *et al.* 2010), and by infiltration (Bakker 2000), for freshwater lenses in dunes or in islands (Ghassemi *et al.* 2000, Eeman *et al.* 2011), for anisotropy (Abarca *et al.* 2007, Bakker 2014), for taking mixing into account (Pool and Carrera 2011, Werner 2017), for sea tidal effects (Levanon *et al.* 2016), and for pumping management (Park and Aral 2004, Mantoglou and Papantoniou 2008).

This is not the place to discuss the individual or combined effects of the main assumptions underlying this concept of sharp interface in more detail. Currently, development of numerical groundwater models makes it possible to simulate any of the practical and specific problems with the required level of complexity. If needed, simple or complex simulations can be considered that account for the density dependence of the flow and transport equations, mixing and reactive transport, tidal effects, heterogeneities, and transient conditions (Watson *et al.* 2010). Most of the proposed software have been validated using well-known synthetic "benchmark" case studies, such as those of Henry (1964a, 1964b) and Huyakorn *et al.* (1987) for seawater intrusion and the Elder problem for free convection phenomena (Elder 1967).

10.3 Coupled density dependent groundwater flow and solute transport equations

As mentioned above, density dependent flow and transport equations can be needed to accurately describe a wide variety of hydrogeological situations. This can be of critical importance for groundwater production and seawater intrusion management in coastal zones, as well as other contexts, such as those linked to gas or waste storage projects in salt formations or the precipitation of evaporates in arid zones.

Density dependent flow equations

A density dependent groundwater flow, also named *density-driven groundwater flow,* occurs where and when the flow pattern is influenced "significantly" by the spatial distribution of groundwater density differences combined with the temporal variations in the groundwater density (i.e., under transient conditions). Theoretically, the groundwater density can be dependent on the pressure, temperature, and solute concentration (see Equation 4.62) $\rho = \rho(p, T, C)$. In most hydrogeological problems, the pressure dependency can be neglected in favor of the temperature and concentration effects. Furthermore, the spatial distribution and variation of density are not the main objectives of studies dedicated to contamination by heat (see Chapter 11) or solutes in the groundwater. Varying density or density differences do not necessarily have a significant influence on flow if the advection resulting from the head (or pressure) gradient is very strong. The opposite situation is found where a slight density variation could have an important influence and change the main flow patterns of a preexisting groundwater flow driven by a very low head gradient. The term buoyancy-induced groundwater flow is found in the literature with the buoyancy force to describe any density difference $\Delta\rho$ multiplied by g (Turner 1973, Cushman-Roisin and Beckers 2011). As explained by Holzbecher (1998), it seems that the term *density-driven* groundwater flow includes any buoyancy effect, and the contrary is not especially true.

As heat transfer issues are treated in Chapter 11, including thermally induced groundwater density variations (see Section 11.2), the salinity induced density dependence is mainly treated hereunder. The coupled density dependent groundwater flow and solute transport equations have been discussed by many authors in the literature (among others, Hassanizadeh 1986, Kolditz *et al.* 1997, Holzbecher 1998, Ackerer *et al.* 1999, Simmons *et al.* 2001, Diersch and Kolditz 2002, Graf and Simmons 2009, Hamann *et al.* 2015).

When discussing the Bernoulli equation (Equation 4.16) that gives the piezometric head by $h = z + p/\rho g$ in Chapter 4, it was mentioned that for density reasons, piezometric head values can be compared only if their groundwater is considered to have the same temperature and salt content. Then, by the hydraulic conductivity definition $K = k\rho g/\mu$ (Equation 4.27), the density can play an important role with the dynamic viscosity. The generalized 3D Darcy's law (similar to Equation 4.28) can thus be written as:

$$q = -\frac{k}{\mu} \cdot (\nabla p + \rho g \nabla z) \tag{10.7}$$

where k is the intrinsic permeability tensor (m²) [L²], μ is the dynamic viscosity (kg/(m · s)) [ML⁻¹T⁻¹], g is the acceleration of gravity (m/s²) [LT⁻²], ρ is the density of water (kg/m³) [ML⁻³], and p is the pressure in N/m² or Pascal (Pa) [ML⁻¹T⁻²].

Note that the "buoyancy term" $(\rho g \nabla z)$ appears explicitly in Equation 10.7 as a function of the groundwater pressure. However, pressure is not the only variable in the equation, as we also have $\rho = \rho(p, T, C)$, and $\mu = \mu(T, C)$. Entering Darcy's law into the steady-state water mass conservation equation leads to the following 3D equation (Equation 4.49):

$$\nabla \cdot \left(\rho \frac{k}{\mu} \cdot (\nabla p + \rho g \nabla z) \right) + \rho q' = 0 \tag{10.8}$$

where q' (s⁻¹) [T⁻¹] is the water flow rate that is withdrawn ($q' < 0$) or injected ($q' > 0$) per unit volume of medium. As mentioned in Chapter 4, the different terms of these equations are expressed in kg/(m³s) [ML⁻³T⁻¹], the mass of the flowing water per unit volume of porous medium per second.

Under transient conditions, Equation 10.8 has a second member and is written as:

$$\nabla \cdot \left(\rho \frac{k}{\mu} \cdot (\nabla p + \rho g \nabla z) \right) + \rho q' = \frac{\partial (\rho n)}{\partial t} \tag{10.9}$$

Under all the previously described conditions, p was the only variable, as constant values were assumed for ρ and μ. If the porous medium was additionally considered as nondeformable, constant values would also be considered for n and the components of k. More information can be found in Chapter 6 (Section 6.2) about consolidation in compressible porous media and its coupling with groundwater flow via the specific storage coefficient (see also Section 4.10). Let us assume here that the k components can be considered as constant and that the water compressibility β_w (Equation 4.63) is no more negligible compared to the volume compressibility; thus, the *specific storage coefficient* (or specific storativity) in m^{-1} $[L^{-1}]$ is expressed in a way similar to that in Equation 4.76, with $S_s = \rho g (\alpha + n \beta_w)$.

Equation 10.9 can then be written as:

$$\nabla \cdot \left(\rho \frac{k}{\mu} \cdot (\nabla p + \rho g \nabla z) \right) + \rho q' = \frac{1}{g} S_s \frac{\partial p}{\partial t} \tag{10.10}$$

where the different terms are expressed in $kg/(m^3 s)$ $[ML^{-3}T^{-1}]$, the mass of the flowing water per unit volume of porous medium per second. Note that a rotational flow component arises when $\nabla \rho$ is not parallel to the gravity vector **g**.

Density dependent solute transport equation

As mentioned in Chapter 8 (Section 8.2), a *mass concentration* (C) in kg/kg (in practice, often in mg/kg $= ppm$ (one part per million) or $\mu g/kg = ppb$ (one part per billion) [-]) must be used in equations if a variable groundwater density can be expected, particularly $\rho(C)$ due to a high salt content. If we consider Equation 8.51, the main variable C^v must now be replaced by $\rho(C)C$, and the following density-dependent transport equation is obtained:

$$R \frac{\partial (\rho(C)C)}{\partial t} = - v_a \cdot \nabla(\rho(C)C) + \nabla \cdot (D_h \cdot \nabla(\rho(C)C)) - R\lambda \rho(C)C$$

$$- \frac{q_s}{n_m} (\rho(C)C - \rho(C_s)C_s) \tag{10.11}$$

where $\rho(C_s)C_s$ is the concentration (kg/m^3) $[ML^{-3}]$ multiplied with the density of the water associated with the source/sink flux q_s (i.e., volumetric flow rate per unit volume of the porous medium $(s^{-1})[T^{-1}]$); if $q_s > 0$ (inflow) $\rho(C_s)C_s$ is prescribed by the external conditions, and if $q_s < 0$ (outflow) $\rho(C_s)C_s = \rho(C)C$ automatically because any outflowing groundwater is assigned the local concentration in the control volume.

As shown in Chapter 8, this equation can possibly be developed further to include immobile water effect/matrix diffusion and reactive transport terms.

Constitutive or state equations

The groundwater flow and transport balance equations (Equations 10.10 and 10.11) are coupled with the groundwater density variation as a function of the salt concentration. Theoretically, viscosity can also vary as a result of salinity changes. More generally, the two state equations written for the groundwater density and viscosity evolutions should express $\rho(p, T, C)$ and $\mu(p, T, C)$, respectively. If we consider the density variations with temperature and pressure to be negligible compared to the influence of salinity on density, simplified polynomial or exponential expressions can be used for $\rho(C)$ and $\mu(C)$. Neglecting the density and viscosity variations resulting from pressure changes can easily be acceptable in most hydrogeological problems where fluid pressure changes are limited (e.g., unlike in oil engineering problems). As mentioned in Chapter 4, viscosity is very dependent on the temperature, and density is very dependent on the salt content (see Table 4.4). The temperature dependences of groundwater density and viscosity are discussed in Chapter 11 (see Section 11.2, Figure 11.1). Assuming that the temperature is constant, the groundwater density constitutive equation for groundwater salinization problems can be reduced (Langevin and Guo 2006) to:

$$\rho = \rho_0 + \frac{\partial \rho}{\partial C}(C - C_0) \tag{10.12}$$

where ρ_0 is the groundwater density when the concentration is C_0 (Voss and Souza 1987). The coefficient of density variability $(\partial \rho / \partial C)$ is given as approximately 700 kg/m³. For a seawater intrusion problem, we may consider freshwater with $C_0 \approx 0$, $\rho_0 \approx 1000$ kg/m³, and we may consider seawater reaching a maximum concentration $C = C_{max} \approx 0.035$ and $\rho_{sw} \approx 1025$ kg/m³. In this case, the relative concentration (i.e., mass fraction) C_r is defined[1] as $C_r = C/C_{max}$ and Equation 10.12 becomes:

$$\rho \approx 1000 + 700C \approx 1000 + 700C_{max}C_r \tag{10.13}$$

If $C_r = 1$, a value of 1025 kg/m³ is calculated for the seawater density (ρ_{sw}).

For the $\mu(C)$ in groundwater, the following empirical relation (van der Vorst et al. 1984) fitting the variation is generally chosen:

$$\mu \approx \mu_0(1 + 0.4819C_r - 0.2774C_r^2 + 0.7814C_r^3) \tag{10.14}$$

where $\mu_0 = 1.002 \; 10^{-3}$ is the dynamic viscosity (Pa s) at approximately 20°C (Figure 11.1) and C_r is the salt mass fraction or relative concentration in the groundwater assuming a constant temperature of approximately 20°C (Ackerer et al. 1999).

Boussinesq approximation and usual assumptions for seawater intrusions

For most seawater intrusion problems, the variation of the dynamic viscosity as approximated by Equation 10.14 is neglected.

---

1 The same type of reasoning may be done for any other salinization.

More importantly, the use of the *Boussinesq approximation* leads to simplifications in Equations 10.10 and 10.11. The "generalized Boussinesq approximation" in Equation 10.10 assumes that changes in density can be neglected except in the terms of Darcy's law. The "classical Boussinesq approximation" goes further and also neglects the change in the hydraulic conductivity. Thus, most often, when the latter is adopted it is stated that the only term where the density variation must be accounted for is the buoyancy term (i.e., containing ρg) (Holzbecher 1998). This approximation is usually accepted for steady-state problems with groundwater density changes lower than 2% (i.e., $\Delta\rho/\rho_0 < 2\%$). Munhoven (1992) has demonstrated, on the basis of an error analysis, that this approximation can be accepted for salinity variations up to 5% (more than the usual contrast between freshwater and seawater) under pressure variations not exceeding 3.10^8 Pa (i.e., always the case in hydrogeological applications). Note that the buoyancy term that is not neglected may create a rotational flow component when $\nabla\rho$ is not parallel to the gravity vector **g**.

With this Boussinesq approximation, Equation 10.10 becomes:

$$\nabla \cdot \left[\frac{k}{\mu} \cdot (\nabla p + \rho g \nabla z) \right] + q' = \alpha \frac{\partial p}{\partial t} \tag{10.15}$$

where α is the volume compressibility of the porous medium (Equation 4.58). If the medium is considered to be incompressible ($\alpha \approx 0$), Equation 10.15 becomes a steady-state equation.

For transport, Equation 10.11 becomes:

$$R \frac{\partial C}{\partial t} = - v_a \cdot \nabla C + \nabla \cdot (D_h \cdot \nabla C) - R\lambda C - \frac{q_s}{n_m}(C - C_s) \tag{10.16}$$

where the only density dependent term is $(v_a \cdot \nabla C)$. The coupling between both equations is expressed by Equation 10.12.

References

Abarca, E., Carrera, J., Sánchez-Vila, X. and M. Dentz. 2007. Anisotropic dispersive Henry problem. *Advances in Water Resources* 30: 913–926.

Abdollahi-Nasab, A., Boufadel, M.C., Li, H. and J. Weaver. 2010. Saltwater flushing by freshwater in a laboratory beach. *Journal of Hydrology* 386: 1–12.

Abdoulhalik, A., Ahmed, A. and G.A. Hamill. 2017. A new physical barrier system for seawater intrusion control. *Journal of Hydrology* 549: 416–427.

Ackerer, Ph., Younes, A. and R. Mose. 1999. Modeling variable density flow and solute transport in porous medium: 1. Numerical model and verification. *Transport in Porous Media* 35: 345–373.

Appelo, C.A.J. and D. Postma. 2005. *Geochemistry, groundwater and pollution* (2nd Edition). Amsterdam: Balkema.

Badon-Ghijben, W. 1888. Nota in verband met de voorgenomen putboring nabij Amsterdam (in Dutch). *Tijdschrift van het Koninklijk Instituut van Ingenieurs* (The Hague NL) 9: 8–22.

Bakker, M. 2000. The size of the freshwater zone below an elongated island with infiltration. *Water Resources Research* 36: 109–117.

Bakker, M. 2006. Analytic solutions for interface flow in combined confined and semi-confined, coastal aquifers. *Advances in Water Resources* 29: 417–425.

Bakker, M. 2014. Exact versus Dupuit interface flow in anisotropic coastal aquifers. *Water Resources Research* 50: 7973–7983.

Carabin, G. and A. Dassargues. 1999. Modeling groundwater with ocean and river interaction. *Water Resources Research* 35(8): 2347–2358.

Cushman-Roisin, B. and J.-M. Beckers. 2011. *Introduction to geophysical fluid dynamics. Physical and numerical aspects* (2nd Edition). Int. Geophysics Series 101, Amsterdam: Elsevier Academic Press.

Dagan, G. and D.G. Zeitoun. 1998. Seawater–freshwater interface in a stratified aquifer of random permeability distribution. *Journal of Contaminant Hydrology* 29: 185–203.

de Marsily, G. 1986. *Quantitative hydrogeology: Groundwater hydrology for engineers*. San Diego: Academic Press.

Diersch, H.-J.G. and O. Kolditz. 2002. Variable-density flow and transport in porous media: Approaches and challenges. *Advances in Water Resources* 25: 899–944.

Eeman, S., Leinse, A., Raats, P.A.C. and S.E.A.T.M. van der Zee. 2011. Analysis of the thickness of a fresh water lens and of the transition zone between this lens and upwelling saline water. *Advances in Water Resources* 34: 291–302.

Elder, J. 1967. Transient convection in a porous medium. *Journal of Fluid Mechanics* 27(3): 609–623.

Evans, T.B. and A.M. Wilson. 2016. Groundwater transport and the freshwater–saltwater interface below sandy beaches. *Journal of Hydrology* 538: 563–573.

Ghassemi, F., Alam, K. and K. Howard. 2000. Fresh-water lenses and practical limitations of their three-dimensional simulation. *Hydrogeology Journal* 8: 521–537.

Graf T. and C.T. Simmons. 2009. Variable-density groundwater flow and solute transport in fractured rock: Applicability of the *Tang et al.* [1981] analytical solution. *Water Resources Research* 45: W02425.

Hamann, E., Post, V., Kohfahl, C., Prommer, H. and C.T. Simmons. 2015. Numerical investigation of coupled density-driven flow and hydrogeochemical processes below playas. *Water Resources Research* 51: 9338–9352.

Hassanizadeh, M.S. 1986. Derivation of basic equations of mass transport in porous media: 2. Generalized Darcy's and Fick's laws. *Advances in Water Resources* 9: 207–222.

Henry, H.R. 1964a. Interfaces between salt water and fresh water in coastal aquifers. In *Sea Water in Coastal Aquifers*. US Geological Survey Water-Supply Paper 1613-G: C35–C69.

Henry, H.R. 1964b. Effects of dispersion on salt encroachment in coastal aquifers. In *Sea Water in Coastal Aquifers*. US Geological Survey Water-Supply Paper 1613-G: C70–C84.

Herzberg, A. 1901. Die wasserversorgung einiger Nordseebäder (in German). *Z. Gasbeleucht. Wasserversorg* 44, 45: 815–819, 842–844.

Holzbecher, E. 1998. *Modeling density-driven flow in porous media. Principles, Numerics, Software*. Berlin: Springer.

Huyakorn, P.S., Anderson, P.F., Mercer, J.W. and W.O. White Jr. 1987. Saltwater intrusion in aquifers: Development and testing of a three dimensional finite element model. *Water Resources Research* 23: 293–312.

Klose, S. 2012. Regional hydrogeology and groundwater budget modeling in the arid Middle Drâa Catchment (South-Morocco). PhD dissertation, Bonn: Rheinischen Friedrich-Wilhelms Universität Bonn.

Kolditz, O., Ratke, R., Diersch, H.J. and W. Zielke. 1997. Coupled groundwater flow and transport: 1. Verification of variable density flow and transport models. *Advances in Water Resources* 21(1): 27–46.

Koussis, A.D., Mazi, K., Riou, F. and G. Destouni. 2015. A correction for Dupuit-Forchheimer interface flow models of seawater intrusion in unconfined coastal aquifers. *Journal of Hydrology* 525: 277–285.

Langevin, C.D. and W.X. Guo. 2006. MODFLOW/MT3DMS-based simulation of variable-density ground water flow and transport. *Ground Water* 44(3): 339–351.

Levanon, E., Shalev, E., Yechieli, Y. and H. Gvirtzman. 2016. Fluctuations of fresh-saline water interface and of water table induced by sea tides in unconfined aquifers. *Advances in Water Resources* 96: 34–42.

Lu, C., Chen, Y., Zhang, C. and J. Luo. 2013. Steady-state freshwater–seawater mixing zone in stratified coastal aquifers. *Journal of Hydrology* 505: 24–34.

Luyun, R., Momii, K. and K. Nakagawa. 2011. Effects of recharge wells and flow barriers on seawater intrusion. *Ground Water* 49: 239–249.

Mantoglou, A. and M. Papantoniou. 2008. Optimal design of pumping networks in coastal aquifers using sharp interface models. *Journal of Hydrology* 361: 52–63.

Munhoven, S. 1992. *Modélisation d'un aquifère salé* (in French). MSc thesis, Engineering Faculty. University of Liège.

Oude Essink, G.H.P. 2001. Improving fresh groundwater supply problems and solutions. *Ocean & Coastal Management* 44: 429–449.

Park, C.-H. and M.M. Aral. 2004. Multi-objective optimization of pumping rates and well placement in coastal aquifers. *Journal of Hydrology* 290(1–2): 80–99.

Peck, A.J. and D.R. Williamson. 1987. Effects of forest clearing on groundwater. *Journal of Hydrology* 94: 47–65.

Pool, M. and J. Carrera. 2010. Dynamics of negative hydraulic barriers to prevent seawater intrusion. *Hydrogeology Journal* 18: 95–105.

Pool, M. and J. Carrera. 2011. A correction factor to account for mixing in Ghyben-Herzberg and critical pumping rate approximations of seawater intrusion in coastal aquifers. *Water Resources Research* 47: W05506.

Reichard, E.G., Li, Z. and C. Hermans. 2010. Emergency use of groundwater as a backup supply: Quantifying hydraulic impacts and economic benefits. *Water Resources Research* 46: W09524.

Reilly, T.E. and A.S. Goodman. 1987. Analysis of saltwater upconing beneath a pumping well. *Journal of Hydrology* 89: 169–204.

Sikkema, P.C. and J.C. van Dam. 1982. Analytical formulas for the shape of the interface in a semi-confined aquifer. *Journal of Hydrology* 56: 201–220.

Simmons, C.T., Fenstemaker, T.R. and J.M. Sharp. 2001. Variable-density groundwater flow and solute transport in heterogeneous porous media: Approaches, resolutions and future challenges. *Journal of Contaminant Hydrology* 52: 245–275.

Strack, O.D.L. 1976. A single-potential solution for regional interface problems in coastal aquifers. *Water Resources Research* 12: 1165–1174.

Strack, O.D.L. 1984. Three-dimensional streamlines in Dupuit–Forchheimer models. *Water Resources Research* 20: 812–822.

Strack, O.D.L. and B.K. Ausk. 2015. A formulation for vertically integrated groundwater flow in a stratified coastalaquifer. *Water Resources Research* 51: 6756–6775.

Terry, J.P. and A.C. Falkland. 2010. Responses of atoll freshwater lenses to storm-surge over-wash in the Northern Cook Islands. *Hydrogeology Journal* 18: 749–759.

Turner, J.S. 1973. *Buoyancy effects in fluids*. Cambridge: Cambridge University Press.

van der Vorst, J.M.C., Glasbergen, P., Leijnse, A., Praagman, N. and J. Taat. 1984. *Transport by groundwater of radionuclides released after flooding of a repository in a salt-dome*, Report 84042402. Leidschendam, The Netherlands: Nat. Inst. Of Publ. Health and Environ. Hygiene.

Vandenbohede, A. and L. Lebbe. 2006. Occurrence of salt water above fresh water in dynamic equilibrium in coastal groundwater flow systems. *Hydrogeology Journal* 14: 462–72.

Vandenbohede, A. and L. Lebbe. 2007. Effects of tides on a sloping shore: Groundwater dynamics and propagation of the tidal wave. *Hydrogeology Journal* 15: 645–58.

Violette, S., Boulicot, G. and S.M. Gorelick. 2009. Tsunami-induced groundwater salinization in southeastern India. *CR Geosciences* 341: 339–346.

Voss, C. and W.R. Souza. 1987. Variable density flow and solute transport simulation of regional aquifers containing a narrow freshwater-saltwater transition zone. *Water Resources Research* 23(10): 1851–1866.

Watson, T.A., Werner, A.D. and C.T. Simmons. 2010. Transience of seawater intrusion in response to sea level rise. *Water Resources Research* 46: W12533.

Werner, A.D. 2017. Correction factor to account for dispersion in sharp-interface models of terrestrial freshwater lenses and active seawater intrusion. *Advances in Water Resources* 102: 45–52.

Werner, A.D., Bakker, M., Post, V.E.A., Vandenbohede, A., Lu, C., Ataie-Ashtiani, B., Simmons, C.T. and D.A. Barry. 2013. Seawater intrusion processes, investigation and management: Recent advances and future challenges. *Advances in Water Resources* 51: 3–26.

Yoon, S., Williams, J.R., Juanes, R. and P.K. Kang. 2017. Maximizing the value of pressure data in saline aquifer characterization. *Advances in Water Resources* 109: 14–28.

Chapter 11

Heat transfer in aquifers and shallow geothermy

11.1 Introduction

Heat transfer in geological media, especially in aquifers, is of increasing importance. New demands for renewable energy sources have greatly increased the attention given to this topic. For years, temperature has been measured for hydrological or hydrogeological studies; for example, for detecting groundwater-surface water interactions (Irvine *et al.* 2017a), quantifying flow through individual fractures, and detecting specific flow patterns in aquifers (Anderson 2005). This topic corresponds to a field that is developing quickly as new measurement techniques are made available (among others, Rau *et al.* 2010, Kurylyk and Irvine 2016, Irvine *et al.* 2017b).

For river-aquifer exchange flux assessments, heat tracer techniques are often combined with other hydrochemical and isotopic measurements (Xie *et al.* 2015). Heat has been used as a tracer since the 1960s, constraining inverse problems with additional data for hydraulic conductivity estimates. Now, the development of new sensors, such as the fiber-optic distributed-temperature-sensing (DTS) technology (Selker *et al.* 2006) enables measurements with a high spatial resolution (Hermans *et al.* 2015, Seibertz *et al.* 2016, Shanafield *et al.* 2017). More info about the DTS technology is reported in Hausner *et al.* (2011), van de Giesen *et al.* (2012), and Bense *et al.* (2016). For example, DTS can be useful to assess borehole flow rates with in-well heat tracer tests (Sellwood *et al.* 2015). More generally, the recent improvement of field equipment for measuring temperatures greatly helps the quantification of heat transport parameters in heterogeneous aquifers at a time of increased interest in various geothermal energy systems. Heat transport in geological layers can also be of decisive importance in radioactive waste disposal studies.

The importance of renewable energy in the new water-energy-food nexus is resulting in a huge expansion of the interest on the quantification of heat transfer in partially and fully saturated zones. Site specific values for heat transport properties/parameters are needed to design geothermal reservoirs and heat storage systems in aquifers. They will influence the short , mid , and long-term performances of the systems, as well as the estimated impacts on the groundwater resources. Previously, only deep geothermal resources were considered, but currently, very low enthalpy shallow geothermal systems are considered nearly everywhere for heat pumps and heat storage.

Various shallow geothermal systems involving no specific anomalous temperature gradients will be briefly described (see Section 11.3). This chapter is only an introductory summary as entire books are now dedicated to this topic (see Stauffer *et al.* 2014).

Two main types of shallow geothermy can be considered: Borehole Thermal Energy Storage (BTES) and Aquifer Thermal Energy Storage (ATES) systems (see Section 11.3). When pumping and reinjection is considered in a shallow aquifer (ATES), very specific hydrogeological conditions are required. The hydraulic conductivity should be high enough to allow significant pumping to provide the required energy (in kW) in a given building. Usually, at the considered site, the same zone, the groundwater flow should be limited enough to prevent the migration of the heat plume far away from where it can be used. Additionally, permit considerations should not be neglected, and in many industrialized countries, strict conditions (i.e., the protection of water resources mostly influenced by drinking water regulations) are observed for the pumping and reinjection of water in an aquifer.

11.2 Heat transfer processes, equations, and properties

Physically, it is easy to understand that the main differences in solute transport result from the fact that heat can be transported through both the pore/fissure space and the solid matrix of the medium whereas solutes in groundwater are transported only through the pores and fissures. To build the heat conservation equation in a saturated medium, the heat conduction, heat convection/ advection and heat dispersion fluxes will be considered. The heat balance equation will then be written with the addition of the "divergence" of each heat flux vector under saturated conditions. This equation will be expressed directly under transient conditions, but a steady-state heat transport problem can be considered in some very specific cases.

Heat conduction

A conduction heat flux tends to migrate from high temperature to low temperature areas of any saturated or partially saturated porous/fractured medium. It is important to note that *heat conduction* exists without any fluid flow as a result of a thermal gradient. A linear law of Fourier (equivalent to Darcy's law for saturated groundwater flow) describes the heat transfer by conduction. The conduction heat (or thermal) flux f_{tcond} is expressed in W/m² or J/(s.m²) [MT⁻³] by:

$$f_{tcond} = -\lambda_b \, \mathbf{grad}\, T = -\lambda_b \nabla T \tag{11.1}$$

where λ_b is the heat (or thermal) conductivity in W/(m°K) [MLT⁻³Θ⁻¹] of the bulk porous medium and ∇T the temperature gradient (°K/m) [Θ L⁻¹]. The *bulk thermal conductivity* is generally considered as a scalar corresponding to an isotropic property of the bulk porous medium, and it can be calculated as the weighted geometric mean of the solid and water thermal conductivities:

$$\lambda_b = \lambda_s^{(1-\theta)}\lambda_w^{\theta} \tag{11.2}$$

where θ is the volumetric water content (see Equation 4.13 in Chapter 4, Section 4.4), and λ_s and λ_w the solid and water heat conductivities, respectively. In general, the equivalent bulk thermal conductivities at the considered scale depends on the true geometry of the medium within the given REV (Nield and Bejan 2013). If we consider that the heat conduction occurs more "in parallel" in the solid and water phases, then a weighted arithmetic mean of the water and solid heat conductivities should be used:

$$\lambda_b = (1 - \theta)\lambda_s + \theta\lambda_w \tag{11.3}$$

If, on the other hand, for any reason linked to the geometry and the spatial distribution of water and solid in the REV, we can consider the heat conduction flux to mostly occur "in series" (i.e., through solid and water phases sequentially and repeatedly), and a weighted harmonic mean of the water and solid heat conductivities could be used:

$$1/\lambda_b = \frac{1 - \theta}{\lambda_s} + \frac{\theta}{\lambda_w} \tag{11.4}$$

In general, these two averaged equivalent values provide upper and lower bounds, for the equivalent (bulk) thermal conductivity in an REV. That is the reason why, for practical purposes, a ready estimate can be provided by the weighted geometric mean of λ_s and λ_w (Equation 11.2). Typical values for the *thermal conductivity* in W/(m°K) of different geological media are given in Table 11.1. When using this type of value table,

Table 11.1 Typical thermal conductivities in saturated geological media (Wm^{-1}°K^{-1})

Lithology[a]	λ_b (Wm^{-1}°K^{-1})
Granite and gneiss	3.2–4.4
Basalt	3.0–3.5
Quartzite	4.0–6.5
Shales	1.5–3.5
Schists and slates	1.3–3.0
Limestone and dolomite (karstified)	2.5–4.5
Chalk	1.5–2.5
Sandstone	2.5–5.0
Siltstone	2.0–4.0
Volcanic tuff	1–1.5
Gravels	2.5–4.5
Sands	3.0–5.5
Silts	2.0–4.0
Loams, loess, and clays	2.0–3.0
Air	0.024–0.026
Water	0.57–0.60
Organic matter	0.25–0.40

[a] values for dry conditions (i.e., with air in the pores and fissures). Saturated or partially saturated values will be slightly greater according to the water content and resulting from the difference in the thermal conductivities of water and air. Data obtained from local rocks should be privileged for detailed calculations of geothermal systems (Clauser and Huenges 1995, Eppelbaum *et al.* 2014, and Stauffer *et al.* 2014).

the user must note that, in theory, the thermal properties of rocks are also dependent on the temperature, pressure, porosity, and properties of pore- and fissure-filling fluids and gases (Eppelbaum *et al.* 2014). For a heat transfer problem in shallow aquifers, the temperature and pressure dependences are most often neglected.

Heat advection and convection

Heat transport with fluid (water) movement is *heat advection*, which describes the heat fluxes associated with all groundwater movement (i.e., resulting from both head differences and temperature-induced density differences). Thus, it includes the forced convection or advection resulting from the head gradient and the free or natural convection resulting from density differences. However, for some colleagues, the term advection is used for groundwater flow only resulting from head differences and the term convection is reserved only for groundwater movement resulting from temperature-induced density differences (*natural convection* or *free convection*).

The heat advection flux is expressed in W/m^2 or J/(s.m^2) by:

$$f_{tconv} = \rho_w c_w \, q T \tag{11.5}$$

where ρ_w is the water density in kg/m^3 [ML^{-3}], c_w is the *water heat capacity* in J/(kg°K) [L^{2}T^{-2}Θ^{-1}], $\rho_w c_w$ is the *water volumetric heat capacity* in J/(m^3°K), q is the total water flux vector (m/s) from Darcy's law and the possible temperature effect on the water density and viscosity, and T is the temperature (°K). Usually, free convection is considered to occur in association with a high-temperature heat flux so that the density and viscosity of the groundwater is strongly affected. The dynamic water viscosity is decreased (i.e., water fluidity tends to increase) with higher temperatures (Figure 11.1). This will definitely influence the value of hydraulic conductivity as determined by Equation 4.24 (Chapter 4, Section 4.4). Accordingly, the density change resulting from nonisothermal conditions (Figure 11.1) will influence the hydraulic conductivity to a lesser extent than the viscosity changes. However, the decrease in water density with higher temperatures can create non-Darcian natural convective groundwater flow.

In natural conditions, free convection appears when a high heat flow is expected on the boundary of a domain made of very thick permeable geological formations. Holzbecher and Yusa (1995) defined a *mixed convection dimensionless number M* for investigating, in the considered domain, the importance of the vertical free convection relative to the vertical forced convection:

$$M(T) = \frac{[\rho_{max} - \rho(T)]}{[\rho(T)(\partial h/\partial z)]} \tag{11.6}$$

where $\partial h/\partial z$ is the vertical piezometric gradient. Indeed, the mixed convection number becomes large in situations that are dominated by free convection. It is relatively easy to use this dimensionless number, as its value is only a function of the foreseen maximum groundwater density (ρ_{max}) and vertical piezometric gradient. More details will be given at the end of this section on the use of different dimensionless numbers to detect critical conditions that induce significant density and viscosity changes

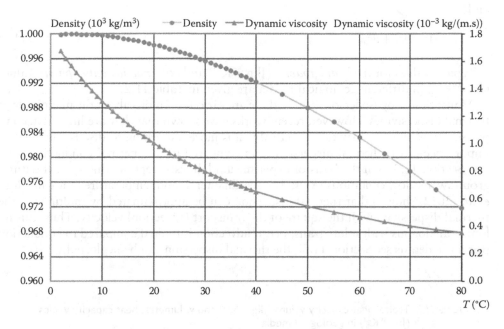

Figure 11.1 Variation of the water density and dynamic viscosity with increasing temperature (Kestin *et al.* 1978).

and their influences on the studied flow and heat transport processes in geothermal problems.

Thermal dispersion

The chosen concept of heat dispersion, similar to the Gaussian solute transport concept (see Section 8.2), describes the spreading of heat not only through the pores and/ or fissures of the medium but also through the solid matrix heterogeneities within the REV. Heat dispersion is known to be lower than solute mechanical dispersion (in relative terms; Bear 1972), even if the dispersion passes through the bulk porous medium (Hopmans *et al.* 2002). The thermal dispersion heat or thermal flux in W/m^2 or J/(s.m^2) is written as:

$$f_{tdisp} = -\rho_b c_b \, \mathbf{D} \cdot \nabla T \tag{11.7}$$

where $\mathbf{D}$ is the thermal dispersion tensor dependent on Darcy's flux q and on the thermal dispersivities (similar to the solute dispersion tensor, see Chapter 8, Section 8.2), ρ_b is the bulk density (kg/m^3) of the porous medium, c_b is the bulk *heat capacity* J/(kg°K) of the porous medium, and $\rho_b c_b$ is the bulk *volumetric heat capacity* J/(m^3°K), calculated as:

$$\rho_b = (1 - \theta)\rho_s + \theta \rho_w \tag{11.8}$$

and

$$c_b = (1 - \theta)c_s + \theta c_w \qquad \qquad (11.9)$$

Typical values for the *heat capacity* ($Jkg^{-1}°K^{-1}$) and *volumetric heat capacity* values ($Jm^{-3}°K^{-1}$) in different geological media are given in Table 11.2.

As mentioned by Anderson (2005), there are various opinions about the magnitude of thermal dispersivities. However, recent works have shown that they are limited in comparison to solute dispersivities. Practically, it is more interesting to discuss the relative importance of the thermal dispersion term compared to the thermal conduction term (see Section 11.3). On the basis of experimental lab tests, representing typical natural groundwater flow conditions ($Re < 2.5$, see Chapter 4) and temperature gradients, Rau *et al.* (2012) showed that thermal dispersion can be approximated by multiplying the thermal dispersivity with the square of the transport (advection) velocity. Thus, this no longer analogous with solute transport. If advection is limited compared to conduction (for more details see Section 11.3), the thermal dispersion can be neglected.

Table 11.2 Typical heat capacity values ($Jkg^{-1}°K^{-1}$) and volumetric heat capacity values ($Jm^{-3}°K^{-1}$) in geological media

Lithology[a]	c_b ($Jkg^{-1}°K^{-1}$)	$\rho_b c_b$ ($Jm^{-3}°K^{-1}$)
granite and gneiss	0.74–$0.79\ 10^3$	2–$2.1\ 10^6$
basalt	0.79–$0.86\ 10^3$	2.1–$2.3\ 10^6$
quartzite	0.70–$0.75\ 10^3$	1.85–$1.95\ 10^6$
shales	0.89–$1.11\ 10^3$	2.5–$3.0\ 10^6$
schists and slates	0.71–$1.11\ 10^3$	2.0–$3.0\ 10^6$
limestone and dolomite	0.80–$0.91\ 10^3$	2.2–$2.5\ 10^6$
chalk	$0.90\ 10^3$	2.2–$2.25\ 10^6$
sandstone	0.74–$0.92\ 10^3$	1.63–$2.2\ 10^6$
siltstone	0.79–$0.88\ 10^3$	2.0–$2.2\ 10^6$
volcanic tuff	$0.2\ 10^3$	0.2–$0.5\ 10^6$
gravels	1.0–$1.5\ 10^3$	1.3–$2.0\ 10^6$
sands	0.80–$0.96\ 10^3$	1.5–$2.1\ 10^6$
silts	0.8–$1.1\ 10^3$	1.7–$2.3\ 10^6$
loams, loess, and clays	1.1–$2.1\ 10^3$	2.3–$4.2\ 10^6$
air	$1.005\ 10^3$	1.25
water	$4.18\ 10^3$	$4.18\ 10^6$
organic matter	$1.93\ 10^3$	$2.51\ 10^6$

[a] Values for dry conditions (i.e., with air in the pores and fissures). Saturated or partially saturated values will be greater according to the water content and resulting from the huge difference in the thermal capacity values of water and air. Data obtained from local rocks should be privileged for detailed calculations of geothermal systems (Clauser and Huenges 1995, Eppelbaum *et al.* 2014, and Stauffer *et al.* 2014).

Heat conservation equation

In the continuum approach, the heat conservation or balance equation can be written for heat transfer in a similar way as the previously described equation for solute transport (see Chapter 8, Section 8.2) as:

$$\frac{\partial \rho_b c_b T}{\partial t} = -\nabla \cdot [\rho_w c_w q T - (\lambda_b + c_b \rho_b \mathbf{D}) \cdot \nabla T] + Q_T \tag{11.10}$$

where Q_T is the heat source (if $Q_T > 0$) or sink (if $Q_T < 0$) term. All terms of this equation are expressed in W/m³ [ML⁻¹T⁻³] (Therrien et al. 2010, Klepikova et al. 2016). Most often, the thermal dispersion term is found to be negligible compared to the thermal conduction term and even more negligible compared to the advection/convection term, particularly in an advection/convection-dominated regime (among others, Hopmans et al. 2002, Constantz et al. 2003, Vandenbohede et al. 2009, Ma and Zheng 2010, Irvine et al. 2015). Then, the equation is written as:

$$\frac{\partial \rho_b c_b T}{\partial t} = -\nabla \cdot [\rho_w c_w q T - \lambda_b \nabla T] + Q_T \tag{11.11}$$

By dividing by the bulk volumetric heat capacity ($\rho_b c_b$), the following equation is found:

$$\frac{\partial T}{\partial t} = -\nabla \cdot \left[\frac{\rho_w c_w}{\rho_b c_b} q T - \frac{\lambda_b}{\rho_b c_b} \nabla T \right] + \frac{Q_T}{\rho_b c_b} \tag{11.12}$$

where each term is expressed in °K/s [ΘT⁻¹]. In Equation 11.12, the coefficient $\lambda_b/(\rho_b c_b)$ is often named *thermal diffusivity* (κ) and is on the order of 10^{-7}–10^{-6} m²/s:

$$\kappa = \frac{\lambda_b}{\rho_b c_b} \tag{11.13}$$

The term $Q_T/(\rho_b c_b)$ (°K/s) can also be understood as the temperature associated with a groundwater flux entering or exiting 1 m³ of the domain. If we compare this to the solute transport equation (see Section 8.2), the thermal diffusivity term is similar to the hydrodynamic dispersion term. Note that the thermal diffusivity is usually 3 or 4 orders of magnitude greater than the molecular diffusion of a solute in groundwater (Anderson 2005).

Dimensionless numbers for assessing the most important processes

Geological heterogeneity can have a greater influence on solute transport than on thermal transport, as the latter occurs through both the pore and fissure fluid and the medium matrix. In relatively low permeability media, the conduction of heat can mask the presence of preferential groundwater and heat paths (Irvine et al. 2015). In aquifers, the ranges of thermal parameters values are usually a few orders of magnitude lower than the variations in the hydraulic conductivities.

In fractured and porous geological media, it is very important to have a reliable picture of the heterogeneous nature of the thermal reservoir for almost all types of enhanced or engineered geothermal systems. The conceptual choices about the thermal processes that are coupled to groundwater flow can be highly site-specific. For example, as reported by Fox et al. (2016), in a fractured rock, the advective/convective heat transport in the fractures can be combined with the 1D heat conduction in the rock matrix perpendicular to the fracture.

Dimensionless numbers were developed in the literature to assess the most important thermal processes that are considered. For solute transport, a *thermal Peclet dimensionless number* is used (Ma et al. 2012) to analyze the ratio between advection and thermal conduction (or thermal conduction + thermal dispersion if the latter is not neglected):

$$Pe = \frac{c_w \, ql}{\lambda_b} \tag{11.14}$$

where l is the characteristic length of the studied problem. However, as shown (Huysmans and Dassargues 2005) for solute transport, a large variety of *Peclet* numbers can be defined according to the chosen characteristic length. According to each specific need, they can be chosen to represent the distance from the heat source, the distance of influence, the distance corresponding to the advective path during a given time step ($v_e \Delta t$), the average particle diameter of the porous medium, the square root of the intrinsic permeability ($\sqrt{k}$), or more "numerically," the grid spacing (Δm) (see Chapter 13) of finite difference cells or finite elements/volumes. Heat transport *Peclet* numbers are lower than solute transport Pe numbers by several orders of magnitude for the same Darcy flow velocities (Rau et al. 2012). They are quite different as heat conduction is also efficient through the solid matrix. As mentioned previously, Rau et al. (2012) indicated that if the *thermal Peclet number* is lower than 0.5, the thermal dispersivity can be neglected.

One of the most discussed question relates to how the natural convection or the buoyancy-driven flow of a fluid heated from below and cooled from above (i.e., the Rayleigh-Bénard convection) is transposed and occurs in real geothermal systems. In a water column (i.e., without any porous or fractured medium) such as a well or a shaft, a *Rayleigh dimensionless number* is defined (Love et al. 2007) as:

$$Ra = \frac{g\beta(\Delta h)^3 \Delta T}{\nu\kappa} \tag{11.15}$$

where β is the volumetric thermal expansion coefficient (°K^{-1}), Δh is the height of the water column in the well (m), ΔT is the temperature difference between the bottom and top of the well or shaft (°K), ν is the kinematic viscosity (m^2/s), and κ is the thermal diffusivity (m^2/s). This is one way to express the ratio between the natural convection and the thermal conduction (+ thermal dispersion if not neglected). The critical value of Ra for the onset of natural convection in a vertical well or shaft mainly depends on the ratio between the well radius and the height of the water column: $\delta = r/\Delta h$, and can be expressed by (Love et al. 2007, Hamm and Bazargan Sabet 2010):

$$Ra_c = \frac{215.6}{\delta^4}(1 + 3.84\delta^2) \tag{11.16}$$

This clearly shows how it may be more convenient to pump groundwater through narrow wells than large shafts to avoid the natural convective mixing of water in the shaft even before the groundwater extraction starts, inducing advection (forced convection). Even if this critical value is calculated for rigid impermeable walls, the value would only change slightly (i.e., be diminished) if a porous medium replaces the solid walls (Love *et al.* 2007).

When advection is also considered, the *mixed convection dimensionless number M* (see Equation 11.6) can be considered as equal to the ratio *Ra/Pe* (Graf and Simmons 2009). This *M* number is useful for determining if it is important to consider natural convection (*M* > 1) with regards to the advection (forced convection). Note that these numbers were developed not only for thermally induced groundwater density variations but also and mostly for determining whether density dependent flow must be considered in solute transport problems involving high concentrations of salt[1] components (Simmons *et al.* 2001; see Chapter 10).

When comparing solute and heat transport, a *Lewis dimensionless number* is defined as the ratio between the thermal diffusivity and the solute mass dispersion:

$$Le = \frac{\lambda_b}{\rho_b c_b D} = \frac{\kappa}{D} \qquad (11.17)$$

where D is the solute hydrodynamic dispersion component in the considered direction (See Section 8.2).

11.3 Hydrogeological methodology for shallow geothermal projects

Deep geothermal projects will not be covered here. They are extensively covered in a specific and specialized research literature.

Introduction to shallow thermal energy storage systems

Geothermal exchange systems with coupled heat pumps target shallow geological layers to provide cooling and heating needs of buildings. They are usually called ground (source) heat pump systems (GSHP or GHP). In typical conditions and without geothermal anomalies, the shallow layers (i.e., at a depth < 200 m) can most often be considered to show a mean temperature corresponding to the annual average local air temperature (Figure 11.2).

Temperature demonstrates very low seasonal variations at depths greater than 5 m in temperate climates. A nearly constant temperature can even be observed at a depth less than 5 m where saturated aquifer conditions are reached at a few meters or centimeters of depth. This results from the high groundwater flux and mixing that rapidly homogenize any temperature change. Note that the temperatures measured in monitoring wells often show more seasonal variations than those in undisturbed geological

---

[1] The term "salt components" is used here to describe all dissolved chemicals in groundwater that significantly change the water density.

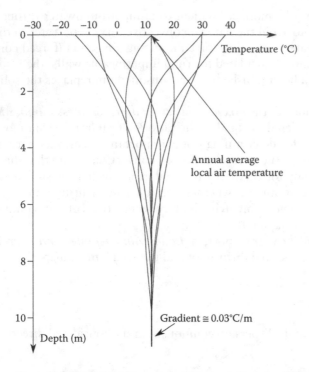

Figure 11.2 Typical cyclic temperature variations in shallow conditions induced by seasonal changes (such as the temperate climatic conditions of Northwest Europe). At a few meters depth, a nearly constant temperature can be found that is equal to the yearly average local air temperature of the location (here 12.5°C). At greater depths, the geothermal gradient induces an increasing temperature gradient of approximately 0.03°C/m.

formations. Also note that, in urban areas, local "heat islands" are observed (Menberg *et al.* 2013, Arola and Korkka-Niemi 2014) with yearly average air temperatures 2°C or 3°C higher than those in the surrounding countryside. Transferring heat from the subsurface to buildings or from buildings to the subsurface can be performed by using two types of systems: closed-loop systems with heat exchangers comprised of water and antifreeze solution circulating in any buried pipes that can be located within a borehole (BTES) (Figure 11.3); and, open systems with a groundwater pumping and reinjection well-doublet (Figure 11.4). In the latter case, a shallow aquifer is needed for the heat/cold extraction through groundwater pumping and the cold/heat storage by injection. Thus, this can be considered as a type of aquifer thermal energy storage (ATES).

As both systems can be used for heating and cooling, they are also named underground thermal energy storage (UTES) systems. The exploitation of shallow geothermal resources is usually considered to represent renewable energy. However, on one hand, the development and multiplication of these systems, particularly in densely populated areas, should not jeopardize the quantity and quality of groundwater resources. On the other hand, the renewability and efficiency of this environmentally friendly resource could be threatened by inadequate operational conditions.

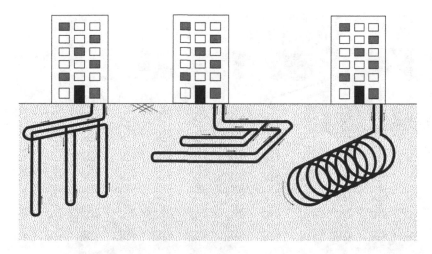

Figure 11.3 Closed-loop geothermal systems with horizontal, vertical, and slinky systems.

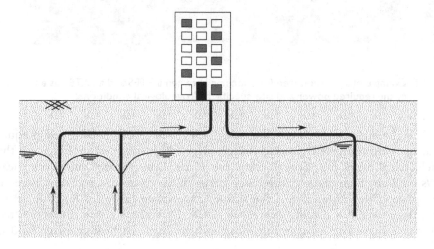

Figure 11.4 Open-loop geothermal system with a well-doublet.

Therefore, the groundwater conditions associated with UTES systems should be well identified, characterized, and simulated to ensure both the efficiency and durability of the installation. An environmental impact assessment should also be performed. These conditions and other parameters, such as the maximum delivered power and the drilling costs, will strongly influence the adopted system. As an example, a decision tree (Figure 11.5) was proposed for the energy conditions and mean drilling costs for Western Europe in 2017. When producing thermal energy with a heat pump, the power P (in W) $[ML^2T^{-3}]$ can be expressed by:

$$P = \frac{Q\Delta T \rho_w c_w}{(1 - (1/COP))} \qquad (11.18)$$

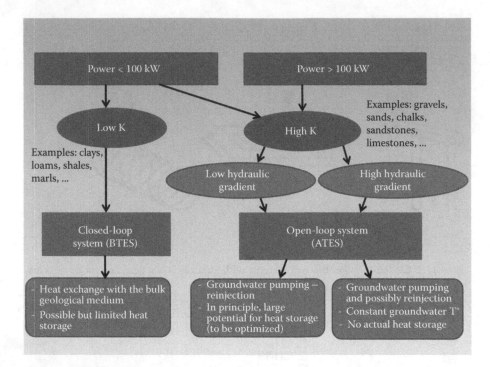

Figure 11.5 Example of a decision tree for choosing between a BTES and an ATES as a function of the required power and site-specific hydrogeological conditions.

where *COP* is the dimensionless *coefficient of performance* of the heat pump equal to the ratio between the useful produced thermal work (heating or cooling) and the required (electrical) work [−], Q is the flow rate of the heat transfer fluid in the heat pump (m³/s), ΔT the temperature difference between the upstream and downstream of the heat pump (°K) [Θ], and c_w is the *water heat capacity* in J/(kg°K) [$L^2 T^{-2} \Theta^{-1}$]. In practice, the power P is usually expressed in kW. By doing so, practitioners often consider the oversimplified assumption that the density of the heat transfer fluid is 10^3 kg/m³.

To produce the needed power, closed-loop systems involve long heat exchange surfaces multiplied in accordance with the number of boreholes or buried exchange pipe loops, while open-loop systems just need to pump more groundwater. Therefore, to meet high power demands, open-loop systems are usually more economically suitable. On the other hand, they have an additional energy cost for pumping and the local hydraulic conductivity can be a clear limitation on the pumped and reinjected flow rate.

In terms of impact, groundwater quality can be affected by temperature changes as a result of their influence on most biochemical and physical processes. Indeed, chemical thermodynamic equilibria can be changed, which has an influence on dissolved minerals and gases. Some of these impacts can also directly influence the future efficiency of the geothermal system and may even compromise its operability. Various possible impacts of anthropogenic groundwater temperature changes are cited in the literature, including (among others): redox processes and associated changes in the microbial communities inducing clogging (Possemiers *et al.* 2014, 2016), acidity changes and matrix dissolution inducing subsidence (García-Gil *et al.* 2016), the increase of TCE solubility inducing increased TCE emissions in source zones (Beyer *et al.* 2016), the mobilization of organic matter and metallic elements (Bonte *et al.* 2013), gas formation (Lüders *et al.* 2016), and lower gas solubility and increased microbial activity leading to anoxic groundwater dependent ecosystems (Griebler *et al.* 2016).

The need for a system with a balanced energy load balanced system to avoid negative effects is increasingly more important to obtain permits. Consequently, more new systems tend to be artificially balanced by other energy sources. In some countries, priority is given to drinking water resources and supplies, so thermal uses (by a BTES or an ATES) may not be permitted at all in aquifers. In Switzerland, some regulations impose a limitation on the induced temperature change in the groundwater (e.g., $\Delta T < 3°C$) at a given distance (e.g., 100 m in the Swiss Federal regulations) from the BTES wells or from the reinjection well (ATES). However, in many urban areas, it remains difficult to assess the local natural or initial groundwater states.

In terms of positive environmental impacts, De Keuleneer and Renard (2015) proposed that shallow geothermal use of groundwater could be helpful to mitigate seawater intrusions in coastal aquifers. Such a system could only work in very specific conditions with the well-doublets of the ATES located between the coast and the main pumping wells with (if possible) the fresh- and cold-water reinjection located on the coastal side (Figure 11.6). The design (location, screened depth, and flow rate) of such a system should be accurately optimized by using density dependent groundwater flow and heat transport modeling to find the best configuration for a specific site.

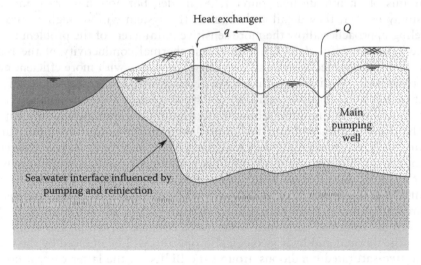

Figure 11.6 Theoretical 2D schema for optimizing an open shallow geothermal system (ATES) to mitigate the seawater intrusion in a pumped unconfined coastal aquifer (De Keuleneer and Renard 2015).

Closed-loop shallow geothermal systems

Horizontal closed-loop systems are usually embedded underground in the first few meters. They are made of loops of pipes and come in series, parallel, basket, or slinky types, requiring a length of 35–60 m of exchanger pipe per kW (Florides and Kalogirou 2008). Thus, they are suitable only for lower power demands, such as for individual houses. Vertical closed-loop systems are more efficient and more common, as they mostly benefit from a constant annual ground temperature. Even if they can be embedded in different civil engineering structures such as concrete piles, foundation slabs, retaining walls, and tunnel structures, they are most often named *borehole heat exchangers* (BHEs). In the most classic case, HDPE (high-density polyethylene) single or double U-pipe loops are installed in one or many borehole(s).

The true heat transfer between the circulating antifreeze fluid and the ground is highly dependent on the grouting material and conditions. This grouting is also useful to prevent contamination in the event of a leak of the circulating fluid. These systems are known to produce (in temperate climate conditions) between 20 W/m and 80 W/m (VDI-4640 2001 in Radioti 2016) and are widely used because they do not need specific hydraulic conductivity conditions. The latter observation is unfortunately also one of the reasons why these systems are sometimes installed without a proper characterization of the local underground conditions leading to mid-term strong decrease in the system efficiency and also to possible significant environmental impacts.

Standard *thermal response tests* (TRTs) are performed to obtain a global borehole thermal resistance and, from its geometric characteristics, the mean global thermal effective conductivity of the system (i.e., the borehole and surrounding underground area; Gehlin 2002, Spitler and Gehlin 2015).

An extensive body of literature can be found about practical recommendations for in situ measurements (duration, insulation, and constant temperature conditions) and interpretations of TRT results. Interpretations were traditionally based on analytical solutions of an infinite line source (ILS) model, but now numerical models are increasingly used. If they detail and simulate the system with enough accuracy, these modeling approaches allow the more sensitive parameters of the problem to be distinguished (Raymond *et al.* 2011). The actual thermal conductivity of the bulk medium is never homogeneous or isotropic. There are zones with more efficient and rapid heat transport, especially resulting from groundwater heat transport. Another weak point of the usual feasibility and impact studies based only on the analytical interpretation of TRTs is that little attention is given to the geochemistry in connection with the possible mid- and long-term imbalances in groundwater temperatures (Florea *et al.* 2017). For future and more realistic studies, fiber-optic distributed-temperature-sensing (DTS) should be used as spatial and temporal high resolution sensor (Seibertz *et al.* 2016, Bense *et al.* 2016) in combination with TRTs in order to get enough spatial and temporal information to constrain the inversion of heat transport models representing the spatially heterogeneous hydrogeological conditions. If BHEs are installed in not fully and permanently saturated conditions, spatial and temporal variations in the thermal and hydrogeological properties will be induced by the variable unsaturated-saturated conditions around the BHEs. In the latter case, it could also be useful to consider the latent heat transfers produced by evaporation and condensation processes (Moradi *et al.* 2016).

As mentioned previously, energy balanced systems minimize negative impacts. If the groundwater flow is very limited (like in aquitards), this balance can be partially achieved by inverse seasonal energy demands (i.e., cold being stored during heating periods and heat being stored during cooling periods). However, artificially balanced systems are often needed. Energy balanced systems are less sensitive to the array configuration of the BHEs and the groundwater flow conditions (Dehkordi *et al.* 2015). An important groundwater flux tends to smooth temperature differences around the BHEs, but the thermal plume moves downgradient.

If the BTES is located in an aquitard (i.e., low hydraulic conductivity and limited groundwater flow), the thermal energy input mainly comes from conduction with the surface and infiltrating rain water. These fluxes are known to be more important than the mean thermal energy earthflow (i.e., estimated between 0.05 and 0.11 W/m^2 according to Pollack *et al.* 1993) to balance the several W/m^2 of the thermal energy exchange system. If the BTES is located in an aquifer (i.e., high hydraulic conductivity and significant groundwater flow), an important thermal flux is transported by the groundwater, so the site-specific hydrogeological conditions, including the groundwater flux, are recognized as the main factors for assessing efficiency and impact (Dehkordi *et al.* 2015).

Open-loop shallow geothermal systems

Open systems (ATES) are characterized by the direct use of groundwater as the heat carrier fluid. They are classically used with groundwater pumping and reinjection well-doublets for heating and air conditioning/cooling purposes with seasonal variations. Ideal hydrogeological contexts involve (Figure 11.5) high hydraulic conductivity with a limited gradient in order to limit the groundwater flow and the associated downgradient temperature plume. The well-doublet location can be optimized as a function of the injection/pumping operations and the considered variation period (i.e., most often seasonal) to maintain an energy balance.

The ATES should lose as little thermal energy as possible in a downgradient plume, which induces side effects in groundwater dependent ecosystems. In particular, in urban aquifers, the overexploitation and conflicting use of thermal and groundwater resources can rapidly become tricky if not rigorously studied in an integrated way. These studies must be based on a clear understanding of the underground system and reliable data about the main parameter values and their heterogeneity (Vienken *et al.* 2015). New research for cost-efficient exploration techniques is currently ongoing (among others, Schwede *et al.* 2014, Wagner *et al.* 2014, Wildemeersch *et al.* 2014, Hermans *et al.* 2015, Klepikova *et al.* 2016) and involve, for example, different hydrogeophysical techniques combined to solute and heat tracer tests (Figure 11.7). For example, to assess the potential for a seasonal ATES in a shallow sandy aquifer in the Brussels region (Belgium), Anibas *et al.* (2016) concluded that, for favorable conditions with a hydraulic conductivity (K) greater than 1.10^{-4} (m/s), it is necessary to have a natural groundwater velocity (v_e) lower than $5\ 10^{-4}$ (m/s) and to pump groundwater only from above or below the aquifer redox boundary to avoid clogging by $Fe(OH)_3$ precipitation. Indeed, these values strongly depend on the considered power demand and the site-specific conditions.

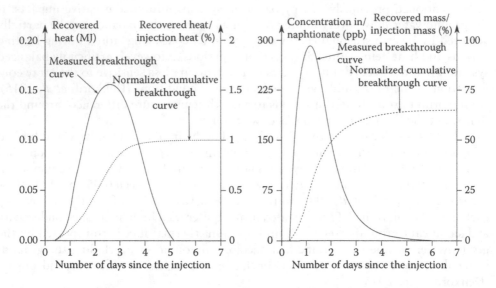

Figure 11.7 Measured breakthrough and normalized cumulative breakthrough curves in a pumping well for naphthionate and temperature as a result of a combined tracer test in an alluvial unconfined aquifer. Heat transfer is delayed compared to the dissolved tracer (Wildemeersch *et al.* 2014).

Forecasting the efficiency of ATES systems is challenging, particularly in terms of the neutral thermal balance. The extracted heat flux should be balanced over a yearly basis mostly by the thermal flux transported by groundwater. Thus, site specific hydrogeological conditions are the main driver for assessing the efficiency and impacts of an ATES, to an even greater degree than they are for a BTES. During an ATES assessment, short-term variations in the energy storage and recovery (i.e., with shorter periods than seasonal changes) can also be considered. When the timing and magnitude of the energy demand does not coincide with renewable production, an ATES could be considered for intraday heat storage and recovery. Indeed, a neutral thermal balance is again an initial guarantee of a limited environmental impact and an increased probability of mid- to long-term efficiency. As ATESs are mainly considered in high-hydraulic conductivity layers, lateral advective heat fluxes are very important. In thermal balance computations, the groundwater flow and associated heat transport conditions prescribed at the lateral external boundaries of the considered zone have a greater influence on the results than they do in the case of a BTES. This is especially true in confined aquifer conditions, even if it has been demonstrated that heat conduction fluxes through vertical boundaries in the over- and underlying aquitards can become significant (Miotlinski and Dillon 2015).

Geothermal systems in old flooded mines

Old flooded mines are increasingly considered for low-enthalpy geothermal exploitation. The cessation of pumping induces the flooding of old works including shafts, galleries, and exploited panels. Depending on the type of abandoned mine, the true geometry of the interconnected network of open galleries and shafts can be highly complex. A high-velocity water flow is expected in this type of network, while low-velocity groundwater flow occurs in less permeable fractured and porous rock massif (Figure 11.8). Logically, warm water is usually expected to be pumped (or injected) in the deep parts of the open network, and cold water is expected to be reinjected (or pumped) in the shallower parts. To assess the geothermal efficiency, it is indeed critical to compute the temperature evolution in the pumping zones to account for mixing with colder waters from the upper parts of the mine. This is particularly sensitive as "pipe-like" flows can occur in the open galleries and shafts. The water temperature inside a vertical shaft may be strongly influenced by convection produced by buoyancy forces as a result of density changes resulting from temperature evolution. Thus, the geothermal gradient is its own driver for further mixing, inducing a temperature drop in the warm water pumping zone (Hamm and Bazargan Sabet 2010). Indeed, possible salinity variations can also influence the water density. Usually, the groundwater salinity increases with depth. Consequently, a partial mitigation of the buoyancy contrast resulting from temperature changes could be expected. If only thermal effects are considered, the mixed convection M *number* (Equation 11.6), the *Rayleigh number* (Equation 11.15), and its critical value (Love *et al.* 2007) for a vertical shaft with infinite hydraulic conductivity and rigid walls (Equation 11.16) can be used for an initial assessment of the expected mixing. The main old shafts may be totally or partially filled by heterogeneous backfill materials, accordingly decreasing

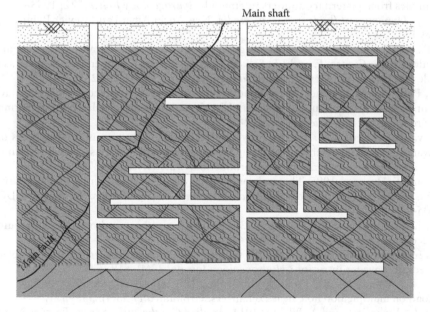

Figure 11.8 Schematic 2D vertical old mine network for low-enthalpy geothermal use: intricate and interconnected network galleries and shafts with high-velocity "pipe-like" water flow (if not refilled with backfill material) and fractured rock massif with low-velocity "porous-like" groundwater flow.

the hydraulic conductivity that must be considered in the shaft. In practice, in most cases, it is probably more efficient to drill new boreholes that reach the deep ancient galleries for pumping warm water. Similarly, a reinjection could then be foreseen in a shallow gallery, thus avoiding the use of old shafts.

In any case, further feasibility investigations must use numerical simulations of variable density flow and coupled heat transport with advanced software that allow additional complex spatial discretization associated with process-based simulations of combined high-velocity "pipe-like" water flows (in the galleries and shafts) and porous/fractured groundwater flow (in the rock matrix).

More generally, the transition toward renewable energy and power supplies necessitates increasingly more projects involving the underground storage of gas (methane, hydrogen, compressed air, and CO_2), water and heat. Hydrogeologists are key actors in these projects. The interactions between storage projects and groundwater will influence the dimensioning for mid- and long-term efficiency, as well as the results of impact study/risk assessment investigations (Kabuth *et al.* 2017).

References

Anderson, M.P. 2005. Heat as a ground water tracer. *Ground Water* 43(6): 951–968.

Anibas, Ch., Kukral, J., Possemiers, M. and M. Huysmans. 2016. Assessment of seasonal Aquifer Thermal Energy Storage as a groundwater ecosystem service for the Brussels-Capital Region: Combining groundwater flow, and heat and reactive transport modelling. *Energy Procedia* 97: 179–185.

Arola, T. and K. Korkka-Niemi. 2014. The effect of urban heat islands on geothermal potential: Examples from quaternary aquifers in Finland. *Hydrogeology Journal* 22: 1953–1967.

Bear, J. 1972. *Dynamics of fluids in porous media*. New-York: American Elsevier Publishing Company Inc.

Bense, V.F., Read, T., Bour, O., Le Borgne, T., Coleman, T., Krause, S., Chalari, A., Mondanos, M., Ciocca, F. and J.S. Selker. 2016. Distributed Temperature Sensing as a downhole tool in hydrogeology. *Water Resources Research* 52: 9259–9273.

Beyer, Ch., Popp, S. and S. Bauer. 2016. Simulation of temperature effects on groundwater flow, contaminant dissolution, transport and biodegradation due to shallow geothermal use. *Environmental Earth Sciences* 75: 1244.

Bonte, M., van Breukelen, B.M. and P.J. Stuyfzand. 2013. Temperature-induced impacts on groundwater quality and arsenic mobility in anoxic aquifer sediments used for both drinking water and shallow geothermal energy production. *Water Research* 47: 5088–5100.

Clauser, C. and E. Huenges. 1995. Thermal conductivity of rocks and minerals. In *Rock physics & phase relations: A handbook of physical constants*, ed. T.J. Ahrens. Washington, DC: AGU, 105–126.

Constantz, J., Cox, M.H. and G.W. Su. 2003. Comparison of heat and bromide as ground water tracers near streams. *Ground Water* 41(5): 647–656.

De Keuleneer, F. and Ph. Renard. 2015. Can shallow open-loop hydrothermal well-doublets help remediate seawater intrusion? *Hydrogeology Journal* 23(4): 619–629.

Dehkordi, S.E., Schincariol, R.A. and B. Olofsson. 2015. Impact of groundwater flow and energy load on multiple borehole heat exchangers. *Groundwater* 53(4): 558–571.

Eppelbaum, L., Kutasov, I. and A. Pilchin. 2014. *Applied Geothermics*, Series: Lecture Notes in Earth System Sciences. Berlin Heidelberg: Springer-Verlag.

Florea, L.J., Hart, D., Tinjum, J. and C. Choi. 2017. Potential impacts to groundwater from ground-coupled geothermal heat pumps in district scale. *Groundwater* 55(1): 8–9.

Florides, G. and S. Kalogirou. 2008. First in situ determination of the thermal performance of a U-pipe borehole heat exchanger, in Cyprus. *Applied Thermal Engineering* 28: 157–163.

Fox, D.B., Koch, D.L. and J.W. Tester. 2016. An analytical thermohydraulic model for discretely fractured geothermal reservoirs. *Water Resources Research* 52: 6792–6817.

García-Gil, A., Epting, J., Ayora, C., Garrido, E., Vázquez-Suñé, E., Huggenberger, P. and A.C. Gimenez. 2016. A reactive transport model for the quantification of risks induced by groundwater heat pump systems in urban aquifers. *Journal of Hydrology* 542: 719–730.

Gehlin, S. 2002. *Thermal Response Test—method, development and evaluation.* PhD diss., Luleå University of Technology, Sweden.

Graf T. and C.T. Simmons. 2009. Variable-density groundwater flow and solute transport in fractured rock: Applicability of the *Tang et al.* [1981] analytical solution. *Water Resources Research* 45: W02425.

Griebler, C., Brielmann, H., Haberer, Ch.M., Kaschuba, S., Kellermann, C., Stumpp, Ch., Hegler, F., Kuntz, D., Walker-Hertkorn, S. and T. Lueders. 2016. Potential impacts of geothermal energy use and storage of heat on groundwater quality, biodiversity, and ecosystem processes. *Environmental Earth Sciences* 75: 1391.

Hamm, V. and B. Bazargan Sabet. 2010. Modelling of fluid flow and heat transfer to assess the geothermal potential of a flooded coal mine in Lorraine, France. *Geothermics* 39(2): 177–186.

Hausner, M.B., Suárez, F., Glander, K.E., van de Giesen, N., Selker, J.S. and S.W. Tyler. 2011. Calibrating single-ended fiber-optic raman spectra distributed temperature sensing data. *Sensors* 11(11): 10859–10879.

Hermans, T., Wildemeersch, S., Jamin, P., Orban, P., Brouyère, S., Dassargues, A. and F. Nguyen. 2015. Quantitative temperature monitoring of a heat tracing experiment using cross-borehole ERT. *Geothermics* 53: 14–26.

Holzbecher, A. and Y. Yusa. 1995. Numerical experiments on free and forced convection in porous media. *Int. J. Heat. Mass Transfer.* 38(11): 2109–2115.

Hopmans, J.W., Simunek, J. and K.L. Bristow. 2002. Indirect estimation of soil thermal properties and water flux using heat pulse probe measurements: Geometry and dispersion effects. *Water Resources Research* 38(1): 7-1–7-13.

Huysmans, M. and A. Dassargues. 2005. Review of the use of Peclet numbers to determine the relative importance of advection and diffusion in low permeability environments. *Hydrogeology Journal* 13(5–6): 895–904.

Irvine, D.J., Simmons, C.T., Werner, A.D. and T. Graf. 2015. Heat and solute tracers: How do they compare in heterogeneous aquifers? *Groundwater* 53(S1): 10–20.

Irvine, D.J., Briggs, M.A., Lautz, L.K., Gordon, R.P., McKenzie, J.M. and I. Cartwright. 2017a. Using diurnal temperature signals to infer vertical groundwater-surface water exchange. *Groundwater* 55: 10–26.

Irvine, D.J., Kurylyk, B.L., Cartwright, I., Bonham, M., Post, V.E.A., Banks, E.W. and C.T. Simmons. 2017b. Groundwater flow estimation using temperature-depth profiles in a complex environment and a changing climate. *Science of the Total Environment* 574: 272–281.

Kabuth, A., Dahmke, A., Beyer, C., Bilke, L., Dethlefsen, F., Dietrich, P., Duttmann, R., Ebert, M., Feeser, V., Görke, U.-J., Köber, R., Rabbel, W., Schanz, T., Schäfer, D., Würdemann, H. and S. Bauer. 2017. Energy storage in the geological subsurface: Dimensioning, risk analysis and spatial planning: the ANGUS+ project. *Environmental Earth Science* 76: 23.

Kestin, J., Sokolov, M. and W.A. Wakeham. 1978. Viscosity of liquid water in the range -8°C to 150°C. *Journal of Physical and Chemical Reference Data* 7(3): 941–948.

Klepikova, M., Wildemeersch, S., Jamin, P., Orban, Ph., Hermans, T., Nguyen, F., Brouyere, S. and A. Dassargues. 2016. Heat tracer test in an alluvial aquifer: Field experiment and inverse modelling. *Journal of Hydrology* 540: 812–823.

Kurylyk, B.L. and D.J. Irvine. 2016. Analytical solution and computer program (FAST) to estimate fluid fluxes from subsurface temperature profiles. *Water Resources Research* 52: 725–733.

Love, A.J., Simmons, C.T. and D.A. Nield. 2007. Double-diffusive convection in groundwater wells. *Water Resources Research* 43(8): W08428.

Lüders, K., Firmbach, L., Ebert, M., Dahmke, A., Dietrich, P. and R. Köber. 2016. Gas-phase formation during thermal energy storage in near-surface aquifers: experimental and modelling results. *Environmental Earth Sciences* 75: 1404.

Ma, R. and Ch. Zheng. 2010. Effects of density and viscosity in modeling heat as a groundwater tracer. *Ground Water* 48(3): 380–389.

Ma, R., Zheng, C., Zachara, J.M. and M. Tonkin. 2012. Utility of bromide and heat tracers for aquifer characterization affected by highly transient flow conditions. *Water Resources Research* 48: W08523.

Menberg, K., Bayer, P., Zosseder, K., Rumohr, S. and P. Blum. 2013. Subsurface urban heat islands in German cities. *Science of the Total Environment* 442: 123–133.

Miotlinski, K. and P.J. Dillon. 2015. Relative recovery of thermal energy and fresh water in aquifer storage and recovery systems. *Groundwater* 53(6): 877–884.

Moradi, A., Smits, K.M., Lu, N. and J.S. McCartney. 2016. Heat transfer in unsaturated soil with application to borehole thermal energy storage. *Vadoze Zone Journal* 15(10): doi:10.2136/vzj2016.03.0027

Nield, D.A. and A. Bejan. 2013. *Convection in porous media*. Springer.

Pollack, H.N., Hurter, S.J. and J.R. Johnson. 1993. Heat flow from the earth's interior: Analysis of the global data set. *Reviews of Geophysics* 31: 267–280.

Possemiers, M., Huysmans, M., Anibas, Ch., Batelaan, O. and J. Van Steenwinkel. 2016. Reactive transport modeling of redox processes to assess $Fe(OH)_3$ precipitation around aquifer thermal energy storage wells in phreatic aquifers. *Environmental Earth Sciences* 75: 648.

Possemiers, M., Huysmans, M. and O. Batelaan. 2014. Influence of aquifer thermal energy storage on groundwater quality: A review illustrated by seven case studies from Belgium. *Journal of Hydrology: Regional Studies* 2: 20–34.

Radioti, G. 2016. *Shallow geothermal energy: effect of in-situ conditions on borehole heat exchanger design and performance*. PhD diss., University of Liège, Belgium.

Rau, G.C., Andersen, M.S. and R.I. Acworth. 2012. Experimental investigation of the thermal dispersivity term and its significance in the heat transport equation for flow in sediments. *Water Resources Research* 48: W03511.

Rau, G.C., Andersen, M.S., McCallum, A.M. and R.I. Acworth. 2010. Analytical methods that use natural heat as a tracer to quantify surface water-groundwater exchange, evaluated using field temperature records. *Hydrogeology Journal* 18(5): 1093–1110.

Raymond, J., Therrien, R., Gosselin, L. and R. Lefebvre. 2011. Numerical analysis of thermal response tests with a groundwater flow and heat transfer model. *Renewable Energy* 36(1): 315–324.

Schwede, R.L., Li, W., Leven, C. and O.A. Cirpka. 2014. Three-dimensional geostatistical inversion of synthetic tomographic pumping and heat-tracer tests in a nested-cell setup. *Advances in Water Resources* 63(0): 77–90.

Seibertz, K.S.O., Chirila, M.A., Bumberger, J., Dietrich, P. and T. Vienken. 2016. Development of in-aquifer heat testing for high resolution subsurface thermal-storage capability characterisation. *Journal of Hydrology* 534: 113–123.

Selker, J.S., Thévanez, L., Huwald, H., Mallet, A., Luxemburg, W., Van de Giesen, N., Stejskal, M., Zeman, J., Westhoff, M. and M.B. Parlange. 2006. Distributed fiber-optic temperature sensing for hydrologic systems. *Water Resources Research* 42: W12202.

Sellwood, S., Hart D.J. and J.M. Bahr. 2015. Evaluating the use of in-well heat tracer tests to measure borehole flow rates. *Groundwater Monitoring & Remediation* 35(4): 85–94.

Shanafield, M., McCallum, J., Cook, P.G. and S. Noorduijn. 2017. Using basic metrics to analyze high-resolution temperature data in the subsurface. *Hydrogeology Journal* 25(5): 1501–1508.

Simmons, C.T., Fenstemaker, T.R. and J.M Sharp. 2001. Variable-density groundwater flow and solute transport in heterogeneous porous media: approaches, resolutions and future challenges. *Journal of Contaminant Hydrology* 52: 245–275.

Spitler, J.D. and S. Gehlin. 2015. Thermal response testing for ground source heat pump systems—an historical review. *Renewable and Sustainable Energy Reviews* 50: 1125–1137.

Stauffer, F., Bayer, P., Blum, Ph., Molino-Giraldo, N. and W. Kinzelbach. 2014. *Thermal use of shallow groundwater.* Boca Raton: CRC Press, Taylor & Francis Group.

Therrien, R., McLaren, R.G., Sudicky, E.A. and S.M. Panday. 2010. *Hydrogeosphere: A three-dimensional numerical model describing fully-integrated subsurface and surface flow and solute transport.* Waterloo, ON: Groundwater Simulations Group, University of Waterloo.

van de Giesen, N., Steele-Dunne, S.C., Jansen, J., Hoes, O., Hausner, M.B., Tyler, S. and J.S. Selker. 2012. Double ended calibration of fiber optic Raman spectra distributed temperature sensing data. *Sensors* 12(5): 5471–5485.

Vandenbohede, A., Louwyck, A. and L. Lebbe. 2009. Conservative solute versus heat transport in porous media during push-pull tests. *Transp. Porous Media* 76(2): 265–287.

VDI 4640. 2001. *Thermal use of the underground—GSHP systems (German guidelines for ground coupled heat pumps, UTES and direct thermal use of the underground).* Part 2, VDI-Verlag: Verain Deutscher Ingenieure, Düsseldorf.

Vienken, T., Schelenz, S., Rink, K. and P. Dietrich. 2015. Sustainable intensive thermal use of the shallow subsurface—A critical view on the status quo. *Groundwater* 53(3): 356–361.

Wagner, V., Li, T., Bayer, P., Leven, C., Dietrich, P. and Ph., Blum. 2014. Thermal tracer testing in a sedimentary aquifer: Field experiment (Lauswiesen, Germany) and numerical simulation. *Hydrogeology Journal* 22(1): 175–187.

Wildemeersch, S., Jamin, P., Orban, P., Hermans, T., Klepikova, M., Nguyen, F., Brouyère, S. and A. Dassargues. 2014. Coupling heat and chemical tracer experiments for estimating heat transfer parameters in shallow alluvial aquifers. *Journal of Contaminant Hydrology* 169(0): 90–99.

Xie, Y., Cook, P.G., Simmons, C.T. and C. Zheng. 2015. On the limits of heat as a tracer to estimate reach-scale river-aquifer exchange flux. *Water Resources Research* 51(9): 7401–7416.

Chapter 12

Methodology for groundwater flow and solute transport modeling

12.1 Introduction and definitions

We have entered an era of "Big Data," "Cloud computing," and "Artificial Intelligence." Any process can be numerically modeled, simulated, optimized, or simply imaged for a better understanding, for automation or for predictions. Modeling tools continuously evolve and improve but a perfect simulation of the reality will remain a dream ... especially in natural sciences. Any type of modeling includes subjective decisions and simplifying assumptions because the true complexity of a natural system is never fully represented and data about properties and variables include uncertainties (Fienen 2013). Natural systems can be described by several coupled processes that involve coupled variables interacting in a nonlinear way. Chapters 12 and 13 of this book present some general considerations about definitions and conceptual models, and then focus on saturated groundwater flow and solute transport models.

A model is, at least, a useful tool for synthesizing in one framework all the available information on a considered system. The model can be used for understanding step-by-step how the considered processes evolve in space and time in the domain. A model could be considered as a computer-aided thinking step in the problem analysis (Molz III 2017). Indeed, the confidence into the model results remains highly dependent on field data. The more the model is constrained on field data, the more reliable the model results become.

Definitions and terminology

A *model* is a simplified representation of a complex reality. A traditional illustration can be given by invoking a road map (Wang and Anderson 1982). The map is a 2D representation of a 3D reality. Simplifications and conventional choices lead to a description of the road network with different colors, signs, and distance information that have as a main aim that the drivers can find their way or estimate the distance still to be traveled. One can observe that these choices and conventions strongly depend on the purpose and the associate scale of consideration. The type of map used for trekking would not be the same as that for driving. This means that the way of conceptualizing the 4D (i.e., spatial dimensions + time) complex reality is strongly dependent on the purpose and the final aim of the study. The scale of consideration is also directly associated, and indeed other critical conceptual choices are following such as the dimensionality, the considered processes (or "a contrario" the neglected processes and influences), the boundary locations and conditions, and the time dependency, among several other factors.

In physical models, flow and solute transport processes can be observed and measured in laboratory columns or tanks containing various types of geological media. Recent developments in various microsensors for measuring pressures, concentrations, temperature, and hydrochemical parameters make possible to consider a series of new detailed experiments allowing accurate measurements under controlled conditions. They can also be very useful for detailed validation of numerical models at the scale considered. Physical models can also make use of existing analogy between equations describing different physical processes (e.g., electric current models for groundwater flow). They were widely used before the development of computers.

Although physical models exist, the most commonly used models are mathematically based. They can be analytical or numerical models. *Analytical models* are mathematical models using exact or approximate analytical solutions to solve the equations describing groundwater flow and solute transport. For example, the Theis solution (See Chapter 5, Section 5.4) is an exact mathematical solution (using the exponential integral function E1) for 2D radially convergent groundwater flow toward a pumping well, which is expressed in radial coordinates with Equation 5.39. To obtain an exact closed-form solution to the flow equation, the Theis solution requires several simplifying assumptions such as homogeneous aquifer properties and uniform aquifer thickness. The simplifying assumptions required by analytical models limit their applicability for natural systems.

Numerical models, on the other hand, are any numerical tools capable of describing the reality with a reliability judged sufficient in relation to the objectives of the study. They can be "black-box" models or process-based models. These latter can be developed in a deterministic or stochastic (probabilistic) framework.

As shown on the general flowsheet in Figure 12.1, the following terminology will be used for a *direct problem* (as opposed to an inverse problem, see below).

- *Stress factors* (or source/sink terms) are considered as *independent variables* of the problem (i.e., entirely resulting from external processes or calculations).
- The *system* (i.e., the geological domain to be simulated) is described by its different *parameters* describing the geometry and the properties. These parameters are chosen according to the simulated processes.
- The *response(s)* which is unique for a deterministic approach and multiple for a stochastic approach, is (are) described by *dependent variables*. Values of these variables are the main expected result of the simulation. They represent the unknowns and prediction of their future values is often the main aim of the study. They are also named the *main variables* of the problem.

An *inverse problem* consists in computing the values of the system parameters (or some of them) from observed values of the dependent variables. This problem will be addressed in more details in Section 12.5 on model calibration.

For a typical groundwater flow problem, the system to model is described by its geometry, its geological structures, and a whole set of hydraulic conductivity and storage values (the latter if transient flow is considered). Stress factors can be recharge

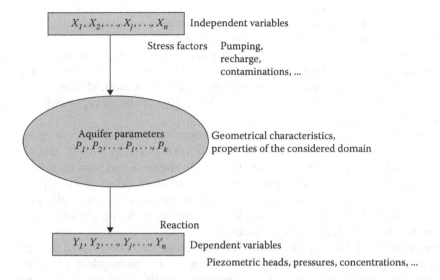

$X_1, X_2, \ldots, X_j, \ldots, X_n$ Independent variables

Stress factors Pumping,
recharge,
contaminations, ...

Aquifer parameters
$P_1, P_2, \ldots, P_l, \ldots, P_k$

Geometrical characteristics,
properties of the considered domain

Reaction

$Y_1, Y_2, \ldots, Y_j, \ldots, Y_n$ Dependent variables

Piezometric heads, pressures, concentrations, ...

Figure 12.1 General schema describing the main terms used in groundwater modeling in a deterministic framework. If a stochastic/probabilistic approach is applied (i.e., using Monte Carlo multiple simulations), the same schema can be used with multiple equally likely sets of parameters, independent variables, and dependent variables.

values or any pumping or reinjection flow rates. Dependent variables (or main variables) are most often the spatially distributed piezometric heads (or the pressure) in the domain.

When a solute transport model is added to the flow model, transport properties (dispersivities, effective transport porosity, diffusion coefficient, degradation constants, partitioning coefficient) are added to the system parameters. Contaminant mass injection or withdrawal could be associated to the source/sink terms as stress factors. Dependent variables are most often the spatially distributed concentrations in the domain.

Note that some *derived dependent variables* can also be calculated from the calculated dependent variables. For example, the groundwater flux can be calculated by multiplying the local first derivative of the main dependent variable (i.e., piezometric head) by the local value of hydraulic conductivity.

In *black-box models*, a set of mathematical equations is developed by empirical or statistical fitting of parameters to reproduce historical records of the main variable. They are referred to as "black-box" models if the equations developed do not physically describe the processes responsible for the observed historical values of the main variable. This means that after the fitting or the "learning stage" of the model, the adjusted values of the parameters have no physical meaning. These models, also named *"data-driven" models* (Anderson *et al.* 2015), have been tried on different hydrogeological problems in particular those deemed too complex for a physically consistent or process-based approach such as, for example, some karstic aquifers. As they do not include any consistent physics, and even if the calibration is made with great care, they can be considered as less reliable for predictions than process-based models. It is particularly the case when nonlinear or coupling effects can be expected (see Section 12.2) and when they will be used with new stresses outside the range of the fitting historical data (e.g., climate change scenarios).

Some authors define *grey-box models* as an intermediate step between black-box and process-based models. An example is given by hydrological models simulating groundwater flow by using bucket reservoirs in series or in parallel: some physics lies in the approach but we are still far from a physically consistent representation of groundwater flow in porous media.

Process-based numerical models consist in solving the appropriate governing equations for the processes considered (flow and solute transport) on a given geological domain with conditions prescribed on the different boundaries and with initial conditions within the domain (i.e., spatial distribution of the initial values of the dependent variable). They can be solved analytically (if simplifications are conceptually acceptable) but most often they are solved numerically using numerical techniques such as finite difference, finite element, or finite volume.

In *deterministic models*, the relationship between cause and effect based on a physical model is unique and deterministic. To a given set of independent variables or stress factors corresponds only one set of dependent variables (Figure 12.1) translating the unique reaction of the system. The philosophy behind using a deterministic approach is that we can rely on a high degree of insight into the processes responsible for the system's response to stress factors. We assume that the calibrated model will provide reliable but unique predictions for any new stress factor even out of range of historically observed values (Konikow and Mercer 1988).

A model is *stochastic or probabilistic* if any of the parameters or stress factors are described by a probabilistic distribution to account for uncertainty. As a model will never exactly match reality, a probabilistic or stochastic approach allows to assess the uncertainty of results. The sources of uncertainty are multiple and of different types. They can be associated to subjective conceptual choices made to simplify the reality into a model (Cooley 2004, Rojas *et al.* 2008, Wildemeersch *et al.* 2014 and many others). They are also embedded in parameters data uncertainty (de Marsily *et al.* 2005, Brunner *et al.* 2012, among many others) and highly parameterized models where parameters value determination (see Section 12.5) represents an ill-posed problem (among others, Carrera and Neuman 1986a, Moore and Doherty 2005, Hill and Tiedeman 2007, Beven 2009). Uncertainty can also result from initial and boundary conditions. For predictions, the uncertainty of the stress factors linked to each simulated scenario can be integrated and accounted for (for example, Rojas *et al.* 2010c, Sulis *et al.* 2012, Goderniaux *et al.* 2015, among many others). A formal stochastic formulation in the partial differential equations for flow and solute transport can be used (see, among others, Dagan 1989, Gelhar 1993, Kitadinis 1997, Zhang 2002, Rubin 2003). In practice, the most commonly-used method is the Monte Carlo simulation with multiple equally-likely realizations of the model parameter sets that are conditioned on the existing data (for example, Vecchia and Cooley 1987, Deutsch and Journel 1998, Huysmans and Dassargues 2006, Tonkin *et al.* 2007, among many others). Multiple simulations are needed and the multiple responses obtained can be statistically treated assuming (most often) Gaussian behavior. The model results are expressed in terms of statistical distributions. The probability distribution for each response is thus based on the statistical distribution of data, parameters, and stress factors.

The aim of Chapters 12 and 13 is mainly to describe the groundwater modeling methodology and the most-commonly used numerical methods, using as a reference the deterministic approach. However, stochastic approaches are increasingly used and they are extensively described and discussed in the specialized literature. These geostatistical aspects and stochastic/probabilistic techniques, together with uncertainty analysis, are summarized for practitioners in Sections 12.6 and 12.7.

Purposes and methodology overview

A numerical model is ideal for integrating in one analytical tool all the knowledge available for a given aquifer or case study. However, modeling is not an end in itself. Modeling can be used to test new theories and hypotheses, to improve our understanding of the concerned processes, to distinguish how different processes interact (Rosbjerg and Madsen 2005), and to make predictions about the future influence of new stress factors. Building an integrated full model representing all possible active processes represents an ill-advised strategy, since the conceptual choices may be too complicated or at least ambiguous. The exact purpose of the model should instead be clearly defined. Will the model be used to make predictions, to help understanding a system, or to exercise the modeler skills? What are the practical questions to be answered through the model's use? Is there a better (i.e., simpler) way to answer these questions? It is indeed not the aim here (or even possible) to enumerate all possible modeling objectives. Except for fundamental scientific purposes, model objectives are most often in line with very practical societal issues related to groundwater quality and quantity, or any human action that could influence or be influenced by groundwater resources.

A general methodological workflow can be described, as shown in Figure 12.2, to illustrate the different steps of a groundwater model construction. These steps include sequentially (among others): definition of the final aim or purpose, conceptual model choices, formulation of the mathematical model, selection/development of a modeling numerical code, model design and data input, calibration and validation, sensitivity analysis during calibration, application (use) of the model for predictions, sensitivity analysis of predictions, analysis of results with respect to the initial question, and writing of a rigorous report.

12.2 Conceptual model

The conceptual model represents the way in which "reality" is simplified in order to be modeled. Simplification is necessary because a complete reproduction of reality is impossible. After a first collection of available data and information about the pursued objectives, one can get a first insight into the physical and hydrogeological behavior of the system. First estimates of the most important relationships can be qualitatively determined and a conceptual model can then be elaborated and formulated to represent the system's functioning. The conceptual model provides a framework for designing, step-by-step, the numerical model (Anderson *et al.* 2015).

Processes to be simulated

Even if models allow to calculate complex interacting processes, the hydrogeological conceptual model remains needed for selection of the most important processes

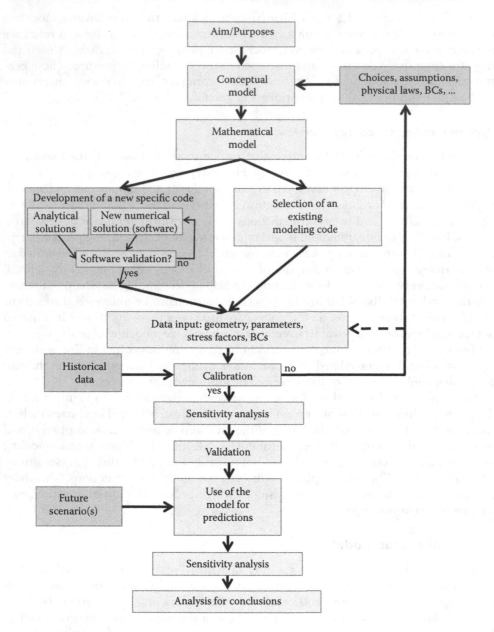

Figure 12.2 General methodological workflow with the different steps of a groundwater model construction.

to be considered at the lab scale, the in situ investigation scale, or the regional scale. Despite some opposing opinions (for example, Chaminé 2015), it is hard to imagine that a multiobjective "total integrated hydrological model" that simulates all coupled processes at all scales could ever be developed. Scientific projects promising this type of result should be considered as naive or at least misleading.

One should determine which processes and, accordingly, which equations must be solved to answer the questions. Note that the choice could be dependent on the scale of interest. In groundwater systems, many coupled processes can potentially affect the expected simulation results. A description of the interactions between different processes is shown in Figure 12.3. The schema illustrates most of the coupling and associated nonlinear effects (i.e., interactions inducing changes in parameters of the system, themselves inducing a change of the main variable, which in turn will influence the parameter) acting as soon as groundwater flow exists. Some of the most important and studied processes are listed below. Some of them were described in more in detail in previous chapters of this book.

Groundwater flow and solute transport coupling (see Chapter 10)

Solute transport is closely dependent on groundwater flow and, in case of transport with high salt concentrations, it could be coupled back to flow resulting from the density effect (i.e., density dependent groundwater flow).

Groundwater flow and geomechanical effects coupling (see Chapter 6)

Changes in groundwater pressures modifies effective stress in the solid part of the medium, inducing geomechanical effects such as consolidation and possible compaction, which can in turn give rise to land subsidence (i.e., induced land subsidence). Consolidation also modifies hydraulic conductivity and porosity, which in turn influences groundwater flow.

Groundwater flow and heat transport (see Chapter 11)

Similar to solute transport, heat transport in groundwater may change groundwater viscosity and density influence groundwater flow and transport conditions.

Groundwater flow, solute transport, and physicochemical reactions

Groundwater flow and contaminant transport may, for specific cases, induce a decrease or increase of the porosity and hydraulic conductivity of the medium due to precipitation and reactive dissolution processes, respectively. These modifications can further modify flow conditions and solute transport.

Other coupling can be invoked on the basis of the processes mentioned in Figure 12.3, such as thermal effects influencing physicochemical reactions and vice-versa, physicochemical reactions influencing the geomechanical behavior, thermal effects influencing the geomechanical behavior, and so on. In strictly rigorous and fundamental terms, one could consider that any groundwater flow could induce effects on all the mentioned processes and each of them in turn could exert feedback on flow and on all the other processes. In practice, given the question, the scale of consideration, and the available data, the most influencing processes should be selected to simulate the "real world" with an acceptable reliability. In most models, only one process or two coupled processes are considered. If more than two fully-coupled processes must be

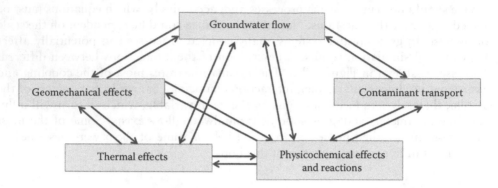

Figure 12.3 In natural groundwater systems, many processes are directly or indirectly coupled to groundwater flow conditions.

considered, the complexity of the model increases and there is a need for specific software that are more often related to scientific expertise than to common engineering practice. An example is the type of model generally used to study the impact of future radioactive waste disposal where all the cited processes (Figure 12.3) are concerned and coupled.

From this first and very important step in the conceptual choices, the equations describing the considered processes are selected. Accordingly, the main parameters together with the dependent and independent variables are identified.

Note that "*poorly justified assumptions can potentially discredit an entire groundwater model*" (Peeters 2017). Each assumption must be identified and its effect on predictions discussed openly in any reporting.

Revisiting the relationship between data, models, and decision-making, Ferré (2017) recently proposed to build model ensembles "*in a way that minimizes the natural multiple-narrative approach that underlies most decision-making processes*" (see Sections 12.6 and 12.7).

Parsimony or complexity

As a consequence of the choices mentioned above, the question of the needed complexity level is clearly an open debate (Wildemeersch 2012). Several papers, such as Hill (2006) and Gómez-Hernández (2006) among others, have addressed the merits and pitfalls of simplicity versus complexity. Nature is infinitely complex. Thus, any process-based groundwater model rapidly becomes complex and, at the same time, remains uncertain. Knowing that different models (i.e., in structure and parametrization) can reproduce similar observed behaviors (i.e., principle of equifinality, Beven and Freer 2001), *complexity* in groundwater modeling should be considered through the use of stochastic approaches conditioned on the available data (Gómez-Hernández 2006). This gives rise to numerous uncalibrated results whose likelihood depends on their capacity to reproduce historical data and observed parameter values within a predefined heterogeneity. Likelihood is assessed using methods such as GLUE (Generalized Likelihood Uncertainty Estimation, Beven and Binley 1992), BMA

(Bayesian Model Averaging, Hoeting *et al.* 1999), MLBMA (Maximum Likelihood Bayesian Model Averaging, Neuman 2003), or GLUE-BMA combination (Rojas *et al.* 2008, 2010a).

If, on the other hand, by *parsimony,* the model is kept as simple as possible (Hill 2006), only the main characteristics and processes are considered in direct relation to the specific prediction purposes. Complexity could be introduced in a stepwise fashion, introducing interacting processes shown to have a significant influence on the results. In this step-by-step progression from simple to complex, an important requirement is to preserve refutability and transparency. Refutability requires that each chosen hypothesis can be tested. Transparency implies that the modeled processes remain understandable. The aim is to increase our objective understanding with a moderate complexity in the model with regards to what is "a priori" known and understood (the *prior,* see Sections 12.6 and 12.7). This can be referred to as *elegant simplicity* (Ward 2005, Schwartz *et al.* 2017). This debate will probably be open for a long time, but with increasing computer simulation efficiency and "cloud computing" (for example, Kurtz *et al.* 2017), we may hope that computing power will no longer be an issue. However, to determine if a simple model provides reliable results, its results should be compared to results from a more complex one (de Marsily *et al.* 2005).

Steady-state versus transient simulations

All natural processes are transient and it would be hard to identify a steady-state natural process. It is also true in hydrogeology, owing to the transient nature of groundwater recharge and other possible source terms (e.g., variable contaminant mass input) of the model. However, practically, to reduce conceptually a real problem to a steady-state problem could save valuable time and energy when this important assumption is acceptable in relation to the existing data and the question to be answered. For example, the steady-state simulation of a worst-case scenario (i.e., clearly conservative with respect to reality) can be envisaged while providing an answer with an unquantified safety (with respect to the question). A justification could be that very few transient data are available for an adequate transient calibration. Indeed, an adequate calibration under transient conditions (see Section 12.5) is needed for producing reliable transient predictions.

For example, a very classical conceptual framework consists in modeling transient solute transport based on a groundwater flow field assumed as steady-state. This allows to simulate the temporal evolution of a solute plume in an assumed steady-state flow field. However, such simplification of the reality should be duly justified as for all other conceptual choices.

If steady-state conditions are chosen, a difficulty may arise in choosing the calibration target values (i.e., calculated from the historical observation data) for the dependent variable and how to assess the steady-state equivalent values for the stress-factors. For example, which piezometric map should be used to calibrate a steady-state groundwater flow model? An averaged map during a given period? The same question arises for calibration on observed discharge rates. In these delicate conceptual choices, one must be consistent (e.g., choose the same period for averaging stress factors and calibration target observed values) and transparent in justifying the chosen assumptions.

Dimensionality of the model

The choice of considering 1D, 2D horizontal, 2D vertical, quasi-3D, or a 3D model could be critical. This depends strongly on the expected influence of the true heterogeneity of the domain, but also on the spatial significance of the stress factors included in the scenarios to be predicted.

1D models

The most simplified approach is indeed a 1D approach assuming that we know and thus impose the groundwater flow direction. This approach is sometimes chosen for computing solute transport along a groundwater flow streamline. A simplified 1D vertical groundwater flow and solute transport model is often considered in the partially saturated zone. This is often recognized as a fair hypothesis given the mostly vertical infiltration and the heavy computational burden of a full 3D nonlinear calculation.

2D horizontal models

Using a 2D horizontal groundwater flow and solute transport is a common approach based on the assumption that, at the regional scale, groundwater flow can often be considered as mostly horizontal. This is known as the Dupuit or Dupuit-Forchheimer assumption (see Chapter 4, Sections 4.11 and 5.4). Equations 4.90 and 4.93 are used for 2D horizontal confined and unconfined flow conditions, respectively. In subhorizontal layered geological contexts, the vertical hydraulic conductivity is often assumed lower than the horizontal conductivity. In fact, at the regional scale, the groundwater pathways are far more horizontal than vertical. The vertical conductivity has thus little influence on the results. Haitjema (2016) mentioned that the Dupuit-Forchheimer assumption is valid when the horizontal groundwater path is longer than $5\sqrt{K_h / K_v}$ times the aquifer thickness, with K_h and K_v being the horizontal and vertical hydraulic conductivity, respectively. However, the assumption becomes rapidly unacceptable if the heterogeneous or structural geological conditions tends to locally force flow to occur vertically in the layer, even after a long horizontal pathway (Figure 12.4). In the case of an aquifer containing horizontal sublayers with contrasted horizontal hydraulic conductivities, the question arises whether to distinguish several horizontal layers or to consider a single layer with equivalent average properties (see Chapter 4, Section 4.5). The choice will depend on the questions to be answered and on the scale of consideration. If the scenarios and detailed expected answers are anticipated to be influenced by the sublayers contrasted parameters, it requires to simulate the processes in detail (i.e., at the scale of the sublayers): a multilayer approach can be adopted (Figure 12.5).

Quasi-3D models

In layered aquifers, most 3D models are actually 3D multilayer models or "quasi-3D models" made of superimposed horizontal 2D layers. Confining low-permeability beds between two more permeable layers (Figure 12.6) are not explicitly included but rather represented by a "vertical conductance" (i.e., vertical hydraulic conductivity

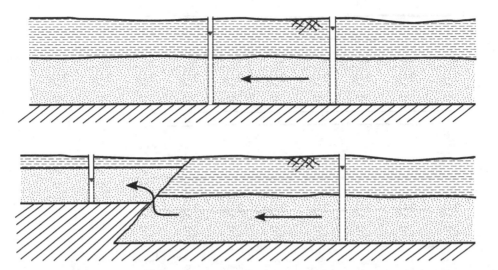

Figure 12.4 For the above situation, a regional 2D horizontal groundwater flow can be considered as nearly independent of the vertical hydraulic conductivity. On the other
hand (below), geological/structural features can drastically change the situation
forcing groundwater to flow along a vertical limited path that will influence strongly
all the regional results.

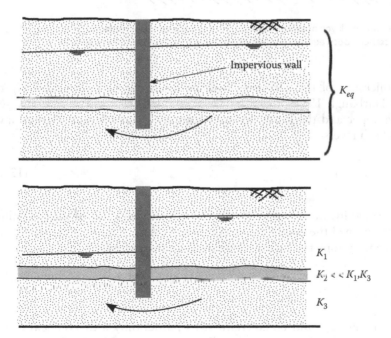

Figure 12.5 The simulated effect of an impervious wall (i.e., the new stress factor) will not be
the same with a single layer 2D horizontal approach using an equivalent hydraulic
conductivity than with a multilayer approach.

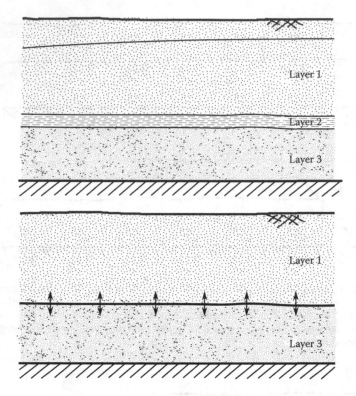

Figure 12.6 A confining layer between two aquifers is conceptually replaced by a "vertical conductance" between both aquifers allowing to calculate a vertical flux.

divided by the thickness of the confining bed) between the two permeable layers (McDonald and Harbaugh 1988). In the quasi-3D model, the vertical groundwater flux between the upper and the lower layers is computed according to Darcy's law applied between both layers:

$$q_v = \frac{K_v}{b}(h_{upper} - h_{lower}) \tag{12.1}$$

where K_v is the vertical hydraulic conductivity and b is the thickness of the confining layer (K_v/b is often named the conductance coefficient), h_{upper} and h_{lower} are the time and space dependent piezometric heads respectively in the upper and lower aquifers, respectively.

2D vertical models

In some circumstances, it could be convenient to model only a 2D vertical cross section (profile model). In this approach, all groundwater flow (and transport) components not located in that vertical section are neglected. One therefore

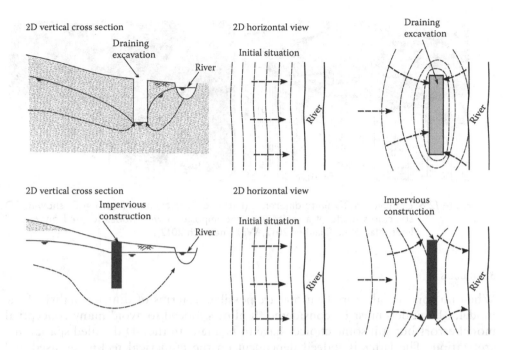

Figure 12.7 Examples of 2D vertical models often used in relation with civil engineering works. On the right, the view in the horizontal plane illustrates how groundwater flow components not located in the cross section may occur at the extremities of the engineered structure.

only considers groundwater flow streamlines in a chosen vertical plane, and assumes that the direction of the streamline would never be affected by the scenarios to be calculated as predictions. The thickness of the vertical profile is set by default to one meter. Equation 4.84 is the used 2D vertical flow equation. Many examples can be found in civil engineering applications where the main groundwater flow component is assumed orthogonal to the structure length (Figure 12.7). The groundwater flux is computed under a linear impervious structure (e.g., a dam or any impervious structure) or toward a draining structure (e.g., excavation or drain) and expressed per meter of length. The flux is then multiplied by the total length of the engineered structure assuming that the geological conditions are similar everywhere. This type of simplifying assumption can be justified for a first assessment when detailed data are not available. However, for real heterogeneous cases and if the length of the linear engineered structure is small with respect to the other dimensions, there are significant groundwater flow components outside of the considered vertical plane. A 3D modeling approach should then be adopted. Other classical examples of 2D vertical conceptual models are the vertical cross sections considered for modeling seawater intrusions in coastal aquifers (at least for the early steps of the modeling; see Section 10.2).

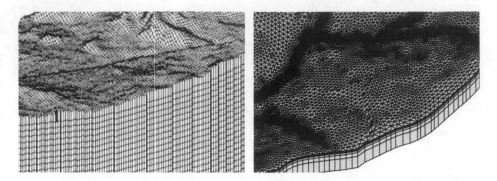

Figure 12.8 Examples of 3D finite difference (left) and finite element (right) grids showing 3D structures made of a multilayer superimposition with nodes located on vertical lines (Madioune Diakher 2012, Wildemeersch 2012).

3D models

When no conceptual simplifications are possible in terms of dimensionality of the model, 3D models must be considered. It allows indeed to avoid many conceptual model errors but still some choices have to be made in the 3D detailed spatial discretization. The latter is indeed dependent on the numerical technique used and associated type of mesh (i.e., finite difference blocks or finite elements; see Chapter 13). In hydrogeology, most discretized 3D models can be qualified as multilayer models. Most software (except dedicated specific tools) only allow to superimpose several model layers with nodes located along vertical lines. Therefore, any local grid refinement needed for one layer is propagated vertically to all other layers (Figure 12.8).

Conceptual choices for fractured and karst media

As mentioned in Chapter 4, if the medium is fractured or karstified, a model based on the EPM (Equivalent Porous Media) conceptual approach may not always be applicable. Other specific conceptual models have therefore been developed for flow and transport, such as the discrete fracture approach or the double continuum approach.

Groundwater flow and solute transport in fractured media

Flow and solute transport can be computed in interconnected *discrete fractures (DFN—Discrete Fracture Network models)*, using the cubic law (Equation 4.40) for calculating flow in each fracture. This approach can be generalized for partially saturated conditions or for a two-phase flow (Pruess and Tsang 1990). The rock matrix is assumed completely impermeable as flow and transport are simulated only in the fracture network. Fracture surfaces are subdivided into subelements and a spatially-correlated aperture is assigned to each of them. Elemental saturations for the wetting and the nonwetting phases depend on calculated pressures and the prescribed entry pressure relationship (Pruess and Tsang 1990, Therrien *et al.* 2006). Relative permeabilities are obtained and recalculated at each time step. Usually, the specific storage coefficient for a fracture under saturated conditions is assumed to only depend on water compressibility (i.e., the deformation fracture compressibility is neglected). Similarly, solute transport can be calculated for saturated or partially-saturated conditions using Equations similar to 8.43 or 9.21, respectively, adapted for a 2D fracture plane.

The *dual continuum approach* involves two separate continua with a first continuum being a porous medium and a second representing fractures, macropores, conduits that can be found in true fractured and karstified media (Gerke and Van Genuchten 1993, Therrien *et al.* 2006). These two continua are linked by exchange terms computed at each connecting node and after each time step. Equations solved in the first continuum are classical porous medium flow and solute transport equations. Indeed, in the second medium (fractures) the corresponding equations to be solved are expressed taking the specific characteristics of the fractures parameters. Thus, this approach is more elaborated than the double porosity model (i.e., mobile and immobile water porosities) as described in Chapter 8, Section 8.2 for matrix diffusion/immobile water effect. It implies here also a double permeability concept. Contrary to transport including matrix diffusion (or double porosity models), this approach is not restricted to steady-state conditions because it allows transient exchanges between the two interacting continua (Therrien *et al.* 2006). The only general condition to be respected is that fractions of the total porosity occupied by each continuum are given at each location and the sum must be equal to 1.

More details on these specific conceptual models can be found in (among others): Wang and Narasimhan (1985), Berkowitz *et al.* (1988), Sudicky and McLaren (1992), Therrien and Sudicky (1996), Therrien *et al.* (2006), and Diersch (2014).

Groundwater flow and solute transport in karstified systems

Groundwater models for karst aquifers are particularly challenging as the specific multiple-scale heterogeneity of karstic systems requires a lot of characterization data to hope to be able to use a physically based approach. For groundwater flow, the combination of a very slow component through the rock matrix and a very fast component through conduits and open fractures is particularly difficult to be modeled. Transport is indeed very dependent on these flow conditions with preferential pathways. One paradox is that the karst conduits represent only a few percent of the total porosity but have a major influence on flow and transport (Ghasemizadeh *et al.* 2012).

Representing spatially distributed and transient diffuse or point recharge is also conceptually challenging. Is diffuse infiltration through permeable outcrops the main recharge of the system? Or is it, on the other hand, the preferential infiltration points? Detailed data are needed to properly conceptualize recharge by accounting for (among others) stream-losses.

Various modeling approaches have been tested and extensive scientific literature exists on this topic. To summarize, conceptual approaches include EPM (equivalent porous media) based techniques, lumped parameter models, dual porosity, dual porosity and dual permeability approaches (dual continuum), and discrete conduits and open fractures approaches.

In *EPM models,* one considers that similarly to a porous medium, equivalent hydraulic conductivities that are spatially distributed can represent the true heterogeneity induced by interconnected conduits and fractures (among others, Long *et al.* 1982, Pankow *et al.* 1986, Scanlon *et al.* 2003, Bodin *et al.* 2012, Abusaada and Sauter 2013). The lack of precision in the representation of reality strongly depends on the scale at which the problem is considered (Dassargues 1998). At the regional scale and in slightly karstified media, it can be considered conceptually acceptable to represent the heterogeneity by contrasted parameters values for the different cells/elements of the 3D grid. Special finite elements, such as 1D conduit elements or 2D fracture elements, can also be used within the domain discretized with 3D elements (Dassargues *et al.* 1988). For regional aquifers, heterogeneity can also be reflected in storage coefficient values and in transport parameters: EPM-based models can be used for transient flow and for solute transport simulations. On the other hand, at the local scale, in highly karstified aquifers, groundwater flow and contaminant transport are too dependent on preferential flow in the karst conduit network and, consequently, other approaches should be used.

In *lumped parameters models*, a combination of interconnected "black- or grey-box" models that can be mixing cells or linear reservoirs is used for representing entire zones or even subcatchments in the karst network (among others, Barrett and Charbeneau 1997, Brouyère *et al.* 2011). Quite simpler than EPM models, they do not need high data requirements and computation runs are rapid. They can be used for globally reproducing spring discharge (for example, Geyer *et al.* 2008, Fleury *et al.* 2009), but not for piezometric heads simulations. Lumped parameters models may require data from large range flow conditions to be accurately calibrated (Scanlon *et al.* 2003). Mixing cells of lumped models may lead to unreliable solute transport results when compared to the rapid advective transport associated to the preferential flowpaths.

As described above for fractured aquifers, *dual porosity and dual permeability models* (dual continuum approach) allow to consider two separate continua. They are by far more reliable than EPM or lumped models for simulating solute transport. Still, such models require the necessary detailed data to parameterize these two media, which is often difficult to obtain.

As for fractured media (see above), in discrete conduits and open fractures models, many data must be available about conduits and fractures geometries, and hydraulic properties. This parametrization of the conduits network can be envisaged in a deterministic or stochastic context (among others, Eisenlohr *et al.* 1997, Borghi 2013). Practically this is only feasible at the very local scale (Halihan and Wicks 1998, Jeannin 2001). The advantage is that both laminar and turbulent flows could be considered in conduits represented explicitly by 1D, 2D, or 3D elements.

Interactions with surface waters and integrated models

For most groundwater models, recharge is introduced as a boundary condition directly in the saturated zone. The spatial distribution and time variation should be estimated separately with various degrees of complexity and reliability (see Chapter 2). To be more physically-based, some groundwater models include partially saturated flow (and transport) processes for calculating recharge (and its contaminant content) from infiltration at the soil surface. However, none of these approaches allow to simulate the feedback, in terms of water exchange, between the surface and subsurface domains. This feedback can be an important component of the water cycle particularly when groundwater heads are close to land surface, and finally when river-aquifer interactions may occur in both directions. Computation of simultaneous solutions in the surface and subsurface domains enables a better representation of the whole system (among others, Singh and Bhallamudi 1998, VanderKwaak 1999, Panday and Huyakorn 2004, Goderniaux *et al.* 2009, Irvine *et al.* 2012). This approach also permits solutes to be naturally exchanged with water fluxes between the surface and subsurface domains (Therrien *et al.* 2006, Sudicky *et al.* 2008, Schilling 2017). Detailed physically-based and spatially-distributed models that take into account the whole hydrological system are referred to as *integrated models.* They provide theoretically more realistic simulations of groundwater fluxes, including dynamic time varying exchanges with surface water. They are particularly realistic in the context of climate change impact assessments. The fully-integrated surface-subsurface models require a large amount of detailed spatially distributed data characterizing the subsurface and surface domains but also characterizing the evapotranspiration multiple processes. These modeling approaches are considered by some researchers as being over-parametrized. However, if an appropriate sensitivity study (see Section 12.5) is carried out, priorities can be identified to considerably reduce the number of parameters to be calibrated. Despite the continuous and spectacular progress of computing processors, these integrated models remain heavy to compute. Parallel procedures, requiring parallelization of the code, should be preferred. Cloud computing is also a new trend for managing, in an effective and interactive way, data assimilation and computing for such integrated models (Kurtz *et al.* 2017).

Other choices and assumptions

After having decided about processes to be simulated, transient or steady-state conditions, and the dimensionality of the model, fundamental choices still need to be made about the scale, the actual domain that is included in the model, and the location and conditions to be prescribed at the domain boundaries. The latter point is described in Section 12.3.

There could be a series of additional assumptions or decisions to be made during the construction of the model but the main choices described above must be decided before selecting the software to be used. This selection should be done by checking if the processes to be simulated (with their corresponding equations) can actually be solved with the intended numerical tool. Note that the decision process must be done in this way and not the other way around. Too often, conceptual models are designed with oversimplifying assumptions to allow solution by a specific tool (e.g., an already acquired or developed software). For complex coupled processes, a specific model could be needed and accordingly developed by scientific teams and specialized companies. Any new numerical tool must however be validated step-by-step (i.e., from simple to complex) solving a series of benchmark case studies available in the specialized literature that were solved previously with analytical and/or other already validated models. Note that intercomparison modeling exercises (i.e., with different models) are always very delicate. They must be performed with the greatest level of care and rigor, because the slightest change in assumptions, even implicitly embedded in the used software, can create a big difference in results.

12.3 Initial and boundary conditions

Initial conditions and boundary conditions (BCs) must be specified. They will be discussed here for groundwater flow and solute transport problems.

Initial conditions

In general, initial conditions for the main time-dependent variable are needed. In practice, for any transient simulation, initial maps of piezometric heads and solute concentrations are needed for groundwater flow and solute transport simulations, respectively. In the particular case where the boundary conditions, hydraulic parameters and stress factors are assumed to be known, initial boundary conditions could be deduced from inverse calculation (see Section 12.5; Hassane Maina *et al.* 2017).

Initial conditions are also needed for an iterative solution of a steady-state problem. Any nonlinear problem is solved iteratively and thus requires initial conditions. A typical example is given by a 2D horizontal groundwater flow problem in an unconfined aquifer where the transmissivity $(T = Kh)$ is dependent on the main dependent variable (h) (See Chapter 4, Sections 4.8 and 4.11, Equation 4.93). An iterative procedure is used for *linearizing* the problem. It implies a first set of h values to be associated with the K values in the modeled domain for obtaining initial values of T. As soon as a new value for the h field is computed, new values for T can be calculated and so on. Note that this iterative calculation loop is most often located inside the time-step loop in the computational flowsheet.

Other typical examples of nonlinear problems include variable density (Chapter 10) and variable saturation (Chapter 9) flow and transport. Initial values for state variables as groundwater density (and possibly viscosity) are needed for density dependent flow and transport modeling and under nonisothermal conditions. For a variably saturated problem, initial water content (or saturation) must be known and specified for all nodes of the grid.

Boundary conditions

The choices concerning where to locate the domain boundaries and what to impose on these boundaries are obviously linked. In continuum physics, boundary conditions (BCs) should reflect as much as possible the continuity of the simulated processes between the modeled zone and the outside world. If possible, lateral and bottom model boundaries should correspond to a change in physical conditions that is not sensitive or dependent on the varying hydrogeological conditions. If this is not the case, a general rule is to keep the lateral boundaries as far as possible from the particular area where the model results are expected (including predictive scenarios) in order to keep these results significant. BCs for solute transport will be discussed after having described BCs for groundwater flow problems as their choice may be highly dependent on the adopted choice for the first ones.

Groundwater flow boundary conditions

DIRICHLET BOUNDARY CONDITIONS

A *Dirichlet-type BC* or *first-type BC* corresponds to prescribing the value of the main (dependent) variable of the modeled process on the given boundary. For groundwater flow, Dirichlet BCs are prescribed piezometric heads at each node of the given boundary:

$$h(x,y,z,t) = f'(x,y,z,t) \qquad (12.2)$$

where $f'(x, y, z, t)$ is a known function dependent on coordinates x, y, z and also possibly dependent on time (i.e., the prescribed head value at a given location can be different from one time step to another). Where a piezometric head is prescribed, the numerical code calculates the groundwater flux (i.e., its only freedom) with the computed local gradient and the local value of hydraulic conductivity. Examples are given in Figure 12.9. They have been chosen for drawing the attention on important bias that could be introduced using these BCs in some circumstances.

a. If prescribed heads are assigned at the interface boundary between a river and the simulated aquifer (Figure 12.9a), any variation of the river water level (i.e., considered here as the external world for the groundwater model) is therefore assigned to the groundwater model with different values of prescribed heads for

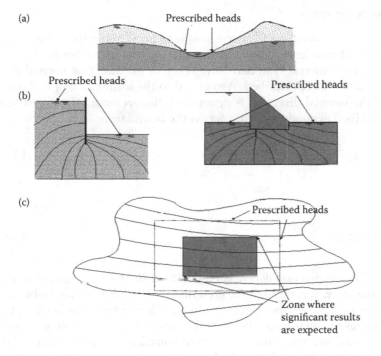

Figure 12.9 Examples of Dirichlet or first-type BCs. Some of these examples are chosen for pointing out issues linked to these BCs in specific cases.

each time step. If variations are frequent and rapid, the time steps must be numerous and short. With this BC, the hydraulic conductivity of the banks and bottom of the river are implicitly assumed so great that any river water level change is immediately transmitted to the aquifer.

b. A typical way of calculating groundwater flux under a civil engineering work (Figure 12.9b) consists in prescribing upgradient and downgradient piezometric heads. If the latter are specified (e.g., at the land surface elevation) a corresponding groundwater flux is computed. Upgradient prescribed heads are well justified for a dam (right side of Figure 12.9b), however, in the case of a slurry wall (left side of Figure 12.9b) it does not take into account the possible decrease of the upgradient heads close to the wall, resulting from groundwater flux under the wall. Note that the lateral (on the left side) upgradient boundary has not been discussed so far.

c. When a local model is considered within a regional aquifer, the temptation is great to prescribe heads on all the boundaries (Figure 12.9c). With that choice, main dependent variable values are prescribed all around so that the model results will be more dependent on the prescribed head values than on the hydraulic conductivity spatial distribution inside the domain. Clearly, for the sake of a reliable calibration and realistic simulation of new scenarios (e.g., with pumping in the modeled zone), prescribed heads should be at least carried over at a larger distance further from the zone where significant predictive results are expected. Alternatively, other types of BCs should be chosen (see next paragraphs).

NEUMANN BOUNDARY CONDITIONS

A *Neumann-type BC* or *second-type BC* corresponds to prescribing the value of the derivative of the main (dependent) variable orthogonal to the given boundary. For groundwater flow, the piezometric head derivative taken in the direction normal (i.e., orthogonal) to the boundary is prescribed. Associated to the local value of hydraulic conductivity, and if the normal direction is denoted by the vector n, Neumann BCs correspond to a prescribed groundwater flux across the boundary:

$$\nabla h \cdot n = \frac{\partial h}{\partial n}(x, y, z, t) = f''(x, y, z, t) \tag{12.3}$$

and

$$K \frac{\partial h}{\partial n}(x, y, z, t) = q''(x, y, z, t) \tag{12.4}$$

where $f''(x, y, z, t)$ is a known function dependent on the coordinates x, y, z and also possibly dependent on time. $q''(x, y, z, t)$ is thus the prescribed normal flux across the boundary (this can be different from one time step to another). Where a flux is prescribed, the numerical code calculates the piezometric head. This type of BC is most often used with $q''(x, y, z, t) = 0$ for describing boundaries that can be considered as impermeable (i.e., a no-flux BC). Examples are given in Figure 12.10. They have been chosen for pointing out some possible bias that could be introduced using Neumann BCs.

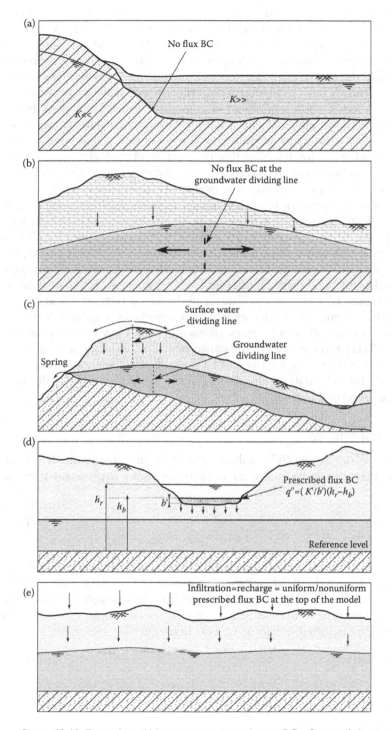

Figure 12.10 Examples of Neumann or second-type BCs. Some of these examples are chosen for pointing out issues linked to these BCs in specific cases.

a. A no-flux Neumann BC is typically used for lateral or bottom boundaries cor-
    responding to low permeability geological formations (Figure 12.10a). A contrast
    greater than 10^3 in hydraulic conductivity is often considered as fully justifying
    this type of assumption. However, for less contrasted K values, it could be interest-
    ing (if the required data are available) to assess a nonzero groundwater flow condi-
    tion by a simple Darcy's law calculation within the low permeability medium.
b. In some circumstances, a no-flux BC could be chosen for describing a groundwater
    divide within a continuous aquifer (Figure 12.10b). Most often used in regional mod-
    eling, at the catchment scale, a null groundwater flow is assumed across a line of high-
    est piezometric head in the aquifer. Note that the location of the highest piezometric
    heads results from nonuniform and variable recharge conditions. If the problem is
    considered under transient conditions, or if new stress factors are considered (e.g.,
    pumping), the highest piezometric heads could move from one location to another. In
    such a case, it is clearly not feasible to move the boundary location at each time step.
c. In the case of Figure 12.10c, the continuous movement of the groundwater divide as
    a function of recharge does not allow using that boundary for specifying a zero flux.
    For that case, other BCs should be chosen specifying that the groundwater outflow
    at the spring is dependent on the local piezometric head gradient (See Cauchy BCs).
d. In the case of Figure 12.10d, a surface water body (i.e., a river or a lake) is not
    interacting with a depressed aquifer except that it induces a greater localized
    recharge. Prescribed flux BC can be assessed (using Darcy's law through the bot-
    tom of the river/lake) and specified in the concerned zone of the modeled aquifer.
e. Any infiltration recharging the aquifer can be prescribed as Neumann BCs. This
    can be uniformly (or not) spatially distributed at the top of the modeled domain.
    Indeed, this can be varied in time.

CAUCHY BOUNDARY CONDITIONS

A *Cauchy-type BC* or *third-type BC* combines Dirichlet and Neumann BC: a linear relation of the piezometric head and its normal derivative is prescribed on the
boundary:

$$a\frac{\partial h}{\partial n}(x,y,z,t) + bh(x,y,z,t) = f'''(x,y,z,t) \tag{12.5}$$

where $f'''(x, y, z, t)$ is a given function dependent on the coordinates x, y, z and also
possibly on time, a and b are coefficients. Prescribing such a BC corresponds to let the
groundwater flux across the boundary being dependent on the calculated piezometric
head and vice-versa. This type of BC can be very general. For example, if we consider
the particular situation described in Figure 12.11a, Equation 12.5 becomes:

$$-K\frac{\partial h}{\partial n} + \frac{k'}{b'}h = \frac{k'}{b'}h_r \tag{12.6}$$

and

$$q'' = -K\frac{\partial h}{\partial n} = \frac{k'}{b'}(h_r - h) \tag{12.7}$$

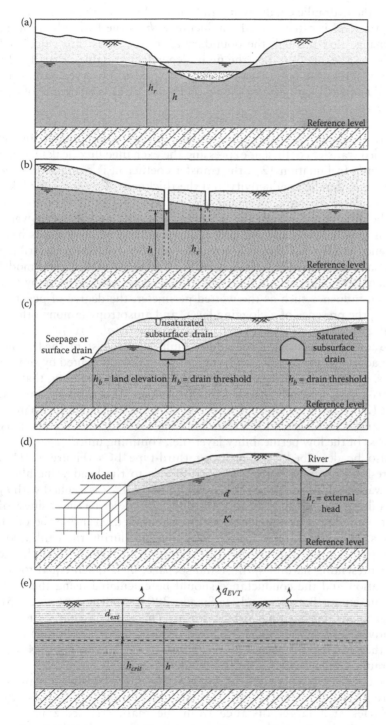

Figure 12.11 Examples of Cauchy or third-type BCs. Some of these examples are chosen for pointing out some dangers of using these BCs in specific cases.

where K' and b' are the hydraulic conductivity and the thickness of the river bottom, respectively, the ratio K'/b' is often named conductance, h_r is the hydraulic head in the river. The exchange flux through the boundary $q''(x, y, z, t)$ is thus calculated as a function of the local value of h (i.e., the main dependent variable). Usually, the computed water flux across the boundary can be compared to observed values for calibrating the conductance. Examples are shown on Figure 12.11 with the use of this type of BCs in different situations.

a. A classical way of conceptualizing groundwater—surface water interactions is to use a simplified transfer coefficient expressing the exchange flux (leakage) as a linear relation of h. In Equation 12.7, the transfer coefficient is expressed as the ratio between the hydraulic conductivity and the thickness of the river (or lake) bottom. The calculated discharge or recharge of the modeled aquifer is thus proportional to the difference between the piezometric head h and the river water level h_r. This is only a rough approximation of the reality. Many authors have published detailed studies for showing how this transfer coefficient describing river seepage conductance should be quantified for different situations and model grids (Harbaugh 2005, Therrien *et al.* 2006). It has been shown that many factors can have an influence such as the wetted perimeter, the degree of penetration of the river, the presence of a clogging layer, and anisotropy (among others, Cousquer *et al.* 2017, Morel-Seytoux *et al.* 2017, Brunner *et al.* 2017).

b. If only a single aquifer is explicitly modeled in a multi-aquifer system, the leakage from the other aquifers through the confining units can be represented by similar linear relations. For the case illustrated in Figure 12.11b, and assuming that the upper aquifer is not significantly affected by this leakage, this can be expressed as in Equation 12.7 where h_r is replaced by h_s the piezometric head in the shallow aquifer. The transfer coefficient is the ratio between the hydraulic conductivity and the thickness of the low permeability layer (i.e., confining unit).

c. Drains can also be described by Cauchy or third-type BCs (Figure 12.11c). The drain threshold or drain bottom h_b corresponds to the head value above which groundwater is drained out of the model. Equation 12.7 is applied with h_b replacing h_r. A drain conductance must be introduced to describe the easiness of groundwater to be drained as a function of the materials used inside the drain and drain walls. In practice, its value is often adjusted during the calibration process of the model comparing modeled discharge to the observed discharge flows. When representing natural springs and seepage, h_b is taken equal to the local land elevation and the conductance should be calculated using the local hydraulic conductivities (Batelaan and De Smedt, 2004, Anderson *et al.* 2015). This is convenient for representing surface drains. The water discharged to the drain is definitively leaving the model. If closed drains (subsurface drains) must be simulated, they can be represented by very local (mostly linear) features with very high hydraulic conductivities inside the model.

d. With reference to the case where a local model is considered within a regional aquifer (Figure 12.9c), Cauchy-type BCs are also an elegant way to prescribe lateral boundaries at a larger distance from the main stressed-zone (i.e., where significant predictive results are expected from the model). Called *general head boundaries* (GHB), they consist in prescribing an "external head" (i.e., not on the true boundary but outside the modeled zone) so that a groundwater flux across the boundary is computed from the difference between this "external head" and the piezometric head on the model boundary using a given conductance (Figure 12.11d). The conductance is calculated from the assessed regional hydraulic conductivities in the zone around the model and the distance to the "external prescribed head." As for the case of Figure 12.9c, "professional judgment is required to decide how far a physical boundary can reasonably be located from the perimeter of the model" (Anderson *et al.* 2015). GHBs have the advantage to avoid useless extension of the model mesh in zones of lower interest (related to the aim of the study) and where data are usually relatively sparse.

e. In arid zones and during very dry and hot periods in temperate zones, evaporation and transpiration (ET) may occur both in the partially saturated zone and also directly in the saturated zone. This is the case if the water table (i.e., the saturated zone) is close to the land surface. Cauchy BCs can be used to represent an evapotranspiration flux leaving the model but dependent on the "depth to water" (i.e., the land surface elevation minus piezometric head). An extinction depth d_{ext} (Anderson *et al.* 2015) corresponding to a critical head h_{crit} can be defined so that ET occurs only if the water table is higher (Figure 12.11e). The vertical discharge water flux corresponding to ET [LT^{-1}] is expressed by:

$$q_{ET} = \frac{R_{ET}}{d_{ext}} (h(x,y,z,t) - h_{crit}(x,y,z,t)) \tag{12.8}$$

where R_{ET} is the maximum evapotranspiration rate in ms^{-1} [LT^{-1}]: the evapotranspirated water column per time unit. The maximum discharge ET flux occurs when h is nearly equal to the land surface elevation (i.e. $(h - h_{crit}) \cong d_{ext}$) and q_{ET} is given a zero value for $h \leq h_{crit}$.

Solute transport boundary conditions

Transport BCs are complementary to groundwater flow BCs. Indeed, prescribing solute mass flux or concentration cannot be done without considering what happens in terms of groundwater flow across the boundary.

DIRICHLET BOUNDARY CONDITIONS

For a solute transport problem, *transport Dirichlet BCs* correspond to prescribed concentrations at each node of the given boundary:

$$C(x,y,z,t) = g'(x,y,z,t) \tag{12.9}$$

where $g'(x, y, z, t)$ is a known function dependent on coordinates x, y, z and also possibly dependent on time (i.e., the prescribed concentration can be different from one time step to another). Where concentration is prescribed, the computed dispersion-diffusion flux is a function of the computed local concentration gradient and the local values of effective diffusion coefficient and dispersion. Examples are given in

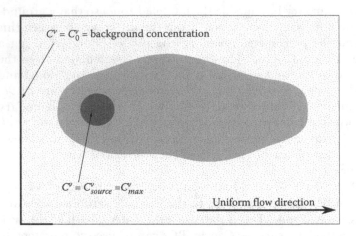

Figure 12.12 Example of Dirichlet transport BCs corresponding to prescribed concentrations for the case of an assumed uniform flow field.

Figure 12.12, pointing out conceptual and numerical issues that can be linked to the first-type transport BCs.

a. It is common to prescribe a zero or a background solute concentration C_0^v at the boundaries located upgradient with respect to the groundwater flow and advection direction (Figure 12.12). This is usually combined with flow BCs allowing an incoming groundwater flux across the boundary (i.e., the water input has a specified concentration). However, precautions must be taken to ensure that this boundary is located far enough from the area where the pollutant plume is expected for all simulated scenarios. If the uncertainty associated with the groundwater flow field is large, then conservative choices (i.e., in terms of distance and locations) should be made.

b. One of the main issues when modeling solute transport for a real contamination case, is to assess the true pollutant source term in Equations 8.43 and 8.44 (expressed as in Equation 8.48: $M_s = q_s C_s^v$). Clearly, in practice, this information is difficult to assess. On the other hand, measured concentrations as function of time are often available at a location considered as corresponding (or very close) to the "contamination source zone." Then, an indirect way of specifying the pollutant input is to prescribe inside the model the (high) observed groundwater concentrations. These C_{max}^v prescribed concentrations will be logically the maximum concentration values of the problem (Figure 12.12). They will be often distributed on different nodes underlying the assumed contamination source zone (i.e., assuming that vertical transport occurred between the land surface and the saturated zone). Unfortunately, prescribing very high concentrations inside a flow field with low preexisting background concentrations will induce unavoidable numerical (artificially high) dispersion resulting from the sudden prescribed concentration gradients. Moreover, serious solute mass conservation problems are usually observed in computed results. This way of introducing the pollutant input in the model is therefore not recommended.

NEUMANN BOUNDARY CONDITIONS

A Neumann-type BC or second-type transport BC corresponds to prescribing the value of the normal derivative of the concentration on a given boundary:

$$\nabla C^v \cdot n = \frac{\partial C^v}{\partial n}(x,y,z,t) = g''(x,y,z,t) \tag{12.10}$$

where $g''(x, y, z, t)$ is a known function dependent on coordinates x, y, z and possibly also on time. In fact, associated to local values of the effective diffusion coefficient and dispersivities, it corresponds to a prescribed hydrodynamic dispersion mass flux across the boundary (see Equation 8.23):

$$n \cdot (-n_m D_h \cdot \nabla C^v) = -n_m D_{h,n} \frac{\partial C^v}{\partial n}(x,y,z,t) = q''(x,y,z,t) \tag{12.11}$$

where $q''(x, y, z, t)$ is the prescribed hydrodynamic dispersion mass flux normal to the boundary (this can be different from one time step to another), $D_{h,n}$ is the dispersion in the direction normal to the boundary.

It is usually hard to assess a nonzero hydrodynamic dispersion mass flux to be prescribed on a boundary. So, this type of BC is essentially used for prescribing a zero-dispersion flux through a boundary by using $\partial C^v/\partial n = 0$. Associated to a zero Neumann BC for flow, it then corresponds to a totally impervious boundary. On the other hand, when combined with a prescribed head (Dirichlet BC for flow), only an advective mass flux is allowed across the boundary. In the case shown in Figure 12.13, both upgradient and downgradient BCs for flow are prescribed heads with the other

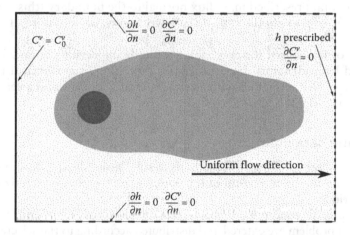

Figure 12.13 On the same example than in Figure 12.12, Neumann transport BCs prescribing a zero-hydrodynamic dispersion mass flux through the boundary ($\partial C^v/\partial n = 0$). Combined with a prescribed head (Dirichlet BC for flow), only an advective mass flux is allowed across the downgradient boundary (on the right-hand side of the figure). Combined with zero Neumann BCs for flow ($\partial h/\partial n = 0$), totally impervious BCs are prescribed.

lateral boundaries being totally impervious (Neumann zero BCs for flow and solute transport). The upgradient transport BC is a prescribed concentration (as discussed in Figure 12.12) but the more delicate choice concerns the downgradient transport BC. Ideally, one would prefer to prescribe neither concentration nor flow. However, in practice a condition must be specified and, although there is no perfect option, the one that introduces the least bias in the simulated plume consists in prescribing a (zero) hydrodynamic dispersion flux on this downgradient boundary. The influence on the simulated plume could be considered as minimum if the boundary is chosen sufficiently far from the zone where significant results are expected.

Cauchy boundary conditions

A *Cauchy-type transport* BC or *third-type* BC corresponds to a prescribed linear combination of the concentration and its first derivative on the given boundary:

$$a\frac{\partial C^v}{\partial n}(x,y,z,t) + bC^v(x,y,z,t) = g'''(x,y,z,t) \tag{12.12}$$

where $g'''(x, y, z, t)$ is a given function dependent on coordinates x, y, z and possibly also on time, a and b are coefficients. Prescribing such a BC corresponds to specify the sum of the advection and hydrodynamic dispersion mass fluxes across the boundary:

$$\boldsymbol{n}\cdot(q C^v - n_m \boldsymbol{D}_h \cdot \nabla C^v) = q_n C^v(x,y,z,t) - n_m D_{h,n}\frac{\partial C^v}{\partial n}(x,y,z,t)$$

$$= q'''(x,y,z,t) \tag{12.13}$$

where $q'''(x, y, z, t)$ is the total prescribed mass flux normal to the boundary (this can be different from one time step to another), q_n is the advection water flux normal to the boundary.

Most often, this type of BC is used for prescribing a totally impervious boundary with zero advective and hydrodynamic dispersion mass flux. Note it is identical to specifying a combination of Neumann BCs with a zero-water flux for flow and a zero-dispersion flux for transport (Figure 12.14).

12.4 Model design and data input

As will be described in Chapter 13, numerical methods to solve groundwater flow and solute transport require spatial discretization of the domain and temporal discretization of the simulated period.

Using the chosen software, the conceptual model is translated into practice by entering the data. All the data for a problem are entered and distributed according to the selected mesh: the spatial discretization. This discretization is chosen according to the geological and geometrical variations, the variations in space of the parameter values, the initial conditions, and the chosen boundaries. If the problem is to be solved for transient conditions, the chosen temporal discretization must account for the time evolution of the stress factors (pumping, injection, recharge, contamination, and so on) and possible identified

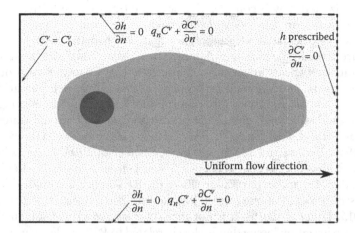

Figure 12.14 On the same example as previously, Cauchy transport BCs prescribing a zero total mass flux through the boundary ($q_nC'-\partial C'/\partial n=0$) are totally equivalent to the combination of zero-flow Neumann BCs and zero-transport Neumann BCs for prescribing totally impervious conditions.

variations in parameters values. An initial state of the system must be entered containing the initial values of the parameters, stress factors, and main variables.

Data required for hydrogeological modeling can be generally classified into four main categories:

a.   the geometry of the system and of the different geological formations
b.   values of the hydrogeological parameters involved in the equations of the simulated processes
c.   the stress factors applied to the system, and
d.   historical data (i.e., measured or observed values) about the main variables of the problem.

Geometry and geology

For defining the geometry of the modeled system, the following data may be required or recommended: maps and geological cross sections showing the horizontal and vertical distribution of the different lithological formations and the 3D boundaries of the system, land surface map with location of the surface water bodies and hydrological surface boundaries, and contour maps for the bottom and top of the aquifers and aquitards levels (or maps of their respective thickness).

Hydrogeological parameters

Values and initial distribution of the hydrogeological parameters must be derived from hard-data measurements and interpretation based on geological and hydrogeological knowledge. It concerns hydraulic conductivity (K) and specific storage coefficient (S_s) for 3D groundwater flow modeling under transient conditions (only hydraulic conductivity is needed for steady-state flow simulations) or transmissivity (T) and storage coefficient

(S) if 2D groundwater horizontal flow is considered. For solute transport, effective diffusion coefficient, effective transport porosity and dispersivities are required for solving a classical ADE. If adsorption, degradation, immobile water effect (matrix diffusion), and reactive transport are considered, then retardation factor, decay constants, matrix diffusion, and reaction rate coefficients are needed, respectively (see Chapter 8, Section 8.2). For most parameters, scale issues should be properly addressed for defining grid equivalent values from field and lab tests with different support scales. This challenging issue has been, for years (and still now), one of the hottest topics in the groundwater modeling literature (among others, Pankow *et al.* 1986, Gómez-Hernández and Gorelick 1989, Paleologos *et al.* 1996, Renard and de Marsily 1997, Chen *et al.* 2003, de Marsily *et al.* 2005, Zhang *et al.* 2006, Fleckenstein and Fogg 2008, Gardet *et al.* 2014, Koch *et al.* 2014, Mahmud *et al.* 2015) and in particular where the REV concept is applied in fractured or slightly karstified rocks (among others, Jourde *et al.* 2007, Rong *et al.* 2013, Ghasemizadeh *et al.* 2015, Liu *et al.* 2016). It is not the scope here to provide detailed information on this "parametrization" issue that is closely related to REV scales, parameters uncertainty, calibration, and inverse modeling and geostatistical assessments from "soft-data" as, for example, hydrofacies heterogeneity. Introductory summaries about these interrelated issues are discussed in Sections 12.5 through 12.7.

Stress factors

The stress factors of the model are the incoming or outgoing fluxes imposed on the system. For a groundwater flow problem, there are, for example, effective infiltration, pumping or injection rates, recharge, or any other specified exchanged fluxes with other water bodies. Similarly, for a solute transport problem, it could be the solute mass fluxes (i.e., water fluxes with associated concentrations) representing the contamination source. In the particular case of a model simulating pump and treat remediation, it could correspond to an outflow of pollutant associated with the pumping of contaminated water. Note that these prescribed stress factors are most often considered as boundary conditions (BCs) of the model (see Section 12.3).

Historical data

For calibration and validation of the model, historical measured and observed data are needed concerning the main problem variables (i.e., piezometric heads and solute concentrations for groundwater flow and solute transport problems, respectively). Also, their first derivatives (e.g., water flux as base-flow of a draining river) can be used. They should be distributed data in the domain and as function of time (for transient models) that will be used for the calibration (or inverse modeling) procedure and for validation. So, maps, cross sections, and graphs showing the values of the main variable as measured at different points in space and at different times are essential. They are taken into account, possibly with different weights, in the calculation of any objective function (see Section 12.5).

12.5 Calibration, validation, sensitivity analysis, and inverse modeling

Change (adaptation) of the parameters values and their spatial distribution for obtaining model results as close to the (observed historical) reality is the main purpose of *calibration. Validation* is usually considered as an additional check using observed historical data that were not used during the calibration. Many publications and text books address this important calibration issue that is a critical part of any reliable groundwater modeling. In this Section, only a summary of this topic is proposed based on the main needs of the model practitioner. This practical synthesis is mostly based on the literature review made by Wildemeersch (2012) in his PhD dissertation.

Optimization

Definitions

Optimization consists in finding the optimum set of parameter values for simulating the reality assumed to be represented by the available historical data. This optimum is measured by using an *objective function* accounting for the discrepancies between observed and computed values of the main variables and/or one or more of their derived variables. This optimization may be obtained by a manual trial-and-error procedure, but nowadays this is most often performed automatically by typically nonlinear regression methods referred to as *inverse modeling*. In some cases, starting with a trial-and-error procedure could be helpful to gain a full understanding of the physical behavior of the simulated system. The automatic inverse modeling procedure is indeed more efficient to produce useful statistics and to reduce subjectivity. The main caveats and issues are however common to both procedures (Carrera *et al.* 2005): the nonuniqueness of the solution (i.e., several sets of parameters lead to a same minimum of the objective function) and the local or global instabilities (i.e., small changes of the historical data lead to huge changes in estimated parameter values). At the other extreme, the insensitivity of simulated results to a parameter change indicates that the information contained in the observations is not sufficient or adequate to help in the parameter optimization (Hill and Tiedeman 2007). It is suggested to introduce prior information on the parameter values to avoid as far as possible an ill-posed inversion (Carrera and Neuman 1986b).

For a chosen mesh, in the most detailed case, optimization can be performed for parameters values in each element or cell, allowing a detailed spatial distribution of the heterogeneity. This *point estimation* is practically not feasible or at least unrealistic in terms of computation (too many unknowns for too little data). In most cases, this is not needed with respect to the available data set. Often, on the basis of geological information, the domain is "a priori" subdivided in fixed zones. This *zonation* reduces largely the parameters values to be optimized but this is a quite rigid and definitive way of prescribing heterogeneity in the simulated domain. *Pilot points* (Certes and de Marsily 1991), is clearly a better alternative consisting in defining parameter values in some (pilot) points distributed in the domain. The set of parameter values introduced at each pilot point is used for further interpolation (i.e., using kriging or any other geostatistical interpolator) over all other nodes of the domain at each iteration of the optimization procedure. This technique is flexible but a relative smoothing of the heterogeneity is to be expected in the results if traditional geostatistical techniques are used for the interpolation. It has been applied by many authors in various hydrogeological contexts (among others, LaVenue *et al.* 2005, Vesselinov *et al.* 2001, Doherty 2003, Moore and Doherty 2005). A combined zonation—pilot point modeling approach can be used also where pilot points are defined in the most characterized part of the domain and classical zonation in other parts (Doppler *et al.* 2007, Batlle-Aguilar *et al.* 2009).

Objective function

Whether this optimization is done by hand or automatically, the most intuitive objective function to be minimized is often taken as a weighted least square:

$$\varphi(b) = \sum_{i=1}^{n} w_i [y_i^{obs} - y_i^{sim}(b)]^2 \tag{12.14}$$

where y_i^{obs} and y_i^{sim} are the ith observed and simulated values, respectively, w_i is the weight attributed to the ith observation, $i = 1, ..., n$ with n being the total number of observations or measurements to be used for constraining the model, b is a vector of order k representing the k parameters values to be calibrated and $\varphi(b)$ is the objective function to be minimized. Note that in the more general case, the n observations may include not only values of the main variables or their derivative variables (e.g., piezometric head or flow-rate for a flow simulation, concentration for a transport simulation), but also prior measures or estimations of parameters values.

This objective function may be generalized to include transient observed and simulated values:

$$\varphi(b) = \sum_{i=1}^{n} w_i \left[\sum_{t=1}^{nt} [y_{i,t}^{obs} - y_{i,t}^{sim}(b)]^2 \right] \tag{12.15}$$

where $t = 1, ...nt$ with nt being the total number of time steps.

Equation 12.14 can also be written as:

$$\varphi(b) = \sum_{i=1}^{n} w_i r_i^2 \tag{12.16}$$

where r_i is the ith residual.

In the literature about inverse modeling the matrix form of this objective function is used:

$$\varphi(b) = [y - y(b)]^T \cdot w \cdot [y - y(b)] = r^T \cdot w \cdot r \tag{12.17}$$

where y is the vector of order n containing the observation values, $y(b)$ is the vector of order n containing the simulated values (each of them dependent on the k parameter values to be fitted), w is the weight matrix with dimensions $n \times n$.

Other performance criteria/objective functions

One can think to adopt specific performance criteria and thus specific objective functions dependent on the type of available data and in function of the purposes of the simulations. However, difficulties are to be expected when different objective functions are considered for different types of data and thus when a multiobjective optimization is pursued.

If performance criteria are measuring the level of agreement between model and historical data, there is however no need to achieve an extreme level of precision. It would be illusory anyway because, resulting from the conceptual assumptions, we would falsify the parameter values to compensate and converge toward the "truth" as perceived by historical data (among others, Beven and Binley 1992, Refsgaard and Henriksen 2004, Rojas *et al.* 2010b, 2010c).

In hydrology, when the main purpose is to reproduce flow rates under transient conditions, the most widely used objective function is the Nash-Sutcliffe criterion (Nash and Sutcliffe, 1970). It compares measured and simulated flow rates at one point in the basin (i.e., most often at an outlet gauging station). It is written as:

$$\varphi_{NS}(b) = 1 - \frac{\sum_{t=1}^{nt} [q_t^{obs} - q_t^{sim}(b)]^2}{\sum_{t=1}^{nt} [q_t^{obs} - \mu^{obs}]^2} \quad \in \,]-\infty, 1]$$
(12.18)

where q_t^{obs} and q_t^{sim} are respectively the observed and simulated flow rates at time step t, nt is the total number of time steps, μ^{obs} is the mean value of observed flow rates. $\varphi_{NS}(b)$ tends to 1 if the match is very good and, it decreases as the fit becomes poorer and becomes negative indicating that the mean value of the measured flow rates (μ^{obs}) would give a better description of the historical data than the simulated values. For groundwater flow models, this Nash-Sutcliffe objective function is used in particular to measure whether the model correctly reproduces baseflow in a watershed as measured in the gaining stream during dry periods (i.e., without rainfall). Many other performance criteria (=objective functions) are provided in the hydrological literature, each having its rationale with respect to the aim of the simulations (e.g., minimizing mass balance errors [Gupta *et al.* 1999], peak flow-rates and peak timing errors [Arico *et al.* 2009]).

For reproducing piezometric heads under transient conditions, the most used objective function is the root mean square error (RMS) in one observation well written as:

$$\varphi_{RMS_h}(b) = \sqrt{\frac{1}{nt} \sum_{t=1}^{nt} [h_t^{obs} - h_t^{sim}]^2}$$
(12.19)

where h_t^{obs} and h_t^{sim} are the observed and simulated piezometric heads at the time step t, respectively, nt is the total number of time steps. $\varphi_{RMS_h}(b)$ is to be minimized for each piezometer. If this is generalized for all the piezometers of the simulated domain, the global objective function is similar to what was proposed in Equation 12.15.

Another example is the case where the main aim is to reproduce the piezometric head variations. The following objective function ($\varphi_{hve}(b)$) is written for minimizing the piezometric head variation error as:

$$\varphi_{hve}(b) = \left(\frac{h_{max}^{sim} \quad h_{min}^{sim}}{h_{max}^{obs} - h_{min}^{obs}} - 1 \right) 100$$
(12.20)

where h_{max}^{sim} and h_{min}^{sim} are the maximum and minimum simulated piezometric heads, respectively, and h_{max}^{obs} and h_{min}^{obs} are the corresponding maximum and minimum observed values, respectively. $\varphi_{hve}(b)$ is expressed in %, it represents the percentage of error and should be minimized.

As indicated previously, if multiple objective functions are used (e.g., considering piezometric heads and flow-rate observations with different specific performance criteria), improvement of one of them can correspond to deterioration of the other(s). A multiobjective optimization becomes rapidly complex (Vrugt *et al.* 2003). It leads to procedure where each performance criterion is weighed and multiple independent optimizations should be performed. Some are concluding that single optimization should be preferred, in particular for complex models with long execution times (Foglia *et al.* 2009).

Weighting

Even when different types of data are available (i.e., observed piezometric heads or flow rates, concentrations, prior estimations of parameters values), a single objective function in the form of Equations 12.14 or 12.15 is the most common choice. In this weighted least square expression, each type of data could have its own weight. The reasons for using different weights can be multiple: (a) to give more relative importance to data that are less uncertain, (b) to produce dimensionless residuals and statistics, and (c) to give more relative importance to data having a regional significance.

As shown in Figure 12.15, weighting the different data is needed to present results in a consistent way.

If the measurement and observation errors can be considered as independent, the weight matrix w, with dimensions $n \times n$ (n being the total number of measured data), is a diagonal matrix. According to the first and second motivations (cited here above), the inverse of the error variance of the considered observation is often proposed for weighing the residuals (Hill and Tiedeman 2007):

$$w_{ii} = \frac{1}{\sigma_{err_i}^2} \qquad (12.21)$$

where $\sigma_{err_i}^2$ is the variance of the errors of the measured ith data. In doing so, the most accurate observations are given the most important weights and values contributing to the objective function are dimensionless. For some observed data, the inaccuracy and uncertainty may increase with the measured value (e.g., flow rates). "Coefficients of variation" (cv_i) can then be used instead of standard deviation of the ith observation errors, so that the weights are written as (Hill and Tiedeman 2007):

$$w_{ii} = \frac{1}{(cv_i y_i^{obs})^2} \qquad (12.22)$$

If needed, it allows to decrease the weight given to high observed values.

Note that another way of thinking could be also adopted. Piezometric heads and concentrations are only discrete measured data (i.e., measured in observation wells). On the other hand, measured flow rates and solute fluxes can be considered as reflecting the behavior of the system on a larger representative volume than just around the measurement point. If this reasoning is followed, more weight should be given, for example, to flow-rate data than to piezometric data during the optimization of a regional groundwater flow model.

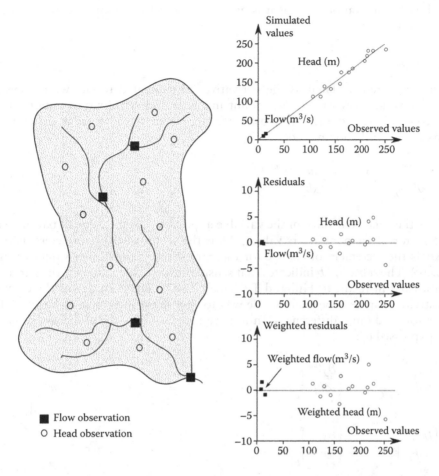

Figure 12.15 If measured (base) flow rates and piezometric heads are available for a simulated watershed, only weighted values of residuals are to be shown for being consistent in terms of units.

Sensitivity analysis

Sensitivity analysis may be seen as an ideal prerequisite tool for performing optimization and thus calibration and inverse modeling procedures. In a sensitivity analysis, the observations-parameters-results interactions are investigated in order to detect which processes and associated parameters are the more influent on the results. Thus, after calibration, the model can be also used for analyzing the prediction sensitivity to various scenarios of changes in stress-factors.

Indeed, we are depending on the conceptual choices made previously in terms of spatial and time discretizations, and parametrization (Refsgaard 1997). To summarize, a sensitivity analysis gives: (a) information about the relation between parameters and observations (i.e., to "feel" how the model behaves), (b) a feedback to the conceptual model (i.e., are the simulated processes sensitive to the chosen parameters?), and

(c) a help for the calibration step, as it is not worth adjusting insensitive parameters to improve calibration.

General sensitivities

Sensitivities (S_{ij}) are calculated as the derivatives of simulated results with respect to the model parameters values. As adjoint methods and sensitivity-equations are relatively complex, sensitivities are most often approximated by a finite difference referred as to "perturbation method":

$$S_{ij} = \left(\frac{\partial y_i^{sim}}{\partial b_j} \right)\bigg|_b \approx \frac{y_i^{sim}(b + \Delta b) - y_i^{sim}(b)}{\Delta b_j} \tag{12.23}$$

where y_i^{sim} is the simulated value of the variable at point i, b_j is the value of parameter j, b is the vector of the parameters values, Δb is the vector of the parameters increments, Δb_j is the increment of change for the parameter j value (other values being equal to zero). The subscript b indicates that sensitivities are calculated for the parameter values listed in vector b (Hill and Tiedeman 2007). The accuracy of the calculated sensitivities can be dependent on the way to choose the increment as backward, central, or forward finite differences. In a more general form, the global sensitivity matrix is expressed by:

$$S = \begin{bmatrix} \left(\dfrac{\partial y_1^{sim}}{\partial b_1} \right)\bigg|_b & \cdots & \left(\dfrac{\partial y_1^{sim}}{\partial b_k} \right)\bigg|_b \\ \vdots & \ddots & \vdots \\ \left(\dfrac{\partial y_n^{sim}}{\partial b_1} \right)\bigg|_b & \cdots & \left(\dfrac{\partial y_n^{sim}}{\partial b_k} \right)\bigg|_b \end{bmatrix} \tag{12.24}$$

where, as stated previously, n is the number of observation points, k is the number of parameters. This sensitivity matrix is also referred as to *Jacobian matrix*, noted J.

Dimensionless scaled sensitivity

Because it is difficult to compare sensitivities when units are different, the *dimensionless scaled sensitivity (dss)* has been proposed (Hill 1992, Hill *et al.* 1998, Hill and Tiedeman 2007):

$$dss_{ij} = \left(\frac{\partial y_i^{sim}}{\partial b_j} \right)\bigg|_b \left| \frac{b_j}{100} \right| \left(100\sqrt{w_{ii}} \right) = \left(\frac{\partial y_i^{sim}}{\partial b_j} \right)\bigg|_b \left| \frac{b_j}{100} \right| \left(\frac{100}{\sigma_i} \right) \tag{12.25}$$

This evaluates the importance of a single observation y_i for the estimation of a single parameter b_j, indicating the amount the simulated value would change given a 1% change in the parameter value. This is expressed as a percent of the observation error standard deviation (σ_i).

Composite scaled sensitivities

More useful are the *composite scaled sensitivities* (css_j) as they are providing the importance of observations as a whole for each parameter j. css_j are calculated as follows (Hill 1992, Anderman *et al.* 1996, Hill and Tiedeman 2007):

$$css_j = \sqrt{\frac{\sum_{i=1}^{n}(dss_{ij})^2}{n}\bigg|_b} \tag{12.26}$$

where $j = 1, ...k$ for the parameters values. They give a measure of the information provided by the entire set of observation data for estimation of one single parameter j. Thus, one single value for each parameter is obtained, which is very useful for comparing from one parameter to another. Largest values are found for parameters for which the data set provides the most information. This is particularly useful for selecting only a few parameters to be optimized in the inverse modeling procedure. As stated previously, introduction of weights in the definition of dss_{ij} and css_j allowed to obtain dimensionless values. A simple example of application of these sensitivity definitions can be found in Hill and Tiedeman (2007) and a nice practical case-study considered for an integrated surface water-groundwater model is shown by Goderniaux *et al.* (2015).

The same type of sensitivity analysis can be performed with regards to simulated predictions of the model. Considering stress-factors changes as model parameters, one can obtain the composite scaled sensitivities and easily detect the more influent scenarios on the prediction results.

Correlation between model parameters

Before selecting the parameters to be optimized in priority in the inverse modeling procedure, one should check if some of them are highly correlated. A parameter correlation coefficient (pcc_{jl}) is proposed by Hill and Tiedeman (2007) as follows:

$$pcc_{jl} = \frac{Cov(b_j, b_l)}{\sqrt{Var(b_j)Var(b_l)}} \quad j = 1, ...k \quad l = 1, ...k \tag{12.27}$$

where $Cov(b_j, b_l)$ is the covariance between two parameters b_j and b_l, $Var(b_j)$ and $Var(b_l)$ are the variances of each of the parameters. It indicates the degree of correlation for a given couple of parameters. If a $pcc > 0.95$ is found, a high correlation is revealed between the two considered parameters. Note that this is not unusual to introduce some of the stress-factors as parameters to be optimized. A typical example is given by the full correlation existing between the hydraulic conductivity of an assumed homogenous aquifer and the assumed uniform recharge in steady-state flow simulations. Increasing recharge or decreasing hydraulic conductivity have the same effect on the calculated piezometric heads so that the best optimized solution for parameter values is not unique. An infinity of coupled values (recharge, hydraulic conductivity) is possible and consequently the model could be highly inaccurate for future predictions without adding other types of observation data. For example, it is advised by many authors (among others, Poeter and Hill 1997, Hill and Tiedeman 2007, Wildemeersch *et al.* 2014) to use discrete (e.g., piezometric heads) and more global information (baseflow rates) or to combine concentration and/or temperature data (for example, Bravo *et al.* 2002, Sanford *et al.* 2004, Niswonger *et al.* 2005, Klepikova *et al.* 2016).

One step forward consists in simulating directly a set of predictions for different scenarios (i.e., changing conceptual choices, parameters values, and stress factors). Then, the sensitivity of predicted results (i.e., the forecast) allows us to detect what is creating the largest forecast uncertainty (White 2017). Thus, not only historical data are used (as for model calibration), but also the uncertainty of prediction scenarios, for influencing the conceptual and main parameters choice to be considered for inversion.

Inverse modeling

Inverse modeling consists thus in minimizing an objective function $\varphi(b)$ (Equation 12.17) that will provide optimized values for the parameters (b). This minimum is found solving the normal equations obtained by expressing that the derivative of the objective function with respect to the parameters is equal to zero:

$$\frac{\partial}{\partial b}[y - y(b)]^T \cdot w \cdot [y - y(b)] = 0 \tag{12.28}$$

where the vector 0 is a vector with k components (number of parameters to be optimized) equal to zero.

If the direct problem to be solved is considered as linear, this is written under the form:

$$y(b) = X \cdot b \tag{12.29}$$

where X is a matrix with dimensions ($n \times k$) containing the system responses to the set of excitation, $y(b)$ is the vector of dimension n (number of observations) and b is the vector of dimension k (number of parameters). The objective function (Equation 12.17) can be written as:

$$\varphi(b) = [y - X \cdot b]^T \cdot w \cdot [y - X \cdot b] \tag{12.30}$$

If the number of observations (n) is larger than the number of parameters (k), the solution of the problem (i.e., solving Equation 12.28 and using Equation 12.29) is mathematically unique and equal to:

$$b = (X^T \cdot w \cdot X)^{-1} \cdot X^T \cdot w \cdot y \tag{12.31}$$

Gauss-Newton gradient method

Unfortunately, most of the groundwater systems are nonlinear. A linearization technique is thus required for using Equation 12.31 for finding the optimal parameters values (*b*). Without going into too much detail, a linearization is obtained applying the Taylor's theorem:

$$y(b) \approx y(b_0) + J(b - b_0) \qquad (12.32)$$

where $J = S$ is the Jacobian matrix or sensitivity matrix as expressed in Equation 12.24, b_0 is only slightly different from *b*. This approximation tends to equality with b_0 tending to *b*.

The weighted objective function (Equation 12.17) written for $y(b_0) + J(b - b_0)$ in place of $y(b)$ can be written as:

$$\varphi(b) = [y - y(b_0) - J(b - b_0)]^T \cdot w \cdot [y - y(b_0) - J(b - b_0)] \qquad (12.33)$$

In fact, Equations 12.30 and 12.33 are similar when replacing y and b of Equation 12.30, respectively, by $[y - y(b_0)]$ and $[b - b_0]$ of Equation 12.33. The matrix J replacing the matrix X.

Solving this system, the parameter upgrade vector $[b - b_0]$ is obtained from the discrepancy between the observed values y and their simulated equivalent $y(b_0)$. This is the main principle of the inverse modeling procedure. The parameter upgrade is calculated with:

$$[b - b_0] = (J^T \cdot w \cdot J^{-1} \cdot J \cdot w \cdot [y - y(b_0)] \qquad (12.34)$$

This equation is similar to Equation 12.31 replacing y and b respectively by $[y - y(b_0)]$ and $[b - b_0]$ and matrix X by matrix J. Note that the variance-covariance matrix ($V(b)$) is expressed by:

$$V(b) = s_{calc}^2 (J^T \cdot w \cdot J) \qquad (12.35)$$

where s_{calc}^2 is the calculated error variance giving the variance of the observation values embodied in the vector y.

This method is referred to as the *Gauss-Newton gradient method*. Starting from a set of initial values for the parameters, the optimal parameters values are obtained iteratively. In a space of k dimensions, the followed path of the evolution of the parameters vector could be represented (as shown for two parameters in Figure 12.16). Sequentially, at each iteration, the calibration is evaluated, sensitivities are calculated, the problem is linearized, a minimum of the linear problem is found, and new parameters are defined.

Modified method

Without entering into detailed considerations, this Gauss-Newton gradient method must additionally be modified by (a) scaling (i.e., to avoid too contrasted values in

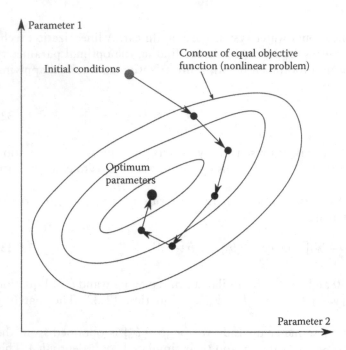

Parameter 1

Initial conditions

Contour of equal objective
function (nonlinear problem)

Optimum
parameters

Parameter 2

Figure 12.16 From initial parameter values, the Gauss-Newton gradient method optimizes
iteratively the parameter values for reaching a minimum of the objective function
(Doherty 2005).

the matrix *J*), (b) the introduction of a Marquard-Levenberg parameter (i.e., changing the direction of the increment to move toward the steepest descent direction, and
(c) damping (i.e., to prevent overshoot). Further detailed information can be found
about this modified method (referred to as the *Gauss-Newton-Levenberg-Marquard
method*) in Doherty (2005), Poeter *et al.* (2005), and Hill and Tiedeman (2007).

Local minima and global minimum

The *Gauss-Newton-Levenberg-Marquard method* is based on local sensitivities.
Thus, the inverse model can converge in reality toward a local minimum of the objective function. The iterative process may then be stuck in this local minimum. Skahill
and Doherty (2006) introduced, in the inversion code PEST, the possibility of withdrawing the most insensitive parameter when the improvement of the last iterations
becomes too poor. The parameter vector (*b*) is recalculated without this parameter
to check if it works better or not. If this is not the case, a second (most insensitive)
parameter is withdrawn, and so on, with the same procedure. This is referred to as
the "temporary parameter immobilization strategy" allowing to freeze some parameters and consequently to explore a larger spectrum of the parameter space for the
remaining ones.

This technique highlights that it could also be more efficient to perform a sensitivity analysis beforehand to directly withdraw some insensitive parameters for reducing
accordingly the vector *b* dimension.

Theoretically, another alternative to ensure the convergence toward a "global minimum" of the objective function, could consist in using global optimization methods (rather than local optimization methods as described here above). More details can be found in (among others) Moore *et al.* (2010) and Keating *et al.* (2010). However, it requires more computational time.

Nonuniqueness and regularization

Another issue is clearly the nonuniqueness of the solution when the number of parameter values to be estimated exceeds the number of observations (Figure 12.17). This is an ill-posed problem. Indeed, this is the main reason for avoiding too much complexity and to keep the model as simple as possible following in that way the *parsimony* principle (Hill 2006; see Section 12.2).

Introducing additional constraints (i.e., additional observations, as for example, baseflow data added to piezometric heads in a transient groundwater flow model), allows generally to avoid this kind of issue. Adding constraints can also be obtained when a zonation (i.e., with homogeneity in each zone) is prescribed before inversion or when predefined relations are used between some parameters values. More rigorously, a "regularization" technique (Hunt *et al.* 2007) can be used where the objective function is penalized when the predefined constraints are violated. This is referred to as a Tikhonov regularization (Tikhonov and Arsenin 1977) and described in detail in (among others) Doherty (2003) and van den Doel and Ascher (2006).

Combining regularization together with reduction of the parameter space (resulting from insensitive parameters) is indeed recommended for gaining in efficiency for highly parameterized groundwater modeling problems (Tonkin and Doherty 2005, Hunt *et al.* 2007).

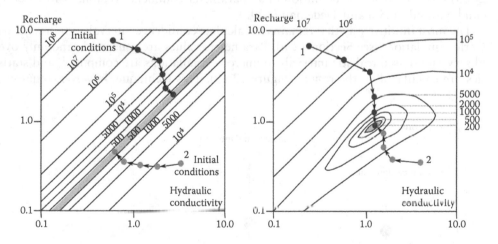

Figure 12.17 Nonuniqueness of the solution of the inverse problem. Depending on the starting parameter values, the optimization leads to a minimum value of the objective function that can be nonunique (left). If additional observations are available they can be used as additional constraints for a convergence toward a unique solution (right) (Hill and Tiedeman 2007).

Uncertainty of predictions

The results of the inverse procedure (or calibration) can be further used to evaluate the uncertainty around predictions. The simplest techniques for quantifying prediction uncertainty are based on inferential statistics. Inferential techniques produce linear or nonlinear confidence intervals for predictions. The standard deviation of predictions is calculated by:

$$s_{y_m^{pred}} = \left[\sum_{j=1}^{k} \sum_{l=1}^{k} \frac{\partial y_m^{pred}}{\partial b_l} \cdot V(b) \cdot \frac{\partial y_m^{pred}}{\partial b_j} \right]^{1/2} \tag{12.36}$$

where y_m^{pred} is the mth prediction and $V(b)$ is the parameter variance-covariance matrix of Equation 12.35. Any linear confidence interval is defined as:

$$y_m^{pred} \pm (\text{critical value}) s_{y_m^{pred}} \tag{12.37}$$

where the "critical value" comes from the considered statistical distribution (i.e., Student-t distribution or others) and is dependent on the degrees of freedom and on the significance level (i.e., α commonly taken equal to 0.05 or 0.10, 5 or 10 percent) (Hill and Tiedeman 2007). This way of calculating confidence intervals of predictions allows to use only sensitivities for calibrated parameters. This simplifies the problem and provides symmetrical confidence intervals (Figure 12.18). However, if the problem is too nonlinear, these intervals could be inaccurate as sensitivities may vary significantly as function of the parameters values dependent on the system state.

Nonlinear confidence intervals (Figure 12.18) are calculated by other methods (for example, Vecchia and Cooley 1987) involving a correction (Christensen and Cooley 2005) resulting from the definition of a "parameter confidence region." More details can be found in Hill and Tiedeman (2007).

The most intuitive way that remains to calculate confidence intervals is to use Monte Carlo simulations (see Section 12.6). Parameters values are generated randomly over observation-based realistic intervals, numerous simulations are computed, and statistics are calculated on the results (Figure 12.19). This technique, however, requires a

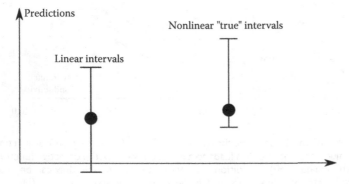

Figure 12.18 Linear and nonlinear confidence intervals of predictions.

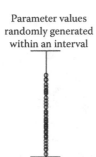
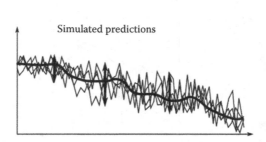

Parameter values
randomly generated
within an interval

Simulated predictions

Figure 12.19 Confidence interval of predictions determined by statistics on Monte Carlo simulations with randomly generated sets of parameters chosen within a determined interval.

large number of parameter sets generated randomly for the calibration and predictions computations. It becomes heavy to handle for models with long execution time.

12.6 Introduction to groundwater geostatistics and probability

It is far beyond the scope of this book to give an overview of the various and numerous applications of geostatistics and stochastic modeling in hydrogeology. The most important notions are introduced as they are needed for many groundwater modeling procedures. In particular for uncertainty assessment of parameters and model predictions, geostatistics and the probabilistic framework are really useful in view of the insufficient knowledge we have about the true heterogeneity of the geological media. Specialized books and dedicated review papers should be consulted for further details and developments (among others, Dagan 1989, Gelhar 1993, Kitadinis 1997, Goovaerts 1997, Deutsch and Journel 1998, Zhang 2002, Rubin 2003, Remy *et al.* 2009, Chilès and Delfiner 1999, Caers 2011, Scheidt *et al.* 2018)

Continuous random geostatistical variables

As mentioned in Section 12.1, in groundwater modeling, the probabilistic or stochastic analysis of data may concern the parameters values and the stress-factors (i.e., including the BCs). More fundamentally, the general structure of a model could also be evaluated within a stochastic or probabilistic approach.

Any data analyzed by geostatistics techniques becomes geostatistical variables. They can be discrete random variables having a limited number of values and continuous random variables for which the number of possible outcomes is infinite (i.e., in line with the continuum assumption of the modeled geological media). However, if we take hydraulic conductivity (K) as an example of continuous random variable, it will never be possible to measure it everywhere, not even at enough locations that would satisfy a model grid. Thus, we have to assess values at locations where there are no measurements (Engesgaard 2003).

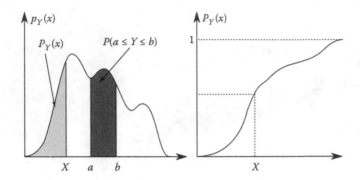

Figure 12.20 Example of a probability density function (left) and of a cumulative distribution function (right) of the random variable Y (Caers 2011).

Thus, contrarily to classical statistics where the analysis is done on different outcomes of a same process, in geostatistics, the geological setting is unique and the outcomes are considered at different locations. This is the *ergodicity assumption* meaning that the spatial distribution of the studied property can be described by a probability distribution function of an ensemble of realizations (de Marsily 1986).

The probability that the random variable Y is less than a given value a is noted:

$$P(Y \leq a) \tag{12.38}$$

The *probability density function* (pdf) ($p_Y(x)$), describes the probabilistic variation of random variable Y evaluated in the point x (Figure 12.20). Accordingly, the probability that the random variable Y is comprised between a and b can be expressed by:

$$P(a \leq Y \leq b) = \int_a^b p_Y(x)dx \tag{12.39}$$

The cumulative distribution function (cdf) ($P_Y(x)$) is defined (Figure 12.20):

$$P_Y(x) = P(Y \leq x) \tag{12.40}$$

and

$$P_Y(x) = \int_{-\infty}^x p_Y(x)dx \quad \text{or} \quad p_Y(x) = \frac{dP_Y(x)}{dx} \tag{12.41}$$

The *expected value* (i.e., equivalent to the *mean* or *average value* if the number of outcomes is large enough) of the continuous variable is written as:

$$E[Y] = \int_{-\infty}^{+\infty} xp_Y(x)dx \tag{12.42}$$

If we know the pdf $p_Y(x)$, we can calculate the integral and find $E[Y]$. For example, if the pdf has a Gaussian (Normal) probabilistic distribution:

$$p_Y(x) = \frac{1}{\sqrt{2\pi\sigma^2}} exp\left[-\frac{1}{2}\left(\frac{x-\mu}{\sigma}\right)^2\right] \tag{12.43}$$

where μ is the population mean (expected value), σ is the population *standard deviation*, σ^2 is the population *variance*, then using Equation 12.42, one can check that $E[Y]$ is equal to μ.

The population variance is defined as follows:

$$Var[Y] = E[(Y - E[Y])^2] = \int_{-\infty}^{+\infty} (x-\mu)^2 p_Y(x)dx \tag{12.44}$$

using Equation 12.43, this can be calculated (Caers 2011) as:

$$Var[Y] = \int_{-\infty}^{+\infty} (x-\mu)^2 \frac{1}{\sqrt{2\pi\sigma^2}} exp\left[-\frac{1}{2}\left(\frac{x-\mu}{\sigma}\right)^2\right](x)dx = \sigma^2 \tag{12.45}$$

As mentioned in Chapter 4, hydraulic conductivity and transmissivity values are often assumed to be log-normally distributed. Thus $log(K)$ or $ln(K)$ are often assumed to follow a Gaussian distribution. This assumption based on observations (for example, Freeze 1975, Hoeksema and Kitadinis 1985) is discussed and sometimes refuted by many authors according to specific geological circumstances (for example, Gómez-Hernández and Wen 1998, Wen and Gómez-Hernández 1998, Whittaker and Teutsch 1999, Zinn and Harvey 2003, Lee et al. 2007, Kerrou et al. 2008, Capilla and Llopis-Albert 2009). Some of them propose to use multi-Gaussian models or apply other transforms as, for example, the power transform (Box and Cox 1964) or the normal score transform (Deutsch and Journel 1998, Ringrose and Bentley 2015).

If $Y = lnK$ is assumed to be normally distributed, the corresponding pdf $(p_Y(x) = p_{lnK}(x))$ is written as in Equation 12.43. It leads to the following expressions for the arithmetic, geometric, and harmonic mean of K, respectively:

$$\mu_K = E[K] = exp\left(\mu_{lnK} + \frac{1}{2}\sigma_{lnK}^2\right) \quad \text{(arithmetic mean)} \tag{12.46}$$

$$K_{Gavg} = exp(\mu_{lnK}) \quad \text{(geometric mean)} \tag{12.47}$$

$$K_{Havg} = exp\left(\mu_{lnK} - \frac{1}{2}\sigma_{lnK}^2\right) \quad \text{(harmonic mean)} \tag{12.48}$$

The variance of K can be obtained by (van Leeuwen 2000):

$$\sigma_K^2 = [exp(\sigma_{lnK}^2 - 1)][exp(2\mu_{lnK} + \sigma_{lnK}^2)] \tag{12.49}$$

Probability and Bayesian approach

Probability can be simply the chance of obtaining a given value (i.e., number of successful events divided by the number of trials) but more often in Earth sciences, this is rather an assessment based on prior knowledge. The probability of a given event, noted $P(E)$ is always between 0 and 1. A conditional probability, describing the probability that an event E occurs given that an event F occurred is noted $P(E|F)$. Note that if the two events are not related the conditional probability $P(E|F) = P(E)$. The Bayes rule expresses the following equivalence (Caers 2011):

$$P(E \, and \, F) = P(E|F)P(F) = P(F|E)P(E) \tag{12.50}$$

meaning that the probability of occurrence of both related events E and F can be calculated with these two combinations of conditional and unconditional probabilities. Therefore:

$$P(E|F) = \frac{P(F|E)P(E)}{P(F)} \tag{12.51}$$

In Bayesian statistics, probability is used for dealing with uncertainty. This is very useful in Earth sciences and particularly in hydrogeology. For example, the hydrogeologist perception or *prior* knowledge about the heterogeneity of the hydraulic conductivity field in an aquifer (i.e., based on his geological knowledge) can be expressed as a *prior distribution of probability*. Subsequently, it will be modified (i.e., increasing or decreasing) as more data becomes available for obtaining finally a *posterior distribution probability*. This updating process is described by the Bayes' rule. In inverse groundwater modeling, if we use this way for describing how initial values of the parameters (vector b) are changed during the calibration or inverse procedure resulting from the conditioning on observations (vector y) the Bayesian approach can be summarized as follows:

$$p(b|y) = \frac{p(b)\,p(y|b)}{p(y)} \tag{12.52}$$

where $p(b)$ is the *prior distribution of probability* of the model parameters values, $p(y)$ is the probability distribution of the observed data, $p(y|b)$ is the conditional probability of observing data (also named likelihood) for a given set of parameter values and $p(b|y)$ is the *posterior distribution probability* of the model parameter values expressing the conditional probability of obtaining a set of parameters for a given set of observed data. Thus, the latter provides a quantification of the uncertainty of the fitted parameters values. Equation 12.52 summarizes remarkably the three sources of uncertainties described previously (see Sections 12.1, 12.2, and 12.5): (a) uncertainty on the parameters values $(p(b))$, (b) uncertainty on the observations/data set $(p(y))$, and (c) uncertainty on the reliability of the (physically-based) model (resulting from the conceptual choices) $(p(y|b))$.

As mentioned by Rojas (2009), Equation 12.52 illustrates the well-known concept of *Bayesian learning* showing the modification of a prior opinion about the parameter values (with the associated uncertainty stated by the prior distribution) by calibration and thus evidence provided by the data (Sorensen and Gianola 2002).

The advantage of the Bayesian technique is to provide a complete description of the uncertainty about parameters values (*b*) after having observed the data. Accordingly, statistics about this uncertainty can be produced. The main disadvantage is the fact that $p(b|y)$ is highly dimensional and complex to be simulated. Most often, Monte Carlo methods are used for approximating this probability distribution.

Monte Carlo simulations

In Monte Carlo simulations, a fair (i.e., random) sampling from the assumed known probability distribution is organized. If the number of samples is large, it allows then to build a cumulative distribution function (cdf) that is an approximation of the cdf of the probability distribution from which the sampling was performed. Obtained statistics from the large number of samples can be used to characterize the uncertainty and confidence intervals. This straightforward method has been used extensively in groundwater modeling (among others, Bair *et al.* 1991, Varljen and Shafer 1991, Cooley 1997, van Leeuwen *et al.* 1998, Hunt *et al.* 2001, Yoon *et al.* 2013).

Monte Carlo simulations are CPU demanding. As the probability distribution of the variable is parameterized, a Markov chain Monte Carlo (MCMC) sampler can be used (for example, Gilks *et al.* 1995, Sorensen and Gianola 2002, Robert 2007, Hassan *et al.* 2009, Mariethoz *et al.* 2010). These MCMC methods are more efficient because the samples are restricted to those obeying the desired probability distribution. Many groundwater modeling applications have been published (among others, Fienen *et al.* 2006, Rojas *et al.* 2008, 2010b, 2010c, Keating *et al.* 2010, Laloy *et al.* 2013).

Geostatistics for modeling heterogeneity

Historically, geological flow properties were considered as *regionalized random variables* in the oil industry by Warren and Price (1961) and in aquifers by Matheron (1967). The main idea is that the medium structure resulting from the geological processes can be "translated" or "captured" in the form of a covariance or variogram (de Marsily *et al.* 2005). In fact, the geological nonrandomness entails that values measured close to each other are more alike than values measured further apart (Caers 2011). Spatial relationships exist among the available spatial data whether the values are of the same type or not. In this perspective, understanding and quantifying of the *connectivity* through connectivity metrics is very important. This is directly in relation with upscaling physical parameters honoring the physical processes such as flow or transport (Renard and Allard 2013).

How to describe the spatial continuity and evaluating any type of correlation between any datum value measured in different locations, has been one of the main topic of geostatistics applied in Earth sciences and particularly in oil reservoir engineering and hydrogeology. To keep it relatively simple, two main ways of describing this spatial continuity will be described below: the *bivariate* and the *multiple points* approaches. The interested reader is referred to the following specialized books or publications for more details: Fogg 1986, de Marsily 1986, Isaaks and Srivastava 1989, Goovaerts 1997, Deutsch and Journel 1998, Chilès and Delfiner 1999, Strebelle 2002, Rubin 2003, Caers 2011, and Scheidt *et al.* 2018, among many others.

Bivariate geostatistics: variogram and kriging

COVARIANCE AND VARIOGRAM

To calculate the autocorrelation in space (i.e., in 2D or 3D), between two observation points, one must define a lag and a search direction. Pairs of observation points are collected, each of them has a lag (distance between the two observation points) and a direction (Caers 2011). If variable Y is considered taking a value in each point of generalized coordinates x, a quantification of the spatial correlation is expresses by the *variogram*, defined as (half) of the variance of the difference for two points located at a distance d to each other:

$$\gamma(d) = \frac{1}{2} Var[Y(x) - Y(x+d)] = \frac{1}{2} E[(Y(x) - Y(x+d))^2] \tag{12.53}$$

This expression measures actually the "degree of dissimilarity" between pairs of points. This is a two-points or bivariate approach.

The *covariance* function $C(d)$ is defined:

$$C(d) = E[(Y(x) - m)(Y(x+d) - m)] \tag{12.54}$$

where m is the average of the observations or expected value $E[Y(x)] = E[Y(x+d)]$.

If $|d| = 0$ then the covariance is equal to the variance $Var(Y) = E[(Y(x) - m)^2]$. Using Equation 12.54, Equation 12.53 can now be written:

$$\gamma(d) = Var(Y) - C(d) \tag{12.55}$$

Note that, contrary to the variance, the covariance contains information about the spatial correlation.

In practice the variogram, as expressed in Equation 12.53, has to be estimated on the basis of all available data with the *experimental variogram*:

$$\text{Estimate of } \gamma(d) = \frac{1}{2n(d)} \sum_{i=1}^{n(d)} [y(x_i) - y(x_i + d)]^2 \tag{12.56}$$

where $n(d)$ is the number of pairs found for distance (lag) d, $y(x_i)$ is the measured data at location x_i. As observation data are distributed irregularly, the value of the experimental variogram is approximated for different distance ranges. A theoretical variogram is then fitted to the experimental variogram, which allows assessing the variogram for any distances.

A variogram shows the following main features (Figure 12.21):

- The *correlation length*, also named the *range*, is the distance at which nearly no correlation is expected: in the variogram it corresponds to the distance at which a plateau is reached (if any).
- The *sill* is the value of the variogram level for this plateau (if it exists), in this case, this can be considered equal to the variance of the sample values as the covariance $C(d)$ becomes zero in Equation 12.55.

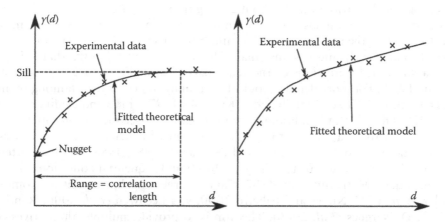

Figure 12.21 Example of variogram, experimental values, fitted theoretical model, nugget, range, and sill (left). It can happen that a variogram does not reach any plateau (right), showing that the property does not have a finite variance.

- The *nugget effect* accounts for the fact that for small d a jump is often observed in experimental variograms, while theoretically the variogram value at $d = 0$ is zero.

Some variograms never reach a clear sill as the experimental values show an increase with distance. The *second-order stationarity* assumption is not valid. This indicates that the property does not have a finite variance and that there is no covariance. As mentioned by Caers (2011), the practical meaning of a variogram is the measurement of a "*geological distance.*" Two observation points can be at a short (Euclidian) distance but at a large distance from a geological perspective if measured observations at the two points indicate very different geological contexts. Note also that variograms can be directional, with a different correlation length according to the direction.

Usually, four commonly (bounded) theoretical models are used for simulating the variogram (i.e., pure constant nugget effect, spherical, exponential, and Gaussian functions). Positive linear combinations and hence nested variograms are also considered. The fitted theoretical variogram can then be used for two types of application (a) calculation of point estimates at unsampled locations: *kriging*, and (b) simulation of equally likely realizations of the studied variable: *geostatistical simulations*, thus generating unconditional or conditional realizations.

KRIGING

Kriging is known as a spatial linear interpolator using weights λ_i:

$$\hat{y}(x) = \sum_{i=1}^{n} \lambda_i \, y(x_i) \qquad (12.57)$$

where $\hat{y}(x)$ is the estimate of the variable y at coordinates x, x_i denotes locations for n available observations used for the estimation, with $i = 1, \ldots n$. Their number can be decided according to the maximum distance of correlation (correlation length = range of the variogram). Without getting into much details, the weights are calculated to provide an unbiased estimation (i.e., the mean of the estimations must be equal to mean of the real values) and a minimum variance of the estimation (i.e., the mean of $[\hat{y}(x) - y(x)]^2$ must be minimal). However, it is well known that kriging produces a smoothed estimation of the true spatial variability of a given variable (Delhomme 1979). For more details about kriging in hydrogeology see (among many others): Delhomme 1978, de Marsily 1986, Kitadinis 1997, Chilès and Delfiner 1999, Zhang 2002). The need of conditioning the estimates on secondary or soft (interpretative geological) data has led to the development of cokriging (among others, Aboufirassi and Marino 1984, Ahmed and de Marsily 1987, 1988, 1993, Knotters et al. 1995, Gloaguen et al. 2001, Yeh et al. 2002) and sequential conditional kriging (among others, Matheron et al. 1987, Kupfersberger and Blöschl 1995, Gómez-Hernández et al. 1997, Nunes and Ribeiro 2000, van Leeuwen et al. 2000, Rentier et al. 2002, Dassargues et al. 2006). The aim is to provide multiple alternatives of a true unknown field rather than provide the best guess at unsampled locations. These types of technique belong to *geostatistical simulations* of equally likely realizations of the studied variable. They require the assumption that the conditional distribution about any unsampled location follows a Gaussian distribution function (Caers 2011). The kriging mean and variance are good candidates for providing that Gaussian conditional distribution. In sequential Gaussian simulations, all distributions are assumed Gaussian. As this is not often the case, variable transformations could be needed. More details on these techniques can be found in: Deutsch and Journel (1998), Lantuejoul (2002), and Caers (2011). Variograms provide only rough description of the spatial heterogeneity and in particular of the connectivity structure of the parameter fields.

Multiple-points (geo)statistics (MPS)

Based on the idea that geostatistics should integrate more geologically-based information (i.e., "genetic models") to represent the spatial continuity and variability, object or Boolean models and then multiple-point geostatistical approaches were developed (Haldorsen and Damsleth 1990, Holden et al. 1998, Strebelle 2002, Krishan and Journel 2003).

Facies maps are simulated using conditional probability functions calculated from a training image. The latter is derived from outcrop observations, geophysics, image analysis, and expert knowledge to simulate the spatial distribution of the property resulting from complex geological processes (among others, Caers and Zhang 2004, Okabe and Blunt 2007, Hu and Chugunova 2008, Huysmans et al. 2008, Alcolea and Renard 2010, Mariethoz et al. 2010, Caers 2011, Comunian et al. 2011, Hermans et al. 2015). It allows also to account for the role of connectivity for the characterization of heterogeneous porous aquifers or reservoirs (Huysmans and Dassargues 2011, Renard and Allard 2013). Multiple-point geostatistics aims to overcome limitations of the variogram by moving beyond two-point correlations between variables and to obtain (cross) correlation moments at three or more locations at a time (Guardiano and Srivastava 1993, Strebelle 2002). A training image is a conceptual explicit representation of the expected spatial distribution of the property or facies type. The main idea is to borrow geological patterns from these training images and anchor them to the subsurface data domain (Huysmans and Dassargues 2009). Construction of suitable training images is critical. They should be representative of the geological heterogeneity and large enough to be statistically characterized. Training images are bound by the principles of stationarity and ergodicity (Caers and Zhang 2004). More and more complex data can be integrated in the choice of the training images through the use of learning processes. High computing costs are linked to pixel-based training image and MPS algorithms (Pirot 2017). Cloud computing, analogs, and the use of "clever tools" to include data more efficiently are needed (Rezaee *et al.* 2015, Pirot *et al.* 2017) and this research area is in full development.

12.7 Prediction focused approaches based on Bayesian evidential learning

As mentioned previously (see Section 12.5), if a detailed sensitivity approach includes directly the forecast (White 2017), not only historical data are used (as for model calibration), but also the uncertainty of predictions, for influencing the conceptual and main parameters to be considered for inversion. Here, the general workflow of groundwater modeling is changed introducing directly the simulation of prediction scenarios for detecting what are the parameters producing the largest uncertainty in the results. These parameters are then the priority of the inverse procedure. This is an efficient way of keeping only the forecast-sensitive input datasets for multiple conceptual model and parameters analysis (Kikuchi *et al.* 2015).

Recently, new ideas appeared giving less importance to the calibration or inverse modeling procedure. Revisiting the relations between data, models, results, and decision-making (Ferré 2017), the predictive ability of a groundwater model is defined as the first and unique priority in prediction-focused approaches (PFA) developed by Hermans *et al.* (2016). Inversion is replaced by a more straightforward approach focusing on prediction. It consists in finding a direct relationship between the data and the forecast by a Bayesian evidential learning process (Scheidt *et al.* 2018). In Equation 12.52 (see Section 12.6), if we assume that the uncertainty of observed data ($p(y)$) is represented by a constant coefficient (k), the conditional probability $p(y|b)$ of observing data y for a given set of parameter values b (also named likelihood function $L(y|b)$) is equal to the product of $p(b)$ and $p(b|y)$ (i.e., the *prior probability distribution* of the model parameter values multiplying the *posterior probability distribution* of the model parameter values expressing the conditional probability of obtaining a set of parameters for a given set of observed data):

$$p(b|y) = k\,p(b)\,p(y|b) = k\,p(b)\,L(y|b) \tag{12.58}$$

Here, PFA reformulates the Bayesian problem as follows:

$$p(b|y) = k\,p(b)\,L(b|y) \tag{12.59}$$

where $p(b|y)$ is the *posterior distribution probability* of obtaining the prediction results b for a given set of observed data y, $p(b)$ is the *prior distribution of probability* of the prediction results, and $L(b|y)$ is the likelihood function measuring the fit between the observed y and calculated b data. Indeed, the whole difficulty lies in finding this direct relationship between data and forecast in Equation 12.59. This high dimensional problem is reduced through reduction-dimension techniques and machine learning. For more practical details, the interested reader will refer to Scheidt *et al.* (2015), Satija and Caers (2015), Hermans *et al.* (2016), Hermans (2017), and Scheidt *et al.* (2018).

The PFA method is most efficient when the data are informative enough with respect to the predictions and if the sampled surrogate models based on the prior are realistic. Therefore, Bayesian evidential learning includes a global sensitivity analysis to assess the information content of data for the specific prediction and a prior falsification process to ensure that the prior is consistent with the data. However, one must make a significant psychological leap and accept that predictions can be made without calibrating the physically-based model. In any case, more applications of these promising methods are required to better assess their effectiveness and reliability.

References

Aboufirassi, M. and M.A. Marino. 1984. Cokriging of aquifer transmissivities from field measurements of transmissivity and specific capacity. *Journal of the International Association for Mathematical Geology* 16(1): 19–35.

Abusaada, M. and M. Sauter. 2013. Studying the flow dynamics of a karst aquifer system with an equivalent porous medium model. *Groundwater* 51: 641–650.

Ahmed, S. and G. de Marsily. 1987. Comparison of geostatistical methods for estimating transmissivity using data on transmissivity and specific capacity. *Water Resources Research* 23(9): 1717–1737.

Ahmed, S. and G. de Marsily. 1988. Combined use of hydraulic and electrical properties of an aquifer in a geostatistical estimation of transmissivity. *Ground Water* 26(1): 78–86.

Ahmed, S. and G. de Marsily. 1993. Co-kriged estimation of aquifer transmissivity as an indirect solution of inverse problem: A practical approach. *Water Resources Research* 29(2): 521–530.

Alcolea, A. and P. Renard. 2010. Blocking moving window algorithm: Conditioning multiple-point simulations to hydrogeological data. *Water Resources Research* 46: W08511.

Anderman, E., Hill, M. and E. Poeter. 1996. Two-dimensional advective transport in groundwater flow parameter estimation. *Ground Water* 34(6): 1001–1009.

Anderson, M.P., Woessner, W.W. and R.J. Hunt. 2015. *Applied groundwater modeling—Simulation of flow and advective transport.* Amsterdam: Academic Press Elsevier.

Arico, C., Nasello, C. and T. Tucciarelli. 2009. Using steady-state water level data to estimate channel roughness and discharge hydrograph. *Advances in Water Resources* 32(8): 1223–1240.

Bair, E.S., Safreed, C.M. and E.A. Stasny. 1991. A Monte Carlo-based approach for determining traveltime-related capture zones of wells using convex hulls as confidence regions. *Ground Water* 29(6): 849–855.

Barrett, M.E. and R.J. Charbeneau. 1997. A parsimonious model for simulating flow in a karst aquifer. *Journal of Hydrology* 196: 47–65.

Batelaan, O. and F. De Smedt. 2004. SEEPAGE, a new MODFLOW drain package. *Groundwater* 42(4): 576–588.

Batlle-Aguilar, J., Brouyère, S., Dassargues, A., Morasch, B., Hunkeler, D., Hohener, P., Diels, L., Vanbroekhoeven, K., Seuntjens, P. and H. Halen. 2009. Benzene dispersion and natural attenuation in an alluvial aquifer with strong interactions with surface water. *Journal of Hydrology* 361: 305–317.

Berkowitz, B., Bear, J. and C. Braester. 1988. Continuum models for contaminant transport in fractured porous formations. *Water Resources Research* 24(8): 1225–1236.

Beven, K.J. 2009. *Environmental modelling: An uncertain future? An introduction to techniques for uncertainty estimation in environmental prediction.* London: Routledge.

Beven, K. and A.M. Binley. 1992. The future of distributed models: Model calibration and uncertainty prediction. *Hydrological Processes* 6: 279–298.

Beven, K. and J. Freer. 2001. Equifinality, data assimilation, and uncertainty estimation in mechanistic modeling of complex environmental systems. *Journal of Hydrology* 249: 11–29.

Bodin, J., Ackerer, P., Boisson, A., Bourbiaux, B., Bruel, D., de Dreuzy, J.R., Delay, F., Porel G. and H. Pourpak. 2012. Predictive modelling of hydraulic head responses to dipole flow experiments in a fractured/karstified limestone aquifer: Insights from a comparison of five modelling approaches to real-field experiments. *Journal of Hydrology* 454–455: 82–100.

Borghi, A. 2013. *3D stochastic modeling of karst aquifers using a pseudo-genetic methodology*, PhD diss., Switzerland: Université de Neuchâtel.

Box, G.E.P. and D.R. Cox. 1964. An analysis of transformations. *Journal of the Royal Statistical Society: Series B* 26: 211–243, discussion 244–252.

Bravo, H.R., Jiang, F. and R.J. Hunt. 2002. Using groundwater temperature data to constrain parameter estimation in a groundwater flow model of a wetland system. *Water Resources Research* 38(8): 1153.

Brouyère, S., Wildemeersch, S., Orban, Ph., Leroy, M., Couturier, J. and A. Dassargues. 2011. The Hybrid Finite-Element Mixing-Cell method: a candidate for modelling groundwater flow and transport in karst systems, In *Proc. H2Karst, 9th Conference on Limestone Hydrogeology*, eds. C. Bertrand, N. Carry, J. Mudry, M. Pronk and F. Zwahlen, 79–82. Besançon: University of Besançon.

Brunner, P., Doherty, J. and C.T. Simmons. 2012. Uncertainty assessment and implications for data acquisition in support of integrated hydrologic models. *Water Resources Research* 48: W07513.

Brunner, P., Therrien, R., Renard, P., Simmons, C.T. and H.J. Hendricks Franssen. 2017. Advances in understanding river-groundwater interactions. *Reviews of Geophysics*, 55: 818–854.

Caers, J. 2011. *Modeling uncertainty in the Earth sciences.* Hoboken (NJ): Wiley-Blackwell.

Caers, J. and T. Zhang. 2004. Multiple-point geostatistics: a quantitative vehicle for integrating geologic analogs into multiple reservoir models. In *Integration of outcrop and modern analog data in reservoir models*, AAPG memoir 80: 383–394, Tulsa (OK): AAPG.

Capilla, J.E. and C. Llopis-Albert. 2009. Gradual conditioning of non-Gaussian transmissivity fields to flow and mass transport data: 1. Theory. *Journal of Hydrology* 371(1): 66–74.

Carrera, J., Alcolea, A., Medina, A., Hidalgo, J. and L. Slooten. 2005. Inverse problem in hydrogeology. *Hydrogeology Journal* 13(1): 206–222.

Carrera, J. and S.P. Neuman. 1986a. Estimation of aquifer parameters under transient and steady state conditions: 1. Maximum likelihood method incorporating prior information. *Water Resources Research* 22(2): 199–210.

Carrera, J. and S.P. Neuman. 1986b. Estimation of aquifer parameters under transient and steady state conditions: 2. Uniqueness, stability, and solution algorithms. *Water Resources Research* 22(2): 211–227.

Certes, C. and G. de Marsily. 1991. Application of the pilot point method to the identification of aquifer transmissivities. *Advances in Water Resources* 14(5): 284–300.

Chaminé, H.I. 2015. Water resources meet sustainability: New trends in environmental hydrogeology and groundwater engineering. *Environmental Earth Sciences* 73: 2513–2520.

Chen, Y., Durlofsky, L.J., Gerritsen, M. and X.H. Wen. 2003. A coupled local-global upscaling approach for simulating flow in highly heterogeneous formations. *Advances in Water Resources* 26: 1041–1060.

Chilès, J.P. and P. Delfiner. 1999. *Geostatistics: Modeling spatial uncertainty.* New York: John Wiley & Sons.

Christensen, S. and R.L. Cooley. 2005. *User guide to the UNC process and three utility programs for computation of nonlinear confidence and prediction intervals using MODFLOW-2000,* US Geological Survey Techniques and Methods Report 2004-1349, Reston (VA): USGS.

Comunian, A., Renard, P. and J. Straubhaar. 2011. 3D multiple-point statistics simulation using 2D training images. *Computers and Geosciences* 40: 49–65.

Cooley, R.L. 1997. Confidence intervals for ground-water models using linearization, likelihood, and bootstrap methods. *Ground Water* 35(5): 869–880.

Cooley, R.L. 2004. *A theory for modeling groundwater flow in heterogeneous media.* USGS Professional Paper 1679.

Cousquer, Y., Pryet, A., Flipo, N., Delbart, C. and A. Dupuy. 2017. Estimating river conductance from prior information to improve surface-subsurface model calibration. *Groundwater* 55: 408–418.

Dagan, G. 1989. *Flow and transport in porous formations.* New York: Springer.

Dassargues, A. 1998. Application of groundwater models in karstic aquifers, In *Karst Hydrology,* Eds. Ch. Leibundgut, J. Gunn and A. Dassargues, IAHS Publication 247: 7–14. Wallinford: IAHS Press.

Dassargues, A., Radu, J.P. and R. Charlier. 1988. Finite elements modelling of a large water table aquifer in transient conditions. *Advances in Water Resources* 11(2): 58–66.

Dassargues, A., Rentier, C. and M. Huysmans. 2006. Reducing the uncertainty of hydrogeological parameters by co-conditional simulations: lessons from practical applications in aquifers and in low permeability layers. In *Calibration and Reliability in Groundwater Modelling: From Uncertainty to Decision Making,* IAHS Publication 304: 3–9. Wallinford: IAHS Press.

de Marsily, G. 1986. *Quantitative hydrogeology: Groundwater hydrology for engineers.* San Diego: Academic Press.

de Marsily, G., Delay, F., Gonçalvès, J., Renard, Ph., Teles, V. and S. Violette. 2005. Dealing with spatial heterogeneity. *Hydrogeology Journal* 13: 161–183.

Delhomme, J.P. 1978. Kriging in the hydrosciences. *Advances in Water Resources* 1(5): 251–266.

Delhomme, J.P. 1979. Spatial variability and uncertainty in groundwater flow parameters: A geostatistical approach. *Water Resources Research* 15(2): 269–280.

Deutsch, C.V. and A.G. Journel. 1998. *GSLIB geostatistical software library and user's guide.* New-York: Oxford University Press.

Diersch, H-J.G. 2014. *Feflow—Finite element modeling of flow, mass and heat transport in porous and fractured media.* Heidelberg: Springer.

Doherty, J. 2003. Ground water model calibration using pilot points and regularization. *Ground Water* 41(2): 170–177.

Doherty, J. 2005. *PEST—Model-independent parameter estimation—User manual,* (5th Edition). Brisbane: Watermark Numerical Computing.

Doppler, T., Franssen, H.-J.H., Kaiser, H.-P., Kuhlman, U. and F. Stauffer. 2007. Field evidence of a dynamic leakage coefficient for modeling river-aquifer interactions. *Journal of Hydrology* 347: 177–187.

Eisenlohr, L., Bouzelboudjen, M., Király, L. and Y. Rossier. 1997. Numerical versus statistical modelling of natural response of a karst hydrogeological system. *Journal of Hydrology* 202(1–4): 244–262.

Engesgaard, P. 2003. *Groundwtaer flow and solute transport.* Lecture notes. Denmark: University of Copenhagen.

Ferré, T. 2017. Revisiting the relationship between data, models, and decision-making. *Groundwater* 55(5): 604–614.

Fienen, M.N. 2013. We speak for the Data. *Groundwater* 51(2): 157.

Fienen, M.N., Luo, J. and P.K. Kitanidis. 2006. A Bayesian geostatistical transfer function approach to tracer test analysis. *Water Resources Research* 42(7): W07426.

Fleckenstein, J.H. and G.E. Fogg. 2008. Efficient upscaling of hydraulic conductivity in heterogeneous alluvial aquifers. *Hydrogeology Journal* 16: 1239–1250.

Fleury, P., Ladouche, B., Conroux, Y., Jourde, H. and N. Dörfliger. 2009. Modelling the hydrologic functions of a karst aquifer under active water management—the Lez spring. *Journal of Hydrology* 365(3–4): 235–243.

Fogg, G.E. 1986. Groundwater flow and sand body interconnectedness in a thick, multiple-aquifer system. *Water Resources Research* 22(5): 679–694.

Foglia, L., Hill, M.C., Mehl, S.W. and P. Burlando. 2009. Sensitivity analysis, calibration, and testing of a distributed hydrological model using error-based weighting and one objective function. *Water Resources Research* 45: W06427.

Freeze, R.A. 1975. A stochastic-conceptual analysis of one-dimensional groundwater flow in nonuniform homogeneous media. *Water Resources Research* 11(5): 725–741.

Gardet, C., Le Ravalec, M. and E. Gloaguen. 2014. Multiscale parameterization of petrophysical properties for efficient history-matching. *Mathematical Geosciences* 46(3): 315–336.

Gelhar, L.W. 1993. *Stochastic subsurface hydrology.* Englewood Cliffs (NJ): Prentice Hall.

Gerke, H.H. and M.T. Van Genuchten. 1993. A dual-porosity model for simulating the preferential movement of water and solutes in structured porous media. *Water Resources Research* 29(2): 305–319.

Geyer, T., Birk, S., Liedl, R. and M. Sauter. 2008. Quantification of temporal distribution of recharge in karst systems from spring hydrographs. *Journal of Hydrology* 348(3–4): 452–463.

Ghasemizadeh, R., Hellweger, F., Butscher, C., Padilla, I., Vesper, D., Field, M. and A. Alshawabkeh. 2012. Review: Groundwater flow and transport modeling of karst aquifers, with particular reference to the north coast limestone aquifer system of Puerto Rico. *Hydrogeology Journal* 20: 1441–1461.

Ghasemizadeh, R., Yu, X., Butscher, C., Hellweger, F., Padilla, I. and A. Alshawabkeh. 2015. Equivalent Porous Media (EPM) simulation of groundwater hydraulics and contaminant transport in karst aquifers. *PLoS ONE* 10(9): e0138954.

Gilks, W.R., Richardson, S. and D. Spiegelhalter. 1995. *Markov chain Monte Carlo in practice.* Boca Raton: CRC press.

Gloaguen, E., Chouteau, M., Marcotte, D. and R. Chapuis. 2001. Estimation of hydraulic conductivity of an unconfined aquifer using cokriging of GPR and hydrostratigraphic data. *Journal of Applied Geophysics* 47(2): 135–152.

Goderniaux, P., Brouyère, S., Fowler, H.J., Blenkinsop, S., Therrien, R., Orban, Ph. and A. Dassargues. 2009. Large scale surface—subsurface hydrological model to assess climate change impacts on groundwater reserves. *Journal of Hydrology* 373: 122–138.

Goderniaux, P., Wildemeersch, S., Brouyère, S., Therrien, R. and A. Dassargues. 2015. Uncertainty of climate change impact on groundwater reserves. *Journal of Hydrology* 528: 108–121.

Gómez-Hernández, J.J. 2006. Complexity. *Ground Water* 44(6): 782–785.

Gómez-Hernández, J.J. and S.M. Gorelick. 1989. Effective groundwater model parameter values: Influence of spatial variability of hydraulic conductivity, leakance and recharge. *Water Resources Research* 25(3): 405–419.

Gómez-Hernández, J.J., Sahuquillo, A. and J.E. Capilla. 1997. Stochastic simulation of transmissivity fields conditional to both transmissivity and piezometric data, I, Theory. *Journal of Hydrology* 203: 162–174.

Gómez-Hernández, J.J. and X-.H. Wen. 1998. To be or not to be multi-Gaussian? A reflection on stochastic hydrogeology. *Advances in Water Resources* 21(1): 47–61.

Goovaerts, P. 1997. *Geostatistics for natural resources evaluation*. New York, Oxford: Oxford University Press.

Guardiano, F. and M. Srivastava. 1993. Multivariate geostatistics: beyond bivariate moments. In *Geostatistics-troia*, ed. A. Soares. Dordrecht: Kluwer, 133–144.

Gupta, H., Sorooshian, S. and P. Yapo. 1999. Status of automatic calibration for hydrologic models: Comparison with multilevel expert calibration. *Journal of Hydrologic Engineering* 4(2): 135–143.

Haitjema, H. 2016. Horizontal flow models that are not. *Groundwater* 54(5): 613.

Haldorsen, H.H. and E. Damsleth. 1990. Stochastic modeling. *Journal of Petroleum Technology* 42: 404–412.

Halihan, T. and C.M. Wicks. 1998. Modeling of storm responses in conduit flow aquifers with reservoirs. *Journal of Hydrology* 208: 82–91.

Harbaugh, A.W. 2005. *MODFLOW-2005: The U.S. Geological Survey modular ground-water model—The ground-water flow process*. U.S. Geological Survey Techniques and Methods 6-A16.

Hassan, A.E., Bekhit, H.M. and J.B. Chapman. 2009. Using Markov Chain Monte Carlo to quantify parameter uncertainty and its effect on predictions of a groundwater flow model. *Environmental Modelling & Software* 24(6): 749–763.

Hassane Maina, F., Delay, F. and P. Ackerer. 2017. Estimating initial conditions for groundwater flow modeling using an adaptive inverse method. *Journal of Hydrology* 552: 52–61.

Hermans, T. 2017. Prediction-focused approaches: An opportunity for hydrology. *Groundwater* 55(5): 683–687.

Hermans, T., Nguyen, F. and J. Caers. 2015. Uncertainty in training image-based inversion of hydraulic head data constrained to ERT data: Workflow and case study. *Water Resources Research* 51(7): 5332–5352.

Hermans, T., Oware, E.K. and J. Caers. 2016. Direct prediction of spatially and temporally varying physical properties from time-lapse electrical resistance data. *Water Resources Research* 52(9): 7262–7283.

Hill, M. 1992. *A computer program (MODFLOWP) for estimating parameters of a transient, three-dimensional, ground-water flow model using nonlinear regression*. Open-File Report 91-484, USGS.

Hill, M. 2006. The practical use of simplicity in developing ground water models. *Ground Water* 44(6): 775–781.

Hill, M., Cooley, R. and D. Pollock. 1998. A controlled experiment in ground-water flow model calibration using nonlinear regression. *Ground Water* 44(6): 775–781.

Hill, M.C. and C.R. Tiedeman. 2007. *Effective groundwater model calibration: With analysis of data, sensitivities, predictions, and uncertainty*. Hoboken (NJ): John Wiley & Sons.

Hoeksema, R.J. and P.K. Kitadinis. 1985. Analysis of the spatial structure of properties of selected aquifers. *Water Resources Research* 21(4): 563–572.

Hoeting, J., Madigan, D., Raftery, A. and C. Volinsky. 1999. Bayesian model averaging: A tutorial. *Statistical Science* 14(4): 382–417.

Holden, L., Hauge, R., Skare, O. and A. Skorstad. 1998. Modeling of fluvial reservoirs with object models. *Mathematical Geology* 30: 473–496.

Hu, L.Y. and T. Chugunova. 2008. Multiple-point geostatistics for modeling subsurface heterogeneity: A comprehensive review. *Water Resources Research* 44: W11413.

Hunt, R., Doherty, J. and M. Tonkin. 2007. Are models too simple? Arguments for increased parameterization. *Ground Water* 45(3): 254–262.

Hunt, R.J., Steuer, J.J., Mansor, M.T.C. and T.D. Bullen. 2001. Delineating a recharge area for a spring using numerical modeling, Monte Carlo techniques, and geochemical investigation. *Ground Water* 39(5): 702–712.

Huysmans, M. and A. Dassargues. 2006. Stochastic analysis of the effect of spatial variability of diffusion parameters on radionuclide transport in a low permeability clay layer. *Hydrogeology Journal* 14: 1094–1106.

Huysmans, M. and A. Dassargues. 2009. Application of multiple-point geostatistics on modelling groundwater flow and transport in a cross-bedded aquifer. *Hydrogeology Journal* 17(8): 1901–1911.

Huysmans, M. and A. Dassargues. 2011. Direct multiple-point geostatistical simulation of edge properties for modeling thin irregularly-shaped surfaces. *Mathematical Geosciences* 43(5): 521–536.

Huysmans, M., Peeters, L., Moermans, G. and A. Dassargues. 2008. Relating small-scale sedimentary structures and permeability in a cross-bedded aquifer. *Journal of Hydrology* 361: 41–51.

Irvine, D.J., Brunner, P., Franssen, H. and C.T. Simmons. 2012. Heterogeneous or homogeneous? implications of simplifying heterogeneous streambeds in models of losing streams. *Journal of Hydrology* 424–425: 16–23.

Isaaks, E.H. and R.M. Srivastava. 1989. *An introduction to applied geostatistics.* New York: Oxford University Press.

Jeannin, P.-Y. 2001. Modeling flow in phreatic and epiphreatic karst conduits in the Holloch Cave (Muotatal, Switzerland). *Water Resources Research* 37: 191–200.

Jourde, H., Fenart, P., Vinches, M., Pistre, S. and B. Vayssade. 2007. Relationship between the geometrical and structural properties of layered fractured rocks and their effective permeability tensor: A simulation study. *Journal of Hydrology* 337: 117–132.

Keating, E., Doherty, J., Vrugt, J. and Q. Kang. 2010. Optimization and uncertainty assessment of strongly nonlinear groundwater models with high parameter dimensionality. *Water Resources Research* 46(10): W10517.

Kerrou, J., Renard, P., Franssen, H.J.H. and I. Lunati. 2008. Issues in characterizing heterogeneity and connectivity in non-multiGaussian media. *Advances in Water Resources* 31(1): 147–159.

Kikuchi, C.P., Ferré, T.P.A. and J.A. Vrugt. 2015. On the optimal design of experiments for conceptual and predictive discrimination of hydrologic system models. *Water Resources Research* 51(6): 4454–4481.

Kitadinis, P.K. 1997. *Introduction to geostatistics: Application in hydrogeology.* Cambridge: Cambridge University Press.

Klepikova, M., Wildemeersch, S., Jamin, P., Orban, P., Hermans, T., Nguyen, F., Brouyere, S. and A. Dassargues. 2016. Heat tracer test in an alluvial aquifer: Field experiment and inverse modelling. *Journal of Hydrology* 540: 812–823.

Knotters, M., Brus, D.J. and J.O. Voshaar. 1995. A comparison of kriging, co-kriging and kriging combined with regression for spatial interpolation of horizon depth with censored observations. *Geoderma* 67(3–4): 227–246.

Koch, J., He, X., Jensen, K.H. and J.C. Refsgaard. 2014. Challenges in conditioning a stochastic geological model of a heterogeneous glacial aquifer to a comprehensive soft data set. *Hydrol. Earth Syst. Sci.* 18(8): 2907–2923.

Konikow, I. F. and J.M. Mercer. 1988. Groundwater flow and transport modelling. *Journal of Hydrology* 100(2): 379–409.

Krishan, S. and A.G. Journel. 2003. Spatial connectivity: From variograms to multiple-point measures. *Mathematical Geology* 35(8): 915–925.

Kupfersberger, H. and G. Blöschl. 1995. Estimating aquifer transmissivities—on the value of auxiliary data. *Journal of Hydrology* 165: 85–99.

Kurtz, W., Lapin, A., Schilling, O.S., Tang, Q., Schiller, E., Braun, T., Hunkeler, D., Vereecken, H., Sudicky, E., Kropf, P., Franssen, H-J.H. and P. Brunner. 2017. Integrating hydrological modelling, data assimilation and cloud computing for real-time management of water resources. *Environmental Modelling & Software* 93: 418–435.

Laloy, E., Rogiers, B., Vrugt, J.A., Mallants, D. and D. Jacques. 2013. Efficient posterior exploration of a high-dimensional groundwater model from two-stage Markov chain Monte Carlo simulation and polynomial chaos expansion. *Water Resources Research* 49(5): 2664–2682.

Lantuejoul, C. 2002. *Geostatistical simulation*. Berlin Heidelberg: Springer Verlag.

LaVenue, A., RamaRao, B., De Marsily, G. and M. Marietta. 2005. Pilot point methodology for automated calibration of an ensemble of conditionally simulated transmissivity fields 2. Application. *Water Resources Research* 31(3): 495–516.

Lee, S.Y., Carle, S.F. and G.E. Fogg. 2007. Geologic heterogeneity and a comparison of two geostatistical models: Sequential Gaussian and transition probability-based geostatistical simulation. *Advances in Water Resources* 30(9): 1914–1932.

Liu, R., Li, B., Jiang, Y. and N. Huang. 2016. Mathematical expressions for estimating equivalent permeability of rock fracture networks. *Hydrogeology Journal* 24(7): 1623–1649.

Long, J.C.S., Remer, J.S., Wilson, C.R. and P.A. Witherspoon. 1982. Porous media equivalents for networks of discontinuous fractures. *Water Resources Research* 18(3): 645–658.

Madioune Diakher, H. 2012. *Etude hydrogéologique du système aquifère du horst de Diass en condition d'exploitation intensive (bassin sédimentaire sénégalais): apport des techniques de télédétection, modélisation, géochimie et isotopie* (in French). *PhD diss.*, University of Liège, Belgium.

Mahmud, K., Mariethoz, G., Baker, A. and A. Sharma. 2015. Integrating multiple scales of hydraulic conductivity measurements in training image-based stochastic models. *Water Resources Research* 51: 465–480.

Mariethoz, G., Renard, P. and J. Caers. 2010. Bayesian inverse problem and optimization with iterative spatial resampling. *Water Resources Research* 46(11): W11530.

Matheron, G. 1967. *Eléments pour une théorie des milieux poreux (in French) [Elements for a theory of porous media]*. Paris: Masson.

Matheron, G., Beucher, H., de Fouquet, C., Galli, A., Guerillot, D. and C. Ravenne. 1987. Conditional simulation of the geometry of fluvio-deltaic reservoirs. *Society of Petroleum Engineers (SPE)* 16753-MS, Dallas: SPE.

McDonald, M.G. and A.W. Harbaugh. 1988. *A modular three-dimensional finite-difference ground-water flow model*. U.S. Geological Survey Techniques of Water-Resources Investigations, Book 6, Reston(VA): USGS.

Molz III, F.J. 2017. The development of groundwater modelling: The end of an era. *Groundwater* 55(1): 1.

Moore, C. and J. Doherty. 2005. Role of the calibration process in reducing model predictive error. *Water Resources Research* 41(5): W05020.

Moore, C., Wöhling, T. and J. Doherty. 2010. Efficient regularization and uncertainty analysis using a global optimization methodology. *Water Resources Research* 46(8): W08527.

Morel-Seytoux, H., Miller, C.D., Miracapillo, C. and S. Mehl. 2017. River seepage conductance in large-scale regional studies. *Groundwater* 55(3): 399–407.

Nash, J.E. and J.V. Sutcliffe. 1970. River flow forecasting through conceptual models part I — A discussion of principles. *Journal of Hydrology* 10(3): 282–290.

Neuman, S. 2003. Maximum likelihood Bayesian averaging of uncertain model predictions. *Stochastic Environmental Research and Risk Assessment* 17(5): 291–305.

Niswonger, R.G., Prudic, D.E., Pohl, G. and J. Constantz. 2005. Incorporating seepage losses into the unsteady streamflow equations for simulating intermittent flow along mountain front streams. *Water Resources Research* 41: W06006.

Nunes, L.M. and L. Ribeiro. 2000. Permeability field estimation by conditional simulation of geophysical data. In *Calibration and Reliability in Groundwater Modelling*, ed. F. Stauffer, W. Kinzelbach, K. Kovar and E. Hoehn, IAHS Publication 265: 117–123, Wallingford: IAHS Press.

Okabe, H. and M.J. Blunt. 2007. Pore space reconstruction of vuggy carbonates using micro-tomography and multiple-point statistics. *Water Resources Research* 43(12): W12S02.

Paleologos, E.K., Neuman, S.P. and D. Tartakovsky. 1996. Effective hydraulic conductivity of bounded, strongly heterogeneous porous media. *Water Resources Research* 32: 1333–1341.

Panday, S. and P.S. Huyakorn. 2004. A fully coupled physically-based spatially distributed model for evaluating surface/subsurface flow. *Advances in Water Resources* 27: 361–382.

Pankow, J.F., Johnson, R.L., Hewetson, J.P. and J.A. Cherry. 1986. An evaluation of contaminant migration patterns at two waste disposal sites on fractured porous media in terms of the equivalent porous medium (EPM) model. *Journal of Contaminant Hydrology* 1(1): 65–76.

Peeters, L.J.M. 2017. Assumption hunting in groundwater modeling: Find assumptions before they find you, *Groundwater*: doi:10.1111/gwat.12565.

Pirot, G. 2017. Using training images to build model ensembles with structural variability, *Groundwater* 55(5): 656–659.

Pirot, G., Linde, N., Mariethoz, G. and J.H. Bradford. 2017. Probabilistic inversion with graph cuts: Application to the Boise hydrogeophysical research site. *Water Resources Research* 53: 1231–1250.

Poeter, E. and M. Hill. 1997. Inverse models: A necessary next step in ground-water modeling. *Ground Water* 35(2): 250–260.

Poeter, E., Hill, M., Banta, E. and S. Mehl. 2005. *UCODE_2005 and six other computer codes for universal sensitivity analysis, calibration and uncertainty evaluation.* Techniques and Methods 6-A11. Reston (VA): USGS.

Pruess, K. and Y.W. Tsang. 1990. On two-phase relative permeability and capillary pressure of rough-walled rock fractures. *Water Resources Research* 26(9): 1915–1926.

Refsgaard, J.C. 1997. Parameterisation, calibration and validation of distributed hydrological models. *Journal of Hydrology* 198: 69–97.

Refsgaard, J.C. and H.J. Henriksen. 2004. Modelling guidelines—terminology and guiding principles. *Advances in Water Resources* 27: 71–82.

Remy, N., Boucher, A. and J. Wu. 2009. *Applied geostatistics with SGeMS.* Cambridge: Cambridge University Press.

Renard, P. and D. Allard. 2013. Connectivity metrics for subsurface flow and transport. *Advances in Water Resources* 51: 168–196.

Renard, P. and G. de Marsily. 1997. Calculating equivalent permeability: A review. *Advances in Water Resources* 20(5-6): 253–278.

Rentier, C., Bouyère, S. and A. Dassargues. 2002. Integrating geophysical and tracer test data for accurate solute transport modelling in heterogeneous porous media. In *Groundwater Quality 2001.* eds. S.F. Thornton and S.E. Oswald, IAHS Publication 275: 3–10. Wallingford: IAHS Press.

Rezaee, H., Marcotte, D., Tahmasebi, P. and A. Saucier. 2015. Multiple-point geostatistical simulation using enriched pattern databases. *Stochastic Environmental Research and Risk Assessment* 29: 893–913.

Ringrose, Ph. and M. Bentley. 2015. *Reservoir model design: A practitioner's guide.* Springer.

Robert, C. 2007. *The Bayesian choice: From decision-theoretic foundations to computational implementation.* New York: Springer Science & Business Media.

Rojas, R. 2009. *Uncertainty analysis in groundwater modelling: An integrated approach to account for conceptual model uncertainty.* PhD diss., Katholieke Universiteit Leuven, Belgium.

Rojas, R., Batelaan, O., Feyen, L. and A. Dassargues. 2010a. Assessment of conceptual model uncertainty for the regional aquifer Pampa del Tamarugal—North Chile. *Hydrology and Earth System Sciences* 14: 171–192.

Rojas, R., Feyen, L., Batelaan, O. and A. Dassargues. 2010b. On the value of conditioning data to reduce conceptual model uncertainty in groundwater modelling. *Water Resources Research* 46(8): W08520.

Rojas, R., Feyen, L. and A. Dassargues. 2008. Conceptual model uncertainty in groundwater modeling: Combining generalized likelihood uncertainty estimation and Bayesian model averaging. *Water Resources Research* 44: W12418.

Rojas, R., Kahundeb, S., Peeters, L., Batelaan, O. and A. Dassargues. 2010c. Application of a multi-model approach to account for conceptual model and scenario uncertainties in groundwater modelling. *Journal of Hydrology* 394: 416–435.

Rong, G., Peng, J., Wang, X., Liu, G. and D. Hou. 2013. Permeability tensor and representative elementary volume of fractured rock masses. *Hydrogeology Journal* 21(7): 1655–1671.

Rosbjerg, D. and H. Madsen. 2005. Concept of hydrologic modelling. In: *Encyclopedia of Hydrological Sciences*, ed. M.G. Anderson, 2061–2080, Chichester (UK): John Wiley & Sons.

Rubin, Y. 2003. *Applied stochastic hydrogeology.* New York: Oxford University Press.

Sanford, W.E., Plummer L.N., McAda, D.P., Bexfield, L.M. and S.K. Anderholm. 2004. Hydrochemical tracers in the Middle Rio Grande basin, USA: 2. calibration of a groundwater model. *Hydrogeology Journal* 12: 389–407.

Satija, A. and J. Caers. 2015. Direct forecasting of subsurface flow response from non-linear dynamic data by linear least squares in canonical functional principal component space. *Advances in Water Resources* 77: 69–81.

Scanlon, B.R., Mace, R.E. Barrett, M.E. and B. Smith. 2003. Can we simulate regional groundwater flow in a karst system using equivalent porous media models? Case study, Barton Springs Edwards aquifer, USA. *Journal of Hydrology* 276: 137–158.

Scheidt, C., Li, L. and J. Caers. 2018. *Quantifying uncertainty in subsurface systems.* Chichester, UK: Wiley-Blackwell.

Scheidt, C., Renard, P. and J. Caers. 2015. Prediction-focused subsurface modeling: Investigating the need for accuracy in flow-based inverse modeling. *Mathematical Geosciences* 47(2): 173–191.

Schilling, O. 2017. *Advances in characterizing surface water—groundwater interactions: combining unconventional data with complex, fully-integrated models,* PhD diss., University of Neuchatel, Switzerland.

Schwartz, F.W., Liu, G., Aggarwal, P. and C.M. Schwartz. 2017. Naïve simplicity: The overlooked piece of the complexity-simplicity paradigm. *Groundwater*: doi:10.1111/gwat.12570.

Singh, V. and S.M. Bhallamudi. 1998. Conjunctive surface-subsurface modeling of overland flow. *Advances in Water Resources* 21: 567–579.

Skahill, B.E. and J. Doherty. 2006. Efficient accommodation of local minima in watershed model calibration. *Journal of Hydrology* 329: 122–139.

Sorensen, D. and D. Gianola. 2002. *Likelihood Bayesian, and MCMC methods in quantitative genetics*, volume I. New York: Springer-Verlag.

Strebelle, S. 2002. Conditional simulation of complex geological structures using multiple point statistics. *Mathematical Geology* 34(1): 1–22.

Sudicky, E.A., Jones, J.-P., Park, Y.-J., Brookfield, E.A. and D. Colautti. 2008. Simulating complex flow and transport dynamics in an integrated surface-subsurface modeling framework. *Geosciences Journal* 12(2): 107–122.

Sudicky, E.A. and R.G. McLaren. 1992. The Laplace transform Galerkin technique for large-scale simulation of mass transport in discretely-fractured porous formations. *Water Resources Research* 28(2): 499–514.

Sulis, M., Paniconi, C., Marrocu, M., Huard, D. and D. Chaumont. 2012. Hydrologic response to multimodel climate output using a physically based model of groundwater/surface water interactions. *Water Resources Research* 48: W12510.

Therrien, R., McLaren, R.G., Sudicky, E.A. and S.M. Panday. 2006. *Hydrogeosphere.* Waterloo, Canada: Groundwater Simulations Group, University of Waterloo.

Therrien, R. and E.A. Sudicky. 1996. Three-dimensional analysis of variably saturated flow and solute transport in discretely-fractured porous media. *Journal of Contaminant Hydrology* 23(1-2): 1–44.

Tikhonov, A. and V. Arsenin. 1977. *Solutions for ill-posed problems.* Hoboken (NJ): John Wiley & Sons.

Tonkin, M. and J. Doherty. 2005. An hybrid regularized inversion methodology for highly parameterized environmental models. *Water Resources Research* 41(10): W10414.

Tonkin, M.J., Tiedeman, C.R., Ely, M.D. and M.C. Hill. 2007. *OPR-PPR, a computer program for assessing data importance to model predictions using linear statistics.* USGS, Techniques and Methods TM-6E2, Reston (VA): USGS.

Van den Doel, K. and U. Ascher. 2006. On level set of regularization for highly ill-posed distributed parameter estimation problems. *Journal of Computational Physics* 216(2): 707–723.

VanderKwaak, J. 1999. *Numerical simulation of flow and chemical transport in integrated surface-subsurface hydrologic systems.* PhD diss., University of Waterloo, Ontario, Canada.

van Leeuwen, M. 2000. *Stochastic determination of well capture zones conditioned on transmissivity data.* PhD diss., Imperial College of Science, University of London, London.

van Leeuwen, M., Butler, A.P., te Stroet, C.B.M. and J.A. Tompkins. 2000. Stochastic determination of well capture zones conditioned on regular grids of transmissivity measurements. *Water Resources Research* 36(4): 949–957.

van Leeuwen, M., te Stroet, C., Butler, A.P. and J.A. Tompkins. 1998. Stochastic determination of well capture zones. *Water Resources Research* 34(9): 2215–2223.

Varljen, M.D. and J.M. Shafer. 1991. Assessment of uncertainty in time-related capture zones using conditional simulation of hydraulic conductivity. *Ground Water* 29(5): 737–748.

Vecchia, A.V. and R.L. Cooley. 1987. Simultaneous confidence and prediction intervals for nonlinear regression models with application to a groundwater flow model. *Water Resources Research* 23(7): 1237–1250.

Vesselinov, V., Neuman, S. and W. Illman. 2001. Three-dimensional numerical inversion of pneumatic cross-hole tests in unsaturated fractured tuff 2. Equivalent parameters, high resolution stochastic imaging and scale effects. *Water Resources Research* 37(12): 3019–3041.

Vrugt, J.A., Gupta, H.V., Bastidas, L.A., Bouten, W. and S. Sorooshian. 2003. Effective and efficient algorithm for multiobjective optimization of hydrologic models. *Water Resources Research* 39(8): 1214.

Wang, H.F. and M.P. Anderson. 1982. *Introduction to groundwater modeling: Finite difference and finite element methods.* San Diego (CA): Academic Press.

Wang, J.S.Y. and T.N. Narasimhan. 1985. Hydrologic mechanisms governing fluid flow in a partially saturated, fractured, porous medium. *Water Resources Research* 21(12): 1861–1874.

Ward, D. 2005. The simplicity cycle: Simplicity and complexity in design. *Defense Acquisition, Technology, and Logistics* 34(6): 18–21.

Warren, J.E. and H.S. Price. 1961. Flow in heterogeneous porous media. *Society of Petroleum Engineers Journal* I: 153–169.

Wen, X.H. and J.J. Gómez-Hernández. 1998. Numerical modeling of macrodispersion in heterogeneous media: A comparison of multi-Gaussian and non-multi-Gaussian models. *Journal of Contaminant Hydrology* 30(1): 129–156.

White, J. 2017. Forecast first: An argument for groundwater modeling in reverse. *Groundwater* 55(5): 660–664.

Whittaker, J. and G. Teutsch. 1999. Numerical simulation of subsurface characterization methods: Application to a natural aquifer analogue. *Advances in Water Resources* 22(8): 819–829.

Wildemeersch, S. 2012. *Assessing the impacts of technical and structure choices on groundwater model performance using a complex synthetic case*. PhD diss., Belgium: University of Liège.

Wildemeersch, S., Goderniaux, P., Orban, P., Brouyère, S. and A. Dassargues. 2014. Assessing the effects of spatial discretization on large-scale flow model performance and prediction uncertainty. *Journal of Hydrology* 510: 10–25.

Yeh, T.C.J., Liu, S., Glass, R.J., Baker, K., Brainard, J.R., Alumbaugh, D. and D. LaBrecque. 2002. A geostatistically based inverse model for electrical resistivity surveys and its applications to vadose zone hydrology. *Water Resources Research* 38(12): 1278.

Yoon, H., Hart, D.B. and S.A. McKenna. 2013. Parameter estimation and predictive uncertainty in stochastic inverse modeling of groundwater flow: Comparing null-space Monte Carlo and multiple starting point methods. *Water Resources Research* 49(1): 536–553.

Zhang, D. 2002. *Stochastic methods for flow in porous media*. San Diego (CA): Academic Press.

Zhang, Y., Gable, C.W. and M. Person 2006. Equivalent hydraulic conductivity of an experimental stratigraphy: Implications for basin-scale flow simulations. *Water Resources Research* 42(5): W05404.

Zinn, B. and C. F. Harvey. 2003. When good statistical models of aquifer heterogeneity go bad: A comparison of flow, dispersion, and mass transfer in connected and multivariate Gaussian hydraulic conductivity fields. *Water Resources Research* 39(3): 1051.

Chapter 13

Main principles of numerical techniques used in groundwater modeling

13.1 Introduction and terminology

The aim of this chapter is to provide the reader with basic numerical techniques for the solution of the groundwater flow and transport equations. Most of the developments are described and demonstrated in saturated and 2D horizontal conditions. Generalization in 3D is not an issue but just heavier to handle in the equations. Foundations about modeling techniques will allow the reader to understand what is numerically done, for gaining confidence using commercial software. The interested and more specialized reader will also be prepared to make the next steps with more advanced techniques dealing with partially saturated conditions, density dependent flow and solute transport, heat transport, and other related processes. The main numerical techniques applied in groundwater modeling as Finite Differences (FD), Finite Elements (FE), and Finite Volumes (FV) are summarized. It is far beyond the scope of this chapter to give an exhaustive description of those techniques that are described in dedicated books.

As mentioned previously, analytical solutions (even approximated) are only limited to very simplified situations. They are usually not applicable for finding solutions in most cases implying irregular boundaries, heterogeneity translated in spatial variability of the parameters, nonlinearities and coupling, stress-factors represented more accurately than just points, and lines or integrals over an interval. Computer-based numerical models have been developed for decades taking advantage of the fast developments in computer technologies improving continuously processors, storage capacities, and parallel and cloud computing facilities (Kropf *et al.* 2014, Hayley 2017, Kurtz *et al.* 2017).

Using numerical techniques allows to transform the *partial differential equations (PDEs)* containing continuous variables describing the process to be simulated, in a system of numerous *linear algebraic equations* that contain discrete variables. The obtained system of equations is then solved by applying numerical analysis techniques involving most often iterative procedures more efficient than matrix inversion. The solution consists in the values of the main variable(s) in discrete locations of the simulated domain that are the result of the domain *spatial discretization*. If the problem is transient, the timescale is also discretized in time steps. Once the solution is found in the n discrete nodes and for all time steps, interpolations can be calculated in space and time to obtain the simulated values in any point and at any time.

A few specific definitions are useful for the user of numerical techniques (Bear and Cheng 2010, Diersch 2014).

The *truncation (or approximation) error* denotes the difference between any value approximated by the chosen numerical technique and the exact value (i.e., due to the approximation method). The *roundoff error* is due to the fact that any simulated value is rounded up by the computer according to the chosen degree of precision. *Truncation* and *roundoff* errors are usually added and named *numerical errors* due to *numerical approximations* of the actual exact value.

In the process of finding iteratively a solution, the *convergence* of a solution denotes that the computed values converge toward the exact values, especially when the spacing between nodes is decreasing. This term should not be confused with the *stability*. This expresses that the numerical errors (truncation + roundoff) should not increase in the solution computation within one time step or from one time step to the next ones.

Conservativity is preserved if the used equations express balance equations for the considered processes. The numerical solutions must preserve and satisfy balance equations at the local as well as at the global scales (Diersch 2014).

Physical consistency (Paniconi and Putti 2015) is depending on the conceptual choices to simplify the reality for an efficient modeling (see Chapter 12). *Numerical consistency* is checked if for decreasing mesh increments and time steps the truncation error is tending to zero.

Accuracy is the general property describing how far the modeling errors are low. Indeed, modeling errors are made of truncation and roundoff errors but also and mostly of conceptual and calibration errors (see Chapter 12).

The *resolution* of a model refers to the smallest increment or decrement of variable value that can be calculated by the model.

13.2 Numerical techniques for groundwater flow modeling

Many numerical methods are used for groundwater modeling. The two most important methods are the Finite Difference Method (FDM) for its simplicity of the underlying mathematics and its use in well-known commercial software, and the Finite Element Method (FEM) for its geometrical flexibility well adapted to complex geological problems. As will be described further, the Finite Volume Method (FVM) is inherently a FEM using low-order elements (with linear relations) (Diersch 2014).

Finite difference method (FDM)

Steady-state groundwater flow

Let's suppose that the variations of piezometric head describe a space and time continuous function $h(x, t)$ (with x the generalized coordinates). A 1D spatial approximation of the gradient by a finite difference is written:

$$\frac{\partial h}{\partial x} \approx \frac{h(x + \Delta x) - h(x)}{\Delta x} \tag{13.1}$$

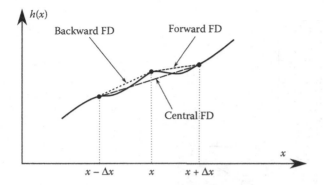

Figure 13.1 Forward, backward, and central finite difference (FD) approximations of the $h(x)$ local gradient.

where $h(x)$ and $h(x + \Delta x)$ are values of the function at two discrete locations distant of Δx (Figure 13.1). The approximation is better for smaller Δx, and it is called a *forward FD approximation*. If $-\Delta x$ is taken in place of Δx, a *backward FD approximation* is found. As shown in Figure 13.1, a *central FD approximation* can also be defined as:

$$\frac{\partial h}{\partial x} \approx \frac{h(x + \Delta x) - h(x - \Delta x)}{2\Delta x} \tag{13.2}$$

HOMOGENEOUS CONDITIONS

To start with very simple mathematics, let's assume a steady-state 2D saturated horizontal groundwater flow in isotropic and homogeneous conditions. Equation 4.54 is then simplified in:

$$\frac{\partial}{\partial x}\left(T \frac{\partial h}{\partial x}\right) + \frac{\partial}{\partial y}\left(T \frac{\partial h}{\partial y}\right) + q'' = 0 \tag{13.3}$$

and if we assume no stress factor ($q'' = 0$), Equation 13.3 becomes the Laplace equation:

$$\frac{\partial^2 h}{\partial x^2} + \frac{\partial^2 h}{\partial y^2} = 0 \tag{13.4}$$

Let's remind that the definition of a partial derivative of a variable $h(x)$ in function of x is as follows:

$$\frac{\partial h}{\partial x} = \lim_{\Delta x \to 0}\left(\frac{h(x + \Delta x) - h(x)}{\Delta x}\right) \tag{13.5}$$

2D view

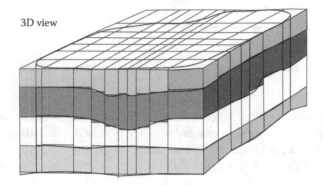

3D view

Figure 13.2 Finite difference discretization in 2D and 3D.

This limit ($\Delta x \to 0$) can be approximated by awarding an arbitrary small value to Δx. In fact, the spatial discretization into small cells of an orthogonal grid (Figure 13.2) allows to calculate easily this approximation in all the modeled domain. As the approximation is better when Δx is small, the cells are made smaller in areas where more accuracy is needed. In most cases, and particularly in the well-known MODFLOW code (McDonald and Harbaugh 1988) the nodes are selected in the center of the cells: the FDM is called "Block Centered Finite Difference" (BCFD) method. In 2D, the nodes are numbered sequentially and can then be referred to as nodes (i, j) with a piezometric head h_{ij} (column i, raw j). In 3D, k layers are added and a given piezometric head is noted h_{ijk} (column i, raw j, layer k).

In 2D, the continuous function $h(x, y)$ can be expanded into a Taylor series in the positive x direction (forward finite difference):

$$h(x + \Delta x) = h(x) + \Delta x \frac{\partial h(x)}{\partial x} + \frac{(\Delta x)^2}{2!} \frac{\partial^2 h(x)}{\partial x^2} + \frac{(\Delta x)^3}{3!} \frac{\partial^3 h(x)}{\partial x^3} + \cdots + \frac{(\Delta x)^n}{n!} \frac{\partial^n h(x)}{\partial x^n}$$

(13.6)

If Δx is small enough, the terms of third order and greater can be omitted because each successive term is of smaller and smaller magnitude:

$$h(x + \Delta x) \approx h(x) + \Delta x \frac{\partial h(x)}{\partial x} + \frac{(\Delta x)^2}{2!} \frac{\partial^2 h(x)}{\partial x^2}$$

(13.7)

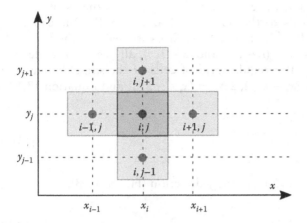

Figure 13.3 A grid detail centered on the considered cell i, j in the reference system (x, y).

This approximation is of the *second order* (or second order accurate) as all neglected terms are of order $(\Delta x)^3$ or higher. In a similar way, a backward finite difference is obtained if Δx is replaced by $-\Delta x$. Considering the piezometric head $h_{ij}(x, y)$ at the node i, j (Figure 13.3), the forward and backward finite difference can be written respectively:

$$h_{i+1j} \approx h_{ij} + (x_{i+1} - x_i)\frac{\partial h}{\partial x} + \frac{(x_{i+1} - x_i)^2}{2}\frac{\partial^2 h}{\partial x^2} \tag{13.8}$$

and

$$h_{i-1j} \approx h_{ij} + (x_{i-1} - x_i)\frac{\partial h}{\partial x} + \frac{(x_{i-1} - x_i)^2}{2}\frac{\partial^2 h}{\partial x^2} \tag{13.9}$$

Rearranging Equations 13.8 and 13.9 gives:

$$\frac{h_{i+1j} - h_{ij}}{(x_{i+1} - x_i)} \approx \frac{\partial h}{\partial x} + \frac{(x_{i+1} - x_i)}{2}\frac{\partial^2 h}{\partial x^2} \tag{13.10}$$

and

$$\frac{h_{i-1j} - h_{ij}}{(x_i - x_{i-1})} \approx -\frac{\partial h}{\partial x} + \frac{(x_i - x_{i-1})}{2}\frac{\partial^2 h}{\partial x^2} \tag{13.11}$$

Adding Equations 13.10 and 13.11, a unique equation providing a value to $\partial^2 h/\partial x^2$ is obtained:

$$\frac{\partial^2 h}{\partial x^2} \approx \frac{2}{(x_{i+1} - x_{i-1})}\left[\frac{h_{i+1j}}{(x_{i+1} - x_i)} - \left(\frac{1}{(x_{i+1} - x_i)} + \frac{1}{(x_i - x_{i-1})}\right)h_{ij} + \frac{h_{i-1j}}{(x_i - x_{i-1})}\right] \tag{13.12}$$

where the coordinates x_{i+1} and x_{i-1} are respectively the x coordinates of the next and previous nodes (with regards to the node i, j) (Figure 13.3). One can easily observe that the right-hand side of Equation 13.12 is expressed in function of the piezometric heads in the nodes (i, j), $(i + 1, j)$ and $(i - 1, j)$, all other things depending only on nodes coordinates. If a regular grid of cells of the same dimension is chosen, then $\Delta x = (x_{i+1} - x_i) = (x_i - x_{i-1})$, $2\Delta x = (x_{i+1} - x_{i-1})$ and Equation 13.12 is simplified in:

$$\frac{\partial^2 h}{\partial x^2} \approx \frac{h_{i+1j} - 2h_{ij} + h_{i-1j}}{(\Delta x)^2} \tag{13.13}$$

A similar equation is found according to the y direction (Figure 13.3):

$$\frac{\partial^2 h}{\partial y^2} \approx \frac{h_{ij+1} - 2h_{ij} + h_{ij-1}}{(\Delta y)^2} \tag{13.14}$$

If we assume $\Delta x = \Delta y = \Delta m$ (i.e., a grid with squared cells of side dimension Δm), the Laplace equation describing our simplified groundwater flow (Equation 13.4) is thus approximated by:

$$\frac{\partial^2 h}{\partial x^2} + \frac{\partial^2 h}{\partial y^2} \approx \frac{(h_{i+1j} + h_{i-1j} + h_{ij+1} + h_{ij-1} - 4h_{ij})}{(\Delta m)^2} = 0 \tag{13.15}$$

For this particular case we obtain for our unknown:

$$h_{ij} = \frac{1}{4}(h_{i+1j} + h_{i-1j} + h_{ij+1} + h_{ij-1}) \tag{13.16}$$

where the calculated piezometric head at a node is equal to the average of the piezometric head values in the four direct neighboring nodes. As many algebraic equations, similar to Equation 13.15, can be written as the model does count internal nodes. On the boundaries, each node corresponds a prescribed equation according to the chosen kind of BC. In total, a system of n equations with n unknown h-values (i.e., one in each node) must be solved.

HETEROGENEITY AND GROUNDWATER CONSERVATION

Consider a cell i, j as represented in Figure 13.4 (i.e., with its four corners A, B, C, D), the conservation law (here in 2D and in steady-state) requires that the sum of the flows into and out of the cells is zero:

$$\int_{AB} T_x \frac{\partial h}{\partial x} dy + \int_{BC} T_y \frac{\partial h}{\partial y} dx + \int_{CD} T_x \frac{\partial h}{\partial x} dy + \int_{DA} T_y \frac{\partial h}{\partial x} dx + Q_{ij} = 0 \tag{13.17}$$

where Q_{ij} is a source term (i.e., positive when water is added to the considered cell from outside).

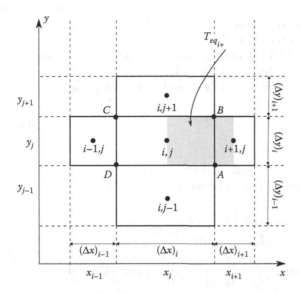

Figure 13.4 The considered cell i, j and its four neighboring cells. In heterogeneous conditions, each cell may be characterized by an individual transmissivity value.

$T_{eq_{i+}}, T_{eq_{j+}}, T_{eq_{i-}}, T_{eq_{j-}}$ are defined respectively as the equivalent transmissivity to be calculated between the cell i, j and its four neighboring cells. Then Equation 13.17 becomes:

$$T_{eq_{i+}}\left(\frac{h_{i+1} - h_{ij}}{x_{i+1} - x_i}\right)(\Delta y)_j + T_{eq_{j+}}\left(\frac{h_{j+1} - h_{ij}}{y_{j+1} - y_j}\right)(\Delta x)_i + T_{eq_{i-}}\left(\frac{h_{i-1} - h_{ij}}{x_{i-1} - x_i}\right)(\Delta y)_j$$

$$+ T_{eq_{j-}}\left(\frac{h_{j-1} - h_{ij}}{y_{j-1} - y_j}\right)(\Delta x)_i + Q_{ij} \approx 0 \tag{13.18}$$

where $(\Delta x)_i$ and $(\Delta y)_j$ are the size of the cell i, j respectively in the x and y directions (Figure 13.4). If a large difference exists between the cell sizes in the x direction and those in the y direction (e.g., $\Delta x \gg \Delta y$) very contrasted coefficients appear in the algebraic linear equation (Equation 13.18). This produces very bad numerical conditions (large roundoff errors due to an ill-conditioned matrix) for solving the whole system of n equations.

Let's assume that the heterogeneity of the geological medium is described by different transmissivity values introduced in each cell (T_{ij} in cell i, j; T_{i+1j} in cell $i + 1, j$; T_{i-1j} in cell $i - 1, j$; T_{ij+1} in cell $i, j + 1$; and T_{ij-1} in cell $i, j - 1$). Equivalent transmissivity values between cells are calculated applying Equation 4.31 (see Chapter 4) taking into account that the flow through the boundary is indeed orthogonal to the heterogeneity boundary:

$$T_{eq_{i+}} = \frac{T_{i+1j}T_{ij}((\Delta x)_{i+1} + (\Delta x)_i)}{(\Delta x)_{i+1}T_{ij} + (\Delta x)_i T_{i+1j}} \tag{13.19}$$

where $(\Delta x)_{i+1}$ is the size of the cell $i + 1, j$ in the x direction. It describes a weighed harmonic mean between the transmissivity values of cell i, j, and cell $i + 1, j$. The weighting is depending on the dimensions of the cells. $T_{eq_{j+}}, T_{eq_{i-}}, T_{eq_{j-}}$ can be found in a similar way using $(\Delta y)_{j+1}$ the size of the cell $i, j + 1$ in the y direction, $(\Delta x)_{i-1}$ the size of the cell $i - 1, j$ in the x direction, and $(\Delta y)_{j-1}$ the size of the cell $i, j - 1$ in the y direction.

If the considered cells are of the same size, Equation 13.19 is simplified in:

$$T_{eq_{i+}} = \frac{2T_{i+1j}T_{ij}}{T_{ij} + T_{i+1j}} \tag{13.20}$$

If all the cells have the same sizes in x and y directions (i.e., a mesh of square cells), then $\Delta x = \Delta y = \Delta m$ and Equation 13.18 can be written:

$$\frac{2T_{i+1j}T_{ij}}{(T_{ij} + T_{i+1j})}\left(\frac{h_{i+1j} - h_{ij}}{\Delta m}\right)\Delta m + \frac{2T_{ij+1}T_{ij}}{(T_{ij} + T_{ij+1})}\left(\frac{h_{ij+1} - h_{ij}}{\Delta m}\right)\Delta m$$
$$\frac{2T_{i-1j}T_{ij}}{(T_{ij} + T_{i-1j})}\left(\frac{h_{i-1j} - h_{ij}}{\Delta m}\right)\Delta m + \frac{2T_{ij-1}T_{ij}}{(T_{ij} + T_{ij-1})}\left(\frac{h_{ij-1} - h_{ij}}{\Delta m}\right)\Delta m + Q_{ij} \approx 0 \tag{13.21}$$

And it could be easily verified that for an isotropic and homogenous medium and without source term ($Q_{ij} = 0$), Equation 13.16 can be calculated back.

Note that the finite difference method can be applied to more complex grids as for example nested mesh with rectangular cells (Figure 13.5). In that particular case, the

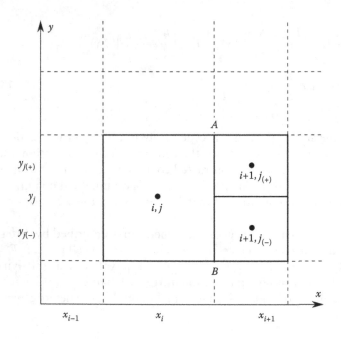

Figure 13.5 In nested meshes, the equivalent transmissivity calculated between cells is more complex, considering individual transmissivity values of more neighboring cells.

calculation of equivalent transmissivity values between cells is more complex. It is done stating the conservation between the considered cell and two neighboring cells (i.e., cells $i + 1$, $j^{(+)}$ and $i + 1$, $j^{(-)}$) instead of only one:

$$\int_{AB} T_x \frac{\partial h}{\partial x} dy =$$

$$\frac{4T_{ij} T_{i+1j^{(+)}} T_{i+1j^{(-)}}}{(T_{ij}T_{i+1j^{(+)}} + T_{i+1j^{(+)}}T_{i+1j^{(-)}} + T_{i+1j^{(-)}}T_{ij})} (h_{i+1j^{(+)}} - h_{ij} + h_{i+1j^{(-)}} - h_{ij}) \tag{13.22}$$

In some particular applications, polygonal cells using Thiessen or Voronoi polygons were adopted. It could be linked to the finite volume method (FVD) (see Section 13.3) as it is a way to calculate the "control volume" of each node of a mesh.

BOUNDARY CONDITIONS

As mentioned previously, as many algebraic equations similar to Equation 13.18 can be written as internal nodes. For nodes located on the boundaries, BCs are used (see Chapter 12, Section 12.3). If a piezometric head is prescribed (Dirichlet BC), a very simple equation of the form $h_{ij} = Cst$ is added. If a flux is prescribed on a boundary (Neumann BC), the corresponding term in Equation 13.18 is replaced by the imposed value. If a flux depending on the piezometric head is prescribed (Cauchy BC), the corresponding term in Equation 13.18 is replaced by the relation linking the piezometric head with the flux.

PRACTICAL RECOMMENDATIONS

From the main characteristics of this numerical method, some practical recommendations can be deduced for the user of a software based on the BCFD technique.

- An initial field of values for the main unknown variable (piezometric head) is needed as an iterative numerical technique is most often used in commercial as scientific codes.
- Accuracy increases with number of cells but portability decreases. A compromise must be found between accuracy and portability of the model.
- Use smaller cells where a steep gradient of the main variable is expected.
- In the spatial discretization, as far as possible, nodes should correspond with pumping wells and observation piezometers: it will make comparison between measured and computed values easier (less uneasy at least).
- Avoid distances between nodes (central points) higher than 1.5 the former one: it is due to the weighted equivalent transmissivity (or hydraulic conductivity in 3D) calculation between cells (see Equation 13.19). If a cell is far longer than the former one, the influence of the transmissivity (or hydraulic conductivity) of the smaller one will be nearly negligible, at least smeared off, while it was perhaps important to have a strong contrast.

- Avoid ratios higher than 1/10 for the cell dimensions, as it would induce highly contrasted coefficients in the equations (see Equation 13.18) producing bad numerical conditions (roundoff errors) for solving the whole system of n equations.
- If possible, boundaries with a prescribed head should preferably correspond to nodes (central points of the cells), while boundaries with a prescribed flux should be better chosen corresponding to sides of the cells (where the flux condition is calculated).

Transient groundwater flow

Keeping the same simplifying assumptions in the 2D groundwater flow equation, but introducing a second member for describing the transient conditions, another way to write Equation 4.90 is:

$$T\left(\frac{\partial^2 h}{\partial x^2} + \frac{\partial^2 h}{\partial y^2}\right) + q'' = S\frac{\partial h}{\partial t} \tag{13.23}$$

with T and S assumed constant in the modeled domain (homogeneity and isotropy), q'' is the water flow rate per unit surface of the geological medium that is withdrawn ($q'' < 0$) or injected ($q'' > 0$) (see Section 4.11). Let's also assume that the spatial discretization corresponds to a uniform grid with square cells ($\Delta x = \Delta y = \Delta m$). The time derivative is replaced by a finite difference:

$$\frac{\partial h}{\partial t} = \frac{h(t + \Delta t) - h(t)}{\Delta t} \tag{13.24}$$

where Δt is the time step. A time step is the result of the time discretization. It is the subdivision of the time line in short periods during which the stress factors on the modeled system remain constant. A smaller time step provides a better estimate of the time derivative. The approximation error caused by the time discretization is directly proportional to the time step size (i.e., the time discretization error is of the first-order). The solution of the problem is computed for each time step and most often results are produced as model output at the end of each time step. In fact, if a new stress factor is started at the beginning of a considered time step, the variable of the system (e.g., piezometric head) evolves during the whole duration of the time step. One can physically observe that this evolution must be faster at the beginning of the time step than at the end (Figure 13.6).

Using the results of Equation 13.15 for approximating the spatial derivatives, and results of Equation 13.24 for the time derivative, Equation 13.23 for the cell i, j is now approximated by:

$$\frac{T}{(\Delta m)^2}(h_{i+1j} + h_{i-1j} + h_{ij+1} + h_{ij-1} - 4h_{ij}) + Q_{ij} = S\frac{h_{ij}(t + \Delta t) - h_{ij}(t)}{\Delta t} \tag{13.25}$$

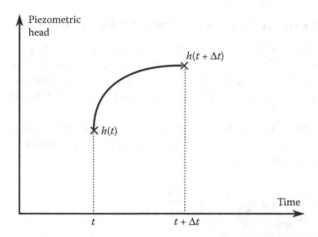

Figure 13.6 Evolution of the piezometric head in one point during a time step in reaction to a given stress factor (i.e., here injection of water in the aquifer). The time evolution is more rapid at the beginning of the time step than at the end.

where Q_{ij} is the water flow rate per unit surface withdrawn (<0) or injected (>0) at the node i, j (e.g., recharge due to infiltration). The actual unknown of Equation 13.25 is $h_{ij}(t + \Delta t)$ and the question now is at what time do we evaluate the values of the piezometric heads (at the node i, j and at the four neighboring nodes) in the left-hand member ?

EXPLICIT TIME INTEGRATION SCHEME

The most intuitive choice is to choose the time t (i.e., at the beginning of the considered time step) for the piezometric head values of the left-hand side member of Equation 13.25. This equation can then be written in a *fully explicit* way:

$$h_{ij}(t + \Delta t) = h_{ij}(t) + \frac{Q_{ij}\Delta t}{S}$$

$$+ \frac{T\Delta t}{(\Delta m)^2 S}(h_{i+1j}(t) + h_{i-1j}(t) + h_{ij+1}(t) + h_{ij-1}(t) - 4h_{ij}(t)) \tag{13.26}$$

The only unknown of this equation is $h_{ij}(t + \Delta t)$ as the piezometric values at time t are all known. An equation similar to Equation 13.26 is written for each node and each of the n equations has only one unknown. This system could eventually be solved equation per equation. Unfortunately, this explicit time integration is not ideal physically and numerically. Physically, as shown schematically on Figure 13.6, calculating $h_{ij}(t + \Delta t)$ from all values taken at the time t is far from the most accurate scheme. Numerically, if the time step is too long, unstability may appear (see example described in Box 13.1). This time integration scheme is only conditionally stable. A criterion should be met to achieve stability in numerical computation.

Box 13.1   Stability of an explicit FD time integration scheme? Example

Equation 13.26 is used on the simplified example of a squared aquifer with pre-scribed heads on all lateral boundaries and a uniform recharge. Detailed data are the following: initial value $h = 10$ m, BCs: $h = 10$ m, recharge: 0.002 m/day, $S = 0.4$ $T = 100$ m^2/day, $\Delta t = 10$ days, $\Delta m = 50$ m. So that $(Q\Delta t)/S = 0.05$ and $(T\Delta t)/((\Delta m)^2 S) = 0.25$.

At the initial time, all h values are 10. For a first time step, all internal nodes have a $h = 10.05$. For a second time step and a third time step, the piezometric head evolves toward logical values due to recharge.

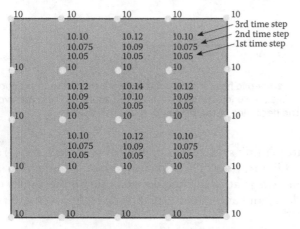

Now, if $\Delta t = 40$ days, $(Q\Delta t)/S = 0.2$ and $(T\Delta t)/((\Delta m)^2 S) = 1$.

It can be observed already at the third time step that strange h values are already obtained ($h = 9.8$ in the aquifer center and $h = 10.6$ in the corners). It is actually showing that the time integration scheme induces unstabilities of the numerical computation.

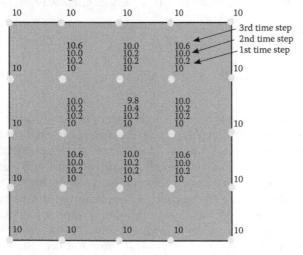

It could be simply found by expressing that the numerical error ε should not increase in the solution from a time step to the next ones (Box 13.2). It is expressed in function of the properties of the aquifer (T and S) and indeed in function of the chosen cell-size (Δm) and the time step (Δt):

$$\frac{T\Delta t}{(\Delta m)^2 S} \leq \frac{1}{4} \tag{13.27}$$

For rectangular cells ($\Delta x \neq \Delta y$) it can be written (Peaceman 1977):

$$\frac{T}{S}\left(\frac{\Delta t}{(\Delta x)^2} + \frac{\Delta t}{(\Delta y)^2}\right) \leq \frac{1}{2} \tag{13.28}$$

Box 13.2   Stability criterion for an explicit FD time integration scheme

Starting from the example of Box 13.1, a very simplified way to find this criterion is proposed considering a zero recharge. In this case, the value of h should converge time step after time step to 10. At a given time t, the worst initial h values are $(10 - \varepsilon)$ in the aquifer center and $(10 + \varepsilon)$ in the four neighboring nodes as the ε will be summed up in the right-hand member of Equation 13.26. In that case, with $(Q\Delta t)/S = 0$ and $(T\Delta t)/((\Delta m)^2 S) = \alpha$, it can be written:

$$h_{ij}(t + \Delta t) = (10 - \varepsilon) + 0 + \alpha(8\varepsilon) \leq (10 - \varepsilon) = h_{ij}(t)$$

and

$(8\alpha - 1)\varepsilon \leq \varepsilon$ so that $\alpha \leq (1/4)$.

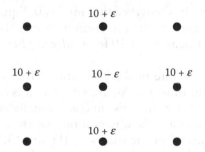

IMPLICIT TIME INTEGRATION SCHEME

At the opposite of the former choice, one may choose that the piezometric head values of the left-hand side member of Equation 13.25 would be taken at time

$(t + \Delta t)$ (i.e., at the end of the considered time step). Then we obtain an implicit equation as the unknown $h_{ij}(t + \Delta t)$ is expressed in function of other (unknown) piezometric heads at the four neighboring nodes. Equation 13.25 is now written in a *fully implicit* way:

$$h_{ij}(t + \Delta t)[1 + 4\alpha] = h_{ij}(t) + \frac{Q_{ij}\Delta t}{S}$$

$$+ \frac{T\Delta t}{(\Delta m)^2 S}(h_{i+1j}(t + \Delta t) + h_{i-1j}(t + \Delta t) + h_{ij+1}(t + \Delta t) + h_{ij-1}(t + \Delta t)) \tag{13.29}$$

If it is apparently mathematically more complex (i.e., one needs the n equations for solving the system of n unknowns), it presents the big advantage to be unconditionally stable (Boxes 13.3 and 13.4). However, physically, calculating $h_{ij}(t + \Delta t)$ from all h values taken at the time $(t + \Delta t)$ is not the best option (even already better than at time t) (Figure 13.6). The error grows with the size of the time step (Bear and Cheng 2010).

CRANK-NICOLSON AND GALERKIN TIME INTEGRATION SCHEMES

If we rewrite Equation 13.25 replacing $(h_{i+1j} + h_{i-1j} + h_{ij+1} + h_{ij-1} - 4h_{ij})$ by a weighted average of those values taken at times t and $(t + \Delta t)$, we obtain:

$$\frac{T}{(\Delta m)^2}(1 - \theta)(h_{i+1j}(t) + h_{i-1j}(t) + h_{ij+1}(t) + h_{ij-1}(t) - 4h_{ij}(t))$$

$$+ \frac{T}{(\Delta m)^2}\theta(h_{i+1j}(t + \Delta t) + h_{i-1j}(t + \Delta t) + h_{ij+1}(t + \Delta t) + h_{ij-1}(t + \Delta t))$$

$$- 4h_{ij}(t + \Delta t) + Q_{ij} = S\frac{h_{ij}(t + \Delta t) - h_{ij}(t)}{\Delta t} \tag{13.30}$$

where θ is called the time integration coefficient. It is observed that for $\theta = 0$, Equation 13.30 reduces to Equation 13.26 for a *fully explicit time integration scheme*. In the same way, if $\theta = 1$, Equation 13.30 reduces to Equation 13.29 for a *fully implicit time integration scheme*.

Now using other values for θ, allows to obtain intermediate situations. If $\theta = 1/2$, average values between h values at time t and at time $(t + \Delta t)$ are adopted: it is called the *Crank-Nicolson time integration scheme*. It remains just unconditionally stable but in terms of accuracy, it is not taking into account the observation that the evolution of h is faster at the beginning than at the end of the time step (Figure 13.5). It means that ideally values at time $(t + \Delta t)$ should receive slightly more weight. That is what is proposed in the *Galerkin time integration scheme* with $\theta = 2/3$. This scheme is also unconditionally stable as it lies in the domain $0.5 \leq \theta \leq 1$ characterizing implicit time integration schemes.

Note that those finite difference (FD) time integration schemes are used in all numerical techniques (i.e., not only FDM, but also FEM, FVM, and the others).

Box 13.3   Stability of an implicit FD time integration scheme? Example

Equation 13.29 is used on the simplified example (as in Box 13.1) of a squared aquifer with prescribed heads on all lateral boundaries and a uniform recharge. Detailed data are the following: initial value $h = 10$ m, BCs: $h = 10$ m, recharge: 0.002 m/day, $S = 0.4$ $T = 100$ m²/day, $\Delta m = 50$ m. The time step is directly chosen at 40 days in order to compare with the unstable situation of the explicit scheme (Box 13.1): $\Delta t = 40$ days, so that $(Q\Delta t/S) = 0.2$ and $(T\Delta t/(\Delta m)^2 S) = 1$.

At the initial time, all h values are 10. After a first time step, $h = 10.2$ at all internal nodes, after a second time step, the piezometric head evolves toward logical values due to the recharge. It will be the same for the next time steps.

It is showing an unconditional stability of the implicit time integration scheme.

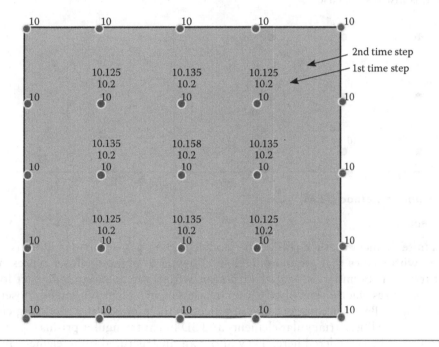

Box 13.4   Implicit FD time integration scheme is always stable

In the same way as in Box 13.2 on the example of Boxes 13.1 and 13.3 with a zero recharge, one can express simply the stability condition: the value of h should converge time step after time step to 10. Taking the same worst initial h values of $(10 - \varepsilon)$ in the aquifer center and $(10 + \varepsilon)$ in the four neighboring nodes, and with $(Q\Delta t/S) = 0$ and $(T\Delta t/(\Delta m)^2 S) = \alpha$ in Equation 13.29, one can find that:

$$h_{ij}(t + \Delta t)[1 + 4\alpha] = (10 - \varepsilon) + 0 + 4\alpha(10 + \varepsilon)$$

and

$$h_{ij}(t + \Delta t) - 10 = \frac{(10 - \varepsilon) + 4\alpha(10 + \varepsilon)}{[1 + 4\alpha]} - 10 \leq \varepsilon$$

So that:

$$10 - \varepsilon + 40\alpha + 4\alpha\varepsilon \leq 10 + \varepsilon + 40\alpha + 4\alpha\varepsilon$$

which is always the case.

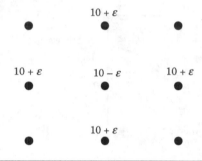

Finite element method (FEM)

General features

In the Finite Element Method (FEM) the modeled domain is subdivided into discrete elements with shapes that are more flexible allowing to obtain a closer representation of irregular boundaries, spatial variations within the domain, and exact locations for the stress-factors and observation measurements. Elements can be chosen of various shapes. Based on the common practices in the groundwater modeling community, only 2D linear triangular elements and 3D linear triangular prisms (i.e., pentahedron) will be considered here. They allow an unstructured finite element mesh (Figure 13.7). The mathematical formulation of the FEM is less straightforward than FDM and it can be established in different ways. Main books or publications where the reader can find more details about FEM applied in groundwater modeling are (among others): Narasimhan *et al.* (1978), Wang and Anderson (1982), Huyakorn and Pinder (1983), Bear and Verruijt (1987), Fitts (2002), Rausch *et al.* (2005), Pinder and Celia (2006), Bear and Cheng (2010), Diersch (2014), and Anderson *et al.* (2015).

In an FE spatial discretization, nodes and elements are numbered separately. An element is defined by its nodes. This implies more topological bookkeeping in terms of nodes locations and numbers of the nodes belonging to each element (Anderson *et al.*

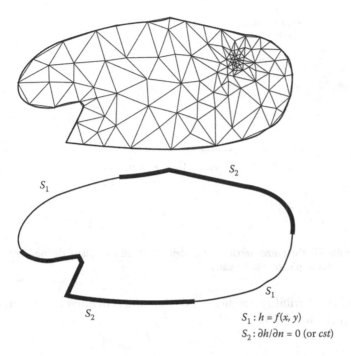

$S_1 : h = f(x, y)$

$S_2 : \partial h/\partial n = 0$ (or cst)

Figure 13.7 Example of a 2D finite element discretization and boundary conditions with a FE mesh. It can easily be adapted to irregular boundaries geometry and be refined locally around a pumping well.

2015). Mesh generation of the whole model may be optimized for reducing the needed memory space allocated during computation (Wang and Anderson 1982).

Local approximation

In each individual element, the continuous field of the variable (i.e., piezometric head) is approximated by interpolation functions, called also *basis functions*. Typically for triangle (2D) or triangular prisms (3D), h is supposed a linear function of the nodes coordinates (linear interpolation).

This linear approximation $\hat{h}(x,y)$ in each FE can be written in 2D:

$$h(x,y) \approx \hat{h}(x,y) = px + qy + r \qquad (13.31)$$

where p, q, and r are the coefficients of a plane equation in the 3D space made of coordinates x, y, and h (Figure 13.8).

It means that the local piezometric field is described in each finite element by a plane. The whole field being approximated by assembling all different planes together

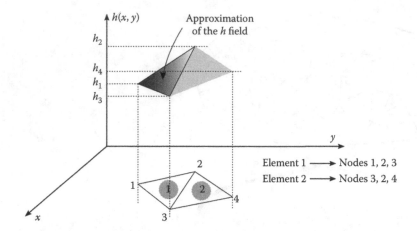

Figure 13.8 In the triangular FE, the piezometric field is approximated by a plane defined by the three nodal discrete piezometric heads.

with some continuity and compatibility conditions. Each plane is defined by the three nodal h values (h_1, h_2, h_3) (Figure 13.8):

$$\begin{cases} h_1 = px_1 + qy_1 + r \\ h_2 = px_2 + qy_2 + r \\ h_3 = px_3 + qy_3 + r \end{cases} \tag{13.32}$$

So that p, q, r can be determined by:

$$\begin{cases} p = \dfrac{h_1 b_1 + h_2 b_2 + h_3 b_3}{\left| x_1(y_2 - y_3) + x_2(y_3 - y_1) + x_3(y_1 - y_2) \right|} = \dfrac{1}{\Delta_D} \sum_{k=1}^{3} b_k h_k \\[3ex] q = \dfrac{h_1 c_1 + h_2 c_2 + h_3 c_3}{\left| x_1(y_2 - y_3) + x_2(y_3 - y_1) + x_3(y_1 - y_2) \right|} = \dfrac{1}{\Delta_D} \sum_{k=1}^{3} c_k h_k \\[3ex] r = \dfrac{h_1 d_1 + h_2 d_2 + h_3 d_3}{\left| x_1(y_2 - y_3) + x_2(y_3 - y_1) + x_3(y_1 - y_2) \right|} = \dfrac{1}{\Delta_D} \sum_{k=1}^{3} d_k h_k \end{cases} \tag{13.33}$$

with $b_1 = y_2 - y_3$, $b_2 = y_3 - y_1$, $b_3 = y_1 - y_2$, $c_1 = x_3 - x_2$, $c_2 = x_1 - x_3$, $c_3 = x_2 - x_1$, $d_1 = (x_2 y_3 - x_3 y_2)$, $d_2 = (x_3 y_1 - x_1 y_3)$, $d_3 = (x_1 y_2 - x_2 y_1)$ and

$$\Delta_D = Det \begin{bmatrix} x_1 & y_1 & 1 \\ x_2 & y_2 & 1 \\ x_3 & y_3 & 1 \end{bmatrix} = \left| x_1 b_1 + x_2 b_2 + x_3 b_3 \right| = 2A$$

where A is the triangular element surface in the 2D horizontal plane (x, y).

The approximated value of the piezometric head in any location (x, y) within the triangular element is:

$$h(x,y) \approx \hat{h}(x,y) = \frac{1}{2A}\sum_{k=1}^{3}b_k h_k x + \frac{1}{2A}\sum_{k=1}^{3}c_k h_k y + \frac{1}{2A}\sum_{k=1}^{3}d_k h_k \qquad (13.34)$$

and it can be written more explicitly in function of the nodal values:

$$h(x,y) \approx \hat{h}(x,y) = \frac{1}{2A}\sum_{k=1}^{3}(b_k x + c_k y + d_k)h_k \qquad (13.35)$$

One can note that in Equation 13.35, the approximation of h is only dependent on the three nodal values and coordinates. Interpolation (basis) functions in the considered triangular element are written:

$$N_k(x,y) = \frac{1}{2A}(b_k x + c_k y + d_k) \qquad (13.36)$$

with b_k, c_k, and d_k as in Equation 13.33. Note also that $N_k = 1$ at node k and $N_k = 0$ at the other two nodes. In the Galerkin method, most used in groundwater modeling, the basis function of the FE is used for the approximation. Equation 13.35 is expressed making use of those basis functions in the considered element:

$$h(x,y) \approx \hat{h}(x,y) = \sum_{k=1}^{3}N_k h_k \qquad (13.37)$$

So, for each element of the whole model, a matrix is obtained expressing the local approximation of the unknown variable within the element. Indeed, accuracy of this approximation depends on the interpolation, thus on the size (and complexity) of the element.

Using Equation 13.35, we may also observe that:

$$\frac{\partial \hat{h}}{\partial x} = \frac{1}{2A}\sum_{k=1}^{3}(b_k)h_k = p \quad \text{and} \quad \frac{\partial \hat{h}}{\partial y} = \frac{1}{2A}\sum_{k=1}^{3}(c_k)h_k = q \qquad (13.38)$$

As mentioned by Wang and Anderson (1982), all the elements which contain the node k form a "patch" around this node. The nodal basis function $N_k(x, y)$ is 1 only at the node k and 0 in all other nodes. Thus, each basis function is pyramidal in shape with its peak located over the node k.

Global minimum energy dissipation

The discrete unknown are the nodal values. All elements of the mesh must be taken together and bound to each other to obtain an assembled global matrix representing the system of n equations to be solved (if n is the total number of nodes). We need an integral approach expressing the *weak formulation* (i.e., a variational form integrating the governing partial differential equation of the process with its BCs and initial conditions) of our problem for obtaining a global continuum balance statement.

Without entering into too much details, there are two ways for expressing that the rate of energy dissipation would be minimized over the problem domain (Wang and Anderson 1982, Diersch 2014). One is based on finding the solution via an equivalent variational problem: most often the minimum of a natural variational functional when this last exists. It is the case when the self-adjoint differential operator is symmetric, but it is not the case (for example) for the solute transport equation (Diersch 2014). The second is the method of weighted residuals applicable to all types of partial differential equations.

In the following, to keep it simple, the reasoning is explained for the case of a steady groundwater flow. The Galerkin method will be briefly summarized as applied in FEM and based on a weighted residual principle that is, in that case, equivalent to the variational principle (Wang and Anderson 1982). The demonstration is simplified to the maximum in order to allow to figure out the general philosophy of the method.

Application to 2D steady-state groundwater flow

Let's assume a steady-state groundwater flow in confined conditions using Equation 13.3 as previously:

$$\frac{\partial}{\partial x}\left(T\frac{\partial h}{\partial x}\right) + \frac{\partial}{\partial y}\left(T\frac{\partial h}{\partial y}\right) + q'' = 0 \tag{13.39}$$

where q'' is a stress factor (in m^3/m^2s) that accounts for recharge (for example). We need to associate at this partial derivative equation (PDE), the three kinds of boundary conditions (BCs). Piezometric heads (Dirichlet BC) and fluxes (Neumann BC) are prescribed on parts of the lateral boundary of the modeled domain shown in Figure 13.7. A piezometric head $h(x, y) = f(x, y)$ is prescribed on parts S_1 of the lateral boundary. A zero-prescribed flux is prescribed $((\partial h / \partial n) = 0)$ on the other parts S_2.[1] The mixed BC (Cauchy BC) can be integrated in Equation 13.39 by introducing an additional term of water flux depending on h (i.e., representing any exchange of flux between the modeled confined aquifer and another layer (or a river; see Equation 12.7):

$$\frac{\partial}{\partial x}\left(T\frac{\partial h}{\partial x}\right) + \frac{\partial}{\partial y}\left(T\frac{\partial h}{\partial y}\right) + q'' + \frac{K'}{d'}(h_r - h) = 0 \tag{13.40}$$

where the ratio K'/d' is the conductance (see Section 12.3), h_r is the hydraulic head in the river or the piezometric head in a neighboring geological layer.

According to the variational principle, a natural quadratic functional can be found (i.e., equivalent to the minimization of the energy dissipation) and minimized for solving the problem (Wang and Anderson 1982, Reddy 1993). This functional is the following:

$$U = \frac{1}{2}\iint_{R}\left[T\left(\frac{\partial h}{\partial x}\right)^2 + T\left(\frac{\partial h}{\partial y}\right)^2 - 2q''h + \frac{K'}{d'}(h^2 - 2hh_r)\right]dxdy \tag{13.41}$$

with $h(x, y) = f(x, y)$ on S_1 and $((\partial h(x,y)) / \partial n = 0)$ on S_2.

---

1   If a nonzero flux is prescribed, an additional boundary integral term will appear in the following demonstration. For the sake of simplicity, a zero flux boundary is assumed here.

From here, the general approach will be explained step-by-step as follows:

- The minimum of the functional (Equation 13.41) provides the approximated solution to the problem (Equation 13.40 applied to the modeled domain with BCs S_1 and S_2).
- The approximated solution is the best for the problem.
- For practical implementation, one can use the interpolation functions (Equation 13.36) for finding the local value of the functional.
- The global minimum of the functional taken at the n nodes of the mesh provides a system of n algebraic equations and n unknowns to be solved.

If the approximated solution $\hat{h}(x,y)$ corresponds to the minimum of Equation 13.41 noted $\hat{U}$, all other solutions, as for example, $h(x,y) = \hat{h}(x,y) + \alpha v(x,y)$ must induce a value $U > \hat{U}$. This is true for any value of $v(x, y)$ but $v(x, y) = 0$ on S_1. If $h(x,y) = \hat{h}(x,y) + \alpha v(x,y)$ is introduced in Equation 13.41, the functional U becomes:

$$U = \hat{U} + \alpha \iint_R \left[T\left(\frac{\partial \hat{h}}{\partial x}\right)\left(\frac{\partial v}{\partial x}\right) + T\left(\frac{\partial \hat{h}}{\partial y}\right)\left(\frac{\partial v}{\partial y}\right) - q''v + \frac{K'}{d'}(\hat{h}v - vh_r) \right] dxdy$$

$$+ \alpha^2 \iint_R \left[T\left(\frac{\partial v}{\partial x}\right)^2 + T\left(\frac{\partial v}{\partial y}\right)^2 \right] dxdy \tag{13.42}$$

That can be written in a summarized way:

$$U = \hat{U} + \alpha A_1 + \alpha^2 A_2 \tag{13.43}$$

Now for proving that $\hat{U}$ is the minimum of U, coefficients A_1 and A_2 must take positive or zero values. It is evident that $A_2 > 0$.

Developing A_1 gives (by expressing the derivative of the product) $\left(v\frac{\partial \hat{h}}{\partial x}\right)$:

$$A_1 = \iint_R \left[\frac{\partial}{\partial x}\left(Tv\frac{\partial \hat{h}}{\partial x}\right) + \frac{\partial}{\partial y}\left(Tv\frac{\partial \hat{h}}{\partial y}\right) \right] dxdy$$

$$- \iint_R v\left[\frac{\partial}{\partial x}\left(T\frac{\partial \hat{h}}{\partial x}\right) + \frac{\partial}{\partial y}\left(T\frac{\partial \hat{h}}{\partial y}\right) + q'' + \frac{K'}{d'}(h_r - \hat{h}) \right] dxdy \tag{13.44}$$

where the second integral of the right-hand side should be equal to zero by Equation 13.40. Then applying the Green theorem transforming surface integrals in contour integrals, the remaining integral of Equation 13.44 is written:

$$A_1 = \int_{S_1} Tv\frac{\partial \hat{h}}{\partial n}dS + \int_{S_2} Tv\frac{\partial \hat{h}}{\partial n}dS \tag{13.45}$$

and using respectively the BCs on S_1 ($v = 0$) and on S_2 ($\partial h/\partial n = 0$), it is evident that $A_1 = 0$.

It is consequently confirmed that $\hat{U}$ is the minimum of U for the approximated solution $\hat{h}(x,y)$.

On a simplified example in Boxes 13.5 and 13.6 it is shown that the calculated $\hat{h}$ minimizing the functional U is well the best approximated solution as possible.

For the problem to be solved (Equation 13.40 and BCs), and considering the leakage flux from the river or from a neighboring layer as zero[2] $((K'/d')(h_r - h) \cong 0)$, the functional can be written (in each element j) as:

$$U_j = \frac{1}{2}\iint\limits_{R}\left[T\left(\frac{\partial h}{\partial x}\right)^2 + T\left(\frac{\partial h}{\partial y}\right)^2 - 2q''h \right]dxdy \tag{13.46}$$

and this integral can be evaluated term-by-term with $U_j = U_j^1 + U_j^2$:

$$U_j^1 = \frac{1}{2}\iint\limits_{R}\left[T\left(\frac{\partial h}{\partial x}\right)^2 + T\left(\frac{\partial h}{\partial y}\right)^2 \right]dxdy \tag{13.47}$$

$$U_j^2 = -\iint\limits_{R} q''h\,dxdy \tag{13.48}$$

Using Equations 13.31 and 13.38 in the element j, we obtain:

$$U_j^1 = \frac{1}{2}\iint\limits_{R} [T(p^2 + q^2)]dxdy \tag{13.49}$$

where the p and q values are expressed only in function of the three nodal coordinates and piezometric heads (see Equations 13.33) of the element j. Further calculation gives:

$$U_j^1 = \frac{1}{2}T_j(p^2 + q^2)\iint\limits_{R} dxdy = \frac{1}{2}T_j(p^2 + q^2)A_j \tag{13.50}$$

where T_j is the transmissivity in the element j, and A_j is (as previously) the surface of element j. The latter is depending only on nodal coordinates.

For calculating Equation 13.48, the approximation of the contribution of the stress factor term in element j, is done using a mean value of h on the element, so that:

$$U_j^2 = -q_j''\iint\limits_{R} h\,dxdy = -q_j''A_j\bar{h} = -q_j''A_j\left(\frac{1}{3}\sum_{k=1}^{3}h_k\right) \tag{13.51}$$

Then, the functional can be written (in each element j) as:

$$U_j = \frac{1}{2}T_j(p^2 + q^2)A_j - \frac{q_j''A_j}{3}\sum_{k=1}^{3}h_k \tag{13.52}$$

---

2 Again, this assumption is chosen for the sake of simplicity. If it is not the case, an additional term would be needed in the next equations.

Box 13.5   Evaluation of the FE solution compared to the analytical solution: The case of the linear interpolation (basis) function

On the very simple 1D homogeneous case illustrated here below, let's compare $\hat{h}(x)$ to the analytical solution. An unconfined aquifer is recharged by a uniform infiltration I and has an impervious boundary on the right side and a prescribed boundary $(h = h_0)$ on the left side. The 1D PDE of the problem is $T(\partial^2 h / \partial x^2) + I = 0$ and the BCs are $x = 0 \rightarrow h = h_0$ and $x = l \rightarrow (\partial h / \partial x) = 0$. The analytical exact solution is easy calculated as:

$$h = h_0 + (Il/T) - (I/2T)x^2$$

If we choose an FEM with linear basis and interpolation functions (most common case), only a linear interpolation can be found between successive nodes. If there is a node at $x = 0$ and the next one at $x = l$, only a linear variation $\hat{h}(x) = h_0 + Cx$ (and $\partial \hat{h} / \partial x = C$) can be found as approximation. The functional is written:

$$U = (1/2) \int_0^l [TC^2 - 2Ih_0 - 2ICx]dx = (1/2)TC^2 l - Ih_0 l - IC(l^2/2)$$

and the minimum is found for

$$(\partial U / \partial C) = TCl - (Il^2/2) = 0 \text{ with } (\partial^2 U / \partial C^2) > 0.$$

So that $C = (Il/2T)$. The approximated solution is thus: $\hat{h}(x) = h_0 + (Il/2T)x$ which is clearly the best possible linear approximation of the reality.

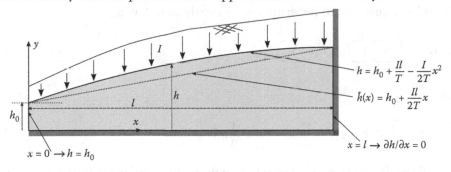

$$h = h_0 + \frac{Il}{T} - \frac{I}{2T}x^2$$

$$\hat{h}(x) = h_0 + \frac{Il}{2T}x$$

$$x = l \rightarrow \partial h/\partial x = 0$$

$$x = 0 \rightarrow h = h_0$$

Introducing the values of p and q as expressed in Equations 13.33, the functional in element j is written:

$$U_j = \frac{1}{4}T_j\frac{1}{2A_j}\left[\left(\sum_{k=1}^{3}b_k h_k\right)\left(\sum_{l=1}^{3}b_l h_l\right) + \left(\sum_{k=1}^{3}c_k h_k\right)\left(\sum_{l=1}^{3}c_l h_l\right)\right]$$

$$- \frac{q_j'' A_j}{3}\sum_{k=1}^{3}h_k \tag{13.53}$$

Box 13.6 **Evaluation of the FE solution compared to the analytical solution: The case of the second-order interpolation (basis) function**

On the same simple example than in Box 13.5, let's calculate $\hat{h}(x)$ with second-order (or quadratic) interpolations. A second-order variation of h must be found as approximation between two successive nodes: $\hat{h}(x) = h_0 + Cx + Dx^2$.

Consequently, $(\partial \hat{h}/\partial x) = C + 2Dx$ and $(\partial \hat{h}/\partial x)^2 = C^2 + 4CDx + 4D^2x^2$ so that U can be written:

$$U = \frac{1}{2}\left[T\left(C^2l + 2CDl^2 + \frac{4}{3}D^2l^3\right) - 2I\left(h_0l + \frac{1}{2}Cl^2 + \frac{1}{3}Dl^3\right)\right]$$

The minimum of U with regards to C and D is found with:

$$\frac{\partial U}{\partial C} = TCl + TDl^2 - \frac{Il^2}{2} = 0 \quad \text{and} \quad \frac{\partial^2 U}{\partial C^2} = Tl > 0$$

$$\frac{\partial U}{\partial D} = TCl^2 + \frac{4}{3}TDl^3 - \frac{Il^3}{3} = 0 \quad \text{and} \quad \frac{\partial^2 U}{\partial D^2} = \frac{4}{3}Tl^3 > 0$$

These two equations allow to find respectively $C = (Il/T)$ and $D = -(I/2T)$. The approximated solution is thus: $\hat{h}(x) = h_0 + (Il/T)x - (I/2T)x^2$ which is clearly the best possible second-order approximation as it is equal to the exact analytical solution.

This last equation can be written more clearly as following:

$$U_j = \frac{1}{2}\sum_{k=1}^{3}\sum_{l=1}^{3}P_{kl}h_kh_l - \sum_{k=1}^{3}Q_kh_k \tag{13.54}$$

with

$$P_{kl} = \frac{T_j}{4A_j}(b_kb_l + c_kc_l) \quad \text{and} \quad Q_k = \frac{1}{3}q_j''A_j \tag{13.55}$$

where P_{kl} coefficients are only dependent on the nodes coordinates and the transmissivity introduced for the jth element, and Q_k coefficients are only dependent on the coordinates, the stress factor value (q''), and the mean piezometric head in the jth finite element.

For the whole modeled domain, in place of $k = 1$ to 3 (or $l = 1$ to 3), the contribution of the n nodes of the m elements is assembled for calculating the total value of the functional U (i.e., obtaining a global continuum statement):

$$U = \frac{1}{2}\sum_{k=1}^{n}\sum_{l=1}^{n}P_{kl}h_kh_l - \sum_{k=1}^{n}Q_kh_k \tag{13.56}$$

where n is the total number of nodes in the FE mesh.

The minimum of this functional is found for $\partial U/\partial h_i = 0$ for $i = 1, \ldots n$. This derivative is calculated term by term:

$$\frac{1}{2}\frac{\partial}{\partial h_i}\left(\sum_{k=1}^{n}\sum_{l=1}^{n}P_{kl}h_k h_l\right) = \frac{1}{2}\left(\sum_{j=1}^{n}P_{ji}h_j + \sum_{j=1}^{n}P_{ij}h_j\right) \tag{13.57}$$

$$\frac{\partial}{\partial h_i}\left(\sum_{k=1}^{n}Q_k h_k\right) = Q_i\frac{\partial h_i}{\partial h_i} = Q_i \tag{13.58}$$

So that, with $P_{ji} = P_{ij}$, the final system of algebraic equations is found:

$$\left(\sum_{j=1}^{n}P_{ij}h_j\right) - Q_i = 0 \tag{13.59}$$

where the h_j are the n unknowns, and the matrix of elements P_{ij} is often called the "rigidity matrix" (in civil engineering applications) or "conductance matrix" (in hydrogeology) containing all information from the coordinates and properties.

It can be written in a matrix form:

$$P \cdot h - Q = 0 \tag{13.60}$$

where P is the conductance matrix, h is the vector (or column matrix) with the n nodal piezometric heads (unknowns), and Q is the vector (or column matrix) containing the boundary conditions and stress factors. A very simple application is illustrated in Box 13.7, showing practically how the matrix is built using contribution from each element and also how BCs are implemented.

Transient groundwater flow

In transient conditions, the time derivative (e.g., see Equation 13.23) is approximated by a finite difference. So all the time integration schemes explained in Section 13.2 can be used. The choice of the time at which the $\sum_{j=1}^{n}P_{ij}h_j$ are expressed is just the same problem as it was for the FDM with the values of the piezometric heads (at the considered node and at its neighboring nodes) in the left-hand member of Equation 13.25. In place of having the system of equations described by Equation 13.60, we have:

$$P \cdot h + S \cdot \frac{\partial h}{\partial t} - Q = 0 \tag{13.61}$$

where S is the matrix describing the storage term, $\partial h/\partial t$ is the vector (or column matrix) with the time derivative of the piezometric heads in the n nodes. The vector h with the n nodal piezometric heads can be expressed (as it was made in Equation 13.30 for FDM):

$$h = (1 - \theta)\,h(t) + \theta h(t + \Delta t) \tag{13.62}$$

Box 13.7 Example of the FE solution: A practical implementation

A steady-state 2D horizontal confined flow in a rectangular aquifer is considered with upstream and downstream prescribed heads BCs respectively of 10 m and 0 m (left and right side of the figure) and no flow BCs elsewhere. The FE spatial discretization is made of four elements and six nodes. Each node has a local number linked to the considered element, and a global number in the whole FE mesh.

$$h_1 = h_2 = 10\,\text{m}, \ h_5 = h_6 = 10\,\text{m}$$

$$T_1 = T_2 = T_3 = T_4 = 2\,\text{m}^2/\text{day}$$

$$h_3 = ? \quad h_4 = ?$$

$$P_{kl} = \frac{T_j}{2\Delta_D}(b_k b_l + c_k c_l)$$

$$\Delta_D = x_1 b_1 + x_2 b_2 + x_3 b_3$$

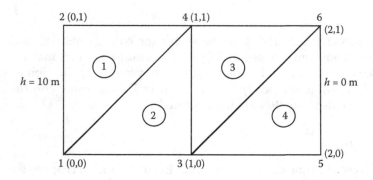

For element 1: with $\begin{cases} b_1 = y_2 - y_3 \\ b_2 = y_3 - y_1 \\ b_3 = y_1 - y_2 \end{cases}$ and $\begin{cases} c_1 = x_3 - x_2 \\ c_2 = x_1 - x_3 \\ c_3 = x_2 - x_1 \end{cases}$

$$x_1 = 0 \quad y_1 = 0 \quad b_1 = 0 \quad c_1 = 1$$

$$x_2 = 0 \quad y_2 = 1 \quad b_2 = 1 \quad c_2 = -1$$

$$x_3 = 1 \quad y_3 = 1 \quad b_3 = -1 \ c_3 = 0$$

$$\Delta_D = -1$$

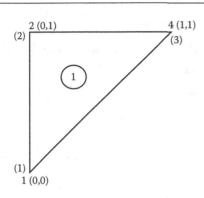

2 (0,1)
(2)
4 (1,1)
(3)
1
(1)
1 (0,0)

The matrix P for element 1: $P^1 = \begin{bmatrix} 1 & -1 & 0 \\ -1 & 2 & -1 \\ 0 & -1 & 1 \end{bmatrix}$ introduced in the global 6×6 matrix P_{ij}:

$$P = \begin{bmatrix} 1 & -1 & . & 0 & . & . \\ -1 & 2 & . & -1 & . & . \\ . & . & . & . & . & . \\ 0 & -1 & . & 1 & . & . \\ . & . & . & . & . & . \\ . & . & . & . & . & . \end{bmatrix}$$

For element 2:

$x_1 = 0 \quad y_1 = 0 \quad b_1 = -1 \quad c_1 = 0$

$x_2 = 1 \quad y_2 = 0 \quad b_2 = 1 \quad c_2 = -1$

$x_3 = 1 \quad y_3 = 1 \quad b_3 = 0 \quad c_3 = 1$

$\Delta_D = 1$

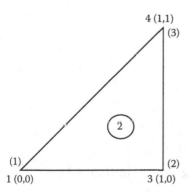

4 (1,1)
(3)
2
(1)
1 (0,0)
(2)
3 (1,0)

The matrix P for element 2: $P^2 = \begin{bmatrix} 1 & -1 & 0 \\ -1 & 2 & -1 \\ 0 & -1 & 1 \end{bmatrix}$ introduced (added) in the global 6×6 matrix P_{ij} gives:

$$P = \begin{bmatrix} 22 & -1 & -1 & 0 & . & . \\ -1 & 2 & . & -1 & . & . \\ -1 & . & 2 & -1 & . & . \\ 0 & -1 & -1 & 2 & . & . \\ . & . & . & . & . & . \\ . & . & . & . & . & . \end{bmatrix}$$

In the same way, for element 3,

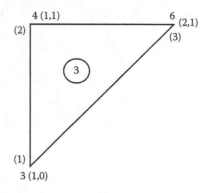

4 (1,1)   6 (2,1)
(2)   (3)

3

(1)

3 (1,0)

The matrix P for element 3: $P^3 = \begin{bmatrix} 1 & -1 & 0 \\ -1 & 2 & -1 \\ 0 & -1 & 1 \end{bmatrix}$ introduced (added) in the global 6×6 matrix P_{ij} gives:

$$P = \begin{bmatrix} 2 & -1 & -1 & 0 & . & . \\ -1 & 2 & . & -1 & . & . \\ -1 & . & 3 & -2 & . & 0 \\ 0 & -1 & -2 & 4 & . & -1 \\ . & . & . & . & . & . \\ . & . & 0 & -1 & . & . \end{bmatrix}$$

For element 4:

The matrix P for element 4: $P^4 = \begin{bmatrix} 1 & -1 & 0 \\ -1 & 2 & -1 \\ 0 & -1 & 1 \end{bmatrix}$ introduced (added) in the global 6×6 matrix P_{ij} gives:

$$P = \begin{bmatrix} 2 & -1 & -1 & 0 & . & . \\ -1 & 2 & . & -1 & . & . \\ -1 & . & 3 & -2 & -1 & 0 \\ 0 & -1 & -2 & 4 & . & -1 \\ . & . & -1 & . & 2 & -1 \\ . & . & 0 & -1 & -1 & 2 \end{bmatrix}$$

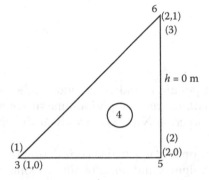

Writing Equations 3 and 4 of the global system with the prescribed BCs gives:

$$\left\{ \begin{array}{l} -h_1 + 0 + 4h_3 - 2h_4 - h_5 + 0 = 0 \\ 0 - h_2 - 2h_3 + 4h_4 - 0 - h_6 = 0 \\ h_1 = h_2 = 10 \\ h_3 = h_4 = 0 \end{array} \right. \quad \rightarrow \quad \left\{ \begin{array}{l} -10 + 4h_3 - 2h_4 = 0 \\ -10 - 2h_3 + 4h_4 = 0 \\ h_1 = h_2 = 10 \\ h_3 = h_4 = 0 \end{array} \right.$$

And the solution ($h_3 = 5$ and $h_4 = 5$) is easy to find. If we try an iterative method (Gauss Seidel) for solving the system with an initial value of $h_3 = 0$:

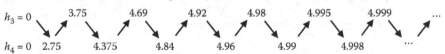

we can observe that it converges toward the same values.

where θ, as previously, is the time integration coefficient. For $\theta = 0$, a *fully explicit time integration scheme* is adopted with the same stability issue than in FDM. If $\theta = 1$, a *fully implicit time integration scheme* is adopted that is unconditionally stable. Intermediate time integration schemes are usually preferred as, for example, those of Crank-Nicolson and Galerkin respectively corresponding to $\theta = (1/2)$ and $= (2/3)$. Note that for solving the system of Equation 13.62, a rearrangement is needed for having all the heads of the time t on the right-hand side and all the heads at the time $t + \Delta t$ on the left-hand side. Note that the matrix P and S must be computed only once as they do not depend on time.

Finite volume method (FVM)

Developed in the 1980s (Patankar 1980, Baliga and Patankar 1983), there are many forms of Finite Volume Methods (FVM) (Chung 2002, Diersch 2014). They are sometimes considered as an extension of FDM able to handle unstructured grids (Narasimhan and Witherspoon 1976, Rausch *et al.* 2005; Figure 13.9). If the spatial discretization is made with triangular elements (2D) or triangular prisms (3D), there are also a lot of similarities with the FEM using similar elements (Fletcher 1988, Idelsohn and Onate 1994). Contrarily to the FDM, FVM approximates the main variable (h) using basis functions in the considered triangular element (as in Equation 13.37):

$$h(x,y) \approx \hat{h}(x,y) = \sum_{k=1}^{3} N_k h_k \qquad (13.63)$$

This approximation depends on the interpolation, and sometimes higher-order functions are used for more accuracy. *Finite volume* refers to the volume surrounding each node point in a mesh with nodal basis function $N_k(x, y) = 1$ only at the node k and 0 in all other nodes.

The difference with FEM, is that conservation law is satisfied considering the control volume with regards to its neighboring volumes and not globally as done in the Galerkin FEM. It means that the balance relies on evaluation of surface integrals on the boundaries (i.e., the conservation must be satisfied across the boundaries of the adjoining control volumes). The main principles of this method often called Control Volume Finite Element Method (CVFEM) can be found in Forsyth *et al.* (1995), Therrien and Sudicky (1996), Pinder and Celia (2006), Therrien *et al.* (2010). Many developments of FVM were done for solving specifically solute transport equations (see Section 13.3).

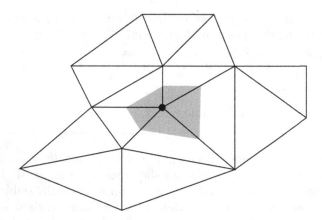

Figure 13.9 If a mesh of triangular elements is chosen (unstructured grid), Thiessen (or Voronoi) polygons can be drawn around each node and a centered finite difference method (FDM) can be applied defining a finite volume for which continuity is expressed with regards to the neighboring other finite volumes as in the FDM (Narasimhan and Witherspoon 1976).

13.3 Numerical techniques for solute transport modeling

The solute transport equation (Equation 8.51) is a partial differential equation (PDE) containing together spatial first and second order partial derivatives representing respectively the advective and the dispersion terms. It also involves a time first order partial derivative. Mathematically, this equation is both elliptic and parabolic and also hyperbolic[3] if the advection term becomes large relative to the dispersion term. It is mostly the case in aquifers.

If we try to solve the solute transport PDE with the same numerical methods as for the groundwater flow PDE, this advection term causes difficulties as a sharp advective front must be solved. Classical FDM and FEM are not efficient for simulating sharp front problems: oscillations are simulated at the location of the sharp front (Figure 13.10), showing unrealistic concentrations (Pinder and Gray 1977). Oscillations can lead to negative concentrations inducing solution failure especially for any associated reactive transport problem. Numerical dispersion (Figure 13.10) may also appear as the approximation by numerical methods tends to smooth any sharp change in the considered variable (i.e., concentration).

Classical numerical methods are modified for solving the solute transport equation. Some are classified as *Eulerian or grid-based methods* using so-called upstream weighting for approximation of the advective term (by FDM and FEM methods). Others are *(mixed) Eulerian-Lagrangian methods* where the partial derivative equation is changed in a substantial derivative equation describing the rate of concentration change (due to all terms except advection) in a Lagrangian coordinate system moving with advection. Also, *Lagrangian* methods are used such as the Random Walk Method, where particles are followed in function of time. Both advection and dispersion are solved with a moving reference system, each particle being labeled by an initial location at the beginning of each time step.

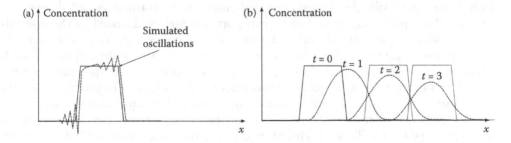

Figure 13.10 Schematic examples of oscillations (a), and numerical dispersion (b), affecting simulated concentration fronts in an advection dominated problem.

---

3 For a variable w an elliptic PDE is typically represented by $(\partial^2 w / \partial x^2) + (\partial^2 w / \partial y^2) = 0$ (as 2D steady-state groundwater flow if $w = h$) and a parabolic PDE by $(\partial w / \partial t) = (\partial^2 w / \partial x^2)$ (as 1D transient groundwater flow if $w = h$). The PDE $(\partial w / \partial t) = (\partial^2 w / \partial x^2) + (\partial^2 w / \partial y^2)$ is both elliptic and parabolic (as 2D transient groundwater flow if $w = h$).

Numerical Peclet and Courant numbers

As numerical problems arise when advection becomes important, the first step is to determine, for each problem to be simulated, what the actual numerical conditions insuring stability and avoiding oscillations and numerical dispersion are.

The dimensionless *Peclet number* (see Chapter 8, Box 8.2) provides a ratio between solute transport by advection and diffusion. It was then modified as the ratio between advection and dispersion (Sauty 1980). So, it can help for determining the nature of the solute transport equation. Accordingly, it is often considered that the equation becomes significantly hyperbolic in nature when $Pe > 2$, inducing a more challenging numerical solution. For grid-based numerical methods, the size of the FDM cells or FEM elements is an important influencing characteristic for simulating accurately sharp advective plumes. Consequently, a numerical *Peclet number* is defined using the typical size of the cells or elements as characteristic length:

$$Pe = \frac{v_a \Delta x}{D} \tag{13.64}$$

where Δx is the typical cell or element dimension (in m) [L]. It is used for assessing the hyperbolic nature of the PDE to be solved on a given grid (spatial discretization). Note that if the problem is considered in 1D along the x direction, and neglecting diffusion with regards to mechanical dispersion, this *Peclet number* can be reduced to:

$$Pe = \frac{v_{a_x} \Delta x}{a_L v_{a_x}} = \frac{\Delta x}{a_L} \tag{13.65}$$

Thus, we should have at least $\Delta x < 2a_L$ in the whole modeled domain for hoping to avoid oscillations with the use of classical grid-based numerical methods (Price *et al.* 1966).

Classical numerical methods are modified for solving the solute transport equation. Some are classified as *Eulerian or grid-based methods* using so-called upstream or central-in-space weighting. Solute transport is largely influenced by the groundwater flow, especially in advection dominated conditions (e.g., in aquifers). As in other physical problems (e.g., waves propagation), we are dealing typically with information being propagated along a certain direction called the *characteristic line*. Starting from the knowledge of this direction, it is logical to give more weight upstream in the spatial approximation. These upwind or upstream schemes are a way to pass information in the direction of the characteristic line. They allow to avoid most of the oscillations. However, they are introducing *numerical dispersion* in the results (Figure 13.10), inducing the spreading/smoothing of sharp concentration fronts (Rausch *et al.* 2005).

Moreover, other constraints arise from the time step for computing solute transport in good conditions. Time integrations schemes can be explicit or implicit (see Section 13.2). Explicit integration schemes are conditionally stable (Box 13.2). In solute transport solved by grid-based methods, whatever the chosen time integration scheme, we need a time step allowing to pass the information slowly enough from a grid cell (element) to the next without losing information. This condition can be expressed by (among others, Daus and Frind 1985, Rausch *et al.* 2005):

$$Cr = \frac{v_a \Delta t}{\Delta x} < 1 \tag{13.66}$$

where Cr is called the *Courant-Friedrich-Levy number* more often called *Courant number* in the groundwater modeling community. Physically, it means that a particle of solute migrating at the velocity v_a should not travel farther then the length of one cell (element). Another time step constraint arises if there is some solute mass decay included in the simulated processes. During a time step, the lost mass due to degradation in a cell (element) could not be greater than the solute mass in the cell at the beginning of the time step. If decay is described by a linear law (Equation 8.34), it gives logically:

$$\Delta t \leq \frac{1}{\lambda} \tag{13.67}$$

So, numerical and dispersion and oscillations can be minimized for small grid sizes and small time steps. For nonreactive solute transport simulations, a small amount of numerical dispersion could be tolerated. As numerical dispersion has the appearance of physical dispersion, it is therefore wise to check the sensitivity of simulated results to longitudinal and transversal dispersivity values (Frind and Germain 1986). It allows to clearly distinguish the relative part of numerical dispersion.

Time integration schemes

As mentioned above, time integrations schemes for solving the solute transport equation can be explicit or implicit (see Section 13.2). As usual, explicit integration scheme, ($\theta < 0.5$) are conditionally stable (Box 13.2) and time integration on the implicit side ($\theta \geq 0.5$) are unconditionally stable. The Crank-Nicolson scheme ($\theta = 0.5$) gives a second-order accuracy (i.e., proportional to $(\Delta t)^2$) and is just unconditionally stable. It means that the reduction of the time step by a factor of 2, reduces the approximation error by a factor of 4. In fact, the time weighting can be combined to different spatial weighting (i.e., upstream weighting) for a variety of different methods. In general, weighting more toward the implicit side will produce less oscillations but more numerical dispersion. This is why the Crank-Nicolson scheme is often adopted as a compromise, with spatial and temporal discretizations adequately chosen in relations to Peclet and Courant constraints.

There are also other schemes for approximating the time derivative. Independently to the choice of θ prescribing at what time the approximation of the spatial derivatives is taken (see, for example, Equation 13.30), a weighting of the finite difference approximation of the time derivative is adopted. The finite difference $(C(t + \Delta t) - C(t))/\Delta t$ is replaced by:

$$\frac{1}{\Delta t}\left[\frac{3}{2}C(t + \Delta t) - 2C(t) + \frac{1}{2}C(t - \Delta t)\right] \tag{13.68}$$

combined with $\theta = 1$ (full implicit) it corresponds to the so-called first-order BDF (Backward Difference Formula) method.

A third-order method is also proposed with the finite difference $(C(t + \Delta t) - C(t))/\Delta t$ replaced by:

$$\frac{1}{\Delta t}\left(\frac{5}{6}C(t + \Delta t) - \frac{2}{3}C(t) - \frac{1}{6}C(t - \Delta t)\right) \tag{13.69}$$

combined with $\theta = (1/3)$ (partially explicit).

Various other high order time integration schemes for approximating the time derivative have been proposed in the specialized scientific literature. Their description falls beyond the scope of this book as they are not so often used.

Eulerian or grid-based methods

For an advection dominated solute transport, the concentration calculated at a given node should be more influenced by the concentration at the upstream node (i.e., with regards to the advection flow) than by concentrations in the other neighboring nodes. It is physically and numerically easy to understand. Accordingly, more weight should be given to upstream values in the finite difference or finite element approximations of the advective term. The other terms of the solute transport PDE are treated by the standard approximations (i.e., similarly to what is done for solving the flow equation, see Section 13.2).

Upwind or upstream methods

If advection is large, with a standard central spatial finite difference scheme, the solution of the advection-dispersion equation (ADE) (i.e., Equation 8.51) may lead to excessive oscillations. Those oscillations in the results are due to the truncation error containing an odd derivative (i.e., third derivative) when approximating the advection term.

A series of upwind or upstream numerical techniques have been developed to decrease oscillations but at the cost of creating numerical dispersion. Indeed, using upstream information induces that the simulated gradients are artificially smoothed, it is numerical dispersion (Figure 13.10). Note that in more general numerical works and books, oscillations are depicted as "dispersive error" and numerical dispersion is depicted as "diffusive error."

The upwind or upstream techniques are relatively similar when applied in the FDM, FVM, and FEM. The discussion here will be done using the finite difference method but it is easily expandable to the other methods. Note that those upwind or upstream techniques require to compute beforehand the advection direction (i.e., groundwater flow direction) for the considered time step. Then the code must adapt its spatial weighting to the new direction.

Two kinds of upwind techniques can be used: central-in-space upwind weighting and upstream weighting. The combination of these methods with different time integration schemes gives rise to a series of different methods. The most popular ones are explained here.

CENTRAL-IN-SPACE UPWIND WEIGHTING

Similarly to Equation 13.2, the central finite difference of the spatial derivative $\partial C / \partial x$ can be written:

$$\frac{\partial C}{\partial x} \approx \frac{C(x + \Delta x) - C(x - \Delta x)}{2\Delta x} \tag{13.70}$$

The assumption of a uniform grid is taken for the sake of simplicity. Using a weighted average of the forward and backward difference approximations, it can be generalized with:

$$\frac{\partial C}{\partial x} \approx (1 - \alpha)\frac{C(x + \Delta x) - C(x)}{\Delta x} + \alpha\frac{C(x) - C(x - \Delta x)}{\Delta x} \tag{13.71}$$

where α is the upwind coefficient, $\alpha \in [0, 1]$. It can be also written:

$$\frac{\partial C}{\partial x} \approx \frac{(1 - \alpha)C(x + \Delta x) - C(x) + \alpha C(x - \Delta x)}{\Delta x} \tag{13.72}$$

and if $\alpha = 1/2$ it is reduced to Equation 13.70 equal to the central finite difference, if $\alpha = 0$ it is reduced to the first term of the right-hand side of Equation 13.71 equal to the forward finite difference, and if $\alpha = 1$ it is reduced to the second term of the right-hand side of Equation 13.71 equal to the backward finite difference. Accordingly, the upwind coefficient α must be chosen larger than 0.5 for creating an upwind weighting.

This central-in-space upwind weighting is combined with the time integration scheme to approximate concentrations values in each node at the end of the time step. It is shown schematically in Figure 13.11 how values at times t and $t + \Delta t$ in the neighboring nodes may contribute to the simulated value $C(x, t + \Delta t)$.

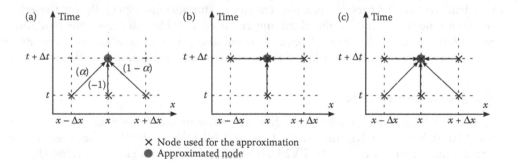

✕ Node used for the approximation
● Approximated node

Figure 13.11 Schematic spatial and temporal view on the nodal contributions to the approximated $C(x, t + \Delta t)$ with a central-in-space upwind weighting combined with respectively (a) an explicit, (b) an implicit, and (c) a Crank-Nicolson time integration scheme. Note that the weight of each nodal contribution is not mentioned for implicit schemes as it depends on the combination of the spatial with the temporal weighting.

UPSTREAM WEIGHTING

In some cases, a higher-order upstream spatial scheme may be required. For example, a second-order upstream scheme is proposed:

$$\frac{\partial C}{\partial x} \approx \frac{1}{2} C(x - 2\Delta x) - 2C(x - \Delta x) + \frac{3}{2} C(x) \tag{13.73}$$

Also, the following third-order upstream scheme is proposed:

$$\frac{\partial C}{\partial x} \approx \frac{1}{6} C(x - 2\Delta x) - C(x - \Delta x) + \frac{1}{2} C(x) + \frac{2}{6} C(x + \Delta x) \tag{13.74}$$

Note that the sum of the weighting factors is always equal to 1.

In Figure 13.12, in a similar way as in Figure 13.11, the contribution of the neighboring nodal values at times $t - \Delta t$, t and $t + \Delta t$ is schematically shown for different spatial weighting (second-order and third-order) combined with different time integration schemes (explicit $\theta = 0$, implicit $\theta = 1$, Crank-Nicolson $\theta = 0.5$, first-order BDF implicit $\theta = 1$, third-order partially explicit $\theta = (1/3)$).

The upwind or upstream techniques described here in the context of the finite difference method (FDM) are fully compatible with the finite volume methods (FVM). Some specific characteristics will be described for application in the finite element method (FEM).

All those upstream techniques reduce oscillations but creating numerical dispersion. It is wise to apply them only if $Pe < 2$ and $Cr < 1$. As mentioned previously, an additional check about the sensitivity to changes in longitudinal and transversal dispersivity values is a good way to assess the relative parts of numerical and physical dispersion in the simulated results.

Note that in 2D and 3D, numerical dispersion may be relatively insignificant compared to physical longitudinal dispersion and, simultaneously, very biasing and troublesome with respect to transverse dispersion (Rausch et al. 2005). This latter effect can lead to large errors for reactive transport simulations especially on the edges of the plume where this artificial mixing creates favorable conditions for oxidation-reduction reactions. In cases of natural or enhanced bioremediation assessments, it can induce a clear overestimation of the degradation (Cirpka et al. 1999).

TVD finite difference method

In the 1990s, a series of solution techniques arose under the generic name of TVD for "Total Variation Diminishing" (Cox and Nishikawa 1991). These techniques can be implemented in FDM-, FVM-, and FEM-based models for solving the solute transport equation in advection dominated problems. TVD schemes are known as more accurate than standard central-in-space weighting and upstream methods for simulating sharp concentration variations. They became mainly known in the groundwater community because a third-order TVD scheme is proposed with the FDM in the MT3D/MT3DMS software (Zheng 1990, Zheng and Bennett 1995, Zheng and Wang 1999), that is one of the most used software for solving solute

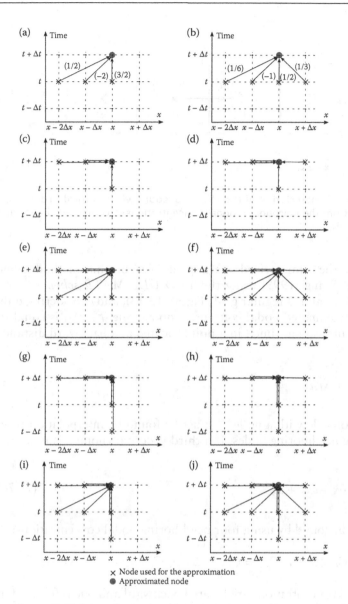

Figure 13.12 Schematic spatial and temporal view on the nodal contributions to the approximated $C(x, t + \Delta t)$ for a second-order (left) and a third-order spatial weighting (right) combined with different time integration schemes (explicit, implicit $\theta = 1$, Crank-Nicolson $\theta = 0.5$, first-order BDF implicit $\theta = 1$. The following cases are shown: (a) second-order spatial upstream and explicit ($\theta = 0$), (b) third-order spatial upstream and explicit ($\theta = 0$), (c) second-order spatial upstream and implicit ($\theta = 1$), (d) third-order spatial upstream and implicit ($\theta = 1$), (e) second-order spatial upstream and Crank-Nicolson ($\theta = 1/2$), (f) third-order spatial upstream and Crank-Nicolson ($\theta = 1/2$), (g) second-order spatial upstream and first-order BDF implicit ($\theta = 1$), (h) third-order spatial upstream and first-order BDF implicit ($\theta = 1$), (i) second-order spatial upstream and third-order partially explicit ($\theta = 1/3$)), (j) third-order spatial upstream and third-order partially explicit ($\theta = (1/3)$).

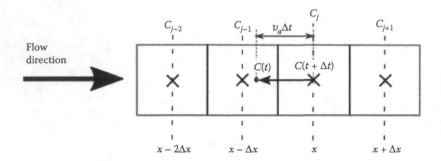

Figure 13.13 Schematic 1D description of the TVD procedure where a third-order interpolation from the neighboring nodal concentrations is computed (Zheng and Wang 1999).

transport. Accordingly, the summarized main principles of this method come mainly from Zheng and Wang (1999) using the TVD *ULTIMATE scheme* developed by Leonard (1988). In an FD regular grid (Figure 13.13), and considering only a 1D advection, the approximated nodal concentration at time $(t + \Delta t)$ is equal to the approximated concentration at time t in a point located at a backward distance equal to $(-v_a\Delta t)$:

$$C(x, t + \Delta t) = C(x - v_a\Delta t, t) \tag{13.75}$$

This point does not coincide with a node. It must be found by interpolation from the concentrations at the neighboring nodes. If a third-order polynomial is used, the 1D form is written:

$$C(x - v_a\Delta t, t) = a + bx + cx^2 + dx^3 \tag{13.76}$$

where a, b, c, and d can be found by using four neighboring nodal concentrations:

$$C_{j+1} = C(x + \Delta x, t) \ C_j = C(x, t) \ C_{j-1} = C(x - \Delta x, t) \ C_{j-2} = C(x - 2\Delta x, t)$$

Note that as two nodal concentrations are located backward and one is forward, it is logical that the desired interpolated point is located backward.

Entering the coefficients values in Equation 13.76, Equation 13.75 can be written as:

$$C_j(t + \Delta t) = C_j(t) - Cr\left[\left(\frac{C_{j+1}(t)}{3} + \frac{C_j(t)}{2} - C_{j-1}(t) + \frac{C_{j-2}(t)}{6}\right)\right.$$
$$\left. - Cr\left(\frac{C_{j+1}(t) - 2C_j(t) + C_{j-1}(t)}{2}\right) + Cr^2\left(\frac{C_{j+1}(t) - 3C_j(t) + 3C_{j-1}(t) - C_{j-2}(t)}{6}\right)\right] \tag{13.77}$$

where $Cr = (v_a \Delta t / \Delta x)$ is the Courant number as defined in Equation 13.66. Using this equation may lead to oscillations in advection dominated problems. The ULTIMATE scheme is applying thus a "flux limiter" making adjustments to calculated concentrations by the polynomial interpolation (Leonard and Niknafs 1990, 1991). This flux limiter is activated when the spatial concentration profile does not show a monotonic evolution. Detailed explanations about this numerical flux limiter are far beyond the scope of this chapter. The interested reader is referred to Zheng and Wang (1999) or to the first references mentioned above.

Note that the presented TVD scheme is explicit, thus subject to the standard stability constraints.

The other terms of the solute transport equation are solved either by an explicit or an implicit procedure. For advection dominated problems, TVD are considered as more accurate than upwind and upstream FD techniques but computationally demanding. They are mostly mass conservative, which is not so often the case for the Lagrangian and Eulerian-Lagrangian techniques.

Finite element upstream methods

In the standard FEM applied to the solute transport equation, the approximation of the concentration is written (similarly to Equation 13.37):

$$C(x,y,z,t) \approx \hat{C}(x,y,z,t) = \sum_{i=1}^{n} C_i(t) N_i(x,y,z) \qquad (13.78)$$

where $N_i(x, y, z)$ is the form or basis functions that are also taken identical to the interpolation functions for the node i in the whole discretized domain with n nodes. This method is called the Bubnov-Galerkin method. For the same reasons than previously, the same problems of oscillations occur for advection dominated problems. In the FEM, the same philosophy for upwind or upstream weighting should be used. For a detailed discussion on the upwind FEM, see Diersch (2014). Without entering in details, an upwind (or upstream) weighting is introduced in the interpolation functions for avoiding oscillations. After having localized the local groundwater flow (advection) direction, asymmetric weighting functions are defined:

$$W_i(x,y,z) = N_i(x,y,z) + \alpha F(x,y,z) \qquad (13.79)$$

where $\alpha F(x, y, z)$ is the upwind form function for the node i, and α the upwind factor taken between 0 and 1. The order of this form function is the upwind order. It is called the Petrov-Galerkin method. Doing so, oscillations are reduced but numerical dispersion (in all directions) is introduced. Consequently, a "streamline" upwind technique has been developed in order to add numerical dispersion only in the longitudinal direction (i.e., advection direction): the *Streamline Upwind Petrov Galerkin* (SUPG) method (Brooks and Hughes 1982).

As previously explained, upwinding is a compromise between accuracy and stability. Dispersion is introduced to avoid oscillations. What can be considered as insidious with this method is that one eradicates the unrealistic but visible oscillations in the results in favor of not so easily detectable numerical dispersion.

Another variant has also been developed where an upwind weighting of the temporal interpolation functions is added and combined to the spatial one. This is the Full Upwind Petrov-Galerkin (FUPG) method (Yu and Heinrich 1986, Franca *et al.* 1992).

As mentioned previously for FDM, all those upstream techniques still require the classical conditions: $Pe < 2$ and $Cr < 1$. If these methods are applied, a check on the simulated dispersion versus the physical dispersion is required.

Eulerian-Lagrangian methods

Until now, the used reference system in which all equations were written was fixed. The partial differential equation describing solute transport in such an Eulerian system can be written as Equation 8.51 (see Chapter 8), that is, a partial derivative equation (PDE):

$$\frac{\partial C^v}{\partial t} = -\frac{v_a}{R} \cdot \nabla C^v + \frac{1}{R} \nabla \cdot (D_h \cdot \nabla C^v) - \lambda C^v - \frac{q_s}{Rn_m}(C^v - C_s^v) \tag{13.80}$$

where the partial derivatives $\partial C^v / \partial t$ is considered respectively with constant x, y, z and accordingly, $\partial C^v / \partial x$ with constant y, z, t, $\partial C^v / \partial y$ with constant x, z, t, and $\partial C^v / \partial z$ with constant x, y, t. Note that if reactive transport is considered, an additional term should be found in Equation 13.80 according to Equation 8.55 in place of Equation 8.51. For the sake of clarity, this term is not considered here, but it will be treated in a specific paragraph about reactive transport modeling (see this section).

In a Lagrangian approach, the reference is moving with advection velocity, so that the Lagrange form of Equation 13.80 is written as an ordinary derivative equation (ODE):

$$\frac{dC^v}{dt} = \frac{1}{R} \nabla \cdot (D_h \cdot \nabla C^v) - \lambda C^v - \frac{q_s}{Rn_m}(C^v - C_s^v) \tag{13.81}$$

where the substantial derivative $(dC^v / dt) = (\partial C^v / \partial t) + (v_a / R) \cdot \nabla C^v$ describes the temporal concentration change along the streamline or "*characteristic* line." The latter is due only to advection with a local velocity (v_a / R) being defined as the "retarded" advection velocity. In Equation 13.81, "the left-hand side is Lagrangian while the right-hand side remains Eulerian" (Bear and Cheng 2010).

Eulerian-Lagrangian methods allow to get rid of the constraint of the $Cr < 1$, and the problem can be solved in two steps. First, advection (i.e., the left-hand side) can be calculated by a particle tracking technique or a method of characteristics that simulates the particle movement in the previously calculated groundwater flow field. Second, all the other terms of Equation 13.80 (i.e., right-hand side of Equation 13.81) are calculated using a standard Eulerian technique. Note that for this second step, as the equation is fully parabolic, all issues linked to solving advective problems have disappeared. This Eulerian-Lagrangian method is proposed in the frame of the FDM in MT3D/MT3DMS (Zheng 1990) and is more and more developed in different FEM-based software with Hybrid Eulerian-Lagrangian methods (HELM).

The main principles of the Eulerian-Lagrangian approach are explained here in the frame of the FDM as used, for example, in MT3D/MT3DMS (Zheng and Wang 1999). The computation of the unknown C^v in each node at time $(t + \Delta t)$ is the result of a two steps process. The first step is thus to track the particles in a Lagrangian approach on the basis of *characteristic lines* or streamlines calculated by a groundwater flow model. Advection is solved separately by the so-called *methods of characteristics* (MOC) for providing an "intermediate" concentration C^{v*} at time $(t + \Delta t)$. The second step will provide the final $C^v(x, y, z, t + \Delta t)$ (i.e., in each node) solving Equation 13.81 by a standard Eulerian technique and using the intermediate concentrations $C^{v*}(x, y, z, t + \Delta t)$.

The substantial derivative of Equation 13.81 can be approximated by finite difference (Zheng and Bennett 1995, Zheng and Wang 1999) as:

$$\frac{dC^v}{dt} \approx \frac{C^v(t + \Delta t) - C^{v*}(t + \Delta t)}{\Delta t} \tag{13.82}$$

It means that at each node, the unknown is approximated by:

$$C^v(t + \Delta t) \approx C^{v*}(t + \Delta t) + \Delta t \left[\frac{1}{R} \nabla \cdot (D_b \cdot \nabla C^v) - \lambda C^v - \frac{q_s}{R n_m}(C^v - C_s^v) \right] \tag{13.83}$$

where the second term of the right-hand side is the contribution to the final solution of our second step (i.e., Eulerian part). Indeed, depending on the chosen time for C^v in this term, an explicit, implicit, Crank-Nicolson or Galerkin time integration scheme can be adopted.

First step: solving advection

As mentioned previously, in order to solve advection, a Lagrangian moving grid is used to obtain the intermediate concentrations C^{v*}. Different methods of characteristics (MOC) can be implemented.

METHOD OF CHARACTERISTICS (MOC)

Initially developed by Garder *et al.* (1964), MOC is a "particle tracking" method that is used in codes as the MOC model (Konikow and Bredehoeft 1978) and the MT3D model (Zheng 1990).

A large number of particles is distributed (randomly or with a desired pattern) on the whole domain. A given concentration and an initial position are assigned to each of these particles. After the considered time step, the particles have moved according to the flow field (i.e., advection only) (Figure 13.14):

$$p_k(t + \Delta t) = p_k(t) + v_a(p_k(t))\Delta t \tag{13.84}$$

where $p_k(t)$ and $p_k(t + \Delta t)$ are respectively the kth particle position at the beginning and at the end of the time step, $v_a(p_k(t))$ is the advection velocity at the position $p_k(t)$. The positions can be tracked forward with this first-order approximation. In each

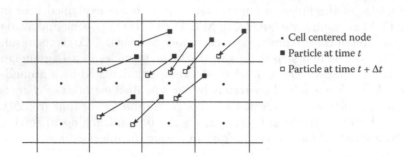

- Cell centered node
- Particle at time t
- Particle at time $t + \Delta t$

Figure 13.14 Schematic illustration of the MOC where particles are tracked forward along the streamlines during each time step for computing an intermediate concentration due only to advection during the considered time step.

cell (FV or FE), an average concentration can be calculated by counting the number of particles lying within the considered cell. If a regular grid is used, this (arithmetic) average can be expressed by:

$$\tilde{C}_i^{v*}(t + \Delta t) = \frac{1}{np_i} \sum_{k=1}^{np_i} C_k^v(t) \tag{13.85}$$

where $\tilde{C}_i^{v*}$ is the concentration at time $(t + \Delta t)$ transported by advection only, $C_k^v(t)$ is the concentration transported by the kth particle between (t) and $(t + \Delta t)$, np_i is the number of particles counted in the cell i at time $(t + \Delta t)$.

More generally, cell (FV or FE) grids are irregular and the average is calculated by weighting the particle concentration by the volume V_i of the origin cell (FV or FE) (Zheng 1993):

$$\tilde{C}_i^{v*}(t + \Delta t) = \frac{1}{\sum_{k=1}^{np_i} V_k} \sum_{k=1}^{np_i} V_k C_k^{v*}(t) \tag{13.86}$$

where V_k is the initial volume of the cell (FV or FE) where the kth particle was generated. The intermediate concentration $C_i^{v*}(t + \Delta t)$ may now be expressed as a linear interpolation between the concentration $\tilde{C}_i^{v*}(t + \Delta t)$ calculated by advection only at time $(t + \Delta t)$ (Equation 13.86) and the concentration at that node at the previous time step $C_i^v(t)$:

$$C_i^{v*}(t + \Delta t) = \omega \tilde{C}_i^{v*}(t + \Delta t) + (1 - \omega)C_i^v(t) \tag{13.87}$$

where ω is a weighting factor that is usually chosen between 0.5 and 1 (i.e., to give more weight to the end of the time step). This intermediate concentration $C_i^{v*}(t + \Delta t)$ can now be used for the second step of the procedure, the computation of the concentration changes due to dispersion, degradation, and sink/source terms, using Equation 13.83. Then the concentrations of all moving particles are updated before starting computation for a new time step.

MOC does not create numerical dispersion even for a large Pe number (Zheng and Wang 1999, Rausch *et al.* 2005). However, errors are coming from the way of interpolating the velocity field from the groundwater flow model. Due to the discrete nature of the particles and the needed counting of particles in each cell after each time step, local mass conservation problems may occur for a given time step, accentuated for cells (FE or FV) with highly irregular shapes. They can be visible on simulated breakthrough curves showing irregular shapes. To limit those problems, a large number of particle is needed increasing rapidly the computing load and memory storage. The method may become too heavy, especially for highly heterogeneous and complex problems involving nonlinearities.

MODIFIED METHOD OF CHARACTERISTICS (MMOC)

To improve the computational efficiency, a modified method was developed (Ewing *et al.* 1983, Cheng *et al.* 1984, Molz *et al.* 1986,) using one particle per cell and a backward particle tracking. One particle located in the centroid of the cell (FE or FV) is followed backward along the streamlines. Indeed, by advection only, the concentration in cell i $C_i^v(t + \Delta t)$ is equal to the concentration that was located in a backtracked position along the local streamline at time t (Figure 13.15). For the particle lying in the center of the ith cell at time $(t + \Delta t)$, calculating advection displacement in a similar way as in Equation 13.84, its position at time t is given by:

$$p_{xyz}(t) = p_i(t + \Delta t) - v_a(p_i(t + \Delta t))\Delta t \tag{13.88}$$

Accordingly, the intermediate concentration (i.e., computed only for advection during the considered time step) is the concentration at that position $p_{xyz}(t)$ at time t:

$$C_i^{v*}(t + \Delta t) = C^v(p_{xyz}(t), t) \tag{13.89}$$

This last concentration value $C^v(p_{xyz}(t), t)$ is calculated using a linear (bilinear in 2D or trilinear in 3D) interpolation of neighboring nodal values at time t.

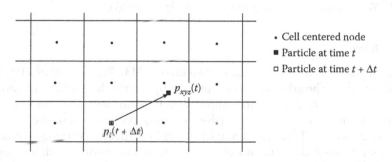

Figure 13.15 Schematic illustration of the MMOC where one particle per cell is tracked backward along the local streamline for computing an intermediate concentration due only to advection.

Memory requirements are reduced, and if a lower order scheme is adopted for the concentration interpolation from nodal values at the previous time step, the method is faster than MOC (Zheng and Wang 1999). MMOC shows the same mass conservation problem than MOC, but the main issue is the numerical dispersion occurring with lower order interpolations of the concentrations in advection dominated problems. Higher-order interpolation schemes, often combined with longer time steps, may lead to better results but induce oscillations when simulating sharp concentration gradients. As previously for MOC, errors can also be introduced depending on the way of interpolating the velocity field from the groundwater flow model.

HYBRID METHOD OF CHARACTERISTICS (HMOC)

For optimizing the choice between MOC and MMOC, a hybrid method of characteristics (HMOC) has been introduced that includes an automatic change of the solution technique in function of the local concentration gradients (Neuman 1981, 1984). Where a steep concentration gradient is found, particles are distributed around and the MOC is applied; elsewhere in the domain, the MMOC is applied. If the steep concentration gradient is disappearing with time (i.e., due to other processes than advection) the forward tracking (MOC) can be automatically stopped and corresponding particles removed (Zheng and Wang 1999).

Second step: solving dispersion and the other terms

As mentioned previously, the second step consists in using the intermediate concentration $C_i^{v*}(t + \Delta t)$ for computation of the concentration changes due to dispersion, degradation, and sink/source terms, using Equation 13.83. Depending on the chosen time for the concentration C^v in the second term of the right-hand side of Equation 13.83, an explicit, implicit, Crank-Nicolson or Galerkin time integration scheme can be adopted. Solving this fully parabolic part of the transport equation does not create any additional numerical issue. At the end of the process, the starting concentrations (for the moving particles) are updated before starting computation for a new time step (first step of the next time step).

There are also other Eulerian-Lagrangian methods called ELLAM for Eulerian-Lagrangian Localized Adjoint Methods (e.g., associated to the FEM) (among others, Wang *et al.* 1999, Younes and Ackerer 2005, Younes *et al.* 2006).

Random walk methods

This method was developed for years (Chandrasekhar 1943, Prickett *et al.* 1981, Uffink 1985) and combines the advection computation by a particle tracking method (see previous paragraphs), with dispersion computation by a random walk process.

Each particle is assigned a mass of solute (they are only present where the solute mass is nonzero; Rausch *et al.* 2005), and particles move by advection along streamlines of the previously computed groundwater flow problem. Diffusion and dispersion are then computed taking the assumption that they can be considered as a random process (Bear and Cheng 2010). A random displacement is added to the advective displacement for each particle taken independently and independently of its previous displacements. If the particles are sufficiently numerous, their random displacements describe a spreading around a mean advection position as described by a Gaussian probabilistic distribution. That can be described by equations similar to Equations 8.24 through 8.26 (written for a uniform 1D flow) as:

$$C^v(x,t) = \frac{C_0^v}{\sqrt{4\pi a_L v_a}} exp\left[-\frac{(x-v_a t)^2}{4a_L v_a t}\right]$$

(13.90)

The random walk method is described synthetically in Chapter 8, Box 8.4. Classical random walk particle tracking methods are working with discrete time steps (discretization of time in given time steps) and variable spatial increment that depends on the velocity field and the random "noise."

Practically, associated to a cell-grid, concentrations at time $t + \Delta t$ are computed by adding mass of the particles lying in each cell (or FV) and dividing by their respective total water volume. Particles are eliminated or added for taking sink/source terms into account. Particles reaching no flow boundaries are "reflected" back. On the prescribed concentration boundaries, particles are added or removed for keeping the considered prescribed concentration.

The analogy between dispersion and the normal distribution implies that the mean advection position ($v_a t$) corresponds to the mean (μ) of the statistic distribution and solute dispersion (i.e., $2a_L v_a t$) corresponds to the variance (σ^2) of the statistic distribution. So that in 1D (along axis x), the position of a tracer is given by (Rausch *et al.* 2005):

$$x = v_a t + Y\sqrt{2a_L v_a t}$$

(13.91)

where Y is the standardized normal distribution with a mean equal to 0 and standard deviation equal to 1 (i.e., produced by a random number generator for normally distributed variables).

Spatially, from the number of particles in the Δx interval (1D), a concentration can be calculated and assigned to the mid-interval at $x + 0.5\Delta x$.

Temporally, the position of a particle at time $t + \Delta t$ can be calculated by applying Equation 13.91:

$$x_p(t + \Delta t) = x_p(t) + v_a(x_p,t)\Delta t + Y\sqrt{2a_L v_a(x_p,t)\Delta t}$$

(13.92)

where $v_a(x_p, t)$ shows explicitly the dependence of the advection velocity on the local and temporal conditions. For nonuniform and time varying advection conditions, the user should choose sufficiently small Δt for assuming that during the time step the advection is uniform locally and constant. Practically, it is recommended to choose advective time steps short enough for having several steps needed for a particle to move across a grid cell (Prickett *et al.* 1981):

$$v_a\Delta t < \frac{1}{5}\Delta x \quad or \quad Cr < \frac{1}{5}$$

(13.93)

Detailed presentation of random walk methods are available in the literature (among others, Uffink 1985, Kinzelbach 1988, Kinzelbach and Uffink 1991, summarized in Rausch *et al.* 2005) with extensions to 2D and 3D cases.

More complex methods are developed for "anomalous" or "non-Fickian" or "non-Gaussian" dispersion behaviors. The random process is declined in time and space: the particle motion occurs with variable spatial increment (ΔL, jumps, and varying length) and variable time increment ($\Delta\tau$, waiting time between two jumps).

If dispersion is considered as temporally Gaussian but spatially not, it is called spatially anomalous. If dispersion is considered as spatially Gaussian but temporally not, it is called temporally anomalous. Dispersion could also be considered as spatially and temporally anomalous. If ΔL and $\Delta\tau$ are independent, the fundamental properties of transport are governed by the asymptotic behavior of the coupled space-time probability distribution function (PDF) $p(\Delta L, \Delta\tau)$ (Berkowitz et al. 2002). The term asymptotic behavior is used to refer to respectively long-distance or long-time behavior (Frippiat and Holeyman 2008).

Methods for simulating anomalous transport are clearly beyond the scope of this book. Let's just mention that continuous time random walk (CTRW) methods (among others, Dentz and Berkowitz 2003, Frippiat and Holeyman 2008, Burnell et al. 2017) have been developed to simulate anomalous transport behaviors in heterogeneous media at different scales (among others, Berkowitz et al. 2006, Le Borgne et al. 2008). Also, multiple time domain random walk methods (TDRW) are developed (among others, Delay et al. 2002) where the transition times are exponentially distributed (Russian et al. 2016). This topic is currently in fast development and new techniques most often based on the random walk method are proposed for simulating non-Gaussian dispersion (among others, Zhang et al. 2016, Benson et al. 2017, Meyer 2017).

Reactive transport modeling

For simulating reactive transport, one must move from a single-species approach toward coupled transport of multiple chemical species including their chemical reactions. This is often crucial to include chemical reactions in a solute transport problem.

The partial differential equation describing saturated groundwater reactive solute transport is Equation 8.55 (see Chapter 8) that is written:

$$R_i \frac{\partial C_i^v}{\partial t} = -v_{a_i} \nabla C_i^v + \nabla \cdot (D_h \cdot \nabla C_i^v) - R_i \lambda_i C_i^v - \frac{q_s}{\theta_i}(C_i^v - C_{s_i}^v)$$

$$+ \frac{1}{\theta_i} \sum_{j=1}^{N_s} S_{ij}(C_1^v, ..., C_n^v) \quad i = 1, ..., N_s \tag{13.94}$$

where, in the "reaction term", S_{ij} is the source/sink term representing the effect of reactions (in kg/m³s)[ML⁻³T⁻¹], θ_i is the groundwater specific volume fraction of the REV where the species i is located. There are as many solute transport equations as the number of considered moving species ($i = 1, ..., n$) but in the "reaction term," there are as many (sub)terms as species being involved in the reactions (N_s). It means that some of them may not be of interest for transport simulations (e.g., species that are known as immobile, as a part of the solid matrix).

The n transport equations are coupled through the $S_{ij}(C_1^v, ..., C_n^v)$ terms. For reactions occurring in the mobile water phase, θ_i are taken equal to the mobile water porosity n_m and the components of v_{a_i} are equal to v_a (i.e., the advection velocity). If all reactions are supposed as occurring in the mobile phase, $N_s = n$ and the system is defined as a *homogeneous reaction system*. On the contrary, and it is mostly the case in the reality, if a part of the involved species is on the solid matrix or in the immobile water, the reaction system is defined as *heterogeneous*, and the species in those immobile phases are considered as just playing a role in the reaction term of Equation 13.94.

A large specialized literature can be found about the development of numerical techniques allowing the simulation of those coupled multicomponent reactive transport systems. Various solutions have been proposed and discussed (among the first, Rubin and James 1973, Rubin 1983, Kirkner and Reeves 1988, Reeves and Kirkner 1988). Many different ways for coupling equations and solving nonlinearities were proposed. It is far beyond the scope of this chapter to describe all the proposed methods.

As mentioned by Walter *et al.* (1994), for a very limited number of involved species, direct methods were used incorporating directly the relevant chemical reactions into the transport equations, the variables (i.e., the concentration of the chemical species) being linked through nonlinear relationships. They were often too demanding in computing resources, and thus "two-steps" methods have been developed giving rise to many possibilities. Transport computation is performed separately from chemical reactions computation. It can be organized sequentially (with internal iterations or not) or in parallel. This latter would be preferred for computational reasons (i.e., parallel processing) while a sequential operator-splitting technique is actually often preferred for chemical coupling consistency. Indeed, the various operator-splitting options have been discussed (among others, Herzer and Kinzelbach 1989, Yeh and Tripathi 1991, Barry *et al.* 2002).

The codes used for computing the two parts of this "two-step" approach are mostly coming from different scientific communities. Only the physical solute transport part must be solved in a spatially connected way (i.e., taking into account coordinates and grid), the chemical part can be solved independently as it depends only on local conditions (i.e., around the concerned node).

An important conceptual choice must be made concerning local chemical equilibrium or not. If the reactions are fast compared to the groundwater flow, the *local equilibrium assumption* (LEA) can be considered as valid. However, quite often it is not always the case and kinetically evolving systems must be considered (Prommer *et al.* 1999). It means that reaction modules with specific kinetically controlled processes are used (Prommer *et al.* 2002, 2003).

Coupling specific chemical reaction codes (as for example PHREEQC-2, Parkhurst and Appelo 1999) with different solute transport codes (as for example MT3DMS, Zheng and Wang 1999) allows to simulate reactive transport with highly versatile capabilities (Prommer and Post 2010). Those developments are crucial for adequate and reliable modeling of natural and enhanced (bio)remediations but also for any other problem involving multiple reactive contaminants (among many others, Prommer *et al.* 2006, Vencelides *et al.* 2007, Cohen *et al.* 2008, Colombani *et al.* 2009, Jung *et al.* 2009, Ng *et al.* 2015).

References

Anderson, M.P., Woessner, W.W. and R.J. Hunt. 2015. *Applied groundwater modeling— Simulation of flow and advective transport*. Amsterdam: Academic Press Elsevier.

Baliga, B.R. and S.V. Patankar. 1983. A control volume finite-element method for two-dimensional fluid flow and heat transfer. *Numerical Heat Transfer* 6(3): 245–261.

Barry, D.A., Prommer, H., Miller, C.T., Engesgaard, P. and C. Zheng. 2002. Modelling the fate of oxidisable organic contaminants in groundwater. *Advances in Water Resources* 25: 899–937.

Bear, J. and A.H.D. Cheng. 2010. *Modeling groundwater flow and contaminant transport*. Dordrecht: Springer.

Bear, J. and A. Verruijt. 1987. *Modeling groundwater flow and pollution*. Dordrecht: Reidel Publishing Company.

Benson, D.A., Aquino, T., Bolster, D., Engdahl, N., Henri, C.V. and D. Fernàndez-Garcia. 2017. A comparison of Eulerian and Lagrangian transport and non-linear reaction algorithms. *Advances in Water Resources* 99: 15–37.

Berkowitz, B., Cortis, A., Dentz, M. and H. Scher. 2006. Modeling non-Fickian transport in geological formations as a continuous time random walk. *Reviews of Geophysics* 44(2): RG2003.

Berkowitz, B., Klafter, J., Metzler, R. and H. Scher. 2002. Physical pictures of transport in heterogeneous media: Advection-dispersion, random-walk, and fractional derivative formulations. *Water Resources Research* 38(10): 1191.

Brooks, A.N. and T.J.R. Hughes. 1982. Streamline upwind/Petrov-Galerkin formulations for convection dominated flows with particular emphasis on the incompressible Navier-Stokes equation. *Computer Methods in Applied Mechanics and Engineering* 32: 199–259.

Burnell, D.K., Hansen, S.K. and J. Xu. 2017. Transient modeling of non-Fickian transport and first-order reaction using continuous time random walk. *Advances in Water Resources* 107: 370–392.

Chandrasekhar, S. 1943. Stochastic problems in physics and astronomy. *Reviews of Modern Physics* 15(1): 1–89.

Cheng, R.T., Casulli, V. and S.N. Milford. 1984. Eulerian-Lagrangian solution of the convection-dispersion equation in natural coordinates. *Water Resources Research* 20(7): 944–952.

Chung, T. 2002. *Computational fluid dynamics*. Cambridge: Cambridge University Press.

Cirpka, O.A., Frind, E.O. and R. Helmig. 1999. Numerical simulation of biodegradation controlled by transverse mixing. *Journal of Contaminant Hydrology* 40(2): 159–182.

Cohen, E.L., Patterson, B.M., McKinley, A.J. and H. Prommer. 2008. Zero valent iron remediation of a mixed brominated ethene contaminated groundwater. *Journal of Contaminant Hydrology* 103(3): 109–118.

Colombani, N., Mastrocicco, N., Gargini, A., Davis, G.B. and H. Prommer. 2009. Modelling the fate of styrene in a mixed petroleum hydrocarbon plume. *Journal of Contaminant Hydrology* 105: 38–55.

Cox, R.A. and T. Nishikawa. 1991. A new total variation diminishing scheme for the solution of advective-dominant solute transport. *Water Resources Research* 27(10): 2645–2654.

Daus, A.D. and E.O. Frind. 1985. An alternating direction Galerkin technique for simulation of contaminant transport in complex groundwater systems. *Water Resources Research* 21(5): 653–664.

Delay, F., Porel, G. and P. Sardini. 2002. Modelling diffusion in a heterogeneous rock matrix with a time-domain Lagrangian method and an inversion procedure. *Comptes Rendus Geoscience* 334: 967–973.

Dentz, M. and B. Berkowitz. 2003. Transport behavior of a passive solute in continuous time random walks and multirate mass transfer. *Water Resources Research* 39(5): 1111.

Diersch, H-J.G. 2014. *Feflow—Finite element modeling of flow, mass and heat transport in porous and fractured media*. Heidelberg: Springer.

Ewing, R.E., Russell, T.F. and M.F. Wheeler. 1983. Simulation of miscible displacement using mixed methods and a modified method of characteristics. In *SPE Reservoir Simulation Symposium*, Dallas (TX): Society of Petroleum Engineers, 12241.

Fitts, Ch. R. 2002. *Groundwater science*. London: Academic Press.

Fletcher, C. 1988. *Computational techniques for fluid dynamics*. Vol.1 and Vol.2, New York: Springer.

Forsyth, P.A., Wu, Y.S. and K. Pruess. 1995. Robust numerical methods for saturated-unsaturated flow with dry initial conditions in heterogeneous media. *Advances in Water Resources* 18(1): 25–38.

Franca, L.P., Frey, S.L. and T.J. Hughes. 1992. Stabilized finite element methods: I. Application to the advective-diffusive model. *Computer Methods in Applied Mechanics and Engineering* 95(2): 253–276.

Frind, E.O. and D. Germain.1986. Simulation of contaminant plumes with large dispersive contrast: Evaluation of alternating direction Galerkin Models. *Water Resources Research* 22(13): 1857–1873.

Frippiat, Ch.C. and A.E. Holeyman. 2008. A comparative review of upscaling methods for solute transport in heterogeneous porous media. *Journal of Hydrology* 362: 150–176.

Garder Jr, A.O., Peaceman, D.W. and A.L. Pozzi Jr. 1964. Numerical calculation of multidimensional miscible displacement by the method of characteristics. *Society of Petroleum Engineers Journal* 4(01): 26–36.

Hayley, K. 2017. The present state and future application of cloud computing for numerical groundwater modeling. *Groundwater*: doi:10.1111/gwat.12555

Herzer, J. and W. Kinzelbach. 1989. Coupling of transport and chemical processes in numerical transport models, *Geoderma* 44(2-3): 115–127.

Huyakorn, P.S. and G.F. Pinder. 1983. *Computational methods in subsurface flow*. New York: Academic Press.

Idelsohn, S. and E. Onate. 1994. Finite volumes and finite elements: Two "good friends". *International Journal for Numerical Methods in Engineering* 37(19): 3323–3341.

Jung, H.B., Charette, M.A. and Y. Zheng. 2009. Field, laboratory, and modeling study of reactive transport of groundwater arsenic in a coastal aquifer. *Environmental Science & Technology* 43(14): 5333–5338.

Kinzelbach, W. 1988. The random walk method in pollutant transport simulation. *Groundwater Flow and Quality Modelling* 224: 227–246.

Kinzelbach, W. and G. Uffink. 1991. The random walk method and extensions in groundwater modelling. In *Transport processes in porous media*, NATO ASI Series, NSSE 202. Springer Netherlands, 761–787.

Kirkner, D.J. and H. Reeves. 1988. Multicomponent mass transport with homogeneous and heterogeneous chemical reactions: Effect of the chemistry on the choice of numerical algorithm: 1. Theory. *Water Resources Research* 24(10): 1719–1729.

Konikow, L.F. and J.D. Bredehoeft. 1978. *Computer model of two-dimensional solute transport and dispersion in ground water*. Washington: US Government Printing Office.

Kropf, P., Schiller, E., Brunner, P., Schilling O., Hunkeler D. and A. Lapin. 2014. Wireless mesh networks and cloud computing for real time environmental simulations. In: *Recent Advances in Information and Communication Technology*, eds. S. Boonkrong, H. Unger and P. Meesad, Advances in Intelligent Systems and Computing, 265: 1–11. Heidelberg: Springer.

Kurtz, W., Lapin, A., Schilling, O.S., Tang, Q., Schiller, E., Braun, T., Hunkeler, D., Vereecken, H., Sudicky, E., Kropf, P., Franssen, H-J. H. and P. Brunner. 2017. Integrating hydrological modelling, data assimilation and cloud computing for real-time management of water resources. *Environmental Modelling & Software* 93: 418–435.

Le Borgne, T., Dentz, M. and J. Carrera. 2008. A Lagrangian statistical model for transport in highly heterogeneous velocity fields. *Physical Review Letters* 101(9): 090601.

Leonard, B.P. 1988. *Universal Limiter for transient interpolation modeling of the advective transport equations: The ULTIMATE conservative difference scheme*, NASA Technical Memorandum 100916 ICOMP-88-11.

Leonard, B.P. and H.S. Niknafs. 1990. Cost-effective accurate coarse-grid method for highly convective multidimensional unsteady flows. In *Computational Fluid Dynamics Symposium on Aeropropulsion*. NASA Conference Publication 3078.

Leonard, B.P. and H.S. Niknafs. 1991. Sharp monotonic resolution of discontinuities without clipping of narrow extrema. *Computer & Fluids* 19(1): 141–154.

McDonald, M.G. and A.W. Harbaugh. 1988. *A modular three-dimensional finite-difference ground-water flow model*. U.S. Geological Survey Techniques of Water-Resources Investigations, Book 6, Reston (VA): USGS.

Meyer, D.W. 2017. Relating recent random walk models with classical perturbation theory for dispersion predictions in the heterogeneous porous subsurface. *Advances in Water Resources* 105: 227–232.

Molz, F.J., Widdowson, M.A. and L.D. Benefield. 1986. Simulation of microbial growth dynamics coupled to nutrient and oxygen transport in porous media. *Water Resources Research* 22(8): 1207–1216.

Narasimhan, T.N. and P.A. Witherspoon. 1976. An integrated finite difference method for analyzing fluid flow in porous media. *Water Resources Research* 12(1): 57–64.

Narasimhan, T.N., Neuman, S.P. and P.A. Witherspoon. 1978. Finite element method for subsurface hydrology using a mixed explicit-implicit scheme. *Water Resources Research* 14(5): 863–877.

Neuman, S.P. 1981. A Eulerian-Lagrangian numerical scheme for the dispersion-convection equation using conjugate space-time grids. *Journal of Computational Physics* 41(2): 270–294.

Neuman, S.P. 1984. Adaptive Eulerian–Lagrangian finite element method for advection-dispersion. *International Journal for Numerical Methods in Engineering* 20(2): 321–337.

Ng, G.-H.C., Bekins, B.A., Cozzarelli, I.M., Baedecker, M.J., Bennett, P.C., Amos, R.T. and W.N. Herkelrath. 2015. Reactive transport modeling of geochemical controls on secondary water quality impacts at a crude oil spill site near Bemidji, MN. *Water Resources Research* 51: 4156–4183.

Paniconi, C. and M. Putti. 2015. Physically based modeling in catchment hydrology at 50: Survey and outlook. *Water Resources Research* 51: 7090–7129.

Parkhurst, D.L. and C.A.J. Appelo. 1999. *User's guide to PHREEQC (Version 2) : a computer program for speciation, batch-reaction, one-dimensional transport, and inverse geochemical calculations*. Water-Resources Investigations Report 99-4259, Denver: USGS.

Patankar, S. 1980. *Numerical heat transfer and fluid flow*. Boca Raton: CRC Press.

Peaceman, D.W. 1977. *Fundamentals of numerical reservoir simulation*, Amsterdam: Elsevier.

Pinder, G.F. and M.A. Celia. 2006. *Subsurface hydrology*. Hoboken (NJ): Wiley & Sons.

Pinder, G.F. and W.G. Gray. 1977. *Finite element simulation in surface and subsurface hydrology*. San Diego (CA): Academic Press.

Price, H.S., Varga, R.S. and J.R. Warren.1966. Application of oscillation matrices to diffusion-convection equations. *Journal of Mathematical Physics* 45: 301–331.

Prickett, T.A., Lonnquist, C.G. and T.G. Naymik. 1981. *A 'random-walk' solute transport model for selected groundwater quality evaluations*. Champaign: Illinois State Water Survey, Bulletin 65.

Prommer, H., and V.E.A. Post. 2010. *PHT3D:Aa reactive multicomponent model for satu-rated porous media*, Version 2.0, User's Manual, http://www.pht3d.org.

Prommer, H., Barry, D.A. and G.B. Davis. 2002. Modelling of physical and reactive processes during biodegradation of a hydrocarbon plume under transient groundwater flow condi-tions. *Journal of Contaminant Hydrology* 59: 113–131.

Prommer, H., Barry, D.A. and C. Zheng. 2003. MODFLOW/MT3DMS-based reactive multi-component transport modelling. *Ground Water* 42(2): 247–257.

Prommer, H., Davis, G.B. and D.A. Barry. 1999. PHT3D—A three dimensional biogeo-chemical transport model for modelling natural and enhanced remediation. In *Proc. Contaminated site remediation: Challenges posed by urban and industrial contaminants*, ed. C.D. Johnston, 351–358, Fremantle, Australia.

Prommer, H., Tuxen, N. and P.L. Bjerg. 2006. Fringe-controlled natural attenuation of phe-noxy acids in a landfill plume: Integration of field-scale processes by reactive transport. *Environmental Science and Technology* 40: 4732–4738.

Rausch, R., Schäfer, W., Therrien, R. and Chr. Wagner. 2005. *Solute transport modelling—An introduction to models and solution strategies*. Berlin-Stuttgart: Gebr.Borntraeger Verlagsbuchhandlung Science Publishers.

Reddy, J.N. 1993. *An introduction to the finite element method* (2nd Edition). New York: McGraw-Hill.

Reeves, H. and D.J. Kirkner. 1988. Multicomponent mass transport with homogeneous and heterogeneous chemical reactions: Effect of the chemistry on the choice of numerical algo-rithm: 2. Numerical results, *Water Resources Research* 24(10): 1730–1739.

Rubin, J. 1983. Transport of reacting solutes in porous media: Relation between mathematical nature of problem formulation and chemical nature of reactions. *Water Resources Research* 19(5): 1231–1252.

Rubin, J. and R.V. James. 1973. Dispersion-affected transport of reacting solutes in saturated porous media: Galerkin method applied to equilibrium-controlled exchange in unidirec-tional steady water flow, *Water Resources Research* 9(5): 1332–1356.

Russian, A., Dentz, M. and P. Gouze. 2016. Time domain random walks for hydrodynamic transport in heterogeneous media. *Water Resources Research* 52(5): 3309–3323.

Sauty, J.P. 1980. An analyse of hydrodispersive transfer in aquifers. *Water Resources Research* 16(1): 145–158.

Therrien, R. and E.A. Sudicky. 1996. Three-dimensional analysis of variably-saturated flow and solute transport in discretely-fractured porous media. *Journal of Contaminant Hydrology* 23(1–2): 1–44.

Therrien, R., McLaren, R.G., Sudicky, E.A. and S.M. Panday. 2010. *HydroGeoSphere: A three-dimensional numerical model describing fully-integrated subsurface and surface flow and solute transport*. User manual. Québec: Université Laval & University of Waterloo.

Uffink, G.J.M. 1985. A random walk method for the simulation of macrodispersion in a strati-fied aquifer. In *IUGG 18th General Assembly, Hamburg August 1983, Proc. Symp.*, IAHS Publication 146: 63–114.

Vencelides, Z., Sracek, O. and H. Prommer. 2007. Modelling of iron cycling and its impact on the electron balance at a petroleum hydrocarbon contaminated site in Hnevice, Czech Republic. *Journal of Contaminant Hydrology* 89: 270–294.

Walter, A.L., Frind, E.O., Blowes, D.W., Ptacek, C.J. and J.W. Molson. 1994. Modeling of multicomponent reactive transport in groundwater: 1. Model development and evaluation. *Water Resources Research* 30(11): 3137–3148.

Wang, H., Dahle, H.K., Ewing, R.E., Espedal, M.S., Sharpley, R.C. and S. Man. 1999. An ELLAM scheme for advection-diffusion equations in two dimensions. *SIAM Journal on Scientific Computing* 20(6): 2160–2194.

Wang, H.F. and M.P. Anderson. 1982. *Introduction to groundwater modelling: Finite difference and finite element methods*, San Diego (CA): Academic Press.

Yeh, G.T. and V.S. Tripathi. 1991. A model for simulating transport of reactive multispecies components: Model development and demonstration. *Water Resources Research* 27(12): 3075–3094.

Younes, A. and P. Ackerer. 2005. Solving the advection-diffusion equation with the Eulerian-Lagrangian localized adjoint method on unstructured meshes and non uniform time stepping. *Journal of Computational Physics* 208(1): 384–402.

Younes, A., Ackerer, P. and F. Lehmann. 2006. A new efficient Eulerian-Lagrangian localized adjoint method for solving the advection-dispersion equation on unstructured meshes. *Advances in Water Resources* 29: 1056–1074.

Yu, C.C. and J.C. Heinrich. 1986. Petrov-Galerkin methods for the time-dependent convective transport equation. *International Journal for Numerical Methods in Engineering* 23(5): 883–901.

Zheng, C. 1990. *MT3D, A Modular Three-Dimensional Transport model for simulation of advection, dispersion and chemical reactions of contaminants in groundwater systems*, Report to the U.S. Environmental Protection Agency Robert S. Kerr Environmental Research Laboratory, Ada, Oklahoma.

Zheng, C. 1993. Extension of the method of characteristics for simulation of solute transport in three dimensions. *Ground Water* 31(3): 456–465.

Zheng, C. and G.D. Bennett. 1995. *Applied contaminant transport modeling: Theory and practice*, New York: John Wiley & Sons.

Zheng, C. and P.P. Wang. 1999. *MT3DMS A modular three-dimensional multispecies transport model for simulation of advection, dispersion and chemical reactions of contaminants in groundwater systems (Release DoD_3.50.A) Documentation and User's guide*. Tuscaloosa (AL): University of Alabama 35487–0338.

Zhang, Y., Meerschaert, M.M. and R.M. Neupauer. 2016. Backward fractional advection dispersion model for contaminant source prediction. *Water Resources Research* 52: 2462–2473.

Index

Note: Page numbers with "f" refer to figures. Page numbers with "t" refer to tables.

Printed in the United States
by Baker & Taylor Publisher Services